The Sharpest Cut

MPS/SIAM Series on Optimization

This series is published jointly by the Mathematical Programming Society and the Society for Industrial and Applied Mathematics. It includes research monographs, textbooks at all levels, books on applications, and tutorials. Besides being of high scientific quality, books in the series must advance the understanding and practice of optimization. They must also be written clearly and at an appropriate level.

Series Volumes

Grötschel, Martin, editor, *The Sharpest Cut: The Impact of Manfred Padberg and His Work*
Renegar, James, *A Mathematical View of Interior-Point Methods in Convex Optimization*
Ben-Tal, Aharon and Nemirovski, Arkadi, *Lectures on Modern Convex Optimization: Analysis, Algorithms, and Engineering Applications*
Conn, Andrew R., Gould, Nicholas I. M., and Toint, Phillippe L., *Trust-Region Methods*

The Sharpest Cut

The Impact of Manfred Padberg and His Work

Edited by
Martin Grötschel
Konrad-Zuse-Zentrum für Informationstechnik Berlin (ZIB)
Berlin-Dahlem, Germany

siam
Society for Industrial and Applied Mathematics
Philadelphia

MPS
Mathematical Programming Society
Philadelphia

10 9 8 7 6 5 4 3 2 1

Library of Congress Cataloging-in-Publication Data

The sharpest cut : the impact of Manfred Padberg and his work / edited by Martin Grötschel.
p. cm. – (MPS/SIAM series on optimization)
Includes bibliographical references and index.
ISBN 978-0-898715-52-1
1. Combinatorial optimization–Congresses. 2. Programming (Mathematics)–Congresses. 3. Combinatorial optimization–Congresses. I. Grötschel, Martin. II. Padberg, M. W. III. MPS-SIAM series on optimization.

QA402.5.S523 2004
519.6'4–dc22

2003067207

Contents

Preface

῾Ο μὴ δαρεὶς ἄνθρωπος οὐ παιδεύεται.
Menandros (c. 342–292 BC)

To celebrate Manfred Padberg's 60th birthday, about 100 mathematicians from 20 different countries around the globe gathered at the Konrad Zuse Zentrum, Berlin, to honor one of the leading men in combinatorial optimization of our times. Twenty-three invited talks were presented at the October 11 to 13, 2001 conference that culminated in a riverboat party on the Spree, circling the center of nightlit Berlin.

The lectures, grouped in this book under the following topics:

- packing, stable sets, and perfect graphs;
- polyhedral combinatorics;
- general polytopes;
- semidefinite programming;
- computation;

touch upon some aspects of Manfred's work. They are enriched with personal reminiscences and anecdotes about encounters with Manfred.

The book contains a short version of Manfred's curriculum vitae and a personal account of his work by Laurence Wolsey. At the end of the book we have included dinner speeches by Egon Balas, Claude Berge, and Harold Kuhn that were read on the boat. The remainder of the book is strictly scientific. The chapters, refereed by the standards of our flagship journal *Mathematical Programming*, present recent results on combinatorial optimization that are closely connected to Manfred's research. Manfred's deep commitment to the geometrical approach to combinatorial optimization can be felt in most of these chapters. His search for increasingly better and computationally efficient cutting planes gave rise to the title of this book.

The Sharpest Cut is about concrete advances in the successful optimization of hard, real-world problems, but it is also mindful of Manfred's descendance from an old family of robber barons of the Sauerland region in Westphalia (Germany). Never mind "sharp" cuts, only the sharpest one is good enough. Give, but do not forget to take. Menandros's sophism "A mensch who has not taken a beating lacks an education" reflects both Manfred's youth in difficult post–World War II times and his pedagogical relation with his students and coworkers. Some have called it very demanding indeed.

The Workshop in Honor of Manfred Padberg was made possible through the financial support of the Konrad Zuse Zentrum (ZIB) and New York University. The conference was organized by the staff of the ZIB under the leadership of Dr. Annegret Wagler, who was also most instrumental in the process of editing this volume. A big and grateful "thank you" to Annegret and everybody who helped.

Berlin,
February, 2003

Martin Grötschel

Part I

Manfred Padberg:

Curriculum Vitae
and
Survey of His Work

Manfred Padberg

Chapter 1

Manfred Padberg

Curriculum Vitae

Personal

1941	Born on October 10 in Bottrop, Germany.
1941–1961	Grew up in Zagreb, Vlotho, Dülmen, Olsberg, Brilon, and Beckum. Up to this time: interests mostly in music, history, Latin, and Greek.

Education

1961–1967	Westfälische Wilhelms Universität, Münster, Germany; M.S. in Mathematics, 1967.
1968–1971	Carnegie-Mellon University, Pittsburgh, U.S.A.; M.S. and Ph.D. in Industrial Administration, 1971.

Positions

1967–1968	Wissenschaftlicher Assistent, Universität Mannheim, Mannheim.
1971–1974	Research Fellow, IIM, Wissenschaftszentrum Berlin, Berlin.
1974–1978	Associate Professor, New York University, New York.
1978–2002	Professor of Operations Research, New York University, New York.
1988–2002	Research Professor, New York University, New York.
2002–	Professor Emeritus, New York University, New York.
1973–2002	Visits: U Waterloo, U Bonn, CMU Pittsburgh, IBM Yorktown, U Münster, INRIA Rocquencourt, CORE Louvain-la-Neuve, EIASM Brussels, U Pisa, IASI Rome, U Grenoble, SUNY Stony Brook, EP Paris, U Augsburg, U Köln.

Selected Honors

1980	Lanchester Prize, Honorable Mention, ORSA.
1983	Lanchester Prize, ORSA.
1985	Dantzig Prize, MPS and SIAM.
1989	Senior U.S. Scientist Research Award, Humboldt-Stiftung.
2000	John von Neumann Theory Prize, INFORMS.

Selected Editorial Activities

1974–1983	Associate Editor, *Mathematical Programming* (also 1988–1993).
1974–1983	Associate Editor, *Mathematical Programming Studies.*
1992–	Associate Editor, *Mathematical and Computer Modelling.*
1995–	Advisory Editor, *TOP: The Spanish Journal on Operations Research.*

Selected Publications

Books

1. *Linear Optimization and Extensions*, Springer-Verlag, Berlin, New York, 1995 (1999, 2nd ed.).
2. *Location, Scheduling, Design and Integer Programming*, with M. Rijal, Kluwer Academic, Boston, MA, 1996.
3. *Linear Optimization and Extensions: Problems and Solutions*, with D. Alevras, Springer-Verlag, Berlin, New York, 2001.

Articles

1. " 'Simple' Zero-One Problems: Set Covering, Matchings and Coverings in Graphs," Man. Sci. Res. Rep. No. 235, Carnegie-Mellon University, Pittsburgh, PA, January 1971.
2. "Equivalent Knapsack-type Formulations of Bounded Integer Linear Programs," *Naval Res. Logistics Quarterly* **19** (1972).
3. "On the Set-Covering Problem," with E. Balas, *Oper. Res.* **20** (1972).
4. "On the Facial Structure of Set Packing Polyhedra," *Math. Program.* **5** (1973).
5. "Perfect Zero-One Matrices," *Math. Program.* **6** (1974).
6. "The Traveling Salesman Problem and a Class of Polyhedra of Diameter Two," with M.R. Rao, *Math. Program.* **7** (1974).
7. "A Note on Zero-One Programming," *Oper. Res.* **23** (1975).
8. "On the Set Covering Problem II: An Algorithm for Set Partitioning," with E. Balas, *Oper. Res.* **23** (1975).
9. "A Note on the Total Unimodularity of Matrices," *Discrete Math.* **14** (1976).
10. "Almost Integral Polyhedra Related to Certain Combinatorial Optimization Problems," *Linear Algebra Appl.* **15** (1976).
11. "Simple Rules for Optimal Portfolio Selection," with E. Elton and M. Gruber, *J. Finance* **31** (1976).
12. "Set Partitioning: A Survey," with E. Balas, *SIAM Rev.* **18** (1976).
13. "On the Complexity of Set Packing Problems," *Discrete Math.* **1** (1977).
14. "On the Traveling Salesman Problem: Theory and Computation," with M. Grötschel, in R. Henn et al. (eds.), *Operations Research and Optimization*, Springer-Verlag, Berlin, New York, 1978.
15. "Covering, Packing and Knapsack Problems," *Discrete Math.* **4** (1979).
16. "On the Symmetric Traveling Salesman Problem I and II," with M. Grötschel, *Math. Program.* **16** (1979).
17. "Null-Eins Entscheidungsprobleme," in M.J. Beckmann, C. Menges, and R. Selten (eds.), *Handwörterbuch der Mathematischen Wirtschaftswissenschaften*, Gabler-Verlag, Wiesbaden, Germany, 1979.
18. "On the Symmetric Traveling Salesman Problem: A Computational Study," with S. Hong, *Math. Program. Studies* **12** (1980).
19. "Solving Large-Scale Symmetric Traveling Salesman Problems to Optimality," with H. Crowder, *Management Sci.* **26** (1980).
20. "$(1, k)$-Configurations and Facets for Packing Problems," *Math. Program.* **18** (1980).

21. "On the Uncapacitated Plant Location Problem I and II," with D. Cho et al., *MOR* **8** (1983).
22. "Odd Minimum Cut-Sets and b-Matchings," with M.R. Rao, *MOR* **7** (1982).
23. "Degree-Two Inequalities, Clique Facets and Biperfect Graphs," with E. Johnson, *Discrete Math.* **16** (1982).
24. "Solving Large-Scale Zero-One Linear Programming Problems," with H. Crowder and E. Johnson, *Oper. Res.* **31** (1983).
25. "Trees and Cuts," with L. Wolsey, *Discrete Math.* **17** (1983).
26. "Valid Linear Inequalities for Fixed Charge Problems," with T. Van Roy and L. Wolsey, *Oper. Res.* **33** (1985).
27. "Polyhedral Aspects of the Traveling Salesman Problem I and II," with M. Grötschel, in E.L. Lawler et al. (eds.), *The Traveling Salesman Problem*, Wiley & Sons, Chichester, New York, 1985.
28. "A Different Convergence Proof of the Projective Method for Linear Programming," *OR Letters* **4** (1986).
29. "Total Unimodularity and the Euler Subgraph Problem," *OR Letters* **7** (1988).
30. "A Polynomial-Time Solution to Papadimitriou and Steiglitz's 'Traps,' " with T.-Y. Sung, *OR Letters* **7** (1988).
31. "The Boolean Quadric Polytope: Some Characteristics, Facets and Relatives," *Math. Program.* B **45** (1989).
32. "An Efficient Algorithm for the Minimum Capacity Cut Problem," with G. Rinaldi, *Math. Program.* **47** (1990).
33. "Facet Identification for the Symmetric Traveling Salesman Problem," with G. Rinaldi, *Math. Program.* **47** (1990).
34. "An Analytical Comparison of Different Formulations of the Traveling Salesman Problem," with T.-Y. Sung, *Math. Program.* B **52** (1991).
35. "Improving the LP-Representation of Zero-One Linear Programming Problems for Branch and Cut," with K. Hoffman, *ORSA J. on Computing* **3** (1991).
36. "A Branch-and-Cut Algorithm for the Resolution of Large-Scale Symmetric Traveling Salesman Problems," with G. Rinaldi, *SIAM Rev.* **33** (1991).
37. "Lehman's Forbidden Minor Characterization of Ideal 0-1 Matrices," *Discrete Math.* **111** (1993).
38. "Solving Airline Crew-Scheduling Problems by Branch-and-Cut," with K. Hoffman, *Management Sci.* **39** (1993).
39. "Order Preserving Assignments," with D. Alevras, *Naval Res. Logistics* **41** (1994).
40. "An Analytical Symmetrization of Max Flow - Min Cut," with T.-Y. Sung, *Discrete Math.* **165/166** (1997).
41. "Optimal Project Selection when Borrowing and Lending Rates Differ," with M. Wilczak, *Math. Comput. Modelling* **29** (1999).
42. "Packing Small Boxes into a Big Box," *Math. Methods Oper. Res.* **52** (2000).
43. "Almost Perfect Matrices and Graphs," *MOR* **26** (2001).
44. "Classical Cuts in Mixed-Integer Programming and Branch-and-Cut," *Math. Methods Oper. Res.* **53** (2001).

Chapter 2

Time for Old and New Faces*

Laurence Wolsey†

2.1 Introduction

It is an impossible task for me to condense Manfred Padberg's work into one short talk. First of all I haven't read or cannot understand all of his papers, and in any case the scientific program is full of talks by specialists on the different areas in which he has made significant contributions.

So I have decided to look back, take a quick overview of some of these areas, and perhaps point out one or two unanswered questions raised along the way, as well as how our viewpoint has changed over the last 30 years. So this will be a look at a variety of familiar old faces or facets, and will finish with a few questions about where some interesting new faces might be found.

2.1.1 Why am I talking?

This is a very legitimate question. All I can say is that both Manfred and I finished our Ph.D.s in the U.S. around 1970 and headed back to Europe. Manfred worked from 1971 to 1974 in Berlin at the International Institute for Management, and I went to Core for nine months. I remember visiting him once in Berlin to give a seminar, and all I can remember is being taken to see the Wall. We were also regularly on call for seminars and workshops that were held at the new Institut für Unternehmungsforschung in Bonn.

I should perhaps point out that when I was asked to prepare this talk by Martin Grötschel about a year ago, he assured me that all the other speakers would be under 50.

---

*This is a written version of the opening talk given at the workshop "The Sharpest Cut."

†CORE and INMA, Université catholique de Louvain, Louvain-la-Neuve, Belgium (wolsey@core.ucl.ac.be).

Taking this as given, it follows that you were all still wearing nappies in the 1970s. Therefore I do not have to tolerate any discussion about what happened during that decade, except with Manfred, and that should be enough to keep the two of us busy arguing for a decade or so.

2.1.2 Terminology

Before trying to make a list of the areas in which Manfred has left his mark, I would like to make a contribution to the terminology used by our community. Here I must say that Manfred and I have never seen eye to eye.

He has introduced one of the most horrible words I have ever seen:

FACETIAL.

What I would like to use is unforgettable, has punch, and will save us all from repetitive strain injury:

PDF.

I leave you to choose what it stands for.

2.1.3 Outline

Now I will try and be serious, so here is a list of the areas that I would like to touch on:

1. Set packing and partitioning;
2. Perfect matrices;
3. Traveling salesman problem (TSP): Theory and computation;
4. Knapsacks, etc.: Theory and computation.

There are several other interesting topics that I will ignore, such as polynomial separation algorithms, pdfs for facility location, pdfs for quadratic 0/1-problems, the ellipsoid algorithm and its consequences, portfolio optimization, and chance-constrained programming.

2.2 Set Packing and Partitioning

This should probably be treated as two topics, but both presumably formed part of Manfred's Ph.D. thesis.

In [22] he developed pdfs for the set packing polytope, showing that clique inequalities are facet-defining and that odd-hole inequalities can be lifted sequentially to produce one or more pdfs. He pointed out that this led to facets with coefficients that were not just 0/1, but that could take all integer values between 0 and $s = \frac{1}{2}(|S| - 1)$, where S is the set of nodes of the odd hole.

This has always been a crucial paper for me in that it explicitly introduced "sequential lifting."

To get an idea of our state of knowledge at the time, I quote: "This, ..., makes it unlikely that an equally elegant and efficient algorithm for the node covering problem can be found as the one Edmonds has developed for the edge-matching problem and which uses implicitly the facets of the matching polyhedron. This observation coincides with the conclusions reached along different lines by Balinski [6] and Karp [20]."

Turning now to set partitioning, Balas and Padberg consider the polytopes

$$P_R = \{x \in R^n_+ : Ax = 1\} \text{ and } P_I = \text{conv}\{x \in Z^n_+ : Ax = 1\}.$$

Based on an observation of Trubin [34] showing that the edges of P_I are all edges of P_R, allowing the possibility of an algorithm moving along edges of P_R from one 0/1-feasible vertex to another, they show that every feasible integer basis has at least $n - m$ adjacent integer bases, and we have the following theorem.

Theorem 2.1 (Balas and Padberg [3]). *If x^1, x^2 are two basic feasible integer solutions with x^1 not optimal, then there exists a sequence of adjacent bases $x^1 = x^{10}, x^{11}, \ldots, x^{1p} = x^2$ such that*
(i) *the basic solutions are integer and feasible;*
(ii) *every row a^T with row sum β that is not a row of A_1 is a copy of a row of A_1*
(iii) *every other row a^T that is not a row of A_1 has a row sum less than β.*

So one sees that both a primal algorithm and the Hirsch conjecture are part of their thinking.

This was followed by several other papers during the 1970s, in particular the surveys [4] and [5].

From this exciting start Padberg then developed this work in several directions, specializing to perfect graphs, moving to another challenging combinatorial 0/1-polytope TSP, and generalizing to more general 0/1-independence systems.

2.3 Perfect Matrices

In [23] Manfred continued his study of pdfs for set packing by studying the incidence matrices of the cliques of perfect graphs. Lovasz had recently proved the weak perfect graph theorem: *A graph is perfect if and only if its complement is perfect*, and, using results of Chvátal and Fulkerson, this implies that a polytope

$$P = \{x \in R^n_+ : Ax \le 1\}$$

is integral if and only if A is the clique matrix of a perfect graph. Manfred studied minimal imperfect graphs and the corresponding matrices, developing a variety of new results, in particular a characterization of perfect matrices in terms of the nonexistence of a class of submatrices.

Definition 2.2. An $m \times n$ $0/1$-matrix A with $m \ge n$ has property $\pi_{\beta,n}$ if
(i) A contains an $n \times n$-nonsingular submatrix A_1 whose row and column sums equal β;
(ii) every row a^T with row sum β that is not a row of A_1 is a copy of a row of A_1;
(iii) every other row a^T that is not a row of A_1 has a row sum less than β.

Theorem 2.3. *Let A be a $0/1$-matrix of size $m \times n$. A is a perfect matrix if and only if A does not contain any $m \times k$ submatrix A' having property $\pi_{\beta,k}$ for $\beta \ge 2$ and $3 \le k \le n$.*

Much more recently he has worked on ideal matrices [26], and he has also returned to the topic of almost perfect matrices [27].

2.4 The Traveling Salesman Problem

2.4.1 Adjacency

Still looking at adjacency, Padberg and Rao [30] show that the diameter of a large class of combinatorial polytopes, including the TSP polytope, is two. They also show that a weak form of the Hirsch conjecture holds for the TSP polytope. Clearly, at the time, it was still thought that small diameter or satisfaction of the Hirsch conjecture might lead to a good algorithm for these problems. Even though this is no longer the case, as all 0/1-polytopes satisfy the Hirsch conjecture [21], it might be time to look again at such questions and see if there is not some stronger (and nontrivial) property leading to polynomial algorithms.

2.4.2 pdfs

As I mentioned earlier, Manfred made regular visits to Bonn, where Korte had an assistant, working on his doctorate, named Martin Grötschel. Somehow this led to a very fruitful collaboration studying the pdfs of the TSP polytope, starting with a 1974 publication [10] in German. In 1975 they had a short note in *Mathematical Programming* on pdfs for the asymmetric TSP, and among others in 1979 two papers on the symmetric TSP (STSP) on (I) *Inequalities*, and (II) *Lifting Theorems and Facets*.

These two papers [12, 13] can be seen as the prototype for all the polyhedral studies over the next 20 years.

In the first paper the following points are dealt with:

- introduction of a new class of "comb" inequalities generalizing Chvátal combs [7];
- the problem that the polytope is not full dimensional;
- analysis of which inequalities lead to the same face;
- counting the number of distinct faces from subtour inequalities and combs;
- calculating the dimension of the polytope;
- showing which trivial inequalities are pdfs;
- studying small polytopes.

In the second paper,

- several lifting theorems are developed showing that, if $ax \leq a_0$ is a pdf for P_n^{TSP}, then $a'x + b'y \leq a'_0$ is a pdf for P_{n+m}^{TSP};
- conditions are derived under which subtour and comb inequalities are pdfs.

In the last paragraph they mention that the Petersen graph and hypohamiltonian graphs also lead to pdfs. In this particular case I have a personal bias: I would call them "pretty dumb facets."

2.4.3 Computation

The first paper mentioning computation for the STSP [11] appeared in 1978 and was quickly followed by many others. In [29] results with a primal algorithm are presented, and in [9] large problems with up to 318 nodes are solved. A few years later Rinaldi spent some time in New York, which led to solution of the 532 city problem by branch-and-cut, followed by [32] in *SIAM Review*, which discusses many interesting aspects of branch-and-cut, including separation. On the topic of separation I must just mention the paper with Rao on separating odd minimum cutsets [31].

2.5 Knapsacks, etc.

Manfred soon realized that lifting could be generalized to independence systems [24], and around 1973 several others got in on the act, developing pdfs for the knapsack problem and further generalizing the ideas of sequential (and then nonsequential) lifting.

One of the challenges we discussed then, which remains to this day, is how to handle two constraints simultaneously. Hammer, Padberg, and Peled [15] give an answer just for the development of logical (two-variable) inequalities, but we know next to nothing about pdfs.

2.5.1 $1 - k$ configurations

For the special knapsack polytope that he called a $1-k$ configuration [25], Manfred showed that he had a complete description of the convex hull.

Written as $\{(x, y) \in \{0, 1\} \times \{0, 1\}^n : kx + \sum_{j=1}^n y_j \geq k\}$, it seemed surprising at the time that this trivially simple set should have so many pdfs:

$$(k - t)x + \sum_{j \in S} y_j \geq k - t \;\; \forall t = 0, 1, \ldots, k - 1, \;\; \forall S : |S| = n - t.$$

Very recently it has been observed [2] that this set and pdf characterization can be viewed as a very special case of the disjunction involving (upper-)monotone polytopes P and Q:

$$\text{Either } x \in P \text{ or } y \in Q.$$

2.5.2 Flow covers

Manfred spent his 1981–82 sabbatical at Core along with Cornuéjols, Conforti, and Hartvigsen. We had just started looking at mixed 0/1-problems and, in particular, the *single node inflow set*

$$\sum_{j \in N} x_j \leq b, \;\; 0 \leq x_j \leq a_j y_j, \;\; y_j \in \{0, 1\}, \;\; j \in N.$$

After several days or weeks, we both arrived in the office one morning—"I've got it," and a new pdf, the flow cover inequality, had been defined [33].

2.5.3 Computation

In 1983 the path-breaking paper of Crowder, Johnson, and Padberg, "Solving large-scale zero-one linear programming problems," appeared in *Operations Research* [8]. Here they showed how the theoretical studies of facets for knapsack polytopes dating from 1974 could be put to use in a general code. They formalized the separation problem for cover inequalities for 0/1-knapsack sets—"Find an inequality of the form $\sum_{j\in C} x_j \leq |C| - 1$ cutting off a fractional point $x^* \in [0, 1]^n$"—as the 0/1-knapsack problem

$$\min \sum_{j\in N}(1 - x_j^*)z_j : \sum_{J\in N} a_j z_j > b,\ z \in \{0, 1\}^n,$$

solved this knapsack problem by a greedy heuristic to find a good cover C, and then sequentially lifted the cover inequality to make it into a pdf (pretty decent facet).

Manfred pursued this work over several years. In [17] a great deal of work was done on preprocessing, in [18] results of a branch-and-cut system for airline crew scheduling are presented, and cuts and branch-and-cut are again discussed in one of his most recent papers [28].

2.6 New Faces, or Whither Branch-and-Cut?

2.6.1 Today

I will start with a few observations about where we are now. Commercial mixed-integer programming systems have discovered cuts in the last two to three years, including not just the lifted knapsack inequalities just mentioned above but also Gomory mixed-integer cuts and mixed-integer rounding (MIR) cuts, which have both been found to be surprisingly effective. Another useful option is that of using model cuts, i.e., a set of constraints that are introduced a priori, but are treated as cuts, so that only those that are tight are kept as part of the active problem. There has also been some progress due to improved primal heuristics and modifications in the tree search strategies.

2.6.2 Tomorrow?

Dual questions, general problems. From the research and development point of view it is natural to ask where the next cuts will come from. One of the commercial software vendors claims to have tried out everything in the literature—one possible candidate is the class of mixing inequalities obtained from MIR inequalities [14]. In developing a branch-and-cut system the question arises of whether to use local or global cuts, and whether to use valid inequalities or pdfs. Manfred has always been a strong advocate of global cuts and pdfs. For the first question, tests to date do not appear conclusive, and the question may not be important. On the second almost everyone would agree that pdfs are best, but every valid inequality is a pdf of some relaxation!

Dual questions, problems with structure. Here it may be of interest to catalog pdfs for small instances of knapsack sets and their generalizations. There is also a variety of mixed-integer programming sets for which both extended formulations and valid inequalities are known. Both a priori reformulations and cutting planes have their advantages and disadvantages, so a better understanding of the trade-offs may be important. Finally, linear program (LP) solvers have become so efficient that one may start to see extended formulations used for cut generation.

Primal and other questions. As the mixed-integer programs (MIPs) we try to solve get larger and larger, a crucial problem is that of finding feasible solutions when the LPs take a long time to solve and the optimal LP solutions contain a large number of integer variables at fractional values. So primal heuristics remain a major challenge. Given these difficulties, the idea of primal algorithms is very attractive, so the recent research in this area, which can be viewed as a major generalization of the work of Balas and Padberg, cited above [3], is being followed with great interest [16]. Another possible idea is to use different formulations for primal and dual—a possible large extended formulation to give a tight dual bound, and a smaller formulation to obtain primal solutions more easily when branching.

In branch-and-cut one important idea is that of treating the "restricted" problem at a node of the tree with all the machinery available. This raises questions such as whether more effort should be spent in preprocessing and reformulating the original problem, whether much more work (preprocessing, cutting, primal heuristics) should be carried out at each node, and whether new branching objects, such as disjunctions, cardinality constraints, and/or aggregate variables, should be developed.

A final algorithmic question concerns the possible use of other relaxations or reformulations. Can the integer programming algorithm of Lenstra [19], which is polynomial for fixed n, or the recent algorithm of Aardal et al. [1], using lattice reformulations, be developed into useful algorithms for some classes of problems?

Solving MIPs also depends on computational speed and modeling languages. In developing a branch-and-cut solver for an MIP with multiple processors, how should the work be divided between processors? Given the development of more powerful modeling languages, should they provide facilities to develop model-dependent heuristics or optimizing strategies?

Final question—How to share knowledge? Further developments of the field to which Manfred has contributed so significantly depend on the development and publication of new results, easy access to these publications, availability of state-of-the-art libraries of test instances for different problem classes, and access to state-of-the-art solvers. So how can the research community, industrial users of integer programming, and the commercial software developers ensure a balanced and mutually beneficial transfer and exchange?

Bibliography

[1] K. Aardal, C.A.J. Hurkens, and A.K. Lenstra. Solving a system of linear diophantine equations with lower and upper bounds on the variables. *Mathematics of Operations Research*, 25(3):427–442, 2000.

[2] E. Balas, A. Bockmayr, N. Psaruk, and L.A. Wolsey. On Unions of Polytopes. Mimeo, 2002.

[3] E. Balas and M.W. Padberg. On the set-covering problem. *Operations Research*, 20(3):1152–1161, 1972.

[4] E. Balas and M.W. Padberg. Set partitioning: A survey. *SIAM Review*, 18:710–760, 1976.

[5] E. Balas and M.W. Padberg. Set partitioning: A survey. In *Proceedings of 1977 Summer School Sogesta, Urbino*, pages 151–210, John Wiley, Chichester, 1979.

[6] M.L. Balinski. On maximum matching, minimum covering and their connections. In H. Huhn, editor, *Proceedings of the Princeton Symposium on Mathematical Programming*, Princeton University Press, Princeton, NJ, 1970.

[7] V. Chvátal. Edmond's polytopes and weakly Hamiltonian graphs. *Mathematical Programming*, 5:29–40, 1973.

[8] H. Crowder, E.L. Johnson, and M.W. Padberg. Solving large-scale zero-one linear programming problems. *Operations Research*, 31:803–834, 1983.

[9] H. Crowder and M.W. Padberg. Solving large-scale symmetric travelling salesman problems to optimality. *Management Science*, 26:495–509, 1980.

[10] M. Grötschel and M.W. Padberg. Zur oberflächenstruktur des travelling saleman polytopen. In *Proceedings in Operations Research*, volume 4, pages 207–211, Physica-Verlag, Würzburg, 1974.

[11] M. Grötschel and M.W. Padberg. On the symmetric travelling saleman problem: Theory and computation. In R. Henn et al., editors, *Optimization and Operations Research*, pages 105–115. Springer-Verlag, Berlin, New York, 1978.

[12] M. Grötschel and M.W. Padberg. On the symmetric travelling salesman problem I: Inequalities. *Mathematical Programming*, 16:265–280, 1979.

[13] M. Grötschel and M.W. Padberg. On the symmetric travelling salesman problem II: Lifting theorems and facets. *Mathematical Programming*, 16:281–302, 1979.

[14] O. Günlük and Y. Pochet. Mixing MIR inequalities for mixed integer programs. *Mathematical Programming*, 90:429–458, 2001.

[15] P.L. Hammer, M.W. Padberg, and U.N. Peled. Constraint pairing in integer programming. *INFOR Canadian Journal of Operations Research*, 13:68–81, 1975.

[16] U. Haus, M. Köppe, and R. Weismantel. The integral basis method for integer programming. *Mathematical Methods in Operations Research*, 53:353–361, 2001.

[17] K.L. Hoffman and M.W. Padberg. Improving LP-representations of zero-one linear programs for branch-and-cut. *ORSA Journal of Computing*, 3:121–134, 1991.

[18] K.L. Hoffman and M.W. Padberg. Solving airline crew scheduling problems by branch-and-cut. *Management Science*, 39:657–681, 1993.

[19] H.W. Lenstra, Jr. Integer programming with a fixed number of variables. *Mathematics of Operations Research*, 8:538–548, 1983.

[20] R.M. Karp. Reducibility and combinatorial problems. In R.E. Miller and J.W. Thatcher, editors, *Complexity of Computer Computations*, pages 85–103. Plenum Press, New York, 1972.

[21] D. Naddef. The Hirsch conjecture is true for 0-1 polytopes. *Mathematical Programming*, 45:109–110, 1989.

[22] M.W. Padberg. On the facial structure of set-packing polyhedra. *Mathematical Programming*, 5:199–215, 1973.

[23] M.W. Padberg. Perfect zero-one matrices. *Mathematical Programming*, 6:180–196, 1974.

[24] M.W. Padberg. A note on 0-1 programming. *Operations Research*, 23:833–837, 1975.

[25] M.W. Padberg. $(1, k)$-configurations and facets for packing problems. *Mathematical Programming*, 18:94–99, 1980.

[26] M.W. Padberg. Lehman's forbidden minor characterization of ideal 0-1 matrices. *Annals of Discrete Mathematics*, 111:409–420, 1993.

[27] M.W. Padberg. Almost perfect matrices and graphs. *Mathematics of Operations Research*, 26:1–18, 2001.

[28] M.W. Padberg. Classical cuts for mixed integer programming and branch-and-cut. *Mathematical Methods of Operations Research*, 53:173–203, 2001.

[29] M.W. Padberg and S. Hong. On the symmetric travelling saleman problem: A computational study. *Mathematical Programming Studies*, 12:78–107, 1980.

[30] M.W. Padberg and M.R. Rao. The travelling salesman problem and a class of polyhedra of diameter two. *Mathematical Programming*, 7:32–45, 1974.

[31] M.W. Padberg and M.R. Rao. Odd minimum cut-sets and b-matchings. *Mathematics of Operations Research*, 7:67–80, 1982.

[32] M.W. Padberg and G. Rinaldi. A branch-and-cut algorithm for the resolution of large-scale symmetric traveling salesman problems. *SIAM Review*, 33:60–100, 1991.

[33] M.W. Padberg, T.J. Van Roy, and L.A. Wolsey. Valid linear inequalities for fixed charge problems. *Operations Research*, 33:842–861, 1985.

[34] V.A. Trubin. On a method of solution of integer linear programming problems of a special kind. *Soviet Mathematics Doklady*, 10:1544–1546, 1969.

Part II

Packing, Stable Sets, and Perfect Graphs

Chapter 3
Combinatorial Packing Problems

*Ralf Borndörfer**

MSC 2000. 90C27
Key words. Packing problems, polyhedral combinatorics

3.1 Introduction

This chapter investigates a certain class of combinatorial packing problems (CPPs) and some polyhedral relations between such problems and the set packing problem (SPP).

Packing constraints are one of the most common problem characteristics in combinatorial optimization. They occur in path packing formulations of vehicle and crew scheduling problems, in Steiner tree packing approaches to VLSI and network design problems, and in coloring models of frequency assignment problems; see [38, 16] for surveys. The pure form of a packing problem is the *SPP* or *stable set problem* in a graph $G = (V, E)$ with node weights w; it asks for a maximum weight set of mutually nonadjacent nodes. This problem has been studied extensively, and deep structural and algorithmic results have been achieved in areas such as antiblocking theory, the theory of perfect graphs, perfect and balanced matrix theory, and semidefinite programming (SDP); see [7, 20, 34, 8] for surveys. There is, in particular, a substantial structural and algorithmic knowledge of the set packing polytope, with many classes of strong and polynomial-time separable inequalities such as odd hole, odd antihole, and orthonormal representation constraints [35, 33, 44, 37, 20].

Several research directions try to translate some of these results into broader settings. A first line investigates generalizations of set packing, such as node packing in hypergraphs

*Konrad-Zuse-Institute for Information Technology Berlin, Takustr. 7, 14195 Berlin, Germany (borndoerfer@zib.de).

[41], independence systems [33, 36, 13, 26], transitive packing [30, 31, 32, 40], and mixed-integer packing [2, 3]. This work aims for a unified polyhedral theory. A second direction is the theory of matrix cuts [27], which generalizes the semidefinite separation machinery that had been developed for the solution of the stable set problem in perfect graphs [20] to arbitrary 0/1-programs. A third direction studies the construction of discrete set packing relaxations [9, 10]; see also [39]. This technique allows us to transfer set packing inequalities and separation algorithms to other combinatorial problems.

Our aim in this chapter is to continue in this general direction. We consider a class of combinatorial optimization problems of packing type where a Dantzig–Wolfe decomposition gives rise to a canonical, yet exponential, set packing formulation, namely, the formulation that one would use in a column generation approach. This alternative formulation allows us, at least in principle, to understand CPPs completely in terms of set packing theory. We show that such Dantzig–Wolfe set packing formulations of CPPs have structural properties that relate them to the original formulation and make them interesting sources of cutting planes.

This chapter consists of two parts. In Section 3.2 we introduce the concept of combinatorial packing. We give two examples of such problems, namely, on packings of two stable sets in bipartite graphs and independent sets in any number of matroids, which are naturally integral. Dantzig–Wolfe set packing formulations of CPPs are discussed in Section 3.3. It is shown that such formulations give rise to cutting planes and that the intersection graphs associated with Dantzig–Wolfe formulations of combinatorial 2-packing problems are perfect.

3.2 Combinatorial Packing

We introduce in this section the notion of combinatorial packing. This concept subsumes a variety of combinatorial optimization problems, among them the Steiner tree packing problem (PST), the multicommodity flow problem (MCFP) with unit capacities, the multiple knapsack problem (MKP), and the coloring problem. It will turn out that, for some problems of this type, namely, the 2-coloring problem in bipartite graphs and the matroid packing problem, the integrality of the individual subproblems carries over to the packing composition.

Consider a family of some number k of combinatorial optimization problems

$$(\mathrm{IP}^i) \qquad \max\ {c^i}^{\mathrm{T}} x^i \text{ subject to } M^i x^i \le b^i,\ 0 \le x^i \le \mathbb{1},\ x^i \in \mathbb{Z}^E, \qquad i = 1, \dots, k, \tag{3.1}$$

on the same ground set E. These are the *individual problems*. Associated with each of them is an *individual polytope* $P^I_{\mathrm{IP}^i} = \mathrm{conv}\{x^i \in \{0,1\}^E \mid M^i x^i \le b^i\}$ and its fractional relaxation $P_{\mathrm{IP}^i} = \{0 \le x^i \le \mathbb{1} \mid M^i x^i \le b^i\}$. An individual problem with the property $P^I_{\mathrm{IP}^i} = P_{\mathrm{IP}^i}$ is called *integral*.

A *packing* is a collection of individual solutions $x^1, \dots, x^k$ of $\mathrm{IP}^1, \dots, \mathrm{IP}^k$, respectively, such that each element of the ground set is contained in at most one solution. The problem of finding a maximum weight packing is the *CPP* associated with the individual problems IP^i, $i = 1, \dots, k$. A CPP with k individual problems is a *(combinatorial) k-packing problem.* The integer programming formulation of a CPP reads

$$
\begin{array}{lrll}
\text{(CPP)} & \max \sum_{i=1}^{k} {c^i}^{\mathrm{T}} x^i \text{ subject to} & & \\
\text{(i)} & M^i x^i \le b^i, & i = 1, \dots, k; & \\
\text{(ii)} & x^i \ge 0, & i = 1, \dots, k; & \\
\text{(iii)} & \sum_{i=1}^{k} x^i \le \mathbb{1}; & & \\
\text{(iv)} & x^i \in \mathbb{Z}^E, & i = 1, \dots, k. &
\end{array}
\tag{3.2}
$$

We call CPP (iii) the *packing constraints.* It will be convenient to use the notation $x^{\mathrm{T}} = ({x^1}^{\mathrm{T}}, \dots, {x^k}^{\mathrm{T}})$ and $c^{\mathrm{T}} = ({c^1}^{\mathrm{T}}, \dots, {c^k}^{\mathrm{T}})$. Likewise, we shall view the ground set of a combinatorial k-packing problem as a disjoint union $\dot{\bigcup} E^i = E^1 \,\dot{\cup}\, \cdots \,\dot{\cup}\, E^k$ of copies of the ground sets of the individual problems, where E^i is the copy of the ground set of problem IP^i. Associated with the CPP are, finally, the *combinatorial packing polytope* and its fractional relaxation

$$
\begin{aligned}
P_{\mathrm{CPP}}^I &= \operatorname{conv}\{x \in \{0,1\}^{\dot{\cup} E^i} \mid \textstyle\sum_{i=1}^{k} x^i \le \mathbb{1},\ M^i x^i \le b^i,\ i = 1, \dots, k\}, \\
P_{\mathrm{CPP}} &= \{x \in [0,1]^{\dot{\cup} E^i} \mid \textstyle\sum_{i=1}^{k} x^i \le \mathbb{1},\ M^i x^i \le b^i,\ i = 1, \dots, k\}.
\end{aligned}
\tag{3.3}
$$

A CPP is *integral* if $P_{\mathrm{CPP}}^I = P_{\mathrm{CPP}}$. If all individual problems as well as CPP itself are integral, we say that CPP is *naturally integral.*

3.2.1 Examples of Combinatorial Packing Problems

The MCFP with unit capacities. This problem involves a *supply digraph* $D_S = (V, A_S)$ and a *demand digraph* $D_D = (V, A_D)$, both on the same nodeset V. We denote an arc from a node s to a node t in these digraphs by st. There are nonnegative weights $w \in \mathbb{Q}_+^{A_S}$ on the arcs A_S of the supply digraph. A *multiflow* is a collection of pairwise arc disjoint directed st-paths in D_S, one for each arc $st \in A_D$ of the demand digraph. The *MCFP* asks for a multiflow of minimum weight [1, 16, 12].

The MCFP is a combinatorial path packing problem. The individual problems are shortest path problems, one for each demand arc $st \in A_D$:

$$
\begin{array}{rll}
\min w^{\mathrm{T}} x^{st} \text{ subject to} & & \\
x^{st}(\delta^+(v)) - x^{st}(\delta^-(v)) = e_s - e_t & \forall v \in V, & \\
0 \le x^{st} \le \mathbb{1}, & & \\
x^{st} \in \mathbb{Z}^{A_S}. & &
\end{array}
\tag{3.4}
$$

Combining the shortest path problems in a CPP adds the packing constraints $\sum_{st \in A_S} x^{st} \le \mathbb{1}$ that model the edge disjointness of the paths.

The PST. This involves a graph $G = (V, E)$, some number k of sets of *terminal* nodes $T^1, \dots, T^k \subseteq V$, and nonnegative edge weights $w^1, \dots, w^k \in \mathbb{Q}_+^E$. The *PST* is to find a collection of Steiner trees $S^1, \dots, S^k$ spanning the terminals $T^1, \dots, T^k$, respectively, such that no two Steiner trees have an edge in common [29, 23, 21, 22, 24]. Note that terminal sets of two nodes will be joined by paths such that the PST subsumes the MCFP.

The PST is a CPP. The individual problems, one for each terminal set T^i, $i = 1, \dots, k$, are Steiner tree problems:

$$\begin{array}{ll} \min\ {w^i}^{\mathrm{T}} x^i \text{ subject to} & \\ \quad x^i(\delta(W)) \geq 1 & \forall W \subseteq V:\ W \cap T^i \neq \emptyset \neq (V \setminus W) \cap T^i, \\ 0 \leq x^i \leq \mathbb{1}, & \\ \quad x^i \in \mathbb{Z}^E. & \end{array} \tag{3.5}$$

Combining the problems in a CPP forces the Steiner trees to be edge disjoint.

The generalized assignment problem (GAP). This deals with a set of *jobs* J to be processed by a set of *machines* I with capacities α^i. There are resource demands a^i_j and profits w^i_j for the assignment of job j to machine i. The *GAP* is to find a maximum profit assignment of jobs to machines [28, 18]. The special case where the resource demands and availabilities do not depend on the machines, i.e., when $a^i = a^k$ and $\alpha^i = \alpha^k$ for all $i, k \in I$, is known as the *MKP* [28, 14, 15].

The GAP models combinatorial packings of job-machine assignments. There is an individual knapsack problem for each of the machines $i \in I$:

$$\max\ {w^i}^{\mathrm{T}} x^i \text{ subject to } {a^i}^{\mathrm{T}} x^i \leq \alpha^i,\ 0 \leq x^i \leq \mathbb{1},\ x^i \in \mathbb{Z}^J. \tag{3.6}$$

The packing constraints forbid assignments of jobs to more than one machine.

The k-coloring problem. This involves a graph $G = (V, E)$ with node weights $w \in \mathbb{Q}^V_+$ and some number $k \in \mathbb{N}$ of colors. The *k-coloring problem* (k-COL) asks for a collection of k mutually disjoint stable sets (color classes) of maximum weight [43].

A combinatorial packing formulation of the k-COL problem is based on k individual stable set problems

$$\max\ w^{\mathrm{T}} x^i \text{ subject to } x^i_u + x^i_v \leq 1 \quad \forall uv \in E,\ 0 \leq x^i \leq \mathbb{1},\ x^i \in \mathbb{Z}^V, \tag{3.7}$$

one for each color $1 \leq i \leq k$. The packing constraints $\sum_{i=1}^k x^i \leq \mathbb{1}$ guarantee that each node can take at most one color.

We finish our list of examples here and remark that, in the same way, graph decomposition problems; constrained path packing problems that arise, e.g., in vehicle routing and duty scheduling; and a variety of other problems are also CPPs.

3.2.2 Natural integrality

The example of the MCFP shows that CPPs can be hard even if all of the individual subproblems are easy and, in particular, even if complete descriptions of the individual polyhedra are explicitly known. There are, however, cases where the integrality of the individual problems carries over to the entire CPP. We give now two examples of CPPs that have this natural integrality property.

The bipartite 2-coloring problem. BIP-2-COL is the special case of the 2-coloring problem where $G = (V, E)$ is a bipartite graph G. The individual problems are two SPPs in this graph G. Their integer programming formulations can be stated as

$$\max\ w^{\mathrm{T}}x^i \text{ subject to } Ax^i \le \mathbb{1}, x^i \ge 0,\ x^i \in \mathbb{Z}^V \qquad (i = 1, 2), \tag{3.8}$$

where $A = A(G)$ denotes the edge-node incidence matrix of G. It is well known (see, e.g., [34, III.1, Corollary 2.9]) that the edge-node incidence matrices of bipartite graphs are totally unimodular. Hence the individual coloring problems are integral.

The integer programming formulation of the entire BIP-2-COL problem reads

$$\begin{array}{rll} \text{(BIP-2-COL)} & \max\ w^{\mathrm{T}}x^1 + w^{\mathrm{T}}x^2 \text{ subject to} & \\ \text{(i)} & x_u^1 + x_v^1 \le 1 & \forall uv \in E, \\ \text{(ii)} & x_u^2 + x_v^2 \le 1 & \forall uv \in E, \\ \text{(iii)} & x_v^1 + x_v^2 \le 1 & \forall v \in V, \\ \text{(iv)} & x_v^1, x_v^2 \ge 0 & \forall v \in V, \\ \text{(v)} & x_v^1, x_v^2 \in \{0,1\}^V & \forall v \in V. \end{array} \tag{3.9}$$

Proposition 3.1. *The BIP-2-COL problem is naturally integral.*

Proof. We show that the constraint matrix of the BIP-2-COL problem is totally unimodular (t.u.). This is easily done by noting that BIP-2-COL can again be seen as an SPP in a larger bipartite graph H. Using the convention to view the ground set of a CPP as a disjoint union of the ground sets of the individual problems, this graph H has as its nodeset the ground set $V^1 \uplus V^2$ of the BIP-2-COL problem, where V^1 is a copy of the nodeset of the first individual coloring problem and V^2 a copy of the second nodeset. For every constraint BIP-2-COL(i) there is an edge u^1v^1 between the first copies u^1 and v^1 of nodes u and v; this edge is a copy of the respective edge uv in the first individual problem. Analogously, there is an edge u^2v^2 between the second copies u^2 and v^2 of nodes u and v for every constraint BIP-2-COL(ii); this edge is a copy of the respective edge uv in the second individual problem. The graph H thus contains two disjoint copies G^1 and G^2 of G, one on the nodes V^1, the other one on the nodes V^2. The only additional edges between these copies come from the constraints BIP-2-COL(iii). There is an edge v^1v^2 that joins the two copies of each original node for every packing constraint.

Let $X \uplus Y$ be a bipartition of the nodes of G. The nodes of H can be partitioned into corresponding copies X^1, Y^1, X^2, and Y^2. Edges run between X^1 and Y^1 (first copy G^1 of G), X^2 and Y^2 (second copy G^2 of G), X^1 and X^2 (packing constraints on the copies of X), and Y^1 and Y^2 (packing constraints on the copies of Y); see Figure 3.1. It follows that $(X^1 \cup Y^2) \cup (X^2 \cup Y^1)$ is a bipartition of H. $\square$

The matroid packing problem. The MPP involves some number k of not necessarily identical matroids on the same ground set E with not necessarily identical nonnegative weights $w^1, \ldots, w^k \in \mathbb{Q}_+^E$. The *MPP* is to find a maximum weight collection of independent sets, one from each matroid, such that no two independent sets intersect on a common element.

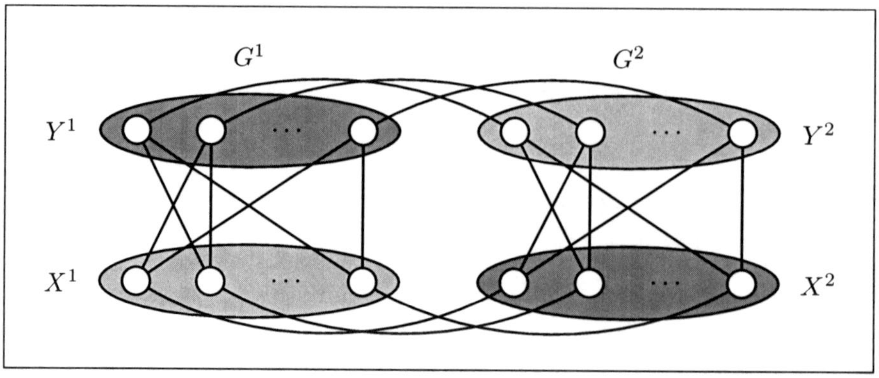

Figure 3.1. *BIP-2-COL problem.*

The MPP can be stated as the following integer program (IP):

$$\begin{array}{llll} \text{(MPP)} & \max \sum {w^i}^{\mathrm{T}} x^i \text{ subject to} & & \\ \text{(i)} & x^i(F) \le r^i(F) & \forall F \subseteq E, & i = 1, \ldots, k, \\ \text{(ii)} & x^i \ge 0, & & i = 1, \ldots, k, \\ \text{(iii)} & \sum x^i \le \mathbb{1}, & & \\ \text{(iv)} & x^i \in \{0, 1\}^E, & & i = 1, \ldots, k. \end{array} \tag{3.10}$$

Here r^i denotes the rank function of matroid i. It is known (see, e.g., [34, Theorem 3.5]) that the individual matroid problems are integral.

Proposition 3.2. *The MPP is naturally integer.*

Proof. The reason for the natural integrality of the MPP is that this problem can be reinterpreted as a matroid intersection problem involving two matroids. Both of these matroids have $E^1 \cup \cdots \cup E^k$ as their ground set. The first matroid is simply the disjoint union of the k individual matroids. The second matroid is also a disjoint union of k matroids, namely, the $|E|$ uniform matroids that are induced by the packing constraints MPP(iii). Consider the packing constraint $\sum_{i=1}^{k} x_e^i \le 1$ for element e. The matroid that is associated with this constraint has as its ground set the set $\{e^1, \ldots, e^k\}$ of copies of the element e. The nontrivial independent sets of this matroid are precisely the one-element sets $\{e^1\}, \ldots, \{e^k\}$. The disjoint union of these $|E|$ uniform matroids forms the second matroid.

By definition, MPP(i) and (ii) are a complete polyhedral description for the first matroid. Trivially, MPP(iii) and (ii) are also a complete polyhedral description of the second matroid. It is, however, well known (see, e.g., [34, III.3, Theorem 5.9]) that the union of two such systems is a complete description of the polytope that is associated with the intersection of two matroids. □

Having seen two examples of naturally integral CPPs, a "converse" question that comes up is whether the integrality of the individual problems is a necessary condition for the natural integrality of a CPP. This is true if the individual problems are down monotone. The following example shows, however, that this is not true in general.

Example 3.3. Consider the combinatorial 2-packing problem

$$\begin{array}{rcrcrcrl} \max x_1 & + & x_2 & + & x_3 & + & x_4 & \text{subject to} \\ x_1 & + & x_2 & & & & & \geq 1, \\ 2x_1 & + & 2x_2 & & & & & \leq 3, \\ & & & & x_3 & + & x_4 & \geq 1, \\ & & & & 2x_3 & + & 2x_4 & \leq 3, \\ & & & & \multicolumn{3}{r}{x_1,\ x_2,\ x_3,\ x_4} & \geq 0, \\ x_1 & & & + & x_3 & & & \leq 1, \\ & & x_2 & & & + & x_4 & \leq 1, \\ & & & & \multicolumn{3}{r}{x_1,\ x_2,\ x_3,\ x_4} & \in \mathbb{Z}. \end{array} \tag{3.11}$$

The individual problems produce the polytopes $P_{\mathrm{IP}^i} = \operatorname{conv}\begin{pmatrix} 0 & \frac{1}{2} & 1 & 1 \\ 1 & 1 & 0 & \frac{1}{2} \end{pmatrix}$, $i = 1, 2$, which have fractional vertices. The entire CPP is, however, integral; its associated polytope is $P_{\mathrm{CPP}} = \operatorname{conv}\begin{pmatrix} 0 & 1 & 1 & 0 \\ 1 & 0 & 0 & 1 \end{pmatrix}^{\mathrm{T}} = P^I_{\mathrm{CPP}}$.

3.3 Dantzig–Wolfe Set Packing Formulations

CPPs give rise to a natural alternative set packing formulation via Dantzig–Wolfe decomposition. This connection creates the possibility of studying CPPs in terms of set packing theory. We show in this section that such Dantzig–Wolfe set packing formulations have interesting structural properties that make them potentially useful sources of cutting planes for CPPs.

Consider a CPP (3.2). Let $M^i \in \{0,1\}^{E \times \mathfrak{V}^i}$ be a matrix whose columns are the incidence vectors of the 0/1-solutions of the individual problem IP^i, $i = 1, \ldots, k$. Let us identify the index $\mathfrak{v} \in \mathfrak{V}^i$ of such a column $M^i_{\cdot\mathfrak{v}}$ with the set associated with that column, i.e., we view $\mathfrak{v}$ as a subset of the ground set E^i whose incidence vector is $M^i_{\cdot\mathfrak{v}}$ (i.e., $\chi^{\mathfrak{v}} = M^i_{\cdot\mathfrak{v}}$).

A Dantzig–Wolfe decomposition subject to the substitutions

$$x^i = M^i \lambda^i, \quad \mathbb{1}^{\mathrm{T}}\lambda^i = 1,\ \lambda^i \geq 0,\ \lambda^i \in \{0,1\}^{\mathfrak{V}^i}, \qquad i = 1, \ldots, k, \tag{3.12}$$

transforms (3.2) into the form

$$\begin{array}{lrll} \text{(XPP)} & \max \sum_{i=1}^k {w^i}^{\mathrm{T}} M^i \lambda^i & \text{subject to} & \\ \text{(i)} & \mathbb{1}^{\mathrm{T}}\lambda^i & = 1, & i = 1, \ldots, k, \\ \text{(ii)} & \sum_{i=1}^k M^i \lambda^i & \leq \mathbb{1}, & \\ \text{(iii)} & \lambda^i & \geq 0, & i = 1, \ldots, k, \\ \text{(iv)} & \lambda^i & \in \{0,1\}^{\mathfrak{V}^i}, & i = 1, \ldots, k. \end{array} \tag{3.13}$$

We call XPP the *Dantzig–Wolfe formulation* associated with CPP. Constraints XPP(i) are the *convexity constraints*, and XPP(ii) are the *packing constraints*. Introducing the notation $\lambda^{\mathrm{T}} = ({\lambda^1}^{\mathrm{T}}, \ldots, {\lambda^k}^{\mathrm{T}})$, $M = (M^1, \ldots, M^k)$, $C = \operatorname{diag}(\mathbb{1}^{\mathrm{T}})$, $w^{\mathrm{T}} = ({w^1}^{\mathrm{T}}, \ldots, {w^k}^{\mathrm{T}})$, and $\mathfrak{V} = \mathfrak{V}^1 \uplus \cdots \uplus \mathfrak{V}^k$, XPP becomes

$$\text{(XPP)} \quad \max\ w^{\mathrm{T}} M\lambda \text{ subject to } C\lambda = \mathbb{1},\ M\lambda \leq \mathbb{1},\ \lambda \geq 0,\ \lambda \in \{0,1\}^{\mathfrak{V}}. \tag{3.14}$$

XPP is closely related to the SPP

$$\text{(SPP)} \quad \max\ w^{\mathrm{T}} M\lambda \text{ subject to } C\lambda \leq \mathbb{1},\ M\lambda \leq \mathbb{1},\ \lambda \geq 0,\ \lambda \in \{0,1\}^{\mathfrak{V}}. \tag{3.15}$$

In fact, XPP arises from SPP by forcing the relaxed convexity constraints $C\lambda \leq \mathbb{1}$ to equality. This is, however, not an essential change. XPP can, e.g., be transformed into the form SPP by adding a suitably large constant $M \cdot \mathbb{1}$ to the objective. As a (modified) packing problem XPP can be restated in graph-theoretical language in terms of the intersection graph $\mathfrak{G} = (\mathfrak{V}, \mathfrak{E})$ that is associated with the constraint matrix $A = \binom{C}{M}$. This graph $\mathfrak{G}$ has a node $\mathfrak{v} \in \mathfrak{V} = \mathfrak{V}^1 \uplus \cdots \uplus \mathfrak{V}^k$ for each individual 0/1-solution. There is an edge $\mathfrak{u}\mathfrak{v}$ for any two individual solutions $\mathfrak{u}$ and $\mathfrak{v}$ that can not be simultaneously contained in a packing. This is the case either when $\mathfrak{u}$ and $\mathfrak{v}$ are both solutions of the same individual problem such that the columns $A_{\cdot\mathfrak{u}}$ and $A_{\cdot\mathfrak{v}}$ intersect on a convexity row, or when $\mathfrak{u}$ and $\mathfrak{v}$ both contain the same element $e \in E$, i.e., $A_{\cdot\mathfrak{u}}$ and $A_{\cdot\mathfrak{v}}$ intersect on the packing row associated with the element e. In terms of $\mathfrak{G}$, XPP is the problem of finding a maximum weight packing in $\mathfrak{G}$ such that each "convexity clique" is covered exactly once. This connection to set packing has polyhedral consequences. Consider the polytopes

$$\begin{aligned} P^I_{\mathrm{XPP}} &= \{\lambda \in \{0,1\}^{\mathfrak{V}} : C\lambda = \mathbb{1},\ M\lambda \leq \mathbb{1}\}, \\ P^I_{\mathrm{SPP}} &= \{\lambda \in \{0,1\}^{\mathfrak{V}} : A\lambda \leq \mathbb{1}\} \end{aligned} \tag{3.16}$$

associated with XPP and SPP and their respective fractional relaxations P_{XPP} and P_{SPP}. The polytope P^I_{SPP} is the set packing polytope associated with $\mathfrak{G}$ and P^I_{XPP} is a face of P^I_{SPP}. The combinatorial packing polytope can be obtained from P^I_{XPP} by projection.

Proposition 3.4. *P^I_{CPP} is the projection of the "extended set packing polytope"*

$$\{x \mid x = \operatorname{diag}(M^i)\lambda,\ \lambda \in P^I_{\mathrm{XPP}}\} \tag{3.17}$$

on the space of the x-variables.

Proposition 3.4 states that all facets of the combinatorial packing polytope are projections of set packing inequalities in some high dimensional space. This means that it is, at least in principle, possible to study CPPs in terms of set packing theory. We remark that such a study is necessary because a Dantzig–Wolfe formulation per se only contains information on the individual problems, not on packings. Namely, Proposition 3.4 implies the following relationship between CPP and XPP (see, e.g., [42, Section 2.3] for essentially the same result).

Corollary 3.5. *Let CPP be a combinatorial packing problem with integral individual problems and let XPP be its Dantzig–Wolfe formulation.*

Then the value of the linear programming relaxation of CPP is equal to the value of the linear programming relaxation of XPP.

For CPPs with integral individual problems such as the MCFP, one can therefore not gain much from just restating the problem in column generation form.

The natural way to exploit Proposition 3.4 algorithmically is by using *lift-and-project* techniques [4, 5]. Suppose we want to check some point $\overline{x}$ for membership in P^I_{CPP}. Suppose also for the moment that we have a complete description $D\lambda \le d$ of P^I_{XPP} at hand. Then, by the Farkas lemma,

$$\begin{aligned} & \overline{x} \notin P^I_{\text{CPP}} \\ \iff & \{\lambda \mid D\lambda \le d, \operatorname{diag}(M^i)\lambda = \overline{x}\} = \emptyset \qquad (3.18) \\ \iff & \exists a, b : a^{\mathrm{T}}D + b^{\mathrm{T}}\operatorname{diag}(M^i) \ge 0,\ a \ge 0,\ a^{\mathrm{T}}d + b^{\mathrm{T}}\overline{x} < 0. \end{aligned}$$

However, as $0 \le a^{\mathrm{T}}D\lambda + b^{\mathrm{T}}\operatorname{diag}(M^i)\lambda \le a^{\mathrm{T}}d + b^{\mathrm{T}}x$ is valid for any $x \in P^I_{\text{CPP}}$, the inequality

$$a^{\mathrm{T}}d + b^{\mathrm{T}}x \ge 0 \qquad (3.19)$$

is a valid inequality for P^I_{CPP} that is violated by $\overline{x}$; such a cut can be determined by solving an appropriate linear program (LP) (involving an additional normalization constraint to bound the recession cone).

Ignoring the technical difficulty of this projection process for the moment, the success of the procedure clearly depends on the quality of the description $D\lambda \le d$ for P^I_{XPP}. Knowledge of a *complete* description of P^I_{XPP} is surely an elusive goal in general. There are, however, significant cases where such a complete description is, in some sense, in fact available.

Proposition 3.6. *The intersection graph associated with the Dantzig–Wolfe formulation of a combinatorial 2-packing problem is perfect.*

Proof. We show that $\mathfrak{G}$ is the complement of a bipartite graph. The nodes of $\mathfrak{G}$ consist of the two sets $\mathfrak{V} = \mathfrak{V}^1 \cup \mathfrak{V}^2$ that correspond to the solutions of the first and the second individual problems, respectively. As there can be only one solution of each individual problem in a packing, the nodes of $\mathfrak{V}^1$ and $\mathfrak{V}^2$ form two cliques in $\mathfrak{G}$. These cliques are joined by the remaining edges, connecting solutions that have elements from the ground set in common; see Figure 3.2. In the complement graph $\overline{\mathfrak{G}}$, the sets $\mathfrak{V}^1$ and $\mathfrak{V}^2$ form two stable sets. Therefore they induce a bipartition in $\overline{\mathfrak{G}}$, and hence $\overline{\mathfrak{G}}$ is perfect. $\square$

Proposition 3.6 shows that all facets of combinatorial 2-packing polytopes are projections of *clique inequalities* [35]. The clique inequalities are subsumed by the larger class of *orthonormal representation constraints* that can be separated in polynomial time [20]. Proposition 3.6 suggests that such separation techniques, combined with lift-and-project

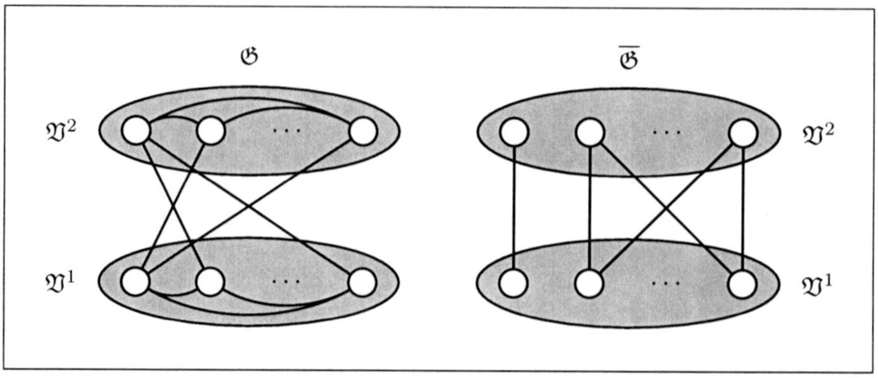

Figure 3.2. *Intersection graph of a combinatorial 2-packing problem.*

methods, are potentially useful tools for the solution of combinatorial 2-packing problems. We remark that such techniques can not, however, lead to polynomial-time algorithms for general combinatorial 2-packing problems, because this class contains NP-hard problems such as the 2-commodity flow problem with unit capacities [17, Problem ND38].

A practical use of lift-and-project cutting planes from Dantzig–Wolfe formulations can not be that one builds up a larger and larger description of P^I_{XPP} in the exponential space $\mathbb{R}^{\mathfrak{V}}$, adding more and more cutting planes and columns. Doing so would be equivalent to a combined column generation and cutting plane approach to CPPs with its well-known difficulties. Instead, we propose to accumulate cutting planes only in the compact original space $\mathbb{R}^{\bigcup E^i}$ and to use the Dantzig–Wolfe formulation solely as a separation tool.

The straightforward way to do this is as follows. Suppose we are given a point $\overline{x}$ to be tested for membership in P^I_{CPP}. The first step is to express $\overline{x}$ as a convex combination of individual solutions in the form $\overline{x}^i = M^i\lambda^i$, $i = 1, \ldots, k$. By Caratheodory's theorem, this can be done in such a way that the resulting multipliers λ^i have at most $|E| + 1$ nonzero components each. We then set up a subproblem of XPP that consists of the columns that appear in these convex combinations, apply whatever separation algorithms we have at hand, and add the resulting cuts. Projecting back, we have to be careful that our cut is dual feasible for the global XPP, i.e., we potentially have to lift a number of additional variables (this can happen because there may be more than one way to express $\overline{x}$ as a convex combination of 0/1-solutions). When this process results in a violated cutting plane for P^I_{CPP}, we add it to our current description of P^I_{CPP}, resolve, and iterate. The procedure that we have just sketched is admittedly expensive, but it points to a possible future algorithmic use of structural results such as Proposition 3.6.

We close this chapter with an example that is supposed to avoid a possible misunderstanding. Proposition 3.6 does *not* make a statement that would relate perfection of the constraint matrix A of a Dantzig–Wolfe formulation or its intersection graph $\mathfrak{G}$ to natural integrality of the original formulation. The obstacle that prevents us from establishing such a connection is that the linear programming relaxation of a Dantzig–Wolfe formulation can have fractional vertices that correspond to integral packings.

Example 3.7. Consider the following CPP with two uniform matroids of rank 2:

$$
\begin{array}{lcccccccccccl}
\max\ x_1^1 & + & x_2^1 & + & x_3^1 & + & x_1^2 & + & x_2^2 & + & x_3^2 & & \text{subject to} \\
x_1^1 & + & x_2^1 & + & x_3^1 & & & & & & & \le & 2, \\
 & & & & & & x_1^2 & + & x_2^2 & + & x_3^2 & \le & 2, \\
x_1^1 & & & & & + & x_1^2 & & & & & \le & 1, \\
 & & x_2^1 & & & & & + & x_2^2 & & & \le & 1, \\
 & & & & x_3^1 & & & & & + & x_3^2 & \le & 1, \\
x_1^1 & & x_2^1, & & x_3^1, & & x_1^2, & & x_2^2, & & x_3^2 & \ge & 0, \\
x_1^1 & & x_2^1, & & x_3^1, & & x_1^2, & & x_2^2, & & x_3^2 & \in & \mathbb{Z}.
\end{array}
\tag{3.20}
$$

By Proposition 3.2 the problem is naturally integral. The Dantzig–Wolfe formulation is

$$
\max\ (0,1,1,1,2,2,2)\lambda^1 + (0,1,1,1,2,2,2)\lambda^2 \text{ subject to}
$$

$$
\begin{pmatrix}
1 & 1 & 1 & 1 & 1 & 1 & 1 & & & & & & & \\
 & & & & & & & 1 & 1 & 1 & 1 & 1 & 1 & 1 \\
 & 1 & & & 1 & 1 & & & 1 & & & 1 & 1 & \\
 & & 1 & & 1 & & 1 & & & 1 & & 1 & & 1 \\
 & & & 1 & & 1 & 1 & & & & 1 & & 1 & 1
\end{pmatrix}
\begin{pmatrix} \lambda^1 \\ \lambda^2 \end{pmatrix}
\begin{array}{l} = 1, \\ = 1, \\ \le 1, \\ \le 1, \\ \le 1, \end{array}
\qquad
\begin{array}{rl} \lambda^1, \lambda^2 & \ge 0, \\ \lambda^1, \lambda^2 & \in \mathbb{Z}^7. \end{array}
\tag{3.21}
$$

The constraint matrix A of this formulation is not perfect. The perfect clique matrix associated with the intersection graph of the Dantzig–Wolfe formulation is

$$
\left(\begin{array}{cccccccccccccc}
1 & 1 & 1 & 1 & 1 & 1 & 1 & & & & & & & \\
 & & & & & & & 1 & 1 & 1 & 1 & 1 & 1 & 1 \\
 & 1 & & & 1 & 1 & & & 1 & & & 1 & 1 & \\
 & & 1 & & 1 & & 1 & & & 1 & & 1 & & 1 \\
 & & & 1 & & 1 & 1 & & & & 1 & & 1 & 1 \\
\hline
 & & & & & & 1 & & & 1 & 1 & 1 & 1 & 1 \\
 & & & & & 1 & & & 1 & & 1 & 1 & 1 & 1 \\
 & & & & & 1 & 1 & & & & 1 & 1 & 1 & 1 \\
 & & & & 1 & & & & 1 & 1 & & 1 & 1 & 1 \\
 & & & & 1 & & 1 & & & 1 & & 1 & 1 & 1 \\
 & & & & 1 & 1 & & & 1 & & & 1 & 1 & 1 \\
 & & & & 1 & 1 & 1 & & & & & 1 & 1 & 1 \\
 & & & 1 & 1 & 1 & 1 & & & & & & 1 & 1 \\
 & & 1 & & 1 & 1 & 1 & & & & & 1 & & 1 \\
 & & 1 & 1 & 1 & 1 & 1 & & & & & & & 1 \\
 & 1 & & & 1 & 1 & 1 & & & & & 1 & 1 & \\
 & 1 & & 1 & 1 & 1 & 1 & & & & & & 1 & \\
 & 1 & 1 & & 1 & 1 & 1 & & & & & 1 & &
\end{array}\right).
\tag{3.22}
$$

This matrix adds 13 missing cliques to A. The clique in the last row contains the highlighted columns of A.

Similarly, one can verify that the 3-packing problem associated with three uniform matroids of rank 2 has an imperfect intersection graph.

Acknowledgements. I would like to thank an anonymous referee for helpful comments and suggestions, among them the idea to investigate the question behind Example 3.3. I would also like to thank Robert Weismantel and Alexander Martin for helpful comments.

Bibliography

[1] R.K. Ahuja, T.L. Magnanti, and J.B. Orlin. *Network Flows*, volume 1 of *Handbooks in Operations Research and Management Science*, chapter IV, pages 211–369. Elsevier Science B.V., Amsterdam, 1989.

[2] A. Atamturk, G.L. Nemhauser, and M.W.P. Savelsbergh. The Mixed Vertex Packing Problem, with G.L. Nemhauser and M.W.P. Savelsbergh, *Mathematical Programming* 89:35–53, 2000.

[3] A. Atamturk, G.L. Nemhauser, and M.W.P. Savelsbergh. Conflict Graphs in Solving Integer Programming Problems, with G.L. Nemhauser and M.W.P. Savelsbergh, *European Journal of Operational Research,* 121:40–55, 2000.

[4] E. Balas. Disjunctive programming. *Annals of Discrete Mathematics*, 5:3–51, 1979.

[5] E. Balas, S. Ceria, and G. Cornuéjols. A lift-and-project cutting plane algorithm for mixed 0-1 programs. *Mathematical Programming*, 58:295–324, 1993.

[6] E. Balas and J. Clausen, editors. *Integer Programming and Combinatorial Optimization*, Proceedings of the 4th International IPCO Conference, Copenhagen, Springer-Verlag, Berlin, 1995.

[7] E. Balas and M.W. Padberg. Set partitioning: A survey. *SIAM Review*, 18:710–760, 1976.

[8] R. Borndörfer. *Aspects of Set Packing, Partitioning, and Covering*. Berichte aus der Mathematik. Shaker, Aachen, 1998. Ph.D. thesis, Technische Universität Berlin. Available at http://www.zib.de/ZIBbib/Publications/.

[9] R. Borndörfer and R. Weismantel. Set packing relaxations of some integer programs. *Mathematical Programming*, 88:425–450, 2000.

[10] R. Borndörfer and R. Weismantel. Discrete relaxations of combinatorial programs. *Discrete Applied Mathematics*, 112(1–3):11–26, 2001.

[11] W.H. Cunningham, S.T. McCormick, and M. Queyranne, editors. *Integer Programming and Combinatorial Optimization*, Proceedings of the 5th International IPCO Conference, Vancouver, British Columbia, Canada, Springer-Verlag, Berlin, 1996.

[12] M. Deza and M. Laurent. *Geometry of Cuts and Metrics*. Springer-Verlag, Berlin, 1997.

[13] R. Euler, M. Jünger, and G. Reinelt. Generalizations of cliques, odd cycles and anticycles and their relation to independence system polyhedra. *Mathematics of Operations Research*, 12(3):451–462, 1987.

[14] C.E. Ferreira. *On Combinatorial Optimization Problems Arising in Computer System Design*. Ph.D. thesis. Available at http://www.zib.de/ZIBbib/Publications/. Technische Universität Berlin, 1994.

[15] C.E. Ferreira, A. Martin, and R. Weismantel. Solving multiple knapsack problems by cutting planes. *SIAM Journal on Optimization*, 6:858–877, 1996.

[16] A. Frank. Packing paths, circuits, and cuts—A survey. In Korte et al., Paths, Flows, and VLSI-Layout. Springer-Verlag, Berlin, 1990, pages 47–100.

[17] M.R. Garey and D.S. Johnson. *Computers and Intractability: A Guide to the Theory of NP-Completeness*. W.H. Freeman and Company, New York, 1979.

[18] E.S. Gottlieb and M.R. Rao. The generalized assignment problem: Valid inequalities and facets. *Mathematical Programming*, 46:31–52, 1990.

[19] R. Graham, M. Grötschel, and L. Lovász, editors. *Handbook of Combinatorics*. Elsevier Science B.V., Amsterdam, 1995.

[20] M. Grötschel, L. Lovász, and A. Schrijver. *Geometric Algorithms and Combinatorial Optimization*. Springer-Verlag, Berlin, 1988.

[21] M. Grötschel, A. Martin, and R. Weismantel. Packing Steiner trees: A cutting plane algorithm and computational results. *Mathematical Programming*, 72:125–145, 1996.

[22] M. Grötschel, A. Martin, and R. Weismantel. Packing Steiner trees: Further facets. *European Journal on Combinatorics*, 17:39–52, 1996.

[23] M. Grötschel, A. Martin, and R. Weismantel. Packing Steiner trees: Polyhedral investigations. *Mathematical Programming*, 72:101–123, 1996.

[24] M. Grötschel, A. Martin, and R. Weismantel. Packing Steiner trees: Separation algorithms. *SIAM Journal on Discrete Mathematics*, 9:233–257, 1996.

[25] B. Korte, L. Lovász, H.J. Prömel, and A. Schrijver, editors. *Paths, Flows, and VLSI-Layout*. Springer-Verlag, Berlin, 1990.

[26] M. Laurent. A generalization of antiwebs to independence systems and their canonical facets. *Mathematical Programming*, 45:97–108, 1989.

[27] L. Lovász and A. Schrijver. Cones of matrices and set-functions and 0-1 optimization. *SIAM Journal on Optimization*, 1:166–190, 1991.

[28] S. Martello and P. Toth. *Knapsack Problems*. John Wiley & Sons, Chichester, U.K., 1990.

[29] A. Martin. *Packen von Steinerbäumen: Polyedrische Studien und Anwendung*. Ph.D. thesis, Technische Universität Berlin, 1992.

[30] R. Müller. On the partial order polytope of a digraph. *Mathematical Programming*, 73(1):31–49, 1996.

[31] R. Müller and A.S. Schulz. The interval order polytope of a digraph. [6], pages 50–64.

[32] R. Müller and A.S. Schulz. Transitive packing. [11], pages 430–444.

[33] G.L. Nemhauser and L.E. Trotter. Properties of vertex packing and independence system polyhedra. *Mathematical Programming*, 6:48–61, 1973.

[34] G.L. Nemhauser and L.A. Wolsey. *Integer and Combinatorial Optimization*. John Wiley & Sons, Inc., New York, 1988.

[35] M.W. Padberg. On the facial structure of set packing polyhedra. *Mathematical Programming*, 5:199–215, 1973.

[36] M.W. Padberg. A note on zero-one programming. *Operations Research*, 23(4):833–837, 1975.

[37] M.W. Padberg. On the complexity of set packing polyhedra. *Annals of Discrete Mathematics*, 1:421–434, 1977.

[38] M.W. Padberg. Covering, packing, and knapsack problems. *Annals of Discrete Mathematics*, 4:265–287, 1979.

[39] M.W. Padberg and T.-Y. Sung. An analytical comparison of different formulations of the travelling salesman problem. *Mathematical Programming*, 52(2):315–357, 1991.

[40] A.S. Schulz. *Polytopes and Scheduling*. Ph.D. thesis. Available at ftp://ftp.math.tu-berlin.de/pub/Preprints/combi/. Technische Universität Berlin, 1996.

[41] Yasuki Sekiguchi. A note on node packing polytopes on hypergraphs. *Operations Research Letters*, 2(5):243–247, 1983.

[42] M. Sol. *Column Generation Techniques for Pickup and Delivery Problems*. Ph.D. thesis, Technische Universität Eindhoven, 1994.

[43] B. Toft. *Colouring, Stable Sets and Perfect Graphs*. [19], chapter 4, pages 233–288.

[44] L.E. Trotter. A class of facet producing graphs for vertex packing polyhedra. *Discrete Mathematics*, 12:373–388, 1975.

Chapter 4

Bicolorings and Equitable Bicolorings of Matrices*

Michele Conforti,† *Gérard Cornuéjols,*‡ *and Giacomo Zambelli*‡

Dedicated to Manfred Padberg.

Abstract. Two classical theorems of Ghouila-Houri and Berge characterize total unimodularity and balancedness in terms of equitable bicolorings and bicolorings, respectively. In this chapter, we prove a bicoloring result that provides a common generalization of these two theorems.

MCS 2000. 90C27, 90C57

Key words. Totally unimodular, balanced, bicoloring, equitable bicoloring

A $0/\pm 1$-matrix is *balanced* if it does not contain a square submatrix with exactly two nonzero entries per row and per column such that the sum of all the entries is congruent to 2 modulo 4. This notion was introduced by Berge [1] for $0/1$-matrices and generalized by Truemper [15] to $0/\pm 1$-matrices.

A $0/\pm 1$-matrix is *bicolorable* if its columns can be partitioned into blue columns and red columns so that every row with at least two nonzero entries contains either two nonzero entries of opposite sign in columns of the same color or two nonzero entries of the same sign in columns of different colors. Berge [1] showed that a $0/1$-matrix A is balanced if

---

*This work was supported in part by NSF grant DMI-0098427 and ONR grant N00014-97-1-0196.

†Dipartimento di Matematica Pura ed Applicata, Università di Padova, Via Belzoni 7, 35131, Padova, Italy.

‡Graduate School of Industrial Administration, Carnegie Mellon University, Schenley Park, Pittsburgh, Pennsylvania 15213-3890 (gc0v@andrew.cmu.edu).

and only if every submatrix of A is bicolorable. Conforti and Cornuéjols [6] extended this result to $0/\pm 1$-matrices. Cameron and Edmonds [3] gave a simple greedy algorithm to find a bicoloring of a balanced matrix. In fact, given any $0/\pm 1$-matrix A, their algorithm finds either a bicoloring of A or a square submatrix of A with exactly two nonzero entries per row and per column such that the sum of all the entries is congruent to 2 modulo 4. Does this algorithm provide an easy test for balancedness? The answer is no, because the algorithm may find a bicoloring of A even when A is not balanced.

A real matrix is *totally unimodular* (t.u.) if every nonsingular square submatrix has determinant ± 1 (note that every t.u. matrix must be a $0/\pm 1$-matrix).

A $0/\pm 1$-matrix A has an *equitable bicoloring* if its columns can be partitioned into red and blue columns so that, for every row of A, the sum of the entries in the red columns differs by at most one from the sum of the entries in the blue columns. Ghouila-Houri [9] showed that a $0/\pm 1$-matrix is t.u. if and only if every submatrix of A has an equitable bicoloring.

A $0/\pm 1$-matrix that is not t.u. but whose submatrices are all t.u. is said to be *almost t.u.* Camion [4] proved the following.

Theorem 4.1 (Camion [4] and Gomory [cited in [4]]). *Let A be an almost t.u. $0/\pm 1$-matrix. Then A is square, $\det A = \pm 2$, and A^{-1} has only $\pm\frac{1}{2}$ entries. Furthermore, each row and each column of A has an even number of nonzero entries and the sum of all entries in A equals 2 modulo 4.*

A nice proof of this result can be found in Padberg [12, 13]. Note that a matrix is balanced if and only if it does not contain any almost t.u. matrix with two nonzero entries in each row. For any positive integer k we say that a $0/\pm 1$-matrix A is *k-balanced* if it does not contain any almost t.u. submatrix with at most $2k$ nonzero entries in each row. Obviously, an $m \times n$ $0/\pm 1$-matrix A is balanced if and only if it is 1-balanced, while A is t.u. if and only if A is k-balanced for some $k \geq \lfloor n/2 \rfloor$. The class of k-balanced matrices was introduced by Conforti, Cornuéjols, and Truemper in [7].

For any integer k we denote by $\mathbf{k}$ a vector with all entries equal to k. For any $m \times n$ $0/\pm 1$-matrix A, we denote by $n(A)$ the vector with m components whose ith component is the number of -1's in the ith row of A.

Theorem 4.2 (Conforti, Cornuéjols, and Truemper [7]). *Let A be an $m \times n$ k-balanced $0/\pm 1$-matrix with rows a^i, $i \in [m]$; b be a vector with entries b_i, $i \in [m]$; and S_1, S_2, S_3 be a partition of $[m]$. Then*

$$\begin{aligned} P(A, b) = \{x \in \mathbb{R}^n : & \ a^i x \leq b_i \text{ for } i \in S_1, \\ & \ a^i x = b_i \text{ for } i \in S_2, \\ & \ a^i x \geq b_i \text{ for } i \in S_3, \\ & \ \mathbf{0} \leq x \leq \mathbf{1}\} \end{aligned}$$

is an integral polytope for all integral vectors b such that $-n(A) \leq b \leq \mathbf{k} - n(A)$.

This theorem generalizes previous results by Hoffman and Kruskal [10] for t.u. matrices, Berge [2] for $0/1$-balanced matrices, Conforti and Cornuéjols [6] for $0/\pm 1$-balanced matrices, and Truemper and Chandrasekaran [16] for k-balanced 0/1-matrices. As an application of Theorem 4.2, consider the SAT problem where, in each clause of a set of CNF clauses, at least k literals must evaluate to true. This SAT problem can be formulated as $Ax \geq \mathbf{k} - n(A),\ x \in \{0,1\}^n$. If the matrix A is k-balanced, it follows from Theorem 4.2 that the polytope $Ax \geq \mathbf{k} - n(A),\ \mathbf{0} \leq x \leq \mathbf{1}$, is integral and therefore the SAT problem can be solved by linear programming.

A 0/±1-matrix A has a *k-equitable bicoloring* if its columns can be partitioned into blue columns and red columns so that

- the bicoloring is equitable for the row submatrix A' determined by the rows of A with at most $2k$ nonzero entries;
- every row with more than $2k$ nonzero entries contains k pairwise disjoint pairs of nonzero entries such that each pair contains either entries of opposite sign in columns of the same color or entries of the same sign in columns of different colors.

Obviously, an $m \times n$ 0/±1-matrix A is bicolorable if and only if A has a 1-equitable bicoloring, while A has an equitable bicoloring if and only if A has a k-equitable bicoloring for $k \geq \lfloor n/2 \rfloor$. The following theorem provides a new characterization of the class of k-balanced matrices, which generalizes the bicoloring results mentioned above for balanced and t.u. matrices.

Theorem 4.3. *A 0/±1-matrix A is k-balanced if and only if every submatrix of A has a k-equitable bicoloring.*

Proof. Assume first that A is k-balanced and let B be any submatrix of A. Assume, up to row permutation, that

$$B = \begin{pmatrix} B' \\ B'' \end{pmatrix},$$

where B' is the row submatrix of B determined by the rows of B with $2k$ or fewer nonzero entries. Consider the system

$$\begin{aligned} B'x &\geq \left\lfloor \frac{B'\mathbf{1}}{2} \right\rfloor, \\ -B'x &\geq -\left\lceil \frac{B'\mathbf{1}}{2} \right\rceil, \\ B''x &\geq \mathbf{k} - n(B''), \\ -B''x &\geq \mathbf{k} - n(-B''), \\ \mathbf{0} \leq x &\leq \mathbf{1}. \end{aligned} \tag{4.1}$$

Since B is k-balanced, $\binom{B}{-B}$ is also k-balanced. Therefore the constraint matrix of system (4.1) above is k-balanced. One can readily verify that $-n(B') \leq \lfloor \frac{B'\mathbf{1}}{2} \rfloor \leq \mathbf{k} - n(B')$ and $-n(-B') \leq -\lceil \frac{B'\mathbf{1}}{2} \rceil \leq \mathbf{k} - n(-B')$. Therefore, by Theorem 4.2 applied with $S_1 = S_2 = \emptyset$, system (4.1) defines an integral polytope. Since the vector $(\frac{1}{2}, \ldots, \frac{1}{2})$ is a solution for (4.1), the polytope is nonempty and contains a 0/1-point $\bar{x}$. Color a column i of B blue if $\bar{x}_i = 1$, red otherwise. It can be easily verified that such a bicoloring is, in fact, k-equitable.

Conversely, assume that A is not k-balanced. Then A contains an almost t.u. matrix B with at most $2k$ nonzero elements per row. Suppose that B has a k-equitable bicoloring. Then such a bicoloring must be equitable since each row has, at most, $2k$ nonzero elements. By Theorem 4.1 B has an even number of nonzero elements in each row. Therefore the sum of the columns colored blue equals the sum of the columns colored red, so B is a singular matrix, a contradiction. □

Given a $0/\pm 1$-matrix A and positive integer k, one can find in polynomial time a k-equitable bicoloring of A or a certificate that A is not k-balanced as follows.

Find a basic feasible solution of (4.1). If the solution is not integral, A is not k-balanced by Theorem 4.2. If the solution is a 0/1-vector, it yields a k-equitable bicoloring, as in the proof of Theorem 4.3.

Note that, as with the algorithm of Cameron and Edmonds [3], a 0/1-vector may be found even when the matrix A is not k-balanced.

Using the fact that the vector $(\frac{1}{2}, \ldots, \frac{1}{2})$ is a feasible solution of (4.1), a basic feasible solution of (4.1) can actually be derived in strongly polynomial time using an algorithm of Megiddo [11].

Bibliography

[1] C. Berge. Sur certain hypergraphes généralisant les graphes bipartis. In *Combinatorial Theory and Its Applications* I. P. Erdös, A. Rényi, and V. Sós, editors, *Colloquia Mathematica Societatis János Bolyai* 4, pages 119–133, North-Holland, Amsterdam, 1970.

[2] C. Berge. Balanced matrices. *Mathematical Programming*, 2:19–31, 1972.

[3] K. Cameron and J. Edmonds. Existentially polytime theorems. In Polyhedral Combinatorics, *DIMACS Series in Discrete Mathematics and Theoretical Computer Science* 1, pages 83–100. American Mathematical Society, Providence, RI, 1990.

[4] P. Camion. Characterization of totally unimodular matrices. *Proceedings of the American Mathematical Society*, 16:1068–1073, 1965.

[5] P. Camion. Caractérisation des matrices unimodulaires. *Cahier du Centre d'Etudes de Recherche Opérationelle*, 5:181–190, 1963.

[6] M. Conforti and G. Cornuéjols. Balanced 0, ±1 matrices, bicoloring and total dual integrality. *Mathematical Programming*, 71:249–258, 1995.

[7] M. Conforti, G. Cornuéjols, and K. Truemper. From totally unimodular to balanced 0, ±1 matrices: A family of integer polytopes. *Mathematics of Operation Research*, 19:21–23, 1994.

[8] D.R. Fulkerson, A.J. Hoffman, and R. Oppenheim. On balanced matrices. *Mathematical Programming Study*, 1:120–132, 1974.

[9] A. Ghouila-Houri. Caractérisations des matrices totalement unimodulaires. *Comptes Rendus de l'Académie des Sciences*, 254:1192–1193, 1962.

[10] A.J. Hoffman and J.B. Kruskal. Integral boundary points of convex polyhedra, in *Linear Inequalities and Related Systems*. H.W. Kuhn and A.W. Tucker, editors, pages 223–246. Princeton University Press, Princeton, NJ, 1956.

[11] N. Megiddo. On finding primal- and dual-optimal bases. *Journal of Computing*, 3:63–65, 1991.

[12] M. Padberg. Characterization of totally unimodular, balanced and perfect matrices, in *Combinatorial Programming: Methods and Applications*. B. Roy, editor, pages 275–284. Reidel, Dordrecht, 1975.

[13] M. Padberg. Total unimodularity and the Euler subgraph problem. *Operations Research Letters*, 7:173–179, 1988.

[14] M. Padberg. *Linear Optimization and Extensions*. Springer-Verlag, Berlin, 1995.

[15] K. Truemper. Alpha balanced graphs and matrices and GF(3)-representability of matroids. *Journal of Combinatorial Theory Series* B, 32:112–139, 1982.

[16] K. Truemper and R. Chandrasekaran. Local unimodularity of matrix-vector pairs. *Linear Algebra and Its Applications*, 22:65–78, 1978.

Chapter 5

The Clique-Rank of 3-Chromatic Perfect Graphs

*Jean Fonlupt**

Dedicated to Manfred Padberg
on the occasion of his 60th birthday.

Abstract. The clique-rank of a perfect graph G introduced by Fonlupt and Sebö is the linear rank of the incidence matrix of the maximum cliques of G. We study this rank for 3-chromatic perfect graphs. We prove that if, in addition, G is diamond-free, G has two distinct colorations. An immediate consequence is that the Strong Perfect Graph Conjecture holds for diamond-free graphs and for graphs with clique number equal to three. The proofs use both linear algebra and combinatorial arguments.

MSC 2000. 05C17, 90C27
Key words. Perfect graphs, chromatic number, linear algebra

5.1 Introduction

A graph G is *perfect* if for each of its induced subgraphs G' the chromatic number $\chi(G')$ equals the clique number $\omega(G')$, i.e., the maximum number of pairwise adjacent nodes in G'. A perfect graph G with clique number equal to ω is ω-chromatic and an ω-coloration of G will refer to a minimum coloration in ω colors. If this minimum coloration is unique, we will say that G is uniquely colorable.

A *hole* is a chordless cycle with at least four nodes. An *antihole* is the complement of a hole; holes or antiholes are called *even* or *odd* according to the parity of their number

*Équipe Combinatoire (C.N.R.S.), Université Paris 6, Paris, France (Jean.Fonlupt@ecp6.jussieu.fr).

of nodes. Berge [1] conjectured that a graph is perfect if and only if it contains no odd hole and no odd antihole as induced subgraph. This conjecture is known as the *Strong Perfect Graph Conjecture*. The "only if" part is trivial but the "if" part remains open after more than 40 years. We shall call a graph *Berge* if it contains no odd hole and no odd antihole as induced subgraph. A graph G is *critically imperfect* if it is not perfect, i.e., $\omega(G) < \chi(G)$, but all its induced subgraphs are perfect. Clearly, odd holes and odd antiholes are critically imperfect, and the hard part of the Strong Perfect Graph Conjecture consists of proving that a critically imperfect graph is either an odd hole or an odd antihole.

Critically imperfect graphs play an important role in the study of perfect graphs, and some of their properties will be reviewed in Section 5.2; for the moment, let us just mention the following theorem proved by Padberg [5].

Theorem 5.1 (Padberg). *If G is critically imperfect, $G - \{v\}$ is uniquely colorable in $\omega(G)$ colors for every node v of G.*

This result suggests that the characterization of uniquely colorable perfect graphs is a crucial question in the study of perfect graphs and more precisely in the study of the Strong Perfect Graph Conjecture. A 2-chromatic perfect graph is uniquely colorable if and only if it is a bipartite connected graph; hence Padberg's theorem implies that for any node v of a critically imperfect graph G the subgraph of $G - \{v\}$ induced by any two color classes is connected and bipartite. (A color class is a subset of nodes that receives the same color in the coloring—this subset induces a stable set of the graph.) This obvious result is well known and not very useful in the study of perfect graphs.

In this chapter we will be interested in uniquely colorable 3-chromatic perfect graphs G. A *diamond* is the graph obtained by deleting an edge from the clique on four nodes; this edge will be called the missing edge of the diamond. A graph with no diamond as induced subgraph will be called *diamond-free*. Our main result is the following theorem.

Theorem 5.2 (Main Theorem). *A diamond-free, 3-chromatic perfect graph with more than three nodes is not uniquely colorable.*

The interest of this result is, firstly, that it is related to the *Forcing Rule Conjecture* formulated by Fonlupt and Sebö [3] (see Section 5.3). The validity of this conjecture for 3-chromatic perfect graphs would provide a simple characterization of uniquely colorable 3-chromatic perfect graphs. The proof of this conjecture will appear in a forthcoming paper but uses (in a more complicated way) arguments similar to those developed in this chapter.

Another interest is that our proof simultaneously uses two kinds of tools, some from linear algebra and others from graph theory. This idea appears in Fonlupt and Sebö [3], where the clique-rank (also called the rank) of a graph was introduced: the rank $r(G)$ of a graph G is the (linear) rank of the incidence matrix of the maximum cliques of G. This matrix, called the matrix associated with G throughout this chapter, plays an important role when the graph G is perfect. In our approach we study this rank and we give an interpretation of linear dependence and linear independence in connection with the cycle structure of the graph G.

A final interest is that this theorem has as an immediate consequence that the Strong Perfect Graph Conjecture is true for diamond-free graphs. The theorem also provides a new and simple proof of the validity of this conjecture for graphs with clique number equal to three. The first result was settled by Parthasaraty and Ravindra [6] and the second by Tucker [8]. The two proofs are independent and more or less technical. Our result provides a kind of unification of these two results. (In the original proof of Tucker for graphs with clique number equal to three the difficult situation occurs precisely when the critically imperfect graph that he considers is diamond-free, but in our approach this case is straightforward.)

The organization of the chapter is as follows. In Section 5.2 we give some notation and review some results related to critically imperfect graphs and the rank $r(G)$ of a perfect graph G. In Section 5.3 we show how our main result is related to the Forcing Rule Conjecture. Sections 5.4 and 5.5 contain preliminary results for the proof of Theorem 5.2 and Section 5.6 is devoted to the proof of this theorem. In Section 5.7 we give a simple and new proof of the validity of the Strong Perfect Graph Conjecture for graphs with clique number equal to three (Tucker's theorem).

5.2 Preliminaries

5.2.1 Notation

A graph G' is included in a graph G ($G' \subseteq G$) if G' is an induced subgraph of G. A 3-clique of G will be called a triangle. An edge st of G extends into a triangle if there is a triangle of G containing s and t; if G is diamond-free and $\omega = 3$, an edge of G extends into at most one triangle.

If P is a path with nodes $p_1, p_2, \dots, p_m$ and edges $p_1p_2, p_2p_3, \dots, p_{m-1}p_m$, we write $P = p_1p_2\dots p_m$ and we borrow the notation used for intervals: $P[p_i, p_j] = p_ip_{i+1}\dots p_j$, $P[p_i, p_j) = p_ip_{i+1}\dots p_{j-1}$, etc.

Let $P = p_1p_2\dots p_j$ and $Q = p_{j+1}p_{j+2}\dots p_m$ be two paths and assume that p_jp_{j+1} is an edge of G; we denote by PQ the path $p_1p_2\dots p_jp_{j+1}p_{j+2}\dots p_m$. If all the nodes of P are distinct, to avoid complicated notation, we will also denote by P the nodeset of the path P; the parity of P is the parity of the edgeset of P.

5.2.2 The rank of a perfect graph

Let G be an ω-chromatic perfect graph and let $\mathbf{A}$ be its associated matrix; i.e., $\mathbf{A}$ is the incidence matrix of the ω-cliques of G (an ω-clique is a clique of size ω). Note that $\mathbf{A}$ has $|V|$ columns and k rows, where k is the number of ω-cliques. Let $\mathbb{R}^V$ be the vector space whose components are indexed by the nodeset V of G and let $\mathbb{1}$ be the column vector of $\mathbb{R}^k$ with all components equal to 1. Consider the following system of linear equations:

$$\mathbf{A}x = \mathbb{1}. \tag{5.1}$$

As mentioned in the introduction, the rank of G, denoted by $r(G)$, is the linear rank of $\mathbf{A}$. Fonlupt and Sebő [3] proved that, for a perfect graph G,

$$\begin{aligned} &r(G) \le |V| - \omega + 1, \\ &r(G) = |V| - \omega + 1 \text{ if and only if } G \text{ is uniquely colorable.} \end{aligned} \tag{5.2}$$

Let y be a row vector in $\mathbb{R}^k$. From the original system (5.1) we can obtain the following equation, which we call *dependence relation*:

$$\langle y\mathbf{A}, x\rangle = \langle y, \mathbb{1}\rangle. \tag{5.3}$$

We will say that y induces the relation (5.3). A linear equation is a dependence relation if and only if any solution of linear system (5.1) is also a solution of this equation. Assume that G is uniquely colorable and let $\chi^1, \chi^2, \ldots, \chi^\omega$ be the set of incidence vectors of the color classes in this coloration. These vectors are linearly independent and by relation (5.2) they generate the vector space of all solutions of linear system (5.1). If two nodes u and v of G are similarly colored,

$$\chi_u^1 - \chi_v^1 = 0,\ \chi_u^2 - \chi_v^2 = 0, \ldots, \chi_u^\omega - \chi_v^\omega = 0.$$

Thus, for any solution x of linear system (5.1),

$$x_u - x_v = 0 \tag{5.4}$$

and equation (5.4) is a dependence relation; hence there exists a row vector $y \in \mathbb{R}^k$ such that

$$\langle y\mathbf{A}, x\rangle = x_u - x_v, \qquad \langle y, \mathbb{1}\rangle = 0. \tag{5.5}$$

If we take now for the ground field the binary field $\mathbf{GF(2)} = \{0, 1\}$ rather than the real field $\mathbb{R}$, the rank of $\mathbf{A}$ may be different, but if $\omega = 3$, Fonlupt and Sebö [3] proved that the rank is the same and given by relation (5.2). In the rest of this chapter we will work with the binary field if $\omega = 3$; in this case the dependence relation associated with the two similarly colored nodes u, v is

$$x_u + x_v = 0 \qquad \text{(modulo 2)} \tag{5.6}$$

and the row vector y that induces the dependence relation (5.6) has all its components equal to 0 or 1.

5.2.3 Critically imperfect graphs

Consider a given class of graphs closed under graph inclusion (for instance, the class of diamond-free graphs). This class satisfies the Strong Perfect Graph Conjecture if it contains no critically imperfect Berge graph—and this statement will be settled if we can prove that any Berge graph in this class violates some properties characterizing critically imperfect graphs. We list below some of these properties, which are used in the next sections of this chapter.

Let ω be the clique number of a graph G, α be its stability number (α is the largest number of pairwise nonadjacent nodes of G), and $\mathbf{A}$ be its associated matrix. If G is Berge and critically imperfect, G satisfies the following properties:

1. G is a Berge graph.
2. $\alpha \geq 3$ and $\omega \geq 3$.
3. $|V| = \alpha\omega + 1$.
4. $\mathbf{A}$ is a square nonsingular matrix and $r(G) = |V|$.

5. Each node of G belongs to ω cliques of size ω.
6. For any node v of G there exists a unique partition of $G - \{v\}$ in ω stable sets (of cardinality α) and each of these stable sets intersects all but one of the ω-cliques containing v (unique colorability of $G - \{v\}$).
7. G is a $(2\omega - 2)$-connected graph (connectivity refers to node connectivity).

The fundamental conditions 3, 4, 5, and 6 were discovered and proved (among other conditions) by Padberg [5] (for condition 3 see also Lovász [4]). Condition 7 was stated by Sebö [7] as a direct application of condition 4 and equation (5.2).

Assume, in addition, that there exists an edge e of G that does not extend into a triangle and that G and $G - e$ belong to our class of graphs. $\mathbf{A}$ is the associated matrix of $G - e$ since $\omega \geq 3$; by condition 4 and relation (5.2) $G - e$ is not perfect. If $G - e$ is not critically imperfect, there exists a proper induced subgraph of $G - e$ that is critically imperfect. Let V' be the set of nodes of this graph, ω' its clique number, and $\mathbf{A}'$ its associated matrix. As G is a Berge graph, this subgraph is not an odd hole; hence $\omega' \geq 3$. This implies that $\mathbf{A}'$ is also the associated matrix of the subgraph of G induced on V'. By condition 4 this subgraph is not perfect and G is not critically imperfect. So, if we want to prove the Strong Perfect Graph Conjecture for our class of graphs, we can replace G with $G - e$. Thus we can also assume that G satisfies the following condition:

8. All the edges of G extend into triangles.

5.3 The Forcing Rule Conjecture

Let $G = G_0$ be a 3-chromatic perfect graph and assume that G contains a diamond induced on the set $\{v, w, s, t\}$ with st as missing edge. In any 3-coloration of G the same color is assigned to s and t. Let G_1 be the graph obtained by identification of s and t. Clearly, any 3-coloration of G_0 induces a 3-coloration of G_1 and vice versa; moreover, G_0 is uniquely colorable if and only if G_1 is uniquely colorable. If G_1 contains a diamond, we can again implement this operation and we obtain a second graph G_2. Reiterating this operation as far as we can, we obtain a sequence of graphs: $G_0, G_1, \ldots, G_k$, and the last graph G_k is diamond-free. If G_k is a triangle, G_k is uniquely 3-colorable and so is G. In a forthcoming paper we will prove the following theorem.

Theorem 5.3. *G is uniquely colorable if and only if G_k is a triangle.*

Note that this theorem settles the Forcing Rule Conjecture stated by Fonlupt and Sebö [3] in the case $\omega = 3$. The main difficulty in the proof of Theorem 5.3 is that the last graph G_k is diamond-free but, in general, is not perfect. However, if our original graph $G = G_0$ is perfect and diamond-free, $G_k = G_0$ and G_k is perfect and diamond-free. So we can see Theorem 5.2 as a special case of Theorem 5.3.

5.4 Some Combinatorial Results

We shall often rely on the following simple lemma.

Lemma 5.4. *Let G be a graph with clique number equal to three and with at least four edges; assume also that G contains a unique clique of size three $\{v_1, v_2, v_3\}$. If all three graphs $G - \{v_1, v_2\}$, $G - \{v_2, v_3\}$, $G - \{v_2, v_3\}$ are connected, then G contains an odd hole.*

Proof. Let Q be a component of $G - \{v_1, v_2, v_3\}$. We may assume that Q is bipartite (else Q contains an odd hole and we are done); thus the set of nodes of Q splits into stable sets S_1 and S_2. Since each $G - \{v_i, v_j\}$ is connected, each of the three nodes v_k must have a neighbor in $S_1 \cup S_2$. Hence two of the three nodes, say v_1 and v_2, must have a neighbor in the same S_i. It follows that the subgraph of G induced by $Q \cup \{v_1, v_2\}$ is not bipartite, and so it contains an odd hole. □

We will assume from now on that G is a 3-chromatic Berge graph. Let S be a set of nodes of G with the same color in an initial 3-coloration of G. S is a stable set and the graph $B = G - S$ is a bipartite graph. We shall denote by E the edgeset of B.

Definition 5.5. A node v of S and a hole H of $B = G - S$ form an *active pair* if the subgraph of G induced by $H \cup \{v\}$ contains an odd number of triangles. The edges of these triangles belonging to H are called *active edges*; finally, a hole H is *active* if it creates an active pair with a node v of S.

Lemma 5.6. *If a node v of S and a hole H of B form an active pair, the subgraph of G induced by $H \cup \{v\}$ contains a unique triangle.*

Proof. Let F be the set of active edges of the subgraph induced by $H \cup \{v\}$ and assume that F is odd and that $|F| > 1$. $H - F$ splits into disjoint chordless paths; if P is one of these paths, the subgraph of G induced on $P \cup \{v\}$ is bipartite and P is even. Hence H and F have the same parity and H is odd, contradicting our assumption that G is Berge. □

In the remainder of this section we will also assume that G is diamond-free. Note that, if there exists an edge e that extends into no triangle, $G - e$ is also Berge and also diamond-free. As G and $G - e$ have the same associated matrix, G is uniquely colorable if and only if $G - e$ is uniquely colorable. Hence, we can assume that each edge of G extends into a triangle and more precisely into a unique triangle since G is diamond-free.

Lemma 5.7. *Assume that each node of G belongs to at least two triangles; let $e = st$ be an active edge of G. The graph $B - \{s, t\}$ is not connected.*

Proof. By Lemma 5.6 there exists a node $v \in S$ and a hole H of B such that $\{v, s, t\}$ is the unique triangle of the subgraph induced by $H \cup \{v\}$. If $B - \{s, t\}$ is connected, there exists in $B - \{s, t\}$ a path P from a neighbor of v distinct from s, t to a node of $H - \{s, t\}$. Taking for P the smallest possible path, we can assume that no internal node of P is adjacent to v. As G is diamond-free, $\{v, s, t\}$ is the unique triangle of the subgraph induced by $H \cup B \cup \{v\}$, but this graph satisfies the hypothesis of Lemma 5.4 and contains an odd hole, which is impossible if G is Berge. □

5.5 Dependence Relations

We shall assume throughout this section that G is Berge, diamond-free, 3-chromatic, and uniquely colorable; note that G is also perfect. S, B, E are as defined in Section 5.4. As in Section 5.2, we will study the linear system (5.1), $\mathbf{A}x = \mathbb{1}$, over the binary field $\mathbf{GF(2)}$. Note that each triangle contains an edge in E and that each edge of E extends into a unique triangle. Hence we can consider that the rows of $\mathbf{A}$ are indexed by the edges of B and that the vectors y that induce a dependence relation are the incidence vectors of subsets of E. A dependence relation may be written

$$\langle \mu, x \rangle = \sum_{v \in V} \mu_v x_v = \mu_0, \qquad \mu_v \in \{0, 1\} \text{ for all } v \in V, \qquad \mu_0 \in \{0, 1\}. \tag{5.7}$$

Definition 5.8. We say that relation (5.7) is an *Eulerian relation* if

$$\mu_v = 0 \text{ for all } v \in V - S.$$

Note that all the dependence relations (5.6), $x_u + x_v = 0$ for all $u, v \in S$, are Eulerian relations.

Lemma 5.9. *A dependence relation $\langle \mu, x \rangle = \langle y\mathbf{A}, x \rangle = \langle y, \mathbb{1} \rangle$ is an Eulerian relation if and only if y is the incidence vector of the edgeset F of an Eulerian subgraph of B.*

Proof. Let y be the incidence vector of a subset $F \subset E$; y induces a dependence relation

$$\langle \mu, x \rangle = \langle y\mathbf{A}, x \rangle = \langle y, \mathbb{1} \rangle.$$

Let v be a node of B; μ_v is congruent (modulo 2) to the number of edges of F incident to v. Thus $\mu_v = 0$ for all $v \in V - S$ if and only if the subgraph $(V - S, F)$ of B is Eulerian. □

Definition 5.10. An Eulerian relation is a *hole relation* if it is induced by the incidence vector h of a hole H of B.

Lemma 5.11. *The set of Eulerian relations is generated by the set of hole relations.*

Proof. A classical basic result in graph theory (see for instance Bondy and Murty [2]) states that the incidence vectors of the edges of the holes of a graph generate (over the field $\mathbf{GF(2)}$) the set of incidence vectors of the edges of the Eulerian subgraphs of this graph. Hence let y be the incidence vector of the edgeset F of an Eulerian subgraph of B; there exists a set of holes $H^1, H^2, \ldots, H^k$ with incidence vectors $h^1, h^2, \ldots, h^k$ such that

$$y = h^1 + h^2 + \cdots + h^k.$$

It follows immediately that

$$\begin{cases} \langle y\mathbf{A}, x \rangle &= \langle h^1\mathbf{A}, x \rangle + \langle h^2\mathbf{A}, x \rangle + \cdots + \langle h^k\mathbf{A}, x \rangle, \\ \langle y, \mathbb{1} \rangle &= \langle h^1, \mathbb{1} \rangle + \langle h^2, \mathbb{1} \rangle + \cdots + \langle h^k, \mathbb{1} \rangle. \end{cases} \tag{5.8}$$

This establishes our claim. □

We will finish this section with a final lemma.

Lemma 5.12. *Assume that h is the incidence vector of the edgeset of a hole H of B and let $\langle \mu, x\rangle = \langle h\mathbf{A}, x\rangle = \langle h, \mathbb{1}\rangle$ be the corresponding hole relation.*

1. $\langle h, \mathbb{1}\rangle = 0$.
2. $\mu \neq 0$ *if and only if H is an active hole.*
3. *An active hole contains an even number of active edges and at least two active edges.*
4. *For any $v \in S$ there exists a hole H of B that forms an active pair with v.*

Proof.

1. h has an even number of components equal to 1 and $\langle h, \mathbb{1}\rangle = 0$.
2. For any $v \in S$, $\mu_v = 0$ if and only if the subgraph of G induced on $H \cup \{v\}$ has an even number of triangles.
3. If x is the incidence vector of the stable set S, x is a solution of linear system (5.1) and satisfies the relation $\langle \mu, x\rangle = \sum_{v\in S} \mu_v = 0$.
4. In equation (5.8) let us set $\mu^1 = h^1\mathbf{A}, \mu^2 = h^2\mathbf{A}, \ldots, \mu^k = h^k\mathbf{A}$. Assume that the dependence relation $\langle \mu, x\rangle = 0$ is the equation $x_u + x_v = 0$ for v and some other node u of S. Since $1 = \mu_v = \mu_v^1 + \mu_v^2 + \cdots + \mu_v^k$, at least one of the coefficients $\mu_v^1, \mu_v^2, \ldots, \mu_v^k$ is equal to 1. Hence v and one of the holes $H^1, H^2, \ldots, H^k$ form an active pair. □

5.6 Proof of the Main Theorem

Our proof will proceed by induction on the number of nodes $|V|$ of G. Let G be a diamond-free, 3-chromatic perfect graph with more than three nodes.

If $|V| = 4$, G obviously has two distinct 3-colorations.

Assume now that $|V| > 4$ and that every 3-chromatic induced subgraph of G has two distinct 3-colorations. If B is not connected, B has two distinct 2-colorations and G is not uniquely colorable.

If there exists a node $v \in S$ that belongs to one triangle, the variable x_v appears in only one row of linear system (5.1), $\mathbf{A}x = \mathbb{1}$. When we delete this row from $\mathbf{A}$, we obtain the matrix associated with the graph $G - \{v\}$ and

$$r(G) = r(G - \{v\}) + 1,$$
$$r(G - \{v\}) < (|V| - 1) - 2$$

by relation (5.2) and our induction hypothesis.

Hence, $r(G) < |V| - 2$ and G is not uniquely colorable.

So, we can assume that B is connected and that each node of S belongs to at least two triangles. As the color assigned to S plays no special role, we may also assume that each node of B belongs to at least two triangles and has at least two neighbors in B. This implies that, if B is not 2-connected, there exists an induced 2-connected graph B' of B and a node r of B' such that $B' - r$ is a component of $B - r$. B' is a *block* of B and r is the root of B'. If B is 2-connected, B itself is a block with no root.

The proof will now proceed by contradiction: we will prove that, if G is uniquely colorable, G contains an odd hole.

Let st be an active edge of B (Lemma 5.12, statement 4 ensures that there are at least $|S|$ active edges). By Lemma 5.7 the set $\{s, t\}$ disconnects B. Let W be the nodeset of a component of $B - \{s, t\}$. A hole of B lies either in the graph induced by $W \cup \{s, t\}$ or in the graph $B - W$.

If all the active holes belong to $B - W$, the equations $x_u + x_v = 0$ for all $u, v \in S$ are still dependence relations of the linear system (5.1) associated with $G - W$ and, in any coloration of $G - W$, the same color will be assigned to all the nodes of S. But $B - W$ is a connected bipartite graph and uniquely colorable in two colors. Thus $G - W$ is uniquely colorable, which is impossible by our induction assumption. Note that by the same argument we can assert that each block of B contains at least one active edge.

Thus consider a block B' of B and an active edge with endnodes s, t in B' and let us prove first the following claim.

Claim 1. *One of the two nodes s, t is the root of B'.*

Proof. Assume the contrary and let W be the nodeset of a component of $B - \{s, t\}$ that does not contain the root of B'. We can also assume that st is chosen among all possible candidates so that $|W|$ is minimum. We know that the subgraph of B induced on $W \cup \{s, t\}$ contains an active hole H and an active edge $s't'$ distinct from st by Lemma 5.12, property 3. Our choice for s and t ensures that $B - \{s\}$ and $B - \{t\}$ are connected. Hence, $B - (W \cup \{s\})$ and $B - (W \cup \{t\})$ are also connected and eventually contain the root of B' if $B' \neq B$. Since $st \neq s't'$, at least one of these two subgraphs is included in a connected component of $B - \{s', t'\}$; hence there exists a component W' of $B - \{s', t'\}$ that does not contain the root and such that $W' \subset W$, contradicting our choice of W. □

Claim 1 shows that B is not 2-connected and that $B' \subset B$. Let H be an active hole of B'; by Claim 1 H has two active edges rt and rt' (recall that r is the root of B'). $H - \{r\}$ is a chordless path $P = p_1 p_2 \dots p_n$ and none of the edges of P is active (again by Claim 1). Hence, if for some node $v \in S$ the graph induced on $P \cup \{v\}$ contains one triangle, it contains an even number of triangles by Lemma 5.6. It follows immediately that there exists a subchain $L = [p_i p_{i+1} \dots p_j]$, $1 < i < j < n$, of P and two nodes v_1, v_2 of S such that, in the graph induced on $L \cup \{v_1, v_2\}$, the adjacent nodes of v_1 are p_i, p_j and the adjacent nodes of v_2 are p_i, p_{i+1}. Since G is diamond-free, $i + 1 < j$. Note also that no node of L is adjacent to r.

Let B'' be the induced graph obtained from B by deleting all the nodes of $B' - \{r\}$. This graph is connected and all the active edges have at least one endnode in B''. Thus the graph induced on $B'' \cup \{v_1, v_2\}$ is connected and there exists in this graph a chordless path Q from v_1 to v_2. No node of Q is adjacent to a node of L and $\{v_2, p_i, p_{i+1}\}$ is the unique triangle of the graph induced on $L \cup Q \cup \{v_1, v_2\}$. By Lemma 5.4 this graph contains an odd hole, which is impossible if G is Berge.

Corollary 5.13. *Diamond-free graphs satisfy the Strong Perfect Graph Conjecture.*

Proof. Let G be a diamond-free graph and assume that G satisfies conditions 1, 2, and 6 of Subsection 5.2.3. So G is a Berge graph, $\omega \geq 3$, and, if v is a node of G, $G - \{v\}$ is uniquely colorable in ω colors. But, if we consider the subgraph G' of $G - \{v\}$ induced on three classes of colors, Theorem 5.2 asserts that G' is not uniquely 3-colorable and therefore $G - \{v\}$ is not uniquely colorable in ω colors, a contradiction. □

5.7 A New Proof of Tucker's Theorem

Let G be a graph with clique number equal to three and assume that G satisfies all the conditions listed in Subsection 5.2.3. If G is diamond-free, G is not critically imperfect.

So, let us assume that G contains a diamond with nodes v, w, s, t and with st as the missing edge. Denote by $T_1 = \{v, w, s\}$, $T_2 = \{v, w, t\}$ the two triangles containing v and w. By condition 5 of Subsection 5.2.3 each node of G belongs to three triangles. So, the third triangle containing v (resp., w) will be called $T_3 = \{v, s_3, t_3\}$ (resp., $T_4 = \{w, s_4, t_4\}$).

Consider in $G - \{v, w\}$ a path P from s to t, with distinct nodes (but not necessarily a chordless path); let us prove first the following claim.

Claim 2. *If the subgraph of G induced on P is bipartite, P is odd if and only if both triangles T_3, T_4 are included in $P \cup \{v, w\}$.*

Proof. If P is odd, the subgraph induced in G by $P \cup \{v\}$ (resp., $P \cup \{w\}$) is an odd cycle and contains a triangle since G is Berge; but the only possible triangle is T_3 (resp., T_4) if the subgraph of G induced on P is bipartite.

Assume now that P is even and that at least one of the two triangles T_3, T_4 is included in $P \cup \{v, w\}$; we can suppose that this triangle is T_3 and that s_3 appears before t_3 in the description of P from s to t. $P[s, s_3]$ and $P[t_3, t]$ are even since G is Berge (consider the holes $\{v\} \cup P[s, s_3]$ and $\{v\} \cup P[t_3, t]$). Thus P is odd, a contradiction. □

In the unique 3-coloration of $G - \{v\}$ the nodes s, w, t receive distinct colors by condition 6 of Subsection 5.2.3. Take for path P a chordless path from s to t in the connected bipartite graph induced by the set of nodes with the same color as s or t. P is odd and by Claim 2 P can be written

$$P = p_1 \dots p_i \dots p_j \dots p_k \text{ with } 1 \le i \le j < k,\ p_i = s_3,\ p_{i+1} = t_3,\ p_j = s_4,\ p_{j+1} = t_4.$$

$i = j$ implies that $\{v, w, p_i, p_{i+1}\}$ is a clique of size 4, which contradicts our definition of G. $i + 1 = j$ implies that v belongs to a fourth triangle, $\{v, w, p_{i+1}\}$, which again is impossible, so $j > i + 1$.

Let $\{q, p_{i+1}, p_{i+2}\}$ be a triangle of G extending the edge (p_{i+1}, p_{i+2}). As $G - \{v, p_{i+1}, p_j\}$ is connected by condition 7 of Subsection 5.2.3, there exists in this graph a path Q from q to a node r adjacent to some node $P[p_1, p_i] \cup P[p_{j+1}, p_k]$ and we can assume, without loss of generality, that r is adjacent to some node of $P[p_1, p_i]$. Note that $w \notin Q$ since the adjacent nodes of w in $G - \{v, p_{i+1}, p_j\}$ belong to $P[p_1, p_i] \cup P[p_{j+1}, p_k]$. We may also consider that Q is as short as possible with respect to this assumption and therefore no internal node of Q is adjacent to some node of $P[p_1, p_i] \cup P[p_{j+1}, p_k]$. The subgraph of G induced on $P[p_1, p_j] \cup Q - \{p_{i+1}\}$ is connected and there exists in this subgraph a chordless path R from p_1 to p_j. The path $P' = RP[p_{j+1}, p_k]$ contains $\{p_j, p_{j+1}\}$ but not $\{p_{i+1}\}$; hence the subgraph induced on $P' \cup \{v, w\}$ contains the triangle T_4 but not T_3. By Claim 2 the graph induced on P' is not bipartite, else G contains an odd hole. Therefore P'

contains at least one triangle. Our definition of Q implies that this triangle contains r and is contained in $\{r\} \cup P[p_j, p_k]$. By symmetry there also exists a triangle containing r and contained in $\{r\} \cup P[p_1, p_{i+1}]$.

If $\{r\} \cup P[p_1, p_{i+1}] \cup P[p_j, p_k]$ contains three triangles, $q = r$, and r belongs to a fourth triangle $\{r, p_{i+1}, p_{i+2}\}$, which is impossible.

Let us assume now that the unique triangle of $\{r\} \cup P[p_1, p_{i+1}]$ is also included in $\{r\} \cup P[p_1, p_i]$. Let l be the largest subscript such that p_l is adjacent to r. Note that $l \geq j+1$. The subgraph of G induced on $P[p_1, p_i] \cup P[p_l, p_k] \cup \{v, r\}$ satisfies the conditions of Lemma 5.4 and contains an odd hole, which contradicts our assumption that G is a Berge graph. Thus $\{r, p_i, p_{i+1}\}$ and for similar reasons $\{r, p_j, p_{j+1}\}$ are triangles.

To finish our proof, consider the subgraph of G induced on

$$\{v, w, s, t, r, p_i, p_{i+1}, p_j, p_{j+1}\}.$$

This subgraph has at most nine nodes and is a proper subgraph of G by conditions 2 and 3 of Subsection 5.2.3. In any coloration, r, v should get the same color (consider the diamond $\{v, r, p_i, p_{i+1}\}$) and r, w should get the same color (consider the diamond $\{w, r, p_j, p_{j+1}\}$). But v and w are adjacent nodes and cannot be colored by the same color. Hence this subgraph is not 3-chromatic and is not perfect. G is not critically imperfect.

Acknowledgement. The author wishes to thank an anonymous referee for helping him to substantially improve a first version of this paper.

Bibliography

[1] C. Berge. Färbung von Graphen, deren sämtliche bzw. deren ungerade Kreise starr sind. Wissenschaftliche Zeitschrift der Martin-Luther-Universität Halle-Wittenberg, 10:114–115, 1961.

[2] J.A. Bondy and U.S.R. Murty. *Graph Theory with Applications*, MacMillan, London, 1976.

[3] J. Fonlupt and A. Sebö. On the clique rank and the coloration of perfect graphs. In Proceedings of the First International Conference on Integer Programming and Combinatorial Optimization, R. Kannan and W.R. Pulleyblank, editors, Mathematical Programming Society, University of Waterloo Press, Waterloo, Canada, 1990.

[4] L. Lovász. A characterization of perfect graphs. *Journal of Combinatorial Theory*, 13:95–98, 1972.

[5] M. Padberg. Perfect zero-one matrices. *Mathematical Programming*, 6:180–196, 1974.

[6] K.R. Parthasaraty and G. Ravindra. The validity of the graph conjecture for $(K_4 - e)$-free graphs. *Journal of Combinatorial Theory. Series* B, 42:313–318, 1987.

[7] A. Sebö. The connectivity of minimal imperfect graphs. *Journal of Graph Theory*, 23:77–85, 1996.

[8] A.C. Tucker. Critical perfect graphs and perfect 3-chromatic graphs. *Journal of Combinatorial Theory. Series* B, 23:143–149, 1977.

Chapter 6

On the Way to Perfection: Primal Operations for Stable Sets in Graphs*

Claudio Gentile,† Utz-Uwe Haus,‡ Matthias Köppe,‡ Giovanni Rinaldi,† and Robert Weismantel‡

Manfred Padberg is the scientific father, or the scientific grandfather, or the scientific great grandfather of each of the five authors. This chapter is dedicated to him on the occasion of his 60th birthday.

Abstract. In this chapter some operations are described that transform every graph into a perfect graph by replacing nodes with sets of new nodes. The transformation is done in such a way that every stable set in the perfect graph corresponds to a stable set in the original graph. These operations can be used in an augmentation procedure for finding a maximum weighted stable set in a graph. Starting with a stable set in a given graph, one defines a simplex-type tableau whose associated basic feasible solution is the incidence vector of the stable set. In an iterative fashion, nonbasic columns that would lead to pivoting into nonintegral basic feasible solutions are replaced by new columns that one can read off

*First, second, third, and fifth authors supported by the European DONET program TMR ERB FMRX-CT98-0202. Fifth author supported by a Gerard-Hess-Preis and grant WE 1462 of the Deutsche Forschungsgemeinschaft. Second and third authors supported by grants FKZ 0037KD0099 and FKZ 2495A/0028G of the Kultusministerium of Sachsen-Anhalt.

†Istituto di Analisi dei Sistemi ed Informatica "Antonio Ruberti"—CNR, Roma, Italy (gentile@iasi.rm.cnr.it, rinaldi@iasi.rm.cnr.it).

‡Otto-von-Guericke-Universitaet Magdeburg, Department for Mathematics/IMO, Germany (haus@imo.math.uni-magdeburg.de, mkoeppe@imo.math.uni-magdeburg.de, weismant@imo.math.uni-magdeburg.de).

from special graph structures such as odd holes, odd antiholes, and various generalizations. Eventually, either a pivot leading to an integral basic feasible solution is performed, or the optimality of the current solution is proved.

MSC 2000. 90C10, 90C27, 05C60, 05C70

Key words. Stable set problem, perfect graphs, primal integer programming

6.1 Introduction

The *stable set problem* (or *node packing problem*) is one of the most studied problems in combinatorial optimization. It can be defined as follows: Let (G, c) be a weighted graph, where $G = (V, E)$ is a graph with $n = |V|$ nodes and $m = |E|$ edges and $c \in \mathbb{R}_+^V$ is a node function that assigns a weight to each node of G. A set $S \subseteq V$ is called *stable* if its nodes are pairwise nonadjacent in G. The problem is to find a stable set S^* in G of maximum weight $c(S^*) = \sum_{v \in S^*} c_v$. The value $c(S^*)$ is called the *c-weighted stability number* $\alpha_c(G)$ of the graph G.

This problem is equivalent to maximizing the linear function $\sum_{v \in S} c_v x_v$ over the *stable set polytope* P_G, the convex hull of the incidence vectors of all the stable sets of G. Thus linear programming techniques can be used to solve the problem, provided that an explicit description of the polytope is given. It is nowadays well known that, the stable set problem being NP-hard, it is very unlikely that such a description can be found for instances of arbitrary size. Moreover, even if a partial description is at hand, due to the enormous number of inequalities, it is not obvious how to turn this knowledge into a useful algorithmic tool.

Despite these difficulties, the literature in combinatorial optimization of the last 30 years abounds with successful studies where nontrivial instances of NP-hard problems were solved with a cutting plane procedure based on the generation of strong cuts obtained from inequalities that define facets of certain polytopes.

The idea of using facet-defining inequalities in a cutting plane algorithm was proposed by Padberg in [20] and pursued in many other of his papers. His contribution goes much beyond the advances in the knowledge of the stable set problem and its polytope, as it influenced the developments of the following three decades in polyhedral combinatorics and computational combinatorial optimization.

The basic integer linear programming formulation of the problem is obtained by adding the integrality requirement on the variables to the following system:

$$\begin{aligned} x_u + x_v &\leq 1 \quad \text{for each edge } (u, v) \text{ in } G, \\ x_v &\geq 0 \quad \text{for each node } v \in V. \end{aligned} \tag{6.1}$$

Such a system is called the *edge-node formulation* and provides a relaxation of P_G that has been studied in depth in [20], where it is proved that its solutions have values in the set $\{0, 1/2, 1\}$.

A set $Q \subseteq V$ is called a *clique* if its nodes are pairwise adjacent in G. In [20] it is proved that for every clique Q of G the *clique inequality*

$$\sum_{v \in Q} x_v \leq 1$$

defines a facet of P_G as long as Q is maximal with respect to set inclusion. If in (6.1) instead of one inequality per edge we have a clique inequality per maximal clique, we obtain the *clique formulation*

$$\sum_{v \in Q} x_v \leq 1 \quad \text{for each clique } Q \text{ in } G,$$
$$x_v \in \{0, 1\} \quad \text{for each node } v \in V,$$

which provides a tighter relaxation of P_G.

Let $C \subseteq V$ be a set of nodes such that $G[C]$, the subgraph of G induced by C, is a cycle of odd length. If the cycle is chordless, it is called an *odd hole*, and the inequality

$$\sum_{v \in C} x_v \leq \frac{|C| - 1}{2}$$

is called an *odd-hole inequality*. This inequality was proved in [20] to define a facet of $P_{G[C]}$. In the same paper a *sequential lifting procedure* is described that turns an odd-hole inequality, and actually any inequality defining a facet of the polytope associated with a subgraph of G, into a facet-defining inequality for P_G.

After the work of Padberg, several other results were produced on the structure of the stable set polytope. Among the facet-defining inequalities that were characterized we mention the *antihole inequalities* introduced by [19]; their definition is as for the hole inequalities, except that the subgraph induced by C is not a chordless cycle but its complement (a so-called odd antihole). For a list of references to further facet-defining inequalities for which a characterization is known we refer to, e.g., Borndörfer [5].

It is not a trivial task to exploit this vast amount of knowledge on the stable set polytope to devise an effective cutting plane algorithm that is able to solve nontrivial instances of large size. Among the few attempts, we mention those of Nemhauser and Sigismondi [18] and Balas et al. [1]. Unlike the case of other NP-hard problems, polyhedrally based cutting plane algorithms for the stable set problem have not yet shown their superiority over alternative methods. On the other hand, several approaches have been tried to solve difficult instances. For a collection of papers on algorithms for the stable set problem and for a recent survey on the subject, see [16] and [4], respectively.

The cutting plane procedure mentioned before has a "dual flavor," in the sense that the current solution is infeasible until the end, when feasibility and hence optimality is reached. A primal cutting plane procedure was first proposed by Young [22]: One starts with an integral basic feasible solution, then either pivots leading to integral solutions are performed or cuts are generated that are satisfied by the current solution at equality. Padberg and Hong [21] were the first to propose a similar primal procedure based on strong polyhedral cutting planes. These kinds of algorithms produce a path of adjacent vertices of the polytope associated with the problem.

A profound study of the vertex adjacency for the polytope of the set partitioning problem was produced by Balas and Padberg [2]. They provided the theoretical background for the realization of a primal algorithm that produces a sequence of adjacent vertices of the polytope, ending with the optimal solution. Their basic technique was to replace a column of the current simplex tableau with a set of new columns in order to guarantee the next pivot to lead to an integral basic feasible solution.

These ideas were generalized to the case of general integer programming by Haus, Köppe, and Weismantel [14] who called their method the "integral basis method." This method requires neither cutting planes nor enumeration techniques. In each major step the algorithm either returns an augmenting direction that is applicable at the given feasible point and yields a new feasible point with better objective function value or provides a proof that the point under consideration is optimal. This is achieved by iteratively substituting one column with columns that correspond to irreducible solutions of a system of linear diophantine inequalities. A detailed description of the method is given in the paper [14].

This chapter provides some graph-theoretical tools for a primal algorithm for the stable set problem in the same vein as the work of Balas and Padberg and of Haus, Köppe, and Weismantel.

The cardinality of the largest stable set of a graph $G = (V, E)$ is called the *stability number* of G and is denoted by $\alpha(G)$. The minimum number of cliques of G whose union coincides with V is called the *clique covering number* of G and is denoted by $\overline{\chi}(G)$. A graph G is *perfect* if and only if $\alpha(G') = \overline{\chi}(G')$ for all subgraphs G' of G induced by subsets of its nodeset V. For the fundamentals on perfect graphs and balanced matrices and on their connections, which will be used throughout the chapter, we refer to, e.g., [6].

A graph is perfect if and only if its clique formulation defines an integral polytope. Moreover, for perfect graphs the stability number can be computed in polynomial time [11]; thus the separation problem for P_G is also polynomially solvable in this case. Therefore one can devise a primal cutting plane algorithm for the stable set problem for perfect graphs. We start, for example, with the edge formulation and with a basic feasible solution corresponding to a stable set. Then we perform simplex pivots until we either reach optimality or produce a fractional solution. In the latter case we add clique inequalities to the formulation that make the fractional solution infeasible, we step back to the previous (integral) basic feasible solution, and we iterate.

Suppose now that the graph is not perfect. We assume that at hand is a graph transformation that in a finite number of "steps to perfection" transforms the original graph into a possibly larger graph that is perfect. Then it may be possible to apply again the previous primal cutting plane procedure as follows: As soon as the fractional solution cannot be cut off by clique inequalities, because other valid inequalities for P_G would be necessary, we make one or more "steps to perfection" until the clique formulation of the current graph makes the fractional solution infeasible. This procedure eventually finds an optimal stable set in the latest generated graph. It can be used to solve the original problem as long as the graph transformation is such that the optimal stable set in this graph can be mapped into an optimal stable set in the original graph.

This procedure provides a motivation for this chapter where in Section 6.2 we define valid transformations that have the desired properties mentioned above; in Section 6.3 we translate the graph transformations into algebraic operations on the simplex tableaux; finally, in Section 6.4 we give some properties of the proposed transformations that may be useful when the primal algorithm sketched above is implemented.

6.2 Valid Graph Transformations

Throughout this section, we will denote by $G^0 = (V^0, E^0)$ and c^0 the graph and node-weight function of the original weighted stable set problem, respectively. The purpose of this section is to devise several types of transformations $(G, c) \mapsto (G', c')$ with the property

$\alpha_c(G) = \alpha_{c'}(G')$, i.e., transformations maintaining the weighted stability number. After a sequence of these transformations, a perfect graph G^* with a node-weight function c^* will be produced. In perfect graphs the stability number can be computed in polynomial time [11]. Moreover, the c^*-weighted stable set problem in G^* can be solved with linear programming over the maximal-clique formulation of G^*.

Typically, one is interested not only in the weighted stability number of a graph but also in a stable set where the maximum is attained. Thus, once the c^*-weighted stable set problem in G^* is solved, one would like to recover a corresponding maximum c^0-weighted stable set in the original graph G^0. For this purpose we shall attach a *node labeling* $\sigma: V \to 2^{V^0}$ to each graph $G = (V, E)$. This labeling assigns a stable set $\sigma(v) \subseteq V^0$ in the original graph to each node $v \in V$. The label of a node also determines its weight by the setting

$$c(v) = \sum_{u \in \sigma(v)} c^0(u) \quad \text{for } v \in V. \tag{6.2}$$

Thus, each node represents a partial stable set configuration in the original graph. For brevity of notation, we shall define $\sigma^0: V^0 \to 2^{V^0}$ by $\sigma^0(v) = \{v\}$ for $v \in V^0$.

Now, given a stable set $S \subseteq V$ in G with labeling σ, we intend to reconstruct a stable set $S^0 \subseteq V^0$ in G^0 by

$$S^0 = \bigcup_{s \in S} \sigma(s). \tag{6.3}$$

For this to work, we need to impose some properties on a labeling.

Definition 6.1 (valid labeling). Let $G = (V, E)$ be a graph. A mapping $\sigma: V \to 2^{V^0}$ is called a *valid node labeling* of G (with respect to G^0) if the following conditions hold:

(a) For $v \in V$, $\sigma(v)$ is a nonempty stable set in G^0.

(b) For every two distinct nodes $u, v \in V$ with $\sigma(u) \cap \sigma(v) \neq \emptyset$, the edge (u, v) is in E, i.e., nodes with nondisjoint labels cannot be in the same stable set.

(c) Let $u, v \in V$ be distinct nodes. If there exists an edge $(u^0, v^0) \in E^0$ with $u^0 \in \sigma(u)$ and $v^0 \in \sigma(v)$, then the edge (u, v) belongs to E.

Note that for a valid labeling σ the union in equation (6.3) is disjoint and gives a stable set in G^0.

Lemma 6.2. *Let σ be a valid labeling of a graph $G = (V, E)$ and let $c: V \to \mathbb{R}_+$ be defined by (6.2). Let S be a stable set in G. Then $S^0 = \bigcup_{s \in S} \sigma(s)$ is a stable set in G^0 with $c^0(S^0) = c(S)$.*

Proof. Assume that $u^0, v^0 \in S^0$ are distinct nodes with $(u^0, v^0) \in E^0$. There exist $u, v \in S$ such that $u^0 \in \sigma(u)$ and $v^0 \in \sigma(v)$. If $u \neq v$, condition (c) of Definition 6.1 implies that $(u, v) \in E$, thus S is not stable in G, contrary to the assumption. Otherwise, if $u = v$, the set $\sigma(u)$ is not stable in G^0, contradicting condition (a) of Definition 6.1. Hence, S^0 is a stable set in G^0.

Finally, note that the union $\bigcup_{s\in S}\sigma(s)$ is disjoint due to condition (b) of Definition 6.1. Therefore, $c(S) = \sum_{s\in S} c(s) = \sum_{s\in S}\sum_{s^0\in\sigma(s)} c^0(s^0) = c^0(S^0)$. □

Definition 6.3 (faithful labeling). Let σ be a valid labeling of a graph $G = (V, E)$ with respect to G^0 and let $c\colon V \to \mathbb{R}_+$ be defined by (6.2). We call σ a *faithful labeling* of G if for every stable set S in G that has maximum weight with respect to c, the stable set $S^0 = \bigcup_{s\in S}\sigma(s)$ in G^0 has maximum weight with respect to c^0. A *faithfully labeled graph* (G, c, σ) is a weighted graph (G, c) with a faithful labeling σ.

Definition 6.4 (valid transformation). A *valid graph transformation* is a transformation that turns a faithfully labeled graph (G, c, σ) into a faithfully labeled graph (G', c', σ').

Throughout this chapter, we shall make use of only a simple type of valid graph transformation, which can be characterized by the following lemma.

Lemma 6.5. *Let $G = (V, E)$ be a graph with a faithful labeling $\sigma\colon V \to 2^{V^0}$ with respect to G^0. Let $G' = (V', E')$ be a graph with a valid labeling $\tau\colon V' \to 2^V$ with respect to G such that, for every stable set S in G, there is a stable set S' in G' with $S = \bigcup_{s'\in S'}\tau(s')$. Then $\sigma'\colon V' \to 2^{V^0}$, defined by $\sigma'(v') = \bigcup_{v\in\tau(v')}\sigma(v)$ for $v' \in V'$, is a faithful labeling of G' with respect to G^0, and $(G, c, \sigma) \mapsto (G', c', \sigma')$ is a valid graph transformation.*

Proof. Obviously, σ' is a valid labeling of G' with respect to G^0. Now let S' be a stable set in G' that has maximum weight with respect to c'. Let $S = \bigcup_{s'\in S'}\tau(s')$. Since τ is a valid labeling of G' with respect to G, the set S is stable in G, and we have

$$c'(S') = \sum_{s'\in S'}\sum_{s^0\in\sigma'(s')} c^0(s^0) = \sum_{s'\in S'}\sum_{s\in\tau(s')}\sum_{s^0\in\sigma(s)} c^0(s^0) = \sum_{s'\in S'}\sum_{s\in\tau(s')} c(s) = c(S). \tag{6.4}$$

Suppose that there is a stable set $\tilde{S}$ with $c(\tilde{S}) > c(S)$. Then there exists a stable set $\tilde{S}'$ in G' with $\tilde{S} = \bigcup_{s'\in\tilde{S}'}\tau(s')$. Since (6.4) also holds when S' and S are replaced with $\tilde{S}'$ and $\tilde{S}$, respectively, we have $c'(\tilde{S}') > c'(S')$, which is a contradiction to the assumption. Hence, σ' is a faithful labeling of G' with respect to G^0. □

We first consider a very simple transformation. Take an odd path of nodes in G,

$$P = (v_1, v_2, \ldots, v_{2l}, v_{2l+1}),$$

that together with the edge (v_{2l+1}, v_1) forms an odd hole (see Figure 6.1). Let S be a stable set in G with $v_1 \in S$. Since there are at most l elements of S in P, there exists an index i such that both v_{2i} and v_{2i+1} are not in S. Therefore, if we replace v_1 with l pairwise adjacent copies $w_1, \ldots, w_l$, where w_i is adjacent to both v_{2i} and v_{2i+1} for $i = 1, \ldots, l$, it is not difficult to see that any stable set in G corresponds to a stable set in the new graph. The advantage of applying such an operation is, as will be made clear in the following, that in the new graph the odd hole has disappeared. This observation motivates the following definition.

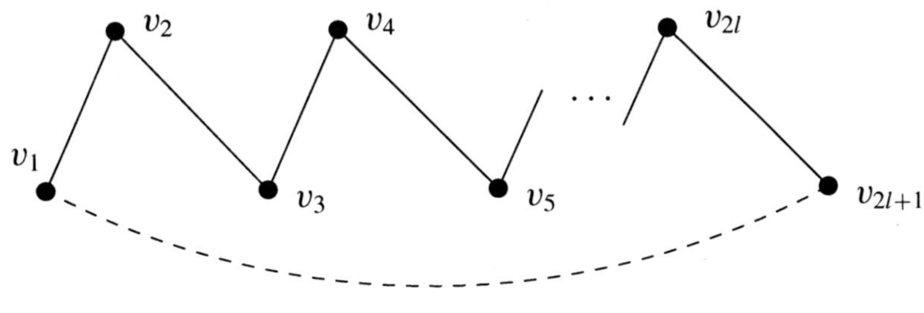

Figure 6.1. *An odd path of nodes.*

The set of all nodes in G adjacent to a node v is denoted by $N_G(v)$.

Definition 6.6 (node-path substitutions). Let $G = (V, E)$ be a graph with a valid node labeling $\sigma\colon V \to 2^{V_0}$. For some $l > 0$ let

$$P = (v_1, v_2, \dots, v_{2l}, v_{2l+1})$$

be a sequence of nodes of V such that v_i is adjacent to v_{i+1} for $i = 1, \dots, 2l$. We call P an *odd path of nodes*; see Figure 6.1. A *node-path substitution* along P that transforms a graph G with a valid labeling into a graph G' with a labeling σ' is obtained in the following way:

- Replace v_1 with the clique W of new nodes defined by

$$W = \begin{cases} \{w_1, \dots, w_l\} & \text{if } v_1 \text{ and } v_{2l+1} \text{ are adjacent in } G, \\ \{w_1, \dots, w_l, t\} & \text{otherwise.} \end{cases}$$

- For $w \in W$ connect w to all nodes of $N_G(v_1)$.
- For $i \in \{1, \dots, l\}$ connect w_i to both v_{2i} and v_{2i+1}, then set $\sigma'(w_i) = \sigma(v_1)$.
- Connect node t (if it is present in W) to v_{2l+1} and all nodes of $N_G(v_{2l+1})$, then set $\sigma'(t) = \sigma(v_1) \cup \sigma(v_{2l+1})$.

The following definition gives a generalization of the node-path substitution.

Definition 6.7 (clique-path substitutions). Let $G = (V, E)$ be a graph with a valid node labeling $\sigma\colon V \to 2^{V_0}$. For some $l > 0$, let

$$P = (Q_1 = \{v_1\}, Q_2, \dots, Q_{2l+1})$$

be a sequence of cliques of G such that $Q_{i,i+1} := Q_i \cup Q_{i+1}$ is a clique in G and $Q_i \cap Q_{i+1} = Q_{2l+1} \cap Q_1 = \emptyset$ for all $i \in \{1, \dots, 2l\}$. We call P an *odd path of cliques*. Let

$$R = \{\, v \in Q_{2l+1} : v \text{ is not adjacent to } v_1 \text{ in } G \,\}. \tag{6.5}$$

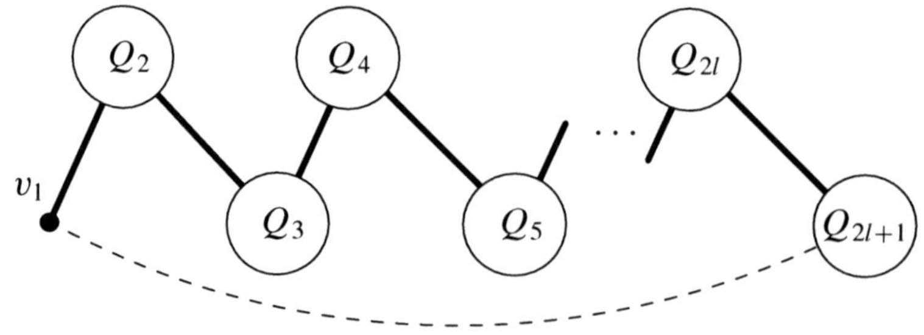

Figure 6.2. *An odd path of cliques.*

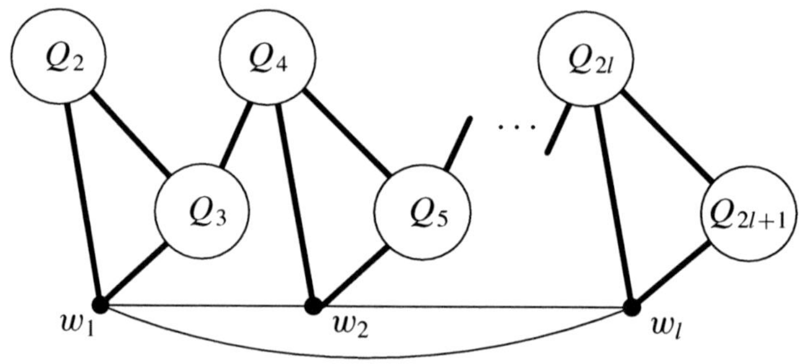

Figure 6.3. *Clique-path substitution with $R = \emptyset$.*

A *clique-path substitution* along P that transforms a graph G with a valid labeling into a graph G' with a labeling σ' is obtained in the following way:

- Replace v_1 with the clique of new nodes
$$W = \{w_1, w_2, \dots, w_l\} \cup \{ t_r : r \in R \}.$$
- For $w \in W$ connect w to all nodes of $N_G(v_1)$.
- For $i \in \{1, \dots, l\}$ connect w_i to all the nodes of $Q_{2i,2i+1}$, then set $\sigma'(w_i) = \sigma(v_1)$.
- For $r \in R$ connect t_r to r and all the nodes of $N_G(r)$, then set $\sigma'(t_r) = \sigma(v_1) \cup \sigma(r)$.

In Figure 6.2 an odd path of cliques is shown. In Figure 6.3, the graph G' that is obtained by the clique-path substitution for the case $R = \emptyset$ is shown. Note that, to unclutter the picture, some edges have been omitted; in fact all nodes w_i are connected with the nodes in Q_2 and $Q_{2l+1} \setminus R$.

Definition 6.7 does not require the cliques Q_i to be pairwise disjoint; nonconsecutive cliques may share nodes. Obviously, when $|Q_i| = 1$ for $i = 2, \dots, 2l + 1$, Definition 6.7 reproduces Definition 6.6. The labeling σ' obtained in a clique-path substitution is a valid labeling; indeed, it turns faithful labelings into faithful labelings.

Proposition 6.8. *Clique-path substitutions are valid graph transformations.*

Proof. Let σ be a faithful labeling of G. We shall make use of Lemma 6.5 to show that σ' is a faithful labeling of G'.

To this end, let $\tau\colon V' \to 2^V$ be defined by $\tau(w_i) = \{v_1\}$ for $i \in \{1, \ldots, l\}$, $\tau(t_r) = \{v_1, r\}$ for $r \in R$, and $\tau(v) = \{v\}$ otherwise. Since v_1 is not adjacent to r for $r \in R$, we have that τ is a valid labeling of G' with respect to G. Moreover, it is easy to see that $\sigma'(v') = \bigcup_{v \in \tau(v')} \sigma(v)$ for $v' \in V'$.

Now let S be a stable set in G. If $v_1 \notin S$, the set S is stable in G' as well. Suppose that $v_1 \in S$. Since v_1 is connected to all nodes of the clique Q_2, we have that $Q_2 \cap S = \emptyset$. If also $Q_3 \cap S = \emptyset$, the set $S' = S \setminus \{v_1\} \cup \{w_1\}$ is stable in G'. Otherwise, $Q_4 \cap S$ is empty, and we can repeat the argument until we reach the end of the path P. If, finally, $Q_{2l+1} \cap S = \emptyset$, the set $S' = S \setminus \{v_1\} \cup \{w_l\}$ is stable in G'. Otherwise, since the nodes in $Q_{2l+1} \setminus R$ are adjacent to v_1 in G, there is a node $r \in R \cap S$. Thus, $S' = S \setminus \{v_1, r\} \cup \{t_r\}$ is stable in G'.

By this construction, for every stable set S in G, we obtain a stable set S' in G' such that $S = \bigcup_{s' \in S'} \tau(s')$. Hence, by Lemma 6.5, σ' is a faithful labeling of G'. □

In certain cases, a clique-path substitution converts an imperfect graph G into a perfect graph G'.

Lemma 6.9. *If G is either an odd hole or an odd antihole, then there exists a node-path substitution such that the resulting graph G' is perfect.*

Proof. Let G be an odd hole with $2k + 1$ nodes denoted by $v_1, \ldots, v_{2k+1}$. Pick node v_1 and consider the path P from v_1 to v_{2k+1} through all nodes. Apply the node-path substitution of node v_1 along P, and call the resulting graph G'. In Figure 6.4 both the original graph G and the transformed graph G' are shown for $k = 2$.

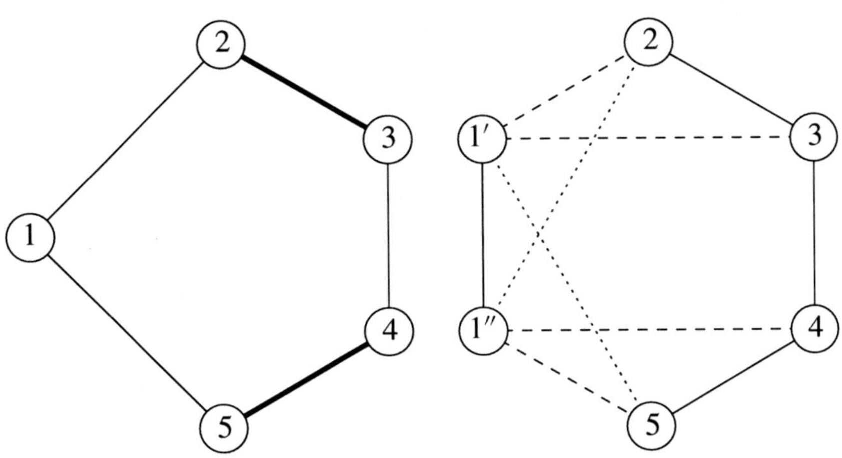

Figure 6.4. *Substitution of node 1 in an odd hole on five nodes.*

Let $W = \{w_1, w_2, \ldots, w_k\}$ be the set of nodes replacing node v_1. Consider the following clique formulation associated with G':

$$\begin{array}{lll} Q_1: & \sum_{i=1}^{k} y_i + x_2 \leq 1, & \\ Q_2: & \sum_{i=1}^{k} y_i + x_{2k+1} \leq 1, & \\ Q_{2i+1}: & y_i + x_{2i} + x_{2i+1} \leq 1 & \text{for } i = 1, \ldots, k, \\ Q_{2i}: & x_{2i-1} + x_{2i} \leq 1 & \text{for } i = 2, \ldots, k, \end{array}$$

where variables y_i correspond to nodes w_i and variables x_i to nodes v_i. We show that the constraint matrix M of such a formulation is balanced. This implies the perfectness of G (see, e.g., [6]).

We consider the row-column bipartite graph of the matrix M (see Figure 6.5), i.e., a graph constructed by taking a node for each row and each column of M and an edge for each nonzero entry of M connecting the nodes corresponding to its row and column. It is well known that M is balanced if and only if the length of all holes of this graph is divisible by four. It is easy to verify that this is the case for M. Hence the matrix is balanced and the claim follows.

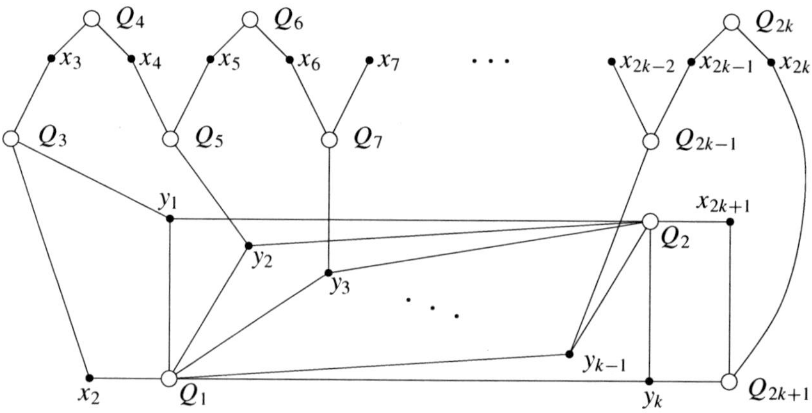

Figure 6.5. *Bipartite graph associated with the clique matrix of graph G'.*

Let G be an odd antihole with $2k + 1$ nodes. Then we number the nodes in such a fashion that the edgeset of the complement $\overline{G}$ of G is the odd hole

$$\{(1, 2), (2, 3), \ldots, (2k, 2k + 1), (2k + 1, 1)\}.$$

Consider the cycle of length five in G:

$$\{(1, 3), (3, 2k + 1), (2k + 1, 2), (2, 4), (1, 4)\}.$$

We select node 1 and perform the node-path substitution along $(1, 3, 2k + 1, 2, 4)$. That is, we replace node 1 with two adjacent nodes $1'$ and $1''$. Node $1'$ will be connected to all neighbors of the old node 1 and to the node $2k + 1$, while node $1''$ will be connected to

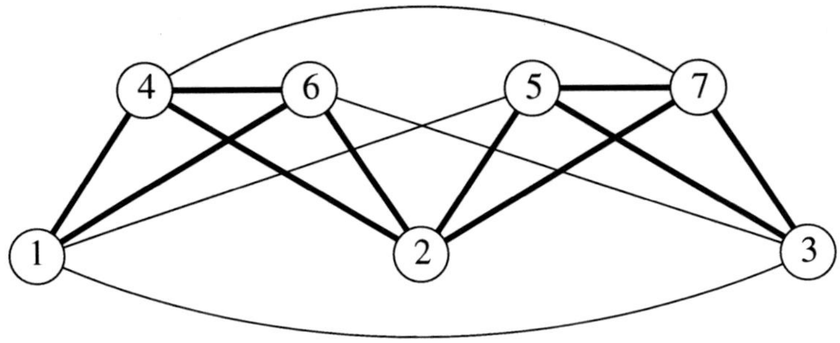

Figure 6.6. *A 7-antihole with a path of cliques.*

the neighbors of the old node 1 and to the node 2. The resulting graph is called G'. Its complement $\overline{G'}$ contains only the simple path

$$\{(1'', 2), (2, 3), \ldots, (2k, 2k+1), (2k+1, 1')\}.$$

Hence, $\overline{G'}$ is perfect. By Lovász's Perfect Graph Theorem [17], G' is perfect, too. □

Note that Lemma 6.9 and the Strong Perfect Graph Conjecture [7] imply that we can transform every minimal imperfect graph into a perfect graph with a single clique-path substitution.[1]

The following example gives an alternative substitution of a node in an antihole structure, which illustrates the more general clique-path substitution.

Example 6.10. Consider an odd antihole $\overline{C}_{2k+1} = (V, E)$ with $2k+1$ nodes, labeled from 1 to $2k+1$. Pick node 1. Then it is easy to verify that the set $V \setminus \{1\}$ can be partitioned into two cliques: Q_{odd} and Q_{even}. The former contains all nodes with an odd label (except node 1); the latter contains all nodes with an even label. So we can consider the following odd path of cliques:

$$P = (\{1\}, Q_{\text{even}} \setminus \{2\}, \{2\}, Q_{\text{odd}} \setminus \{3\}, \{3\}).$$

Figure 6.6 shows the 7-antihole $\overline{C}_7$; the edges of the complete subgraphs induced by the relevant cliques in P are shown with thick lines. The graph G' resulting from the clique-path substitution along P has two new nodes $1'$ and $1''$ replacing node 1; node $1'$ is connected to all nodes but 2, while node $1''$ is connected to all nodes but $2k+1$. The graph G' is shown in Figure 6.7. The edges introduced by the clique-path substitution are drawn with dashed lines, whereas edges merely inherited from node 1 are drawn with dotted lines. The resulting graph is the same as the one obtained by the node-path substitution of Lemma 6.9, hence it is perfect.

For any graph it is easy to construct a finite sequence of clique-path substitutions leading to a graph that is the disjoint union of complete graphs, hence to a perfect graph.

[1]The first version of this paper was written before the Strong Perfect Graph Theorem was proved. Note that the proof of Lemma 6.9 can be simplified by using the Strong Perfect Graph Theorem.

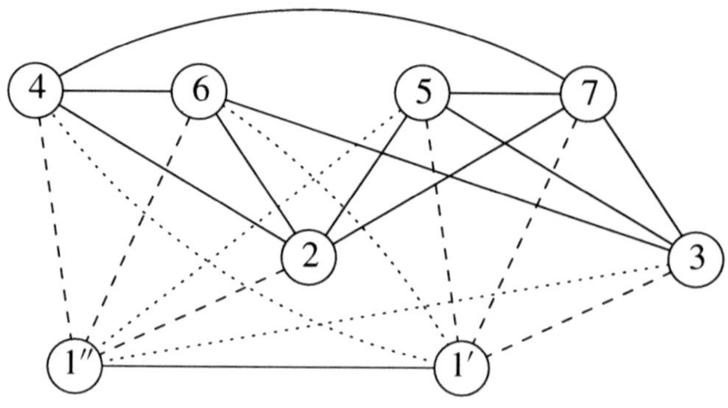

Figure 6.7. *A 7-antihole after a clique-path transformation.*

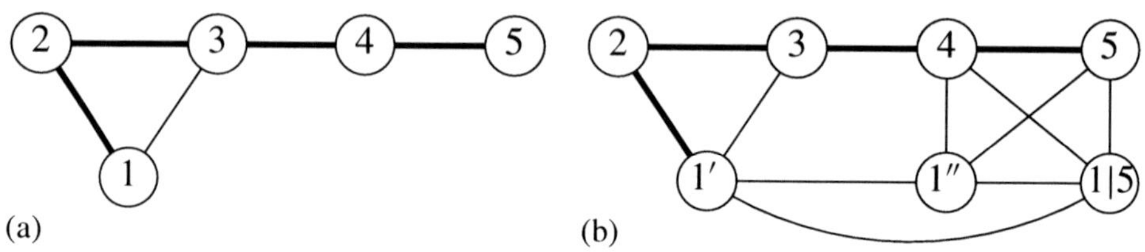

Figure 6.8. *The node-path substitution along the odd path of nodes* $(1, 2, 3, 4, 5)$.

Lemma 6.11 (finiteness). *Let $G = (V, E)$ be a graph. There exists a finite sequence of clique-path substitutions leading to a perfect graph $G^* = (V, E)$.*

Proof. Since clique-path substitutions work within one component, we may assume that G is connected. Let $Q \subseteq V$ be a clique in G that is maximal with respect to inclusion. If $Q = V$, we are done. Otherwise, let $v_1 \in V \setminus Q$ such that $Q_2 := Q \cap N_G(v_1) \neq \emptyset$. Let $Q_3 := Q \setminus Q_2$. Now $P = (v_1, Q_2, Q_3)$ is an odd path of cliques in G. The clique-path substitution along P leads to a graph G', where all the new nodes have been adjoined to the clique Q. We continue with G' and a maximal clique in G' containing the enlarged clique. Since $|V \setminus Q|$ decreases in each step by at least one, the procedure terminates with a complete graph. □

However, we cannot expect that an arbitrarily chosen sequence of clique-path substitutions terminates, as the following example shows.

Example 6.12. Let us consider the graph G^0 shown in Figure 6.8(a). The node-path substitution along the odd path of nodes $(1, 2, 3, 4, 5)$, shown with bold edges, leads to the validly labeled graph G^1 shown in Figure 6.8(b). As G^0 is an induced subgraph of G^1 after renaming node $1'$ to 1, the same node-path substitution can be performed ad infinitum, adding two nodes (copies of $1''$ and $1|5$) in each step.

Remark 6.13 (the struction of Ebenegger et al.). In the paper [9], Ebenegger, Hammer, and de Werra describe a construction that reduces the stability number of a graph by 1;

in the subsequent papers [8, 12, 13] this is called a *struction*. Here, we shall present a variant of the struction that is a valid graph transformation, i.e., it maintains the weighted stability number rather than reducing it. Let $G = (V, E)$ be a graph. Let $v_0 \in V$ be an arbitrary node and let $N(v_0) = \{v_1, \dots, v_p\}$. The idea of the construction is the following: For each stable set S in G not containing v_0 but some of the nodes $v_1, \dots, v_p$, there is a minimum index $i \in \{1, \dots, p\}$ with $v_i \in S$. For each such minimum index i, in G' there is a layer of copies of those nodes $v_i, v_{i+1}, \dots, v_p$ that are not adjacent to v_i in G; these copies are called $v_{i,i}, v_{i,i+1}, \dots, v_{i,p}$. These copies replace the original nodes $v_1, \dots, v_p$. Within one layer, the $v_{i,j}$ inherits all the edges from both v_i and v_j in G, whereas nodes of different layers are all connected by edges. Figure 6.9 illustrates the transformation; note that only edges between adjacent layers have been drawn here and that node v_0, which is connected to all new nodes, has been omitted. The algorithmic idea is to perform a sequence of structions, yielding a graph whose stability number can be computed easily. In the papers [8, 12, 13], it is shown that, for certain classes of graphs, the number of operations necessary is polynomially bounded. In the general case, however, one cannot expect similar results to hold, since the number of nodes in the problem may grow very fast. We are not aware of a thorough computational study of an algorithm based on structions.

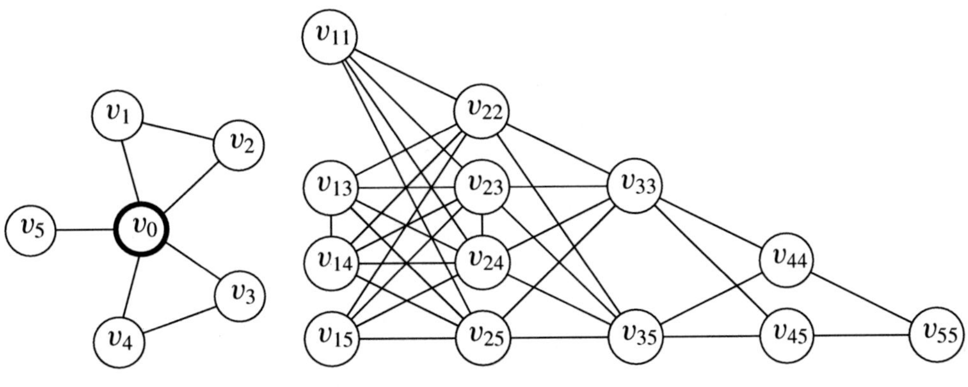

Figure 6.9. *A variant of the struction of a graph.*

Remark 6.14 (comparison to linear programming–based branch-and-bound (B&B) procedures). We use a simple example to illustrate the possible advantage of a method based on valid graph transformations, compared to linear programming–based B&B procedures. For $k \in \{1, 2, \dots\}$ and $l \in \{2, 3, \dots\}$, let the graph C^k_{2l+1} be the disjoint union of k odd holes C_{2l+1}. The maximal clique formulation of the stable set problem in C^k_{2l+1} is

$$
\begin{array}{lll}
\max & \sum_{i=1}^{k} \sum_{j=1}^{2l+1} x_{i,j} & \\
\text{subject to} & x_{i,j} + x_{i,j+1} \le 1 & \text{for } i \in \{1, \dots, k\} \text{ and } j \in \{1, \dots, 2l\}, \\
& x_{i,1} + x_{i,2l+1} \le 1 & \text{for } i \in \{1, \dots, k\}, \\
& x_{i,j} \in \{0, 1\} & \text{for } i \in \{1, \dots, k\} \text{ and } j \in \{1, \dots, 2l+1\}.
\end{array}
\tag{6.6}
$$

The unique optimal solution to the linear programming relaxation of (6.6) is given by $x_{i,j} = \frac{1}{2}$ for $i \in \{1, \ldots, k\}$ and $j \in \{1, \ldots, 2l + 1\}$. A linear programming–based B&B procedure would now select one node variable, $x_{1,1}$, say, and consider the two subproblems obtained from (6.6) by fixing $x_{1,1}$ at 0 and 1, respectively. The graph-theoretic interpretation of this variable fixing is that the copy of C_{2l+1} corresponding to the node variables $x_{1,\cdot}$ is turned into a perfect graph in both branches; see Figure 6.10. Hence the optimal basic solutions to the linear programming relaxations of the subproblems attain integral values there. Since there remain $k - 1$ odd holes in both subproblems, the B&B procedure clearly visits a number of subproblems exponential in k.

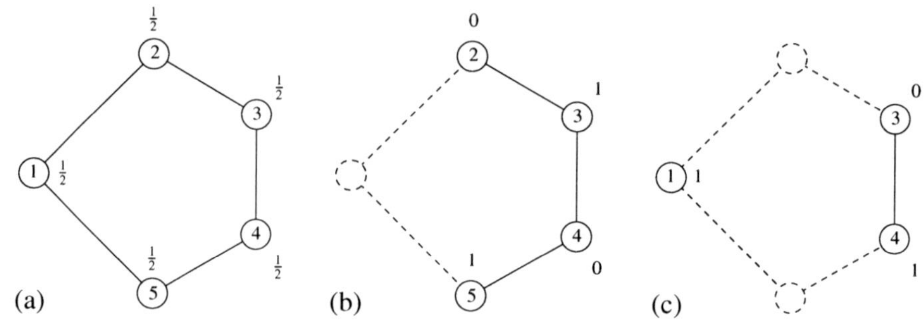

Figure 6.10. *Graph-theoretic interpretation of the branch operation on a fractional node variable in a linear programming–based B&B procedure. The numbers shown beside the nodes are the node variable values in an optimal solution to the linear programming relaxation.* (a) *One of the copies of C_5 in the graph C_5^k.* (b) *Fixing the first node variable at zero.* (c) *Fixing the first node variable at one.*

On the other hand, a method using clique-path substitutions, which performs the whole enumeration implicitly, can turn the graph C_{2l+1}^k into a perfect, validly labeled graph (G, c, σ) by performing only k substitution steps of the type shown in Figure 6.4. The optimal solution to the linear programming relaxation of this formulation is integral, and the corresponding maximum stable set in the original graph C_{2l+1}^k can be computed by means of the node labeling σ.

6.3 Optimizing Over Stable Sets

Since graph transformations transform weighted stable set problems into weighted stable set problems, they can be used as a tool within any optimization algorithm for the stable set problem.

In this section, we will deal with weighted stable set problems in a specific algorithmic framework, namely, in a primal integer programming setting in the vein of works of Balas and Padberg [2] and Haus, Köppe, and Weismantel [14, 15]. It will turn out that the graph transformations discussed in the previous section can be re-interpreted as column operations in an integral simplex tableau.

First we need to fix an integer programming formulation. We note that the problem of finding a maximum stable set in $G = (V, E)$ with respect to the weight function $c \in \mathbb{R}^V_+$ can be formulated as the following integer program (IP):

$$\max c^\top x : \; x_v + x_w + z_{v,w} = 1 \; \text{ for } (v, w) \in E, \; x \in \{0, 1\}^V, \; z \in \{0, 1\}^E. \tag{6.7}$$

Note that in this formulation $z_{v,w}$ is the slack variable of the edge $(v, w) \in E$. A better integer programming formulation is achieved when edges are replaced by cliques of larger size. Let $Q_1, \ldots, Q_k$ be cliques in G that cover all the edges of G, i.e., for every edge $(v, w) \in E$ there is an index $i \in \{1, \ldots, k\}$ such that $\{v, w\} \subseteq Q_i$. Note that this set of cliques may not coincide with all the maximal cliques of the complete clique formulation and that the cliques may not be maximal. Introducing a slack variable z_{Q_i} for each clique Q_i, the weighted stable set problem is formulated as

$$\max c^\top x : \sum_{v \in Q_i} x_v + z_{Q_i} = 1 \; \text{ for } i \in \{1, \ldots, k\}, \; x \geq 0, \; z \geq 0, \tag{6.8}$$

$$x \in \mathbb{Z}^V, \; z \in \mathbb{Z}^k. \tag{6.9}$$

This IP is the starting point of our further investigations. We call (6.8) a *maximal-clique formulation* if all cliques Q_i, $i \in \{1, \ldots, k\}$, are maximal. Moreover, if the set $\{Q_i\}_{i \in \{1,\ldots,k\}}$ includes all maximal cliques of G, (6.8) is called the *complete maximal-clique formulation*.

Now let $S \subseteq V$ be a stable set in G. To construct a basic feasible solution associated with S we select for each of the rows in the program (6.8) a basic variable as follows:

- For every $v \in S$, let i_v be a row index such that $v \in Q_{i_v}$. We select x_v as the basic variable associated with the row i_v of the tableau.

- For each of the remaining clique constraints i, we select the slack variable z_{Q_i} as the corresponding basic variable.

Note that the indices i_v are all distinct, and hence the construction yields a basis corresponding to S. As usual, let B and N denote the sets of basic and nonbasic variables, respectively. We can now rewrite (6.8) in tableau form:

$$y_B + A_B^{-1} A_N y_N = b,$$
$$(y_B, y_N) \in \mathbb{Z}^{n+k}_+.$$

The variables y_j correspond to node variables x_v or slack variables z_{Q_i}. We will henceforth call a tableau obtained by this procedure a *canonical tableau* for S. Realizing the fact that the original objective function was nonnegative on the nodes and zero on the slack variables, the following observation is immediate.

Observation 6.15. Let N be the set of nonbasic variables in a canonical tableau for a stable set S in the graph G. Then the reduced cost of a nonbasic slack variable z_{Q_i} is always nonpositive. The reduced cost of a nonbasic node variable x_k may be zero, negative, or positive.

Example 6.16. Let C_5 be an odd hole on five nodes. The problem of finding a stable set of maximum size in C_5 is formulated as the following IP:

$$\begin{array}{lllllllllll}
\max & x_1 + x_2 + x_3 + x_4 + x_5 \text{ subject to} \\
& x_1 + x_2 & & & & + z_{12} & & & & & = 1, \\
& \quad x_2 + x_3 & & & & & + z_{23} & & & & = 1, \\
& \qquad x_3 + x_4 & & & & & & + z_{34} & & & = 1, \\
& \qquad\quad x_4 + x_5 & & & & & & & + z_{45} & & = 1, \\
& x_1 \qquad\qquad\quad + x_5 & & & & & & & & + z_{15} & = 1.
\end{array}$$

The construction described above for the stable set $S = \{3, 5\}$ in C_5 yields the following tableau for S:

$$\begin{array}{lllllll}
\max & & & & x_1 - z_{23} - z_{45} \text{ subject to} \\
x_3 & & + x_2 & & + z_{23} & & = 1, \\
x_5 & & + x_4 & & & + z_{45} & = 1, \\
z_{34} & & - x_2 + x_4 & & - z_{23} & & = 0, \\
z_{15} & & - x_4 + x_1 & & & - z_{45} & = 0, \\
z_{12} + x_2 & & + x_1 & & & & = 1.
\end{array}$$

Starting from a basic feasible integer solution, the Balas–Padberg procedure [2] and the Integral Basis Method by Haus, Köppe, and Weismantel [14, 15] proceed as follows. As long as nondegenerate integral pivots are possible, such steps are performed, improving the current basic feasible integer solution. When a solution is reached that would only permit a degenerate integral pivot, a "column generation procedure" replaces some nonbasic columns with new "composite" columns, which are nonnegative integral combinations of nonbasic columns in the current tableau. In this way columns are eventually generated that allow nondegenerate integral pivots, or optimality of the current basic feasible solution is proved.

The Balas–Padberg procedure has not proved to be an efficient algorithm for set partitioning problems. Also, the implementation of the Integral Basis Method as described in [14] shows a rather weak computational performance when applied to stable set problems. The reason is that both algorithms generate the composite columns to add to the tableau without making use of the graph-theoretic properties of the problem.

The idea to improve the performance is to use "strong," "combinatorial" composite columns whenever possible, rather than the general composite columns derived from the tableau. This is where clique-path substitutions come into play. In the following, we show how they can be dealt with in the integer programming setting.

Definition 6.17 (odd alternating path of cliques). Let $G = (V, E)$ be a graph and σ a faithful labeling of G, and let the node-weight function $c\colon V \to \mathbb{R}_+$ be defined by (6.2). Fix an integral simplex tableau for a formulation of the c-weighted stable set problem in G. Let x_{v_1} be a nonbasic variable of positive reduced cost, and let $P = (\{v_1\}, Q_2, \ldots, Q_{2l+1})$ be an odd path of cliques in G. Again we denote by R the set

$$R = \{\, v \in Q_{2l+1} : v \text{ is not adjacent to } v_1 \text{ in } G \,\}.$$

For $i \in \{1, \ldots, l\}$ assume there is a nonbasic slack variable z_i for the clique $Q_{2i,2i+1} = Q_{2i} \cup Q_{2i+1}$ and there is $x_{j_i} = 1$ with $j_i \in Q_{2i}$. Moreover, presume that all variables x_v

for $v \in R$ are nonbasic. In this setting we call

$$(\{v_1\}, Q_{2,3}, Q_{4,5}, \dots, Q_{2l,2l+1})$$

an *odd alternating path of cliques*. If $|Q_i| = 1$ for $i \in \{2, \dots, 2l+1\}$, we call it an *odd alternating path of nodes*.

In this setting we are going to remove the column of the nonbasic variable x_{v_1} from the tableau and replace it with nonbasic integral combinations of other columns of the tableau. We shall use the notation $x_v \wedge x_w$ for a new variable associated with a column that is the sum of the columns for x_v and x_w.

Definition 6.18. For an odd alternating path of cliques

$$(\{v_1\}, Q_{2,3}, Q_{4,5}, \dots, Q_{2l,2l+1})$$

we define the corresponding *alternating-path substitution* in the tableau as follows: Substitute x_{v_1} with new binary variables according to the following column operations:

- $u_r = x_{v_1} \wedge x_r$ for all $r \in R$,
- $y_i = x_{v_1} \wedge z_i$ for all $i \in \{1, \dots, l\}$,

where all new variables are nonbasic and 0/1.

Observation 6.19. Let $\bar{x}$ and $\bar{z}$ denote the variables x and z, respectively, in the formulation obtained after the substitution. Then we can map a solution of the new formulation back into a solution of the old formulation via the following relations:

$$x_{v_1} = \sum_{i=1}^{l} y_i + \sum_{r \in R} u_r, \tag{6.10a}$$

$$x_r = \bar{x}_r + u_r \qquad \text{for all } r \in R, \tag{6.10b}$$

$$z_i = \bar{z}_i + y_i \qquad \text{for all } i \in \{1, \dots, l\}. \tag{6.10c}$$

For all other variables we have $x_v = \bar{x}_v$ and $z_i = \bar{z}_i$. In the following we will denote by $\mathcal{F}$ a generic formulation of type (6.8) and by $\mathcal{S}_P(\mathcal{F})$ the formulation obtained by applying an alternating-path substitution along P to $\mathcal{F}$. Moreover, given a formulation $\mathcal{F}' = \mathcal{S}_P(\mathcal{F})$, we will denote by $\mathcal{R}_P(\mathcal{F}')$ the formulation obtained by applying the mapping (6.10) to $\mathcal{F}'$.

Lemma 6.20. *The IP obtained by a sequence of alternating-path substitutions is an integer programming formulation of the stable set problem in G; the optimal solutions translate into the maximum stable sets in G via the iterated mapping (6.10).*

Proof. Let (G', c', σ') denote the labeled graph obtained from (G, c, σ) by performing the clique-path substitution along

$$P = (v_1, Q_2, Q_3, \dots, Q_{2l}, Q_{2l+1}).$$

As in Definition 6.7, let t_r for $r \in R$ and w_i for $i \in \{1, \dots, l\}$ denote the new nodes arising from the substitution. We show that the problem resulting from the above column operations is a formulation of the c'-weighted stable set problem in G'.

The key is to realize that the new variables correspond to the new nodes in the following way:

(i) For $i \in \{1, \dots, l\}$, variable $y_i = x_{v_1} \wedge z_i$ corresponds to the new node w_i.

(ii) For $r \in R$, variable $u_r = x_{v_1} \wedge x_r$ corresponds to the new node t_r.

To verify (i) let $i \in \{1, \dots, l\}$ and note that the original formulation of the c-weighted stable set problem in G implies the following inequalities and equations:

$$\begin{aligned} x_{v_1} + x_v &\le 1 \quad \text{for } v \in N_G(v_1), \\ \textstyle\sum_{v \in Q_{2i,2i+1}} x_v + z_i &\le 1. \end{aligned}$$

By (6.10) we obtain

$$\begin{aligned} \textstyle\sum_{i=1}^{l} y_i + \sum_{r \in R} u_r + \bar{x}_v &\le 1 \quad \text{for } v \in N_G(v_1), \\ \textstyle\sum_{v \in Q_{2i,2i+1}} \bar{x}_v + \sum_{v \in Q_{2i,2i+1} \cap R} u_v + \bar{z}_i + y_i &= 1. \end{aligned}$$

Hence, for all variables x_t corresponding to the neighbors $t \in N_{G'}(w_i)$, as given by Definition 6.7, the new formulation implies an inequality $x_t + y_i \le 1$. The correspondence (ii) can be verified analogously. □

Example 6.21 (Example 6.16 continued). For the stable set problem introduced in Example 6.16, the path

$$P = (1, \{2, 3\}, \{4, 5\})$$

is an alternating path of cliques. The slack variables z_{23} and z_{45} are both nonbasic. Now x_1 is the variable in the tableau with positive reduced cost. Since the edge $(1, 5)$ is present in G, we substitute variable x_1 with the two variables $x_1' = x_1 \wedge z_{23}$ and $x_1'' = x_1 \wedge z_{45}$ corresponding to the two sums of column 1 and the columns associated with z_{23} and z_{45}, respectively. Note that in this example the reduced cost of the two new nonbasic columns is 0. Hence, we have a certificate that S is indeed optimal.

This illustrates the most fortunate situation for our algorithmic framework: Not only has the alternating-path substitution turned the graph G into a perfect graph, as shown in Lemma 6.9, but we also obtain an integral tableau with a linear programming certificate for optimality.

6.4 Properties of Alternating-Path Substitutions

The destruction of one odd hole via our column substitution procedure comes with the cost of needing to enlarge the original stable set problem significantly. One might argue that because of this enlargement of the graph it might algorithmically be more tractable to add just one odd-hole cutting plane to the initial formulation. The drawback of the latter approach is, however, that odd-hole cuts define facets for the stable set polytope associated with a graph that is the odd hole itself. For an arbitrary graph that contains an odd hole as an induced subgraph, lifting becomes necessary to strengthen an odd-hole cut. However, it is not known how to separate each lifted inequality in polynomial time. Therefore, cutting plane procedures apply heuristic techniques (exact or heuristic sequential lifting) to strengthen an odd-hole cut. In contrast, our procedure automatically deals with graph structures where a cut approach would need lifting, as the following example shows.

Example 6.22. Let $G = (V, E)$ be an odd wheel involving $2k$ nodes. The first $2k - 1$ nodes, numbered from 1 to $2k - 1$, form an odd hole. The additional node is called the *hub* and is denoted by h. This node h is adjacent to all nodes on the hole. Associated with such a configuration is an odd-wheel inequality: that can be seen as a lifted odd-hole inequality:

$$\sum_{i=1}^{2k-1} x_i + (k-1)x_h \leq k - 1.$$

We will now show that "destroying" the odd hole by performing a node-path substitution makes the fractional solutions that would be cut by the odd-wheel inequality automatically infeasible. This implies that in this situation a concept of inequality strengthening by lifting is not required for the primal approach. We perform the same graph transformation as in the proof of Lemma 6.9 for the odd hole, i.e., a node-path substitution along the path

$$P = (1, 2, 3, \ldots, 2k-2, 2k-1).$$

The resulting graph G' is illustrated in Figure 6.11 for $k = 3$. Compared to the perfect graph obtained in the proof of Lemma 6.9, G' has only the extra node h, which is connected to all the other nodes. Thus G' is perfect as well.

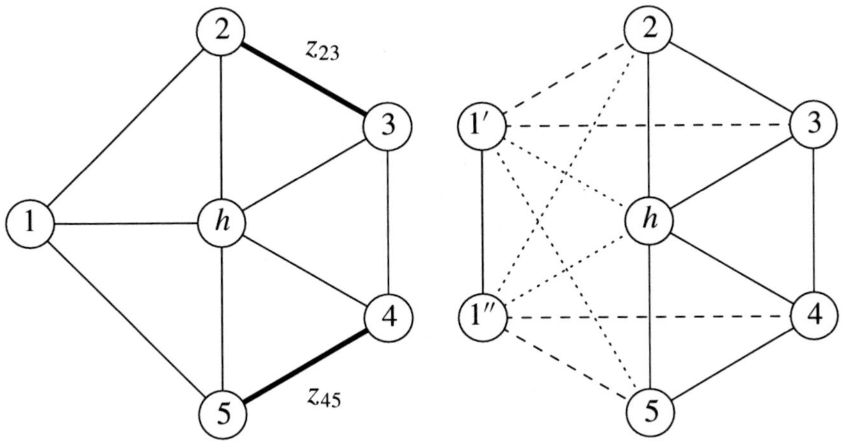

Figure 6.11. *Substitution of node (1) of the odd hole in an odd wheel of size five.*

It has already been pointed out that a clique formulation (6.8) of the weighted stable set problem is much stronger than the node-edge formulation (6.7). In fact, when the maximal cliques are employed, the integrality constraints in (6.9) can be dropped if the underlying graph G is perfect.

Example 6.23 (Example 6.10 continued). Let us again consider the transformation of the odd antihole $\overline{C}_{2k+1}$ carried out in Example 6.10. Let the problem be given in its complete maximal-clique formulation, composed of $2k+1$ cliques of size k, among which are the cliques Q_{odd} and Q_{even} from Example 6.10. Let a basis be fixed such that the node variable x_1 and the clique slack variables $z_{Q_{\text{odd}}}$ and $z_{Q_{\text{even}}}$ are nonbasic. Then the clique-path substitution of Example 6.10 is equivalent to the column operation substituting x_1 with the variables

$$x_{1'} = x_1 \wedge z_{Q_{\text{odd}}}, \qquad x_{1''} = x_1 \wedge z_{Q_{\text{even}}}.$$

It is now easy to verify that in the formulation we obtain in this way, all cliques are maximal in the resulting perfect graph (actually, in this particular example we obtain the complete maximal-clique formulation).

However, this desirable property does not hold in general.

The alternating-path substitution requires that the cliques $Q_{2i} \cup Q_{2i+1}$, $i \in \{1, \dots, l\}$, be present in the formulation $\mathcal{F}$. In the case that these cliques are not maximal, to perform the alternating-path substitution we have to add the corresponding inequalities, which are dominated.

Now we consider the *identification problem* for odd alternating paths of cliques. First we show that an alternating-path substitution "cuts off" the fractional point (x^F, z^F) obtained by a single pivoting step applied to a basic integer solution (x^I, z^I). Moreover, such an alternating path with $R = \emptyset$ can be found in polynomial time if it exists.

Definition 6.24. Let (x^I, z^I) be a basic integer solution and let v_i be a nonbasic variable. If the basic solution obtained by pivoting in x_{v_1} is fractional, we call it a *fractional neighbor* of (x^I, z^I) and denote it by (x^F, z^F).

Lemma 6.25. *Let $\mathcal{F}$ be a formulation for the graph G, (x^I, z^I) a basic integer solution, and x_{v_1} a nonbasic variable. Assume that pivoting x_{v_1} into the basis would produce a fractional neighbor (x^F, z^F). Consider an alternating-path substitution along $P = (v_1, Q_{2,3}, \dots, Q_{2l,2l+1})$ and the corresponding formulation $\mathcal{F}' = \mathcal{S}_P(\mathcal{F})$. Then the solution (x^F, z^F) is not feasible for the mapping $\mathcal{R}_P(\mathcal{F}')$ on the space of the initial variables (where $\mathcal{R}$ is the mapping defined in Observation 6.19).*

Proof. Let $y_i = x_{v_1} \wedge z_{Q_{2i,2i+1}}$ for $i \in \{1, \dots, l\}$ and $u_r = x_{v_1} \wedge x_r$ for $r \in R$ be the new variables obtained by substitution of v_1. We denote by $\bar{x}$ and $\bar{z}$ the variables x and z, respectively, in the formulation $\mathcal{F}'$. The relations that define the mapping $\mathcal{R}$ are satisfied by all integer solutions. We will show that they are violated by the fractional solution (x^F, z^F), so that $(x^F, z^F) \notin \mathcal{R}_P(\mathcal{F}')$. First, note that the variables z_i for $i \in \{1, \dots, l\}$ and x_r for $r \in R$ are nonbasic both before and after the pivoting operation. From equations (6.10b), (6.10c), and the nonnegativity of all variables, we obtain that the variables $\bar{z}_i$ and y_i for $i \in \{1, \dots, l\}$ and $\bar{x}_r$ and u_r for all $r \in R$ must have value zero in the solution corresponding to (x^F, z^F). But at the same time $x_{v_1} > 0$, that is, at least one of the variables y_i for $i \in \{1, \dots, l\}$ or u_r for all $r \in R$ attains a positive value. This is a contradiction. □

Observation 6.26. Lemma 6.25 is valid as long as the variables z_i for $i \in \{1, \ldots, l\}$ and x_r for $r \in R$ are equal to zero for both the solutions (x^I, z^I) and (x^F, z^F). Therefore these variables do not necessarily have to be nonbasic.

In the following, we shall consider the alternating-path substitution in the case of $R = \emptyset$. In this case the path of cliques is in fact a cycle, so an associated tableau has the following form:

$$\begin{array}{llllll}
\sum_{j\in Q_{1,2}} x_j & & & & & \leq 1,\\
& \sum_{j\in Q_{2,3}} x_j & & & + z_1 & = 1,\\
& & \sum_{j\in Q_{3,4}} x_j & & & \leq 1,\\
& & \ddots & & & \vdots\\
& & & \sum_{j\in Q_{2l,2l+1}} x_j & + z_l & = 1,\\
& & & & \sum_{j\in Q_{2l+1}\cup\{v_1\}} x_j & \leq 1.
\end{array}$$

If we perform the substitution of x_{v_1}, the equations in the above system can be rewritten as follows:

$$\begin{array}{llll}
\sum_{j\in Q_{2,3}} x_j & + \bar{z}_1 & + y_1 & = 1,\\
\ddots & \ddots & \ddots & \\
\sum_{j\in Q_{2l,2l+1}} x_j & + \bar{z}_l & + y_l & = 1,
\end{array}$$

where $y_i = x_{v_1} \wedge z_i$ for all $i \in \{1, \ldots, l\}$, and, therefore, $x_{v_1} = \sum_{i=2}^{l} y_i$ and $z_i = \bar{z}_i + y_i$. By substituting $y_1 = x_{v_1} - \sum_{i=2}^{l} y_i$ in the first equation and then summing up all of them, we obtain the relation

$$\sum_{i=1}^{l}\left(\sum_{j\in Q_{2i,2i+1}} x_j\right) + x_{v_1} + \sum_{i=1}^{l} \bar{z}_i = l. \tag{6.11}$$

If the cliques $Q_1, \ldots, Q_{2l+1}$ are pairwise disjoint, this is a lifted odd-hole inequality resulting from the odd hole of length $2l + 1$ of nodes $v_1, v_2, v_3, v_4, \ldots, v_{2l}, v_{2l+1}$, where v_i can be any node in Q_i for $i \in \{2, \ldots, 2l + 1\}$. This means that performing the alternating-path substitution is at least as strong as adding a lifted odd-hole cut to the problem.

Theorem 6.27. *Let $\mathcal{F}$ be a formulation and let (x^I, z^I) denote a basic integer solution. Suppose that there exists an alternating-path substitution along*

$$P = (v_1, Q_{2,3}, \ldots, Q_{2l,2l+1})$$

such that the inequality corresponding to each of the cliques $Q_{2,3}, \ldots, Q_{2l,2l+1}$ coincides with or is dominated by one of the clique inequalities in $\mathcal{F}$ and $R = \emptyset$. Then one can find such a substitution in polynomial time in the size of $\mathcal{F}$. Moreover, the fractional neighbor (x^F, z^F) that would have been obtained by pivoting x_{v_1} into the basis is infeasible for the formulation $\mathcal{R}_P(\mathcal{S}_P(\mathcal{F}))$.

Proof. Let $\{\hat{Q}_i : i \in K\}$ be cliques corresponding to the inequalities in $\mathcal{F}$. We build a digraph H_0, where each node represents a clique $\hat{Q}_i$ whose corresponding variable is nonbasic. We also have an additional node associated with the column x_{v_1} to be substituted. For each clique $\hat{Q}_i$ with nonbasic slack variable, let x_{j_i} be the unique variable such that $j_i \in Q_i$ and $x^I_{j_i} = 1$ in the present basic integer solution. We have an arc labeled (i, m, k) from clique $\hat{Q}_i$ to clique $\hat{Q}_k$ if there exists an index $m \in K$ such that

$$\hat{Q}_i \cap \hat{Q}_m \neq \emptyset \quad \text{and} \quad \hat{Q}_m \cap \hat{Q}_k \neq \emptyset, \tag{6.12a}$$

$$j_i \notin \hat{Q}_m \quad \text{whereas} \quad j_k \in \hat{Q}_m. \tag{6.12b}$$

The node associated with the variable x_{v_1} is connected to the cliques following the same rules, i.e., there exists an arc labeled $(0, m, k)$ from x_{v_1} to a clique $\hat{Q}_k$ not containing x_{v_1} if there exists an index $m \in K$ such that

$$x_{v_1} \in \hat{Q}_m \quad \text{and} \quad \hat{Q}_m \cap \hat{Q}_k \neq \emptyset, \tag{6.13a}$$

$$j_k \in \hat{Q}_m. \tag{6.13b}$$

For arcs labeled $(k, m, 0)$ from a clique $\hat{Q}_k$ to x_{v_1} the last condition must be modified to read $j_k \notin \hat{Q}_m$.

Suppose there exists a directed cycle in H_0 passing through x_{v_1}:

$$C = (x_{v_1}, (0, m_1, k_1), \hat{Q}_{k_1}, (k_1, m_2, k_2), \hat{Q}_{k_2}, \ldots, \hat{Q}_{k_l}, (k_l, m_{l+1}, 0), x_{v_1}).$$

We can construct an odd path of cliques from C as follows. For $i \in \{1, \ldots, l\}$ let

$$\begin{aligned} Q_{2i} &= \hat{Q}_{m_i} \cap \hat{Q}_{k_i}, \\ Q_{2i+1} &= \hat{Q}_{k_i} \cap \hat{Q}_{m_{i+1}}. \end{aligned}$$

Since $j_{k_i} \in Q_{2i}$ for $i \in \{1, \ldots, l\}$, the cliques $Q_{2i,2i+1} = Q_{2i} \cup Q_{2i+1}$ are tight at (x^I, z^I). Note that each clique $Q_{2i,2i+1}$ coincides with or is dominated by the clique $\hat{Q}_{k_i}$ that is present in the formulation. After introducing nonbasic variables for the slacks of $Q_{2i,2i+1}$ (unless already present), the path

$$P = (v_1, Q_{2,3}, \ldots, Q_{2l,2l+1})$$

is clearly an odd alternating path of cliques (see Figure 6.12). Conversely, every odd alternating path of cliques that coincide with or are dominated by cliques in the current formulation corresponds to a directed cycle C in H_0. Hence, we can find them in polynomial time in the size of $\mathcal{F}$ by detecting directed cycles in the auxiliary digraph H_0. By Lemma 6.25 the fractional solution (x^F, z^F) obtained by pivoting x_{v_1} into the basis is infeasible for $\mathcal{R}_P(\mathcal{S}_P(\mathcal{F}))$. $\square$

Note that the digraph H_0 defined in the proof of Theorem 6.27 can be replaced by a directed multigraph, where two cliques $\hat{Q}_i$ and $\hat{Q}_k$ are connected by parallel arcs (i, m, k) for *each* $m \in K$ satisfying the above conditions.

We have noted before that an alternating-path substitution is equivalent to adding a lifted odd-hole cut (6.11). In a dual-type method, one is interested in finding the cut from a given class that is *most violated* by the current solution. We can solve the analogous problem in our primal setting.

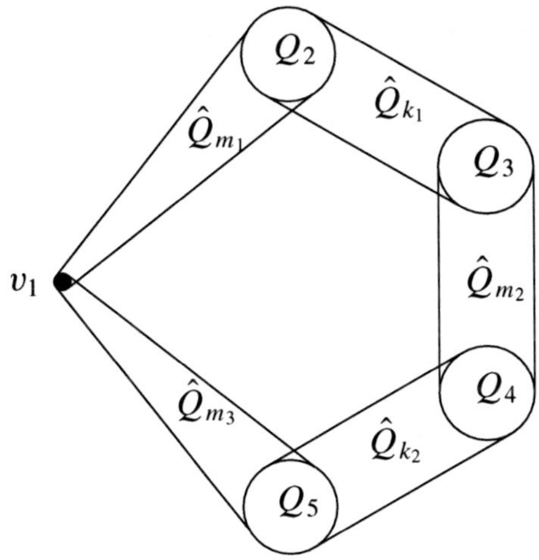

Figure 6.12. *Constructing an odd alternating path of cliques.*

Proposition 6.28. *The problem of finding the alternating-path substitution of the type in Theorem 6.27, whose corresponding inequality (6.11) is the most violated by a fractional neighbor (x^F, z^F) of a basic integral solution (x^I, z^I), can be solved in polynomial time in the size of $\mathcal{F}$.*

Proof. We use the same notation as in the proof of Theorem 6.27. Let H be the graph constructed following the same rules used for H_0, but with nodes associated with all clique inequalities that are tight at (x^I, z^I). Let us now define a digraph H', where the nodes are defined by all those subcliques obtained by the intersections of all the cliques $\hat{Q}_i$ that correspond to the nodes of H with the cliques $\hat{Q}_m$ for all the arcs (i, m, k) and (k, m, i) of H. The graph H' inherits all the arcs of H and has arcs between the pairs of subcliques of the same original clique $\hat{Q}_i$. We define the weight on each arc $e = (Q', Q'')$ as

$$w_e = 1 - \sum_{j \in Q' \cup Q''} x_j^F. \tag{6.14}$$

Note that $w_e \geq 0$ because $Q' \cup Q''$ is a subset of a clique in the formulation. Now the minimum weight directed odd cycle in H' passing through x_{v_1} yields the alternating-path substitution for x_{v_1} corresponding to the most violated constraint (6.11). □

Note that the algorithm described in the proof is in fact a modification of the standard algorithm for separating odd-hole inequalities [10], but in this primal setting we have to deal with directed graphs instead of undirected ones, and every cycle of H' passing through x_{v_1} is odd.

Observation 6.29. If $\mathcal{F}$ contains an inequality for each clique of G of size at most h, then Proposition 6.28 gives an exact polynomial-time primal separation procedure for all the inequalities of type (6.11). Note that, for $h = 3$, inequalities of type (6.11) include all the odd-hole and the odd-wheel inequalities. The standard separation for these inequalities is also possible in polynomial time with a minor change in the procedure given in Proposition 6.28.

Finally, we shall briefly mention a possible way to exploit a solution algorithm for the weighted stable set problem. Suppose we are in the situation where an integral tableau for a maximum weighted stable set is given, but the linear programming certificate for optimality is still missing, as in Example 6.21. The idea is to substitute columns with positive reduced cost along odd alternating paths of cliques until a tableau is obtained where each column has nonpositive reduced cost. In line with this idea, a criterion for finding an alternating-path substitution, an alternative to that of Proposition 6.28, is given by the following.

Problem 6.30 (column substitution problem). Let $G = (V, E)$ be a graph and $S \subseteq V$ a stable set in G. Let (x^I, z^I) be a basic feasible solution corresponding to S in a formulation $\mathcal{F}$. Let x_{v_1} be a nonbasic variable in N with positive reduced cost. Does there exist an odd alternating path of cliques $P = (v_1, Q_{2,3}, \ldots, Q_{2l,2l+1})$ in the given formulation such that the substitution along P yields a tableau where all the new columns have nonpositive reduced cost?

If such an odd path P exists, then we know that we can replace the nonbasic variable x_1 of positive reduced cost with new columns, according to Lemma 6.20, all having nonpositive reduced cost.

Corollary 6.31. *Problem 6.30, restricted to the case $R = \emptyset$, can be solved in polynomial time.*

Proof. We consider a variant of the construction in the proof of Proposition 6.28. After constructing H', we remove every arc $e = (Q', Q'')$ that would give rise to a nonbasic variable of positive reduced cost in the substitution. Then every directed odd cycle in H' passing through v_1 corresponds to an odd alternating path of cliques with the required properties. □

6.5 Conclusions

In this chapter, we have presented graph-theoretical transformations that, at the expense of enlarging a given graph G, can produce a perfect graph, and a weight-preserving map for each of its stable sets to a stable set of G. The graph transformations have a natural analogue in an integral tableau setting and result in replacing a column of the tableau with a set of new columns. These results provide the foundations for a solution algorithm for the maximum weight stable set problem based on a primal simplex method with all integral pivots. Identification procedures that perform these operations in polynomial time in the tableau size are important building blocks for such an algorithm. A number of them are presented in this chapter.

Bibliography

[1] E. Balas, S. Ceria, G. Cornuéjols, and G. Pataki. Polyhedral methods for the maximum clique problem. [16], pages 11–28.

[2] E. Balas and M. Padberg. On the set-covering problem: II. An algorithm for set partitioning. *Operations Research*, 23:74–90, 1975.

[3] C. Berge. Färbung von Graphen, deren sämtliche bzw. deren ungerade Kreise starr sind. Wissenschaftliche Zeitschrift der Martin-Luther-Universität Halle-Wittenberg, 10:114–115, 1961.

[4] I. Bomze, M. Budinich, P.M. Pardalos, and M. Pelillo. The maximum clique problem. In D.-Z. Du and P.M. Pardalos, editors, Volume 4 of *Handbook of Combinatorial Optimization (Supplement Volume* A*)*, pages 1–74. Kluwer Academic Publishers, Dordrecht, 1999.

[5] R. Borndörfer. *Aspects of Set Packing, Partitioning, and Covering*. Ph.D. thesis, Tech. Univ. Berlin, 1998. Published by Shaker-Verlag, Aachen, 1998.

[6] G. Cornuéjols. *Combinatorial Optimization: Packing and Covering*, number 74 in *CBMS-NSF Regional Conference Series in Applied Mathematics* 74. SIAM, Philadelphia, 2001.

[7] M. Chudnovsky, N. Robertson, P. D. Seymour, and R. Thomas. Progress on perfect graphs. *Mathematical Programming*, 97B:405–422, 2003.

[8] D. de Werra. On some properties of the struction of a graph. *SIAM Journal on Algebraic and Discrete Methods*, 5:239–243, 1984.

[9] C. Ebenegger, P.L. Hammer, and D. de Werra. Pseudo-Boolean functions and stability of graphs. In *Algebraic and Combinatorial Methods in Operations Research*, pages 83–97. North-Holland, Amsterdam, 1984.

[10] A.M.H. Gerards and A. Schrijver. Matrices with the Edmonds-Johnson property. *Combinatorica*, 6:365–379, 1986.

[11] M. Grötschel, L. Lovász, and A. Schrijver. *Geometric Algorithms and Combinatorial Optimization*, volume 2 of *Algorithms and Combinatorics*. Springer-Verlag, Berlin, 1988.

[12] P.L. Hammer, N.V.R. Mahadev, and D. de Werra. Stability in CAN-free graphs. *Journal of Combinatorial Theory. Series* B, 38:23–30, 1985.

[13] P.L. Hammer, N.V.R. Mahadev, and D. de Werra. The struction of a graph: Application to CN-free graphs. *Combinatorica*, 5:141–147, 1985.

[14] U.-U. Haus, M. Köppe, and R. Weismantel. A primal all-integer algorithm based on irreducible solutions. *Mathematical Programming*, 96B:205–246, 2003.

[15] U.-U. Haus, M. Köppe, and R. Weismantel. *A Primal All-Integer Algorithm Based on Irreducible Solutions*. Preprint no. 5, Fakultät für Mathematik, Otto-von-Guericke-Universität Magdeburg, 2001. Available online from http://www.math.uni-magdeburg.de/~mkoeppe/art/haus-koeppe-weismantel-ibm-theory-rr.ps.

[16] D.S. Johnson and M.A. Trick, editors. *Clique, Coloring, and Satisfiability: Second DIMACS Implementation Challenge, DIMACS*, volume 26. American Mathematical Society, Providence, RI, 1996.

[17] L. Lovász. Normal hypergraphs and the weak perfect graph conjecture. In *Topics on Perfect Graphs*, pages 29–42. North-Holland, Amsterdam, 1984.

[18] G.L. Nemhauser and G.L. Sigismondi. A strong cutting plane / branch and bound algorithm for node packing. *Journal of the Operational Research Society*, 43:443–457, 1992.

[19] G.L. Nemhauser and L.E. Trotter. Properties of vertex packing and independent system polyhedra. *Mathematical Programming*, 6:48–61, 1973.

[20] M.W. Padberg. On the facial structure of set packing polyhedra. *Mathematical Programming*, 5:199–215, 1973.

[21] M.W. Padberg and S. Hong. On the symmetric travelling salesman problem: A computational study. *Mathematical Programming Studies*, 12:78–107, 1980.

[22] R.D. Young. A simplified primal (all-integer) integer programming algorithm. *Operations Research*, 16:750–782, 1968.

Chapter 7

Relaxing Perfectness: Which Graphs Are "Almost" Perfect?

*Annegret K. Wagler**

Abstract. For all perfect graphs, the stable set polytope STAB(G) coincides with the fractional stable set polytope QSTAB(G), whereas STAB$(G) \subset$ QSTAB(G) holds iff G is imperfect. In the early 1970s, Padberg asked for "almost" perfect graphs. He characterized those graphs for which the difference between STAB(G) and QSTAB(G) is smallest possible. We develop this idea further and define three polytopes between STAB(G) and QSTAB(G) by allowing certain sets of cutting planes only to cut off all the fractional vertices of QSTAB(G). The difference between QSTAB(G) and the largest of the three polytopes coinciding with STAB(G) gives some information on the stage of imperfectness of the graph G. We obtain a nested collection of three superclasses of perfect graphs and survey which graphs are known to belong to one of those three superclasses. This answers the question, Which graphs are "almost" perfect?

MSC 2000. 05C17, 05C69, 90C10, 90C97

Key words. Perfect graph, stable set polytope, relaxations

7.1 Introduction

Berge [1] proposed to call a graph $G = (V, E)$ *perfect* if, for each of its (node-induced) subgraphs $G' \subseteq G$, the chromatic number $\chi(G')$ equals the clique number $\omega(G')$. That is, for all $G' \subseteq G$, we need as many stable sets to cover all nodes of G' as the maximum clique of G' has nodes. (A set $V' \subseteq V$ is a clique (stable set) if the nodes in V' are mutually (non)adjacent; maximum cliques (stable sets) contain a maximal number of nodes.)

*Konrad-Zuse-Institute for Information Technology Berlin, Takustr. 7, 14169 Berlin, Germany (wagler@zib.de). Supported by the Deutsche Forschungsgemeinschaft (Gr 883/9–1).

Berge [1] conjectured two characterizations of perfect graphs. His first conjecture was that a graph G is perfect if and only if the clique covering number $\overline{\chi}(G')$ equals the stability number $\alpha(G')\ \forall G' \subseteq G$ (i.e., that we need as many cliques to cover all nodes of G' as a maximum stable set of G' has nodes). Since complementation transforms stable sets into cliques, we have $\alpha(G) = \omega(\overline{G})$ and $\chi(G) = \overline{\chi}(\overline{G})$, where $\overline{G}$ denotes the complement of G. Hence Berge [1] conjectured and Lovász [18] proved that *a graph G is perfect if and only if its complement $\overline{G}$ is perfect (Perfect Graph Theorem).* The second Berge conjecture concerns a characterization via forbidden subgraphs. It is a simple observation that chordless odd cycles C_{2k+1} with $k \geq 2$, termed *odd holes*, and their complements $\overline{C}_{2k+1}$, called *odd antiholes*, are imperfect. Clearly, each graph containing an odd hole or an odd antihole as subgraph is imperfect as well. Berge conjectured in [1] that *a graph is perfect if and only if it contains neither odd holes nor odd antiholes as subgraphs (Strong Perfect Graph Conjecture).* This conjecture is still open and is one of the most famous conjectures in graph theory.

Padberg [21, 22] asked which graphs are "almost" perfect, i.e., which graphs are imperfect with the property that all of their proper induced subgraphs are perfect. Such graphs are nowadays called *minimally imperfect.* Using this term, the Strong Perfect Graph Conjecture reads as follows: *Odd holes and odd antiholes are the only minimally imperfect graphs.* In order to give a characterization of minimally imperfect graphs (and thereby to verify or falsify the Strong Perfect Graph Conjecture), many fascinating structures of such graphs have been discovered. First, the Perfect Graph Theorem implies that a graph is minimally imperfect if and only if its complement is. Further properties reflecting an extraordinary amount of symmetry of their maximum cliques and stable sets are given by the next two theorems.

Theorem 7.1 (Lovász [18]). *Every minimally imperfect graph G has exactly $\alpha\omega + 1$ nodes and, for every node x of G, the graph $G - x$ can be partitioned into α cliques of size ω and into ω stable sets of size α, where $\alpha = \alpha(G)$ and $\omega = \omega(G)$.*

Theorem 7.2 (Padberg [21]). *Every minimally imperfect graph G on n nodes has precisely n maximum stable sets and precisely n maximum cliques. Each node of G is contained in precisely $\alpha(G)$ maximum stable sets and in precisely $\omega(G)$ maximum cliques. For every maximum clique Q (maximum stable set S) there is a unique maximum stable set S (maximum clique Q) with $Q \cap S = \emptyset$.*

Unfortunately, minimally imperfect graphs are not characterized by these properties but share them with other graphs. Bland, Huang, and Trotter suggested in [2] calling a graph *partitionable* if it satisfies the conditions of Theorem 7.1 for some integers α, ω and verified Theorem 7.2 for all partitionable graphs (see Figure 7.1 for two partitionable graphs that are not minimally imperfect). Thus the class of partitionable graphs contains all potential counterexamples to the Strong Perfect Graph Conjecture. One main interest is, therefore, to find so-called *genuine properties* satisfied by all minimally imperfect graphs but violated by at least one partitionable graph (see [25] for more information on minimally imperfect and partitionable graphs).

Padberg [21, 22] investigated general set packing problems and studied the case when the polyhedron $P(A) = \{x \in \mathbb{R}^n_+ : Ax \leq \mathbb{1}\}$ associated with an $m \times n$ 0/1-matrix A has

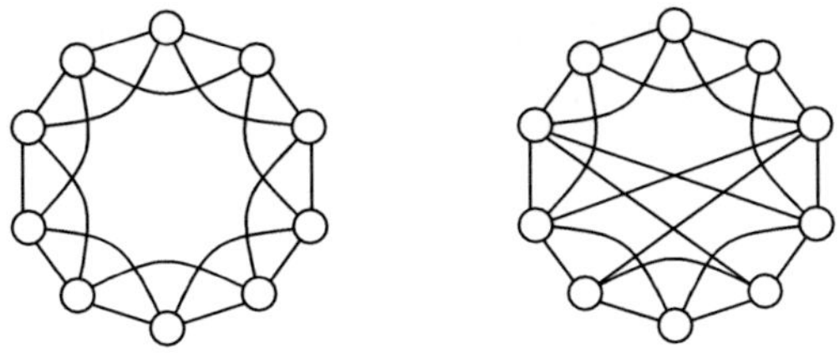

(a) (b)

Figure 7.1. *Examples of partitionable graphs.*

integral vertices only (where $\mathbb{1} = (1, \ldots, 1)$). Padberg proved in [21] that $P(A)$ coincides with $P_I(A)$, the convex hull of integer vertices of $P(A)$, if and only if A is a perfect 0/1-matrix. Translating this result into graph-theoretic terms [21], consider the graph G associated with A, where the nodes of G correspond to the n columns of A and two nodes of G are linked by an edge if the corresponding columns of A have a 1-entry in common. Consequently, A is the clique-node incidence matrix of G and $P(A)$ is the *fractional stable set polytope* QSTAB(G) given by the *nonnegativity constraints*

$$x_i \geq 0 \tag{7.0}$$

for all nodes i of G and by the *clique constraints*

$$\sum_{i \in Q} x_i \leq 1 \tag{7.1}$$

for all cliques $Q \subseteq G$. Furthermore, $P_I(A)$ corresponds to the *stable set polytope* STAB(G), which is defined as the convex hull of the incidence vectors of all stable sets of the graph G. Then the result on perfect 0/1-matrices is as follows.

Theorem 7.3 (Padberg [21]). *STAB(G) = QSTAB(G) if and only if G is perfect.*

Remark 7.4. In [4] Chvátal noted that Theorem 7.1 implies this polyhedral characterization of perfect graphs; further references are Fulkerson [9, 10] and Sachs [27].

If G is an imperfect graph, STAB$(G) \subset$ QSTAB(G) holds and the difference between STAB(G) and QSTAB(G) can be used as a tool in order to decide how far a graph is from being perfect. In this sense Padberg [22, 23] introduced the notion of almost integral polyhedra defined with respect to $m \times n$ 0/1-matrices A: $P(A)$ is called *almost integral* if $P(A)$ possesses at least one fractional vertex, but the polyhedra obtained from $P(A)$ by projecting $P(A)$ into (strictly) lower-dimensional subspaces have integer vertices only. Padberg proved several properties of almost integral polyhedra (see, e.g., Theorem 7.5 below) and showed recently an equivalent version of the Strong Perfect Graph Conjecture in terms of almost integral polyhedra [23]. (Padberg introduced in [23] two kinds of orthogonal projections and proved that the Strong Perfect Graph Conjecture is correct if and only if the studied projections of almost integral polyhedra yield again almost integral polyhedra.)

Theorem 7.5 (Padberg [22]). *If $P(A)$ is almost integral, then the following conditions are simultaneously satisfied. Every fractional vertex has exactly n adjacent integer vertices. $P(A)$ has exactly one fractional vertex. $P_I(A) = P(A) \cap \{x \in \mathbb{R}^n_+ : \sum_{1\le i\le n} x_i \le \overline{\alpha}\}$ with $\overline{\alpha} = \max\{\sum_{1\le i\le n} x_i : x \in P(A)\}$.*

Padberg called a 0/1-matrix A almost perfect if $P(A)$ is almost integral and showed that A is almost perfect if and only if it is the clique-node incidence matrix of an almost perfect (i.e., minimally imperfect) graph. In graph-theoretic terms Theorem 7.5 implies, therefore, the following characterization of minimally imperfect graphs.

Theorem 7.6 (Padberg [21, 22]). *G is minimally imperfect if and only if QSTAB(G) has exactly one fractional vertex (namely, $\frac{1}{\omega(G)}\mathbb{1}$, which is adjacent to the $|G|$ integer vertices coming from the maximum stable sets of G) and STAB(G) = QSTAB$(G) \cap \{x \in \mathbb{R}^{|G|}_+ : \sum_{i\in G} x_i \le \alpha(G)\}$.*

This means that G is minimally imperfect if and only if QSTAB(G) has *precisely one* fractional vertex that can be cut off by *exactly one* cutting plane, namely, the so-called full rank constraint

$$\sum_{i\in G} x_i \le \alpha(G) \tag{7.2}$$

associated with G. The above theorem implies, therefore, a most beautiful nontrivial genuine property that holds *exactly* for all minimally imperfect graphs and for *none* of the other partitionable graphs.

Theorem 7.7 (Padberg [21, 22]). *A partitionable graph G is minimally imperfect if and only if QSTAB(G) has exactly one fractional vertex.*

In the case of minimally imperfect graphs G, the polytope QSTAB(G) is the smallest possible relaxation of STAB(G) and, hence, minimally imperfect graphs are indeed "almost perfect." The next possible relaxation of STAB(G) is the case when QSTAB(G) may have more than one fractional vertex but, again, the full rank constraint is required only as a cutting plane to cut off all those fractional vertices. This led Shepherd [30], inspired by Padberg's results, to the definition of near-perfect matrices: an $m \times n$ 0/1-matrix A is called near-perfect if $P_I(A)$ coincides with $P(A) \cap \{x \in \mathbb{R}^n_+ : \sum_{1\le i\le n} x_i \le \alpha(G)\}$, where G is again the graph with clique-node incidence matrix A. Let FSTAB(G) denote the polytope given by all nonnegativity constraints (7.0), all clique constraints (7.1), and the full rank constraint (7.2). Shepherd [30] called a graph G *near-perfect* if STAB(G) = FSTAB(G). Minimally imperfect graphs are obviously near-perfect. Since there is no requirement that QSTAB(G) has at least one fractional vertex but only that all fractional vertices be cut off by the full rank constraint, perfect graphs are near-perfect, too. Figure 7.2 shows near-perfect graphs that are neither perfect nor minimally imperfect; see Section 7.3 for more examples and considerations of near-perfect graphs.

Following a suggestion of Grötschel, Lovász, and Schrijver [15], one may relax the notion of perfectness further by generalizing clique constraints to other classes of inequalities valid for the stable set polytope and then by investigating all graphs such that their stable set polytope is entirely described by nonnegativity constraints and the inequalities in question.

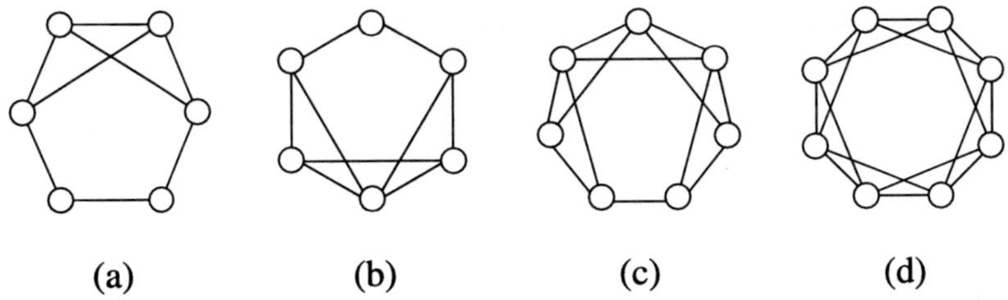

Figure 7.2. *Examples of near-perfect graphs.*

A natural way to generalize both clique constraints and the full rank constraint is to consider *all* 0/1-inequalities, i.e., to investigate the *rank constraints*

$$\sum_{i \in G'} x_i \leq \alpha(G') \tag{7.3}$$

associated with arbitrary induced subgraphs $G' \subseteq G$ (note $\alpha(G') = 1$ holds if G' is a clique). Every rank constraint is obviously valid for the stable set polytope and defines in some cases also a facet (see the next section for examples of graphs G where the full rank constraint is facet-defining). Hence the polytope RSTAB(G) given by all nonnegativity constraints (7.0) and all rank constraints (7.3) is a further relaxation of STAB(G) but contained in FSTAB(G). We define all graphs G with STAB(G) = RSTAB(G) to be *rank-perfect* (i.e., if we need only 0/1-inequalities of the form (7.3) to cut off all fractional vertices of QSTAB(G)). Every perfect, every minimally imperfect, and every near-perfect graph is obviously also rank-perfect. Further classes of rank-perfect graphs are discussed in Section 7.4.

If a rank constraint is associated with a *proper* subgraph $G' \subset G$, then it does not yield a facet of STAB(G) in general, even if $\sum_{i \in G'} x_i \leq \alpha(G')$ is facet-defining for STAB(G'). In the latter case we can determine a facet

$$\sum_{i \in G'} x_i + \sum_{i \in G - G'} a_i x_i \leq \alpha(G') \tag{7.4}$$

of the stable set polytope of the whole graph G by computing appropriate coefficients a_i for all nodes i in $G - G'$ via sequential lifting [20] (see Section 7.2). We call facets of the form (7.4) *weak rank constraints* if the base rank constraint associated with G' is facet-defining for STAB(G'). (This means that a lifted rank constraint $\sum_{i \in G'} x_i + \sum_{i \in G - G'} a_i x_i \leq \alpha(G')$ is a weak rank constraint if an orthogonal projection is the full rank facet of STAB(G').) Clearly, facet-defining rank constraints are weak rank constraints with $a_i = 0$ for $i \in G - G'$. Let WSTAB(G) be the polytope given by all nonnegativity constraints (7.0) and all weak rank constraints (7.4). WSTAB(G) is a further relaxation of STAB(G) but contained in RSTAB(G) (since we allow more general cutting planes than rank constraints only). We define all graphs G with STAB(G) = WSTAB(G) to be *weakly rank-perfect* (see Section 7.5 for classes of weakly rank-perfect graphs). Moreover, the stable set polytope itself is entirely described by all "trivial" facets (7.0) and all "nontrivial" facets of the general form

$$\sum_{i \in G} a_i x_i \leq \alpha(G, a), \tag{7.5}$$

where we interpret the vector $a = (a_1, \ldots, a_n)$ to be a node weighting of G associating the weight a_i to $i \in G$ and denote the weighted graph by (G, a). Furthermore, $\alpha(G, a) =$

$\max\{\sum_{i \in S} a_i : S \subseteq G \text{ stable set}\}$ stands for the weighted stability number. Thus there is no further relaxation of STAB(G) possible beyond WSTAB(G). By the chain of relaxations of STAB(G):

$$\text{STAB}(G) \subseteq \text{WSTAB}(G) \subseteq \text{RSTAB}(G) \subseteq \text{FSTAB}(G) \subseteq \text{QSTAB}(G),$$

we have finally obtained a hierarchy of superclasses of perfect graphs: near-perfect, rank-perfect, and weakly rank-perfect graphs. The difference between QSTAB(G) and the largest polytope coinciding with STAB(G) gives us some information on the stage of imperfectness or answers the following question: Which graphs are more or less "almost perfect"? Our considerations will have a special stress on near-perfect graphs (which are almost "almost" perfect), while we list only known classes of rank-perfect and weakly rank-perfect graphs in Sections 7.4 and 7.5, respectively. We close with some final remarks and open problems in Section 7.6.

7.2 Rank Constraints and Sequential Lifting

Determining the system of facet-defining inequalities of STAB(G), i.e., finding all cutting planes required to cut off the fractional vertices of QSTAB(G), is very difficult in general. Thus one often tries to find classes of valid inequalities for STAB(G) and to investigate when those valid inequalities yield facets of STAB(G). One natural class of valid nontrivial inequalities are rank constraints (7.3) associated with induced subgraphs $G' \subseteq G$. For convenience we often write (7.3) as $x(G', \mathbb{1}) \leq \alpha(G', \mathbb{1})$, $x(G', \mathbb{1}) \leq \alpha(G')$, or just $x(G') \leq \alpha(G')$. The goal is to find out for which subgraphs $G' \subseteq G$ the associated rank constraint $x(G') \leq \alpha(G')$ yields a facet of STAB(G). A first step toward this goal is to identify those graphs G whose stable set polytope has the full rank constraint (7.2)

$$x(G, \mathbb{1}) \leq \alpha(G, \mathbb{1})$$

as a facet (we say that such graphs produce the full rank facet). Padberg showed that this holds if G is a clique [20] or minimally imperfect [21]. Bland, Huang, and Trotter [2] generalized Padberg's result [21] by showing that (7.2) is a facet of STAB(G) for all partitionable graphs G. Webs form a graph class with circular symmetry of their maximum cliques and stable sets that contain many partitionable graphs. A *web* W_n^k is a graph with nodes $1, \dots, n$, where ij is an edge if i and j differ by at most k (i.e., if $|i - j| \leq k$ modulo n); we assume $n \geq 2(k+1)$ in the following since W_n^k is a clique otherwise. W_n^1 is a hole, W_{2k+1}^{k-1} is an odd antihole for $k \geq 2$, and $W_{\alpha\omega+1}^{\omega-1}$ is partitionable with $\alpha = \alpha(W_n^{k-1}) = \lfloor \frac{n}{k} \rfloor$ and $\omega = \omega(W_n^{k-1}) = k$. The partitionable web W_{10}^2 is shown in Figure 7.1(a); the near-perfect graph in Figure 7.2(d) is W_8^2.

Remark 7.8. Webs are also called circulant graphs C_n^k in [5]. Furthermore, graphs $W(n,k)$ with $n \geq 2$, $1 \leq k \leq \frac{1}{2}n$, and $W(n,k) = \overline{W}_n^{k-1}$ were introduced in [32].

Trotter [32] studied the case when the complement of a web, called an *antiweb*, produces the full rank facet; he showed that this happens if and only if the antiweb $\overline{W}_n^{k-1}$ is *prime*, i.e., if k and n are relatively prime. In order to show which webs produce the full rank facet, we need the following result [4]. An edge e of a graph $G = (V, E)$ is α-*critical* if $\alpha(G) < \alpha(G - e)$. We call G *α-connected* if the graph on the same nodeset V having all α-critical edges of G is connected. Chvátal [4] showed that *every α-connected graph produces the full rank facet* (see [29] for a survey and [17, 24] for further results).

Theorem 7.9. *W_n^{k-1} produces the full rank facet if and only if k does not divide n.*

Proof. *If:* Consider the maximum stable set $S_i = \{i, i+k, \ldots, i + (\lfloor \frac{n}{k} \rfloor - 1)k\}$ of W_n^{k-1} (where all indices are taken modulo n). Since k is not a divisor of n, we have $\lfloor \frac{n}{k} \rfloor k < n$. Subtracting $k - i$ from both sides of this relation yields $i + (\lfloor \frac{n}{k} \rfloor - 1)k < i + n - k$, where $i + n - k$ is the last neighbor of i in W_n^{k-1}. Consequently, $(S_i - \{i\}) \cup \{i - 1\}$ is also a maximum stable set of W_n^{k-1} and the edge $i - 1, i$ is, therefore, α-critical for $1 \leq i \leq n$ (where all indices are again taken modulo n). Thus, if k is not a divisor of n, then W_n^{k-1} is α-connected and produces the full rank facet due to Chvátal [4]. □

Only if: In the case that k is a divisor of n, there are only k maximum stable sets in W_n^{k-1} (of size $\frac{n}{k}$). Thus, W_n^{k-1} cannot contain n maximum stable sets whose incidence vectors are affinely independent. □

Remark 7.10. The *If*-part follows the proof of Trotter [32] that W_n^{k-1} produces the full rank facet if k and n are relatively prime. The weaker condition that k is not a divisor of n suffices for the argumentation. (E.g., W_{10}^3 produces the full rank facet but 4 and 10 are not relatively prime.)

Moreover, Edmonds and Pulleyblank [7] established via matching theory that line graphs of 2-connected hypomatchable graphs have the full rank facet: H is called *hypomatchable* if, for all nodes v of H, the subgraph $H - v$ admits a matching (i.e., a set of disjoint edges) meeting all nodes. A graph is 2-connected if it is still connected after removing an arbitrary node. The *line graph* $L(F)$ of a graph F is obtained by taking the edges of F as nodes of $L(F)$ and connecting two nodes in $L(F)$ if and only if the corresponding edges of F are incident. (Note: Matchings of H correspond to stable sets of its line graph $L(H)$ since the line operator transforms nonincident edges of H to nonadjacent nodes of $L(H)$.)

For some cases a sufficient condition is known when a rank constraint $x(G') \leq \alpha(G')$ associated with a *proper* subgraph $G' \subset G$ yields a facet of the stable set polytope of the whole graph G. Padberg [20] showed that clique constraints $x(Q, \mathbb{1}) \leq 1$ are facet-defining for $\text{STAB}(G)$ if and only if Q is an (inclusionwise) maximal clique of G. The results in [7] imply that a rank constraint

$$x(L(H), \mathbb{1}) \leq \frac{|H| - 1}{2} \tag{7.3a}$$

associated with the line graph of a 2-connected hypomatchable graph $H \subseteq F$ is a facet of $\text{STAB}(L(F))$ if and only if H is an induced subgraph of F. In general, a rank constraint associated with a proper subgraph $G' \subset G$ does not need to provide a facet of $\text{STAB}(G)$ even if $\text{STAB}(G')$ admits the full rank facet. This is the case for, e.g., *odd hole constraints*

$$x(C_{2k+1}, \mathbb{1}) \leq k \tag{7.3b}$$

with $C_{2k+1} \subset G$ and for *odd antihole constraints*

$$x(\overline{C}_{2k+1}, \mathbb{1}) \leq 2 \tag{7.3c}$$

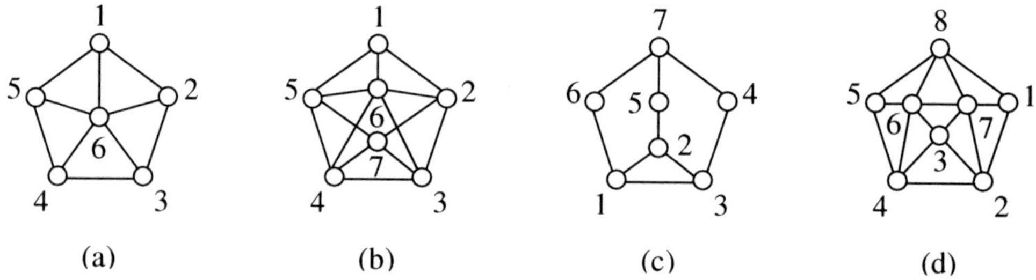

Figure 7.3. *Liftings of an odd hole constraint.*

with $\overline{C}_{2k+1} \subset G$. Figure 7.3(a) shows a graph with an induced C_5 (note a C_5 is both an odd hole and an odd antihole), but the rank constraint associated with this C_5 does not induce a facet of the stable set polytope of the whole graph.

However, rank constraints $x(G') \leq \alpha(G')$ with $G' \subset G$ may be strengthened to a facet of STAB(G) using *sequential lifting*, introduced by Padberg [20], i.e., by determining appropriate lifting coefficients a_i for all nodes i in $G - G'$ such that the right-hand side $\alpha(G')$ of the inequality is still satisfied and that there are $|G|$ stable sets of weight $\alpha(G')$ whose incidence vectors are linearly independent. Every inequality

$$x(G', \mathbb{1}) + x(G - G', a) \leq \alpha(G', \mathbb{1}) \tag{7.4}$$

is a weak rank constraint if it is obtained by lifting a base rank constraint $x(G', \mathbb{1}) \leq \alpha(G', \mathbb{1})$ that is facet-defining for STAB(G'), i.e., if G' produces the full rank facet. The graph G depicted in Figure 7.3(a) yields a weak rank constraint based on an odd hole constraint by using a lifting coefficient not equal to 0, 1 (thus G is not rank-perfect in particular). G consists of an odd hole (nodes $1, \ldots, 5$) and a central node adjacent to all nodes of the odd hole (such graphs are termed *odd wheels*). The C_5 yields 5 stable sets of weight 2 whose incidence vectors are linearly independent. In order to construct the remaining stable set of weight 2 containing the central node 6, we have to choose lifting coefficient $a_6 = 2$. The resulting facet $x(C_5, \mathbb{1}) + 2x_6 \leq 2$ of STAB(G) is a special weak rank constraint called the *odd wheel constraint*:

$$x(C_{2k+1}, \mathbb{1}) + kx_c \leq k, \tag{7.4a}$$

where c is the central node adjacent to all nodes of the odd hole C_{2k+1} and $k \geq 2$. (See Padberg [20] for a general description of how to lift odd hole constraints associated with proper subgraphs to weak rank facets of the whole graph.) Shepherd [31] studied a more general weak rank constraint

$$\sum_{i \leq k} \frac{1}{\alpha(\overline{W}_i)} x(\overline{W}_i) + x(Q) \leq 1 \tag{7.4b}$$

associated with the complete join of prime antiwebs $\overline{W}_1, \ldots, \overline{W}_k$ and a clique Q. The *complete join* of two disjoint graphs G_1 and G_2 is obtained by joining every node of G_1 and every node of G_2 by an edge. E.g., every odd wheel is the complete join of an odd hole and a single node. Note that the support graph of such facets arises by the complete join of graphs that all produce their full rank facet, i.e., we put together disjoint facet blocks. The obtained constraints can be scaled in such a way that they have the form (7.4) with a base rank constraint $x(G') \leq \alpha(G')$ and noninteger coefficients a_i for $i \in G - G'$. In this sense (7.4b) can be seen as a lifted clique constraint.

Shepherd [31] showed that odd antiholes are the only prime antiwebs that occur in complements of line graphs. Thus their stable set polytopes admit weak rank constraints

$$\sum_{i \leq k} x(A_i) + 2x(Q) \leq 2 \tag{7.4c}$$

associated with the complete join of odd antiholes $A_1, \ldots, A_k$ and a clique Q. Cook studied (see [30]) the stable set polytopes of graphs G with $\alpha(G) = 2$. He showed that the inequality

$$x(\tilde{N}(Q)) + 2x(Q) \leq 2 \tag{7.4d}$$

is valid for STAB(G) for every clique Q, where $\tilde{N}(Q)$ is the set of all nodes v of G with $Q \subseteq N(v)$ (note $\tilde{N}(Q) = V(G)$ if $Q = \emptyset$ and $N(v)$ denotes the set of all neighbors of v). Cook showed that (7.4d) is a facet of STAB(G) if and only if no component of $\tilde{N}(Q)$ in the complementary graph $\overline{G}$ is bipartite (see [30]). Since $\alpha(G) = 2$ implies $\omega(\overline{G}) = 2$, a component of $\overline{G}$ is not bipartite if and only if it contains an odd hole. Hence, (7.4d) is a facet if and only if $\tilde{N}(Q)$ is the complete join of subgraphs all containing an odd antihole. Therefore (7.4d) is, as lifting of (7.4c), a special weak rank constraint.

Giles and Trotter [14] studied further weak rank constraints: Consider the webs W_n^{k+1} and W_n^k with $n = 2k(k+2)+1$, where $V(W_n^{k+1}) = \{1, \ldots, n\}$ and $V(W_n^k) = \{1', \ldots, n'\}$. Construct the graph G^k by taking W_n^{k+1} and W_n^k as induced subgraphs and adding the edges $\{i, i'\}, \{i, (i+1)'\}, \ldots, \{i, (i+2k+1)'\}$ for $1 \leq i \leq n$, where all indices are taken modulo n. (The graph G^1 is shown in Figure 7.4; here the odd antihole W_7^2 and the odd hole W_7^1 are emphasized with bold lines.) Then

$$(k+1)\, x(W_n^{k+1}) + k\, x(W_n^k) \leq 2k(k+1) \tag{7.4e}$$

is a facet of STAB(G^k) by [14]. W_n^{k+1} has stability number $2k$ and produces the full rank facet by Theorem 7.9 (since $k+2$ is not a divisor of $n = 2k(k+2)+1$). Hence (7.4e) is a class of weak rank constraints.

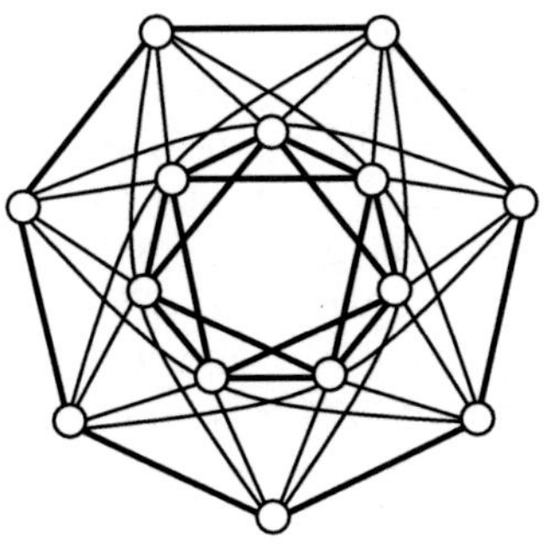

Figure 7.4. *The graph G^1.*

Note that the weak rank facet obtained by lifting may depend on the order in which the nodes are lifted [20]. Hence lifting a base rank constraint may result in *several* weak rank constraints. The graph G in Figure 7.3(b), e.g., contains the 5-wheel from Figure 7.3(a) as induced subgraph, and the associated odd wheel constraint $x(C_5, \mathbb{1}) + 2x_6 \leq 2$ is also a facet of STAB(G). Furthermore, there is another way to lift the rank constraint associated with the C_5 to a facet of STAB(G), namely, by choosing $a_6 = 1$ and $a_7 = 1$ (i.e., STAB(G) also admits the full rank facet).

Finally, STAB(G) may admit nontrivial facets that are not weak rank constraints. The stable set polytope of the graph G in Figure 7.3(c), e.g., has the facet $\sum_{i\leq 6} x_i + 2x_7 \leq 3$, which is not a weak rank constraint: among the nodes of G with coefficient 1, there is no subgraph G' such that $x(G') \leq 3$ is a facet of STAB(G'). (This means that there is no facet-inducing structure of a proper subgraph $G' \subset G$ that we could lift to a facet of STAB(G).) In particular, the graph G in Figure 7.3(c) is an example of a graph that is not weakly rank-perfect. (Checking the stable set polytopes of small imperfect graphs yields that G and $\overline{G}$ are the only two not weakly rank-perfect graphs on up to seven nodes.) The graph G in Figure 7.3(d) is a so-called wedge, introduced in [14]. Wedges are further examples of graphs that are not weakly rank-perfect. The stable set polytope of G has, e.g., the facet $\sum_{i\leq 5} x_i + 2\sum_{6\leq i\leq 8} x_i \leq 3$, which is not a weak rank constraint, either.

Oriolo [19] introduced a new class of inequalities valid for the stable set polytope of every graph. Let $G = (V, E)$ be a graph and $\mathcal{Q}$ be a family of (at least three) maximal cliques of G. Let $k \leq |\mathcal{Q}|$ be an integer, $\lambda(|\mathcal{Q}|, k) = \frac{k-l-1}{k-l}$ with $l = |\mathcal{Q}| - k\lfloor\frac{|\mathcal{Q}|}{k}\rfloor$, and define the following two sets: $I(\mathcal{Q}, k) = \{v \in V : |\{Q \in \mathcal{Q} : v \in Q\}| \geq k\}$ and $O(\mathcal{Q}, k) = \{v \in V : |\{Q \in \mathcal{Q} : v \in Q\}| = k - 1\}$. Oriolo [19] proved that

$$x(I(\mathcal{Q}, k)) + \lambda(|\mathcal{Q}|, k)\, x(O(\mathcal{Q}, k)) \leq \left\lfloor \frac{|\mathcal{Q}|}{k} \right\rfloor \tag{7.5a}$$

is valid for the stable set polytope of every graph G. Furthermore, he showed [19] that (7.5a) is a common generalization of the rank constraints (7.3a) associated with line graphs of 2-connected hypomatchable graphs, the full rank constraints associated with webs W_n^{k-1} where k is not a divisor of n, and the weak rank constraints (7.4e) associated with graphs G^k introduced in [14]. However, it is not known so far whether a *facet* (7.5a) is a weak rank constraint in general.

7.3 Near-Perfect Graphs

The subject of this section is a class of graphs that is, in a polyhedral sense, the smallest superclass of perfect graphs: the class of near-perfect graphs G where only one cutting plane, namely, the full rank constraint, is required to cut off all fractional vertices of QSTAB(G) [30]. This means that for near-perfect graphs G we only have to add the full rank constraint (7.2) to the nonnegativity (7.0) and clique constraints (7.1) in order to arrive at STAB(G). Since there is no requirement that QSTAB(G) has at least one fractional vertex, all perfect graphs are near-perfect in particular (here the full rank constraint is not a facet except in the case of a clique). Hence near-perfect graphs are indeed the closest superclass of perfect graphs.

Minimally imperfect graphs are further examples of near-perfect graphs given by Padberg [21, 22]; see Theorem 7.6. While the characterization of minimally imperfect graphs via the Strong Perfect Graph Conjecture is still open, there is, besides Theorem 7.6, a further polyhedral characterization of minimally imperfect graphs in terms of near-perfection.

Theorem 7.11 (Shepherd [30]). *An imperfect graph G is minimally imperfect if and only if both G and $\overline{G}$ are near-perfect.*

This means that the part of the class of near-perfect graphs that is closed under complementation consists exactly of all perfect and all minimally imperfect graphs. For every partitionable graph G we know that G and $\overline{G}$ produce the full rank facet, by Bland, Huang, and Trotter [2], but at most one of G and $\overline{G}$ is near-perfect. Even more holds.

Theorem 7.12. *A partitionable graph G is minimally imperfect if and only if G is near-perfect.*

Proof. Every minimally imperfect graph is near-perfect, by [21, 22]. We show that a partitionable graph G that is not minimally imperfect cannot be near-perfect either. G properly contains a minimally imperfect subgraph $G' \subset G$ with $\alpha(G') < \alpha(G)$, by [30]. The rank constraint associated with G' yields a nontrivial facet of STAB(G) that is different from a clique facet and the full rank facet of G. □

Hence, we have, in addition to Theorem 7.7, a further nontrivial genuine property that holds *exactly* for all minimally imperfect graphs and for *none* of the other partitionable graphs. This means that if G is partitionable but not minimally imperfect, then QSTAB(G) has at least two fractional vertices, by Theorem 7.7, and at least two cutting planes are required to arrive at STAB(G). (Recall that every partitionable graph G produces the full rank facet by [2], but the full rank facet does not suffice to cut off all fractional vertices of QSTAB(G), by the above Theorem 7.12.)

In order to be near-perfect, an imperfect graph G has obviously to satisfy the condition that every minimally imperfect subgraph of G has the same stability number as G. A further property was conjectured to characterize near-perfect graphs in [30].

Conjecture 7.13 (Shepherd [30]). *A graph G is near-perfect if and only if each lifting of a rank constraint associated with a minimally imperfect subgraph of G yields the full rank facet $x(G) \leq \alpha(G)$.*

Other than perfect and minimally imperfect graphs, no other class is known so far to belong (completely) to the class of near-perfect graphs. In addition to Theorem 7.12, we give characterizations of all the near-perfect graphs in three graph classes. We start with a result from [30] on graphs G with stability number $\alpha(G) = 2$.

Theorem 7.14 (Shepherd [30]). *A graph G with $\alpha(G) = 2$ is near-perfect if and only if the neighborhood of every node of G induces a perfect graph.*

Next we study two classes that contain all odd holes, all odd antiholes, and many partitionable graphs: webs and antiwebs. Recall from Section 7.2 that a web W_n^{k-1} produces the full rank facet if and only if k is not a divisor of n (Theorem 7.9), while the same is true for antiwebs $\overline{W}_n^{k-1}$ if and only if k and n are relatively prime (Trotter [32]). We now determine for which webs and antiwebs the full rank facet is the only facet of the stable set polytope other than facets of type (7.0) and (7.1).

Theorem 7.15. *A web is near-perfect if and only if it is perfect, an odd hole, or W_{11}^2, or if it has stability number two.*

Proof. *If:* The assertion is trivial if W_n^{k-1} is perfect and follows for odd holes from Padberg [21]. In the case $\alpha(W_n^{k-1}) = 2$, we apply Theorem 7.14 due to Shepherd [30]. The neighborhood $N(i)$ of every node i of W_n^{k-1} consists of two disjoint cliques, namely, $\{i-(k-1), \dots, i-1\}$ and $\{i+1, \dots, i+(k-1)\}$, where all indices are taken modulo n. Thus $N(i)$ induces the complement of a bipartite graph and is, therefore, perfect for all nodes i. Hence, W_n^{k-1} is near-perfect by Theorem 7.14 if $\alpha(W_n^{k-1}) = \lfloor \frac{n}{k} \rfloor = 2$ holds.

Checking the stable set polytope of W_{11}^2 explicitly shows that W_{11}^2 is near-perfect, too. (Note: W_{11}^2 has C_7 as only minimally imperfect subgraphs and $\alpha(C_7) = 3 = \alpha(W_{11}^2)$ holds.) □

Only if: W_n^{k-1} is a stable set if $k = 1$ and a hole if $k = 2$, hence either perfect or minimally imperfect and, in the latter case, near-perfect by Padberg [21]. W_{2k}^{k-1} is the complement of the graph consisting of k disjoint edges (recall that we assume $n \geq 2k$ since W_n^{k-1} is a clique whenever $n < 2k$). W_{2k+1}^{k-1} is an odd antihole if $k \geq 2$, hence near-perfect by Padberg [21]. We have to show that, for $k \geq 3$ and $n \geq 2k+2$, the web W_{11}^2 is the only near-perfect web W_n^{k-1} with stability number $\lfloor \frac{n}{k} \rfloor > 2$.

In the case $k \geq 3$ and $n \geq 2k+2$, W_n^{k-1} *properly* contains an odd hole or an odd antihole by Trotter [32]. If one of these odd holes or odd antiholes has a stability number $< \alpha(W_n^{k-1})$, then $\text{STAB}(W_n^{k-1})$ has a nontrivial facet that is associated neither with a clique nor with W_n^{k-1} itself. Hence W_n^{k-1} is near-perfect only if it has stability number two or if it contains only odd holes $W_{n'}^1$ with stability number $\lfloor \frac{n'}{2} \rfloor = \lfloor \frac{n}{k} \rfloor > 2$ but no odd antiholes. We show that W_n^{k-1} with $k \geq 3$ and $n \geq 3k$ has odd holes with stability number $< \alpha(W_n^{k-1})$ except in the case where $k = 3$ and $n = 11$.

Claim 1. *W_n^{k-1} contains odd holes of different lengths if $k = 3, 4$ and $n \geq 24$, if $k = 5$ and $n \geq 27$, or if $k \geq 6$ and $n \geq 5k$.*

Proof of Claim 1. Due to Trotter [32], we have $W_{n'}^1 \subseteq W_n^{k-1}$ if and only if $2n \geq n'k$ and $n \leq n'(k-1)$, i.e., if and only if the following condition (∗) holds

$$\frac{n}{k-1} \leq n' \leq 2\frac{n}{k}. \qquad (*)$$

If $\frac{n}{k-1} + 4 \leq 2\frac{n}{k}$, there exist at least two odd n' that satisfy (∗). Determine n such that

$$\frac{4k(k-1)}{k-2} \leq n$$

holds. We obtain $n \geq 24$ if $k \in \{3, 4\}$, $n \geq 27$ if $k = 5$, and $n \geq 5k$ if $k \geq 6$. Moreover, $\frac{n}{k-1} > 5$ holds in all these cases; hence W_n^{k-1} contains odd holes of different length.

Claim 2. *W_n^{k-1} contains only odd holes with stability number $\lfloor \frac{n}{k} \rfloor \geq 3$ only if $k = 3$ and $n = 11$.*

Proof of Claim 2. Considering W_n^{k-1} containing odd holes of length n' only, we get

$$n' - 2 < \frac{n}{k-1} \leq n' \leq 2\frac{n}{k} < n' + 2$$

by (∗). Replacing n' with $2\lfloor \frac{n}{k} \rfloor + 1$ (since $\lfloor \frac{n'}{2} \rfloor = \frac{n'-1}{2} = \lfloor \frac{n}{k} \rfloor$ is required), we obtain

$$2\left\lfloor \frac{n}{k} \right\rfloor - 1 < \frac{n}{k-1} \leq 2\left\lfloor \frac{n}{k} \right\rfloor + 1 \leq 2\frac{n}{k} < 2\left\lfloor \frac{n}{k} \right\rfloor + 3$$

in order to guarantee $\alpha(W_{n'}^1) = \alpha(W_n^{k-1}) \geq 3$. We first observe that $2\frac{n}{k} < 2\lfloor \frac{n}{k} \rfloor + 3$ is true for all k and n (since $\frac{n}{k} < \lfloor \frac{n}{k} \rfloor + 1$). Further, $2\lfloor \frac{n}{k} \rfloor + 1 \leq 2\frac{n}{k}$ means $\lfloor \frac{n}{k} \rfloor + \frac{1}{2} \leq \frac{n}{k}$ and is fulfilled whenever $ik + \frac{k}{2} \leq n < (i+1)k$ for some i. If $i = 3$, we consider $2\lfloor \frac{n}{k} \rfloor - 1 = 5 < \frac{n}{k-1} = \frac{3k+l}{k-1}$ with $\frac{k}{2} \leq l < k$ and obtain $2k < 5 + l$, which is true only if $k < 4$. If $i = 4$, then $2\lfloor \frac{n}{k} \rfloor - 1 = 7 < \frac{n}{k-1} = \frac{4k+l}{k-1}$ with $\frac{k}{2} \leq l < k$ yields $3k < 7 + l$, which is true only if $k < 3$. If $i = 5$, we only have to check $k = 3, 4$ by Claim 1 (note $5k + \frac{k}{2} > 27$ if $k = 5$), but W_{17}^2, W_{22}^3, and W_{23}^3 all contain a C_9 and a C_{11} (which is implied by (∗)). If $i = 6, 7$, we only have to check $k = 3$ by Claim 1, but we obtain $C_{11}, C_{13} \subseteq W_{20}^2$ and $C_{13}, C_{15} \subseteq W_{23}^2$ by (∗). The case $i \geq 8$ does not have to be checked for any $k \geq 3$ by Claim 1; thus we only have $i = 3$ and $k = 3$ left.

The observation that W_n^{k-1} with $n < 3k$ cannot contain an odd hole different from a C_5 (since $\alpha(W_n^{k-1}) = 2$) finishes the proof. □

Theorem 7.16. *An antiweb is near-perfect if and only if it is perfect, an odd hole, or an odd antihole.*

Proof. In the case that $\overline{W}_n^{k-1}$ is perfect or minimally imperfect, $\overline{W}_n^{k-1}$ is clearly near-perfect. We show that there are no other near-perfect antiwebs. $\overline{W}_n^{k-1}$ is a clique if $k = 1$ and an antihole if $k = 2$. $\overline{W}_n^{k-1}$ consists of k disjoint edges (and is perfect) if $n = 2k$. Trotter [32] has shown that $\overline{W}_n^{k-1}$ contains an odd hole or odd antihole as an induced subgraph if $k \geq 3$ and $n > 2k$. If $n = 2k+1$, then $\overline{W}_n^{k-1}$ is isomorphic to an odd hole. If $n > 2k+1$, then $\overline{W}_n^{k-1}$ *properly* contains an odd hole $\overline{W}_{2k'+1}^{k'-1}$ or an odd antihole, $\overline{W}_{n'}^1$, since $\overline{W}_n^{l-1} \subseteq \overline{W}_n^{k-1}$ implies $l = k$ by [32] and $\overline{W}_{n'}^1 \neq \overline{W}_n^{k-1}$ follows by $k \geq 3$. Then STAB($\overline{W}_n^{k-1}$) has the corresponding (zero-lifted) odd hole or odd antihole facet by Trotter [32]. This facet is different from the full rank constraint associated with $\overline{W}_n^{k-1}$ since the stability number of the odd hole or odd antihole in $\overline{W}_n^{k-1}$ is *strictly* less than $k = \alpha(\overline{W}_n^{k-1})$ (note that $\overline{W}_{n'}^{k-1} \subseteq \overline{W}_n^{k-1}$ implies $n = n'$ by [32] again). Hence $\overline{W}_n^{k-1}$ is not near-perfect if $k \geq 3$ and $n > 2k + 1$. □

7.4 Rank-Perfect Graphs

We now turn to the next superclass of perfect graphs: the class of rank-perfect graphs G, where 0/1-inequalities of the form (7.3):

$$x(G') \leq \alpha(G'),$$

with $G' \subseteq G$ are needed as only nontrivial facets to describe STAB(G). Since clique constraints are special rank constraints (namely, those with $\alpha(G') = 1$), all perfect graphs are rank-perfect in particular. Furthermore, all near-perfect graphs are obviously rank-perfect, too. There are further classes of rank-perfect graphs known.

Chvátal [4] defined graphs G to be *t-perfect* if STAB(G) has rank constraints associated with edges and odd holes as only nontrivial facets. (Note that "t" stands for "trou," the French word for hole, and that every C_{2k+1} with $k \geq 1$ is here considered to be a hole.) Bipartite graphs without isolated nodes are obviously t-perfect. Chvátal conjectured in [4] and Boulala and Uhry proved in [3] that *series-parallel graphs* are t-perfect (these are graphs obtained from disjoint cycle-free subgraphs by repeated application of the following two operations: adding a new edge parallel to an existing edge and subdividing edges, i.e., replacing edges with a path). Further examples of t-perfect graphs are *almost bipartite graphs* (having a node whose deletion leaves the graph bipartite) due to Fonlupt and Uhry [8] and *strongly t-perfect graphs* (having no subgraph obtained from subdividing edges of a K_4 such that all four cycles corresponding to the triangles of the K_4 are odd) due to Gerards and Schrijver [12]. Further investigations of t-perfect graphs without certain subdivisions of K_4 can be found in Gerards and Shepherd [13].

By definition [15], a natural generalization of t-perfect graphs is the class of *h-perfect graphs* (from hole-perfect), where, besides the nonnegativity constraints (7.0), all clique constraints (7.1) and odd hole constraints (7.3b) suffice to describe the associated stable set polytopes. At present, there are no interesting classes of h-perfect graphs known that are not perfect, t-perfect, nor combinations of these. (For combinations see Fonlupt and Uhry [8] and Sbihi and Uhry [28].)

Line graphs are a further class of rank-perfect graphs due to a result of Edmonds and Pulleyblank [7]. Their result implies that the stable set polytopes of line graphs are given by nonnegativity constraints (7.0), clique constraints (7.1), and rank constraints (7.3a) associated with the line graphs of 2-connected hypomatchable graphs. Note that line graphs are a "natural" graph class that is proved to contain rank-perfect graphs only (while near-perfect, t-perfect, and h-perfect graphs are rank-perfect by definition).

It is worth noting that line graphs seem to be a maximal class of rank-perfect graphs. The closest superclass of line graphs consists of all *quasi-line graphs* where the neighborhood of each node partitions into two cliques. (Quasi-line graphs were first investigated by Ben Rebea in his Ph.D. thesis. Tragically, he died shortly after completing his thesis and all the efforts to reorganize and publish his results have been unsuccessful so far.) It is easy to check that, besides all line graphs, each web is a quasi-line graph. We know which webs are near-perfect due to Theorem 7.15. Dahl [6] showed that webs W_n^2 for all $n \geq 4$ are rank-perfect. But there are webs with clique number greater than 4 (e.g., the partitionable web W_{25}^5) whose stable set polytopes have nonrank facets (see Kind [16]). The graphs G^k introduced in [14] are further quasi-line graphs that produce nonrank facets (7.4e). Thus quasi-line graphs are not rank-perfect.

Furthermore, we studied in [33] *critical edges* with respect to perfectness (edges of perfect graphs whose deletion yields an imperfect graph). We investigated the case of deleting critical edges e from perfect line graphs G. Besides 0/1-liftings of rank constraints (7.3a) associated with line graphs of 2-connected hypomatchable graphs, odd wheel constraints (7.4a) associated with 5-wheels as facets of STAB$(G - e)$ also appear; see [33]. Thus deleting edges from line graphs destroys the property of rank-perfection, too. However, the

5-wheel constraint does not appear if we restrict our consideration to line graphs of *bipartite* graphs. Thus $G - e$ might be rank-perfect if G is the line graph of a bipartite graph [33].

7.5 Weakly Rank-Perfect Graphs

This section deals with the class of weakly rank-perfect graphs G where, besides the nonnegativity constraints (7.0), only weak rank constraints (7.4) of the form

$$x(G', \mathbb{1}) + x(G - G', a) \leq \alpha(G', \mathbb{1})$$

are required to describe STAB(G). (Recall that the above inequality is obtained by lifting the base rank constraint associated with $G' \subseteq G$ and that $x(G', \mathbb{1}) \leq \alpha(G', \mathbb{1})$ produces the full rank facet of STAB(G') by the definition of a weak rank constraint.) Since every facet-defining rank constraint $x(G', \mathbb{1}) \leq \alpha(G', \mathbb{1})$ is a weak rank constraint with $a_i = 0$ for $i \in G - G'$, the class of weakly rank-perfect graphs contains all rank-perfect graphs (and, therefore, all near-perfect and all perfect graphs).

One general way to arrive at classes of weakly rank-perfect graphs is as follows: Consider a class of rank-perfect graphs where only nonnegativity constraints and special rank constraints are needed to describe the stable set polytope. Then define the "corresponding" class of weakly rank-perfect graphs by allowing weak rank constraints based on those special rank constraints as the only nontrivial facets of the stable set polytope. E.g., the class of *weakly h-perfect graphs* can be defined that way to contain all graphs whose stable set polytope is given by nonnegativity constraints (7.0), clique constraints (7.1), and lifted odd hole constraints. (See Padberg [20] for a general description of how to lift odd hole constraints to weak rank facets.) The 5-wheel in Figure 7.3(a) and the graph in Figure 7.3(b) are examples of weakly h-perfect graphs that are not h-perfect. (Note that the classes of weakly t-perfect and weakly h-perfect graphs coincide since clique constraints are liftings of edge constraints.)

Two natural graph classes are known to consist of only weakly rank-perfect graphs due to Shepherd [31]: so-called near-bipartite graphs and complements of line graphs. A graph G is *near-bipartite* if removing all neighbors of an arbitrary node leaves the graph bipartite. (That is, $G - N(v)$ can be partitioned into two stable sets for all nodes v of G, and near-bipartite graphs are, therefore, the complements of quasi-line graphs.) The stable set polytope of near-bipartite graphs has facets of type (7.4b):

$$\sum_{i \leq k} \frac{1}{\alpha(\overline{W}_i)} x(\overline{W}_i) + x(Q) \leq 1,$$

associated with the complete join of prime antiwebs $\overline{W}_1, \ldots, \overline{W}_k$ and a clique Q as its only nontrivial facets [31]. The class of near-bipartite graphs contains all complements of line graphs (the nonneighbors of a node v in $\overline{L(F)}$ correspond to the edges incident to the edge v in F, hence to two cliques in $L(F)$ and to two stable sets in $\overline{L(F)}$). Shepherd [31] showed that odd antiholes are the only prime antiwebs that occur in complements of line graphs. Thus the only nontrivial facets of their stable set polytope are weak rank constraints (7.4c) associated with the complete join of odd antiholes and a clique.

We studied in [33] critical edges with respect to perfectness (recall that these are edges of perfect graphs whose deletion yields an imperfect graph). We investigated the case of deleting critical edges e from complements G of perfect line graphs. We showed that odd antiholes are the only minimally imperfect subgraphs of $G - e$ and we showed how to lift the corresponding odd antihole constraints to facets of $\mathrm{STAB}(G - e)$. We were able to prove that these lifted odd antihole constraints are, besides clique constraints (7.1), the only nontrivial facets of $\mathrm{STAB}(G - e)$ if G is the complement of the line graph of a *bipartite* graph. Thus: every graph obtained by deleting a critical edge from the complement of the line graph of a bipartite graph is weakly rank-perfect [33]. That is, deleting edges from complements of line graphs of bipartite graphs leaves the resulting graphs in the same stage of imperfectness as general complements of line graphs; see [33] for more details.

Finally, a description of the facet system of $\mathrm{STAB}(G)$ for all graphs G with $\alpha(G) = 2$ was found (but not published) by Cook; see [30]. He showed that the stable set polytope of graphs G with $\alpha(G) = 2$ is given by nonnegativity constraints (7.0) and weak rank constraints of the form (7.4d):

$$x(\tilde{N}(Q)) + 2x(Q) \leq 2,$$

for every clique Q (recall that $\tilde{N}(Q)$ denotes the set of all nodes v of G with $Q \subseteq N(v)$). That is, graphs G with $\alpha(G) = 2$ are weakly rank-perfect, too. In order to figure out which graphs G with $\alpha(G) = 2$ are rank-perfect, we determine which rank facets may appear. The inequalities (7.4d) can be scaled to have no coefficients different from 0 and 1 only if Q is maximal (then $\tilde{N}(Q) = \emptyset$ follows) or Q is empty (then $\tilde{N}(Q) = V(G)$ follows). Thus, the only possible rank facets are maximal clique facets and the full rank facet. Hence, we have obtained the following: *A graph G with $\alpha(G) = 2$ is near-perfect if and only if G is rank-perfect.*

7.6 Concluding Remarks

For all perfect graphs the stable set polytope coincides with the fractional stable set polytope, whereas $\mathrm{STAB}(G) \subset \mathrm{QSTAB}(G)$ holds if and only if G is imperfect. We used the difference between $\mathrm{STAB}(G)$ and $\mathrm{QSTAB}(G)$ to decide how far away an imperfect graph is from being perfect. For that, we introduced three polytopes that contain $\mathrm{STAB}(G)$ but are contained in $\mathrm{QSTAB}(G)$. The fractional stable set polytope $\mathrm{QSTAB}(G)$ is given by nonnegativity constraints (7.0) and clique constraints (7.1):

$$x(G', \mathbb{1}) \leq 1,$$

for all cliques $G' \subseteq G$. We discussed which additional cutting planes are required to cut off all fractional vertices of $\mathrm{QSTAB}(G)$. We defined $\mathrm{FSTAB}(G)$ to be the polytope where the full rank constraint (7.2) is the only additional cutting plane. Next we defined $\mathrm{RSTAB}(G)$ as the polytope given by nonnegativity constraints (7.0) and all 0/1-inequalities (7.3):

$$x(G', \mathbb{1}) \leq \alpha(G', \mathbb{1}),$$

for arbitrary induced subgraphs $G' \subseteq G$. The last step was to allow in $\mathrm{WSTAB}(G)$ as nontrivial facets more general inequalities of the form (7.4):

$$x(G, a) \leq \alpha(G', \mathbb{1}),$$

where $G' \subseteq G$, with $V(G') \subseteq \{v_i \in V(G) : a_i = 1\}$, and STAB$(G')$ has the full rank facet. Since STAB(G) itself is given by (7.0) and all general inequalities (7.5)

$$x(G\,,a) \leq \alpha(G\,,a),$$

there is no further relaxation of STAB(G) possible beyond WSTAB(G):

$$\text{STAB}(G) \subseteq \text{WSTAB}(G) \subseteq \text{RSTAB}(G) \subseteq \text{FSTAB}(G) \subseteq \text{QSTAB}(G).$$

The difference between QSTAB(G) and the largest of the polytopes coinciding with STAB(G) gives us some information on the stage of imperfectness of the graph G. This answers the following question: Which graphs are "almost" perfect? Closest to perfect graphs are all near-perfect graphs G with STAB(G) = FSTAB(G). The next superclass contains all rank-perfect graphs G with STAB(G) = RSTAB(G). "Less perfect" are all weakly rank-perfect graphs G with STAB(G) = RSTAB(G). The discussion of which graphs are known to belong to one of these superclasses of perfect graphs is summarized in Figure 7.5. For some interesting graph classes strongly related to minimally imperfect graphs, so far we do not know to which of the three superclasses they belong: partitionable graphs, webs, or antiwebs. They are not all near-perfect (see Section 7.3), but there is some hope of proving that antiwebs are all rank-perfect and partitionable graphs and webs are at least weakly rank-perfect.

Furthermore, perfect graphs are closed under complementation, but none of the superclasses of perfect graphs under consideration is: Theorem 7.11 by Shepherd [30] implies this for near-perfect graphs. The 5-wheel is not rank-perfect but its complement is; the wedge depicted in Figure 7.3(d) is not weakly rank-perfect but its complement is.

Finally, other than the perfect graphs, line graphs constitute the only natural class of graphs for which we have a polyhedral description for the stable set polytope for the class as well as for the complementary class. The question of polyhedral descriptions for quasi-line graphs and, more generally, for claw-free graphs (having no node with a stable set of size three in its neighborhood), remains one of the interesting open problems in polyhedral combinatorics. We already know that quasi-line graphs are not rank-perfect; see the web W_{25}^5 and the graphs G^k introduced in [14]. Oriolo [19] conjectured that the only nontrivial facets of the stable set polytope of quasi-line graphs have the form (7.5a), but we do not even know whether these are weak rank constraints. We already know that claw-free graphs are not weakly rank-perfect, since all wedges are claw free but produce facets that are not weak-rank constraints, by Giles and Trotter [14]; see Section 7.2. Pulleyblank and Shepherd [26] showed that all wedges belong to a subclass of claw-free graphs, so-called distance claw-free graphs (where the nodes at distance exactly two from a node do not contain a stable set of size three). Hence, distance claw-free graphs are not weakly rank-perfect, either. But there is a complete description of all rank facet–producing claw-free graphs due to Galluccio and Sassano [11]. They showed that the rank facets of claw-free graphs essentially come from cliques, line graphs of 2-connected hypomatchable graphs, and partitionable webs.

Note added in proof. The author has proved meanwhile that antiwebs are rank-perfect. Chudnovsky, Robertson, Seymour, and Thomas verified in 2002 the Strong Perfect Graph Conjecture after a sequence of remarkable results based on the work of many graph-theoretists.

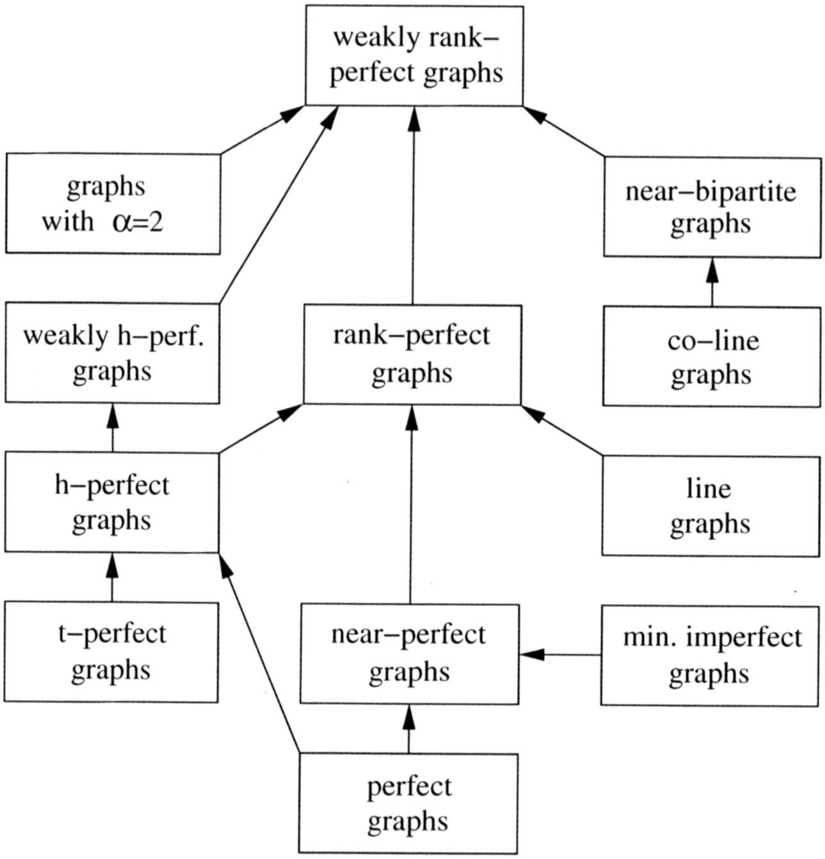

Figure 7.5. *Inclusion relations of the studied graph classes.*

Bibliography

[1] C. Berge. Färbung von Graphen, deren sämtliche bzw. deren ungerade Kreise starr sind. Wissenschaftliche Zeitschrift der Martin-Luther-Universität Halle-Wittenberg, 10:114–115, 1961.

[2] R.G. Bland, H.-C. Huang, and L.E. Trotter. Graphical properties related to minimal imperfection. *Discrete Mathematics*, 27:11–22, 1979.

[3] M. Boulala and J.P. Uhry. Polytope des indépendants dans un graphe série-parallèle. *Discrete Mathematics*, 27:225–243, 1979.

[4] V. Chvátal. On certain polytopes associated with graphs. *Journal of Combinatorial Theory* B, 18:138–154, 1975.

[5] V. Chvátal. On the strong perfect graph conjecture. *Journal of Combinatorial Theory* B, 20:139–141, 1976.

[6] G. Dahl. Stable set polytopes for a class of circulant graphs. *SIAM Journal on Optimization*, 9:493–503, 1999.

[7] J.R. Edmonds and W.R. Pulleyblank. Facets of 1-matching polyhedra. In C. Berge and D.R. Chaudhuri, editors, *Hypergraph Seminar*, pages 214–242. Springer-Verlag, Heidelberg, 1974.

[8] J. Fonlupt and J.P. Uhry. Transformations which preserve perfectness and h-perfectness of graphs. *Annals of Discrete Mathematics*, 16:83–95, 1982.

[9] D.R. Fulkerson. Blocking and antiblocking pairs of polyhedra. *Mathematical Programming*, 1:168–194, 1971.

[10] D.R. Fulkerson. On the perfect graph theorem. In T.C. Hu and S.M. Robinson, editors, *Mathematical Programming*, pages 69–76. Academic Press, New York, 1973.

[11] A. Galluccio and A. Sassano. The rank facets of the stable set polytope for claw-free graphs. *Journal of Combinatorial Theory* B, 69:1–38, 1997.

[12] A.M.H. Gerards and A. Schrijver. Matrices with the Edmonds-Johnson property. *Combinatorica*, 6:403–417, 1986.

[13] A.M.H. Gerards and F.B. Shepherd. The graphs with all subgraphs t-perfect. *SIAM Journal on Discrete Mathematics*, 11:524–545, 1998.

[14] R. Giles and L.E. Trotter. On stable set polyhedra for $K_{1,3}$-free graphs. *Journal of Combinatorial Theory* B, 31:313–326, 1981.

[15] M. Grötschel, L. Lovász, and A. Schrijver. *Geometric Algorithms and Combinatorial Optimization*. Springer-Verlag, Berlin, Heidelberg, New York, 1988.

[16] J. Kind. *Mobilitätsmodelle für zellulare Mobilfunknetze: Produktformen und Blockierung*. Ph.D. thesis, RWTH, Aachen, 2000.

[17] L. Liptá and L. Lováz. Facets with fixed defect of the stable set polytope. *Mathematical Programming* A, 88:33–44, 2000.

[18] L. Lovász. Normal hypergraphs and the weak perfect graph conjecture. *Discrete Mathematics*, 2:253–267, 1972.

[19] G. Oriolo. Clique Family Inequalities for the Stable Set Polytope for Quasi-Line Graphs. *Discrete Applied Mathematics* 132:185–201, 2003.

[20] M.W. Padberg. On the facial structure of set packing polyhedra. *Mathematical Programming*, 5:199–215, 1973.

[21] M.W. Padberg. Perfect zero-one matrices. *Mathematical Programming*, 6:180–196, 1974.

[22] M.W. Padberg. Almost integral polyhedra related to certain combinatorial optimization problems. *Linear Algebra and Its Applications*, 15:69–88, 1976.

[23] M.W. Padberg. Almost perfect matrices and graphs. *Mathematics of Operations Research*, 26:1–18, 2001.

[24] A. Pêcher. *About Facets of the Stable Set Polytope of a Graph*. Rapport no. 2000-12, Université d'Orléans, 2000.

[25] M. Preissmann and A. Sebö. Some aspects of minimal imperfect graphs. In B. Reed and J. Ramirez Alfonsin, editors, *Perfect Graphs*, pages 185–214. Wiley, New York, 2001.

[26] W.R. Pulleyblank and F.B. Shepherd. Formulations for the stable set polytope of a claw-free graph. In G. Rinaldi et al., editors, *Integer Programming and Combinatorial Optimization*, pages 267–279. Librarian CORE, Louvain-la-Neuve, 1993.

[27] H. Sachs. On the Berge conjecture concerning perfect graphs. In R. Guy et al., editors, *Combinatorial Structures and Their Applications*, pages 377–384. Gordon and Breach, New York, 1970.

[28] N. Sbihi and J.P. Uhry. A class of h-perfect graphs. *Discrete Mathematics*, 51:191–205, 1984.

[29] E.C. Sewell. *Stability Critical Graphs and the Stable Set Polytope*. Ph.D. thesis, Cornell University, Ithaca, NY, 1990.

[30] F.B. Shepherd. Near-perfect matrices. *Mathematical Programming*, 64:295–323, 1994.

[31] F.B. Shepherd. Applying Lehman's theorem to packing problems. *Mathematical Programming*, 71:353–367, 1995.

[32] L.E. Trotter. A class of facet producing graphs for vertex packing polyhedra. *Discrete Mathematics*, 12:373–388, 1975.

[33] A.K. Wagler. *Critical Edges in Perfect Graphs*. Ph.D. thesis, Technische Universität Berlin, 2000.

Part III

Polyhedral Combinatorics

Chapter 8
Cardinality Homogeneous Set Systems, Cycles in Matroids, and Associated Polytopes

*Martin Grötschel**

Abstract. A subset $\mathcal{C}$ of the power set of a finite set E is called cardinality homogeneous if, whenever $\mathcal{C}$ contains some set F, $\mathcal{C}$ contains all subsets of E of cardinality $|F|$. Examples of such set systems $\mathcal{C}$ are the sets of all even or of all odd cardinality subsets of E, or, for each uniform matroid, its set of circuits and its set of cycles. With each cardinality homogeneous set system $\mathcal{C}$, we associate the polytope $P(\mathcal{C})$, the convex hull of the incidence vectors of all sets in $\mathcal{C}$. We provide a complete and nonredundant linear description of $P(\mathcal{C})$. We show that a greedy algorithm optimizes any linear function over $P(\mathcal{C})$; we construct, by a dual greedy procedure, an explicit optimum solution of the dual linear program; and we describe a polynomial time separation algorithm for the class of polytopes of type $P(\mathcal{C})$.

MSC 2000. 90C27, 90C57, 52B40, 05B35, 52B55

Key words. Cycles and circuits in matroids, cardinality homogeneous set systems, polytopes, greedy algorithm, polyhedral combinatorics, separation algorithms

8.1 Introduction

Cycles in matroids can be viewed as far-reaching common generalizations of Eulerian subgraphs and cuts of a graph. From an optimization point of view it is of interest to understand the polytopes naturally associated with cycles.

The aim is to develop linear programming techniques for the solution of weighted cycle optimization problems. This chapter contributes to this issue by investigating a class of polytopes, namely, the polytopes associated with cardinality homogeneous set systems,

*Konrad-Zuse-Zentrum für Informationstechnik Berlin, Takustr. 7, 14195 Berlin, Germany (groetschel@zib.de).

which properly contains, e.g., the class of cycle and circuit polytopes associated with uniform matroids.

8.2 Matroids

Good books on matroid theory are [6] and [11]. We follow their notation and terminology to a large extent.

Let E be a finite set. We usually assume that $E = \{1, \ldots, n\}$, $n \geq 1$. A subset $\mathcal{I}$ of the power set 2^E of E is called an *independence system* if $\emptyset \in \mathcal{I}$ and if, whenever $I \in \mathcal{I}$, every subset of I also belongs to $\mathcal{I}$. An independence system $\mathcal{I}$ is called a *matroid* if, whenever $I, J \in \mathcal{I}$ with $|I| < |J|$, there is an element $j \in J \backslash I$ such that $I \cup \{j\} \in \mathcal{I}$. We also write $M = (E, \mathcal{I})$ to give a matroid a name and stress that we are dealing with a matroid $\mathcal{I}$ on the ground set E.

Every set in $\mathcal{I}$ is called *independent* and every set in $2^E \backslash \mathcal{I}$ is said to be *dependent.* The minimal dependent subsets of E are called *circuits* (such sets do not properly contain other dependent sets). Every subset of E that is the disjoint union of circuits is called a *cycle*. For every set $F \subseteq E$, a set $B \subseteq E$ is called a *basis* of F if $B \subseteq F$, $B \in \mathcal{I}$, and F does not contain an independent set B' properly containing B, i.e., B is a maximal independent subset of F.

If $\mathcal{B}$ is the set of bases of the ground set E of a matroid $M = (E, \mathcal{I})$, then $\mathcal{B}^* := \{E \backslash B | B \in \mathcal{B}\}$ is the set of bases of another matroid, denoted by $M^* = (E, \mathcal{I}^*)$ and called the matroid *dual* to M. By construction we have $M^{**} = M$. It is customary to call the bases, circuits, and cycles of M^* the *cobases*, *cocircuits*, and *cocycles* of M.

It is well known that, for any graph $G = (V, E)$, the set of edgesets of its forests forms the system of independent sets of a matroid, the so-called graphic matroid, denoted by $M(G)$. The matroid dual to a graphic matroid is called *cographic* and is denoted by $M(G)^*$. The circuits of a graphic matroid are the edgesets of the *circuits* of the underlying graph G. The cycles are the (not necessarily connected) *Eulerian subgraphs* of G, i.e., the edgesets of all subgraphs with nodes of even degree. The cycles of $M(G)^*$ are the cuts of G, i.e., edgesets of the form $\delta(W) = \{ij \in E \mid i \in W, j \in V \backslash W\}$. The circuits of a cographic matroid are the edgesets of minimal cuts.

Another nice class of matroids is composed of *representable* (or matric) matroids. We choose a field F and an $m \times n$ matrix A with entries from F. A set $I \subseteq E = \{1, \ldots, n\}$ is called independent if the submatrix of A consisting of the columns indexed by I has rank $|I|$, i.e., if the column vectors $A_{.j}$, $j \in I$, are linearly independent in the m-dimensional vector space over F. A matroid that is isomorphic to a matroid of this type is called *representable over* F. A matroid representable over the two-element field **GF(2)** is called *binary*. If M is representable over F, then this also holds for its dual matroid M^*.

There are many equivalent characterizations of binary matroids; see [11], Chapter 10. For instance, we have the following theorem.

Theorem 8.1. *The following statements about a matroid M are equivalent.*

(i) *M is binary.*
(ii) *For any circuit C and any cocircuit C^*, $|C \cap C^*|$ is even.*
(iii) *Every cycle of M is the symmetric difference of distinct circuits of M.*

Graphic matroids (and therefore also cographic matroids) are representable over any field and, hence, they are binary.

One, in many respects, very simple class of matroids comprises the uniform matroids. They are defined as follows. We are given integers $1 \leq k \leq n$. The ground set is $E = \{1, \ldots, n\}$ and every subset with at most k elements is declared to be independent. This matroid is called the *uniform matroid on n elements of rank k* and is denoted by $U_{k,n}$. It has $\binom{n}{k}$ bases (the sets of size k) and $\binom{n}{k+1}$ circuits (the sets of size $k+1$). The cycles of $U_{k,n}$ are the sets of cardinality $i(k+1)$, $0 \leq i \leq \lfloor \frac{n}{k+1} \rfloor$.

8.3 Cycle Polytopes

Polyhedral combinatorics deals with the geometric description of combinatorial problems. Instead of solving a combinatorial problem directly, one associates a polytope with the problem and tries to solve the combinatorial problem as a linear program over this polytope. Two prominent examples are the Chinese postman and the max-cut problems. With respect to these problems, the approach works as follows.

Given a graph $G = (V, E)$ with weights c_e on the edges $e \in E$, we wish to find an Eulerian subgraph of maximum weight. To do this we define the polytope

$$\mathrm{CP}(G) := conv\{\chi^C \in \mathbb{R}^E \mid C \subseteq E \text{ Eulerian subgraph}\},$$

where $\chi^C = (\chi_e^C)_{e \in E}$ denotes the incidence vector of C with $\chi_e^C = 1$ if $e \in C$ and $\chi_e^C = 0$ otherwise. $\mathrm{CP}(G)$ is called the *Chinese postman polytope*. Solving the Chinese postman problem is equivalent to solving the linear programming problem

$$\max\ c^T x, x \in \mathrm{CP}(G).$$

Similarly, given a graph $G = (V, E)$ with weights c_e for all $e \in E$, finding a cut of G with maximum weight is equivalent to maximizing the linear function $c^T x$ over the *cut polytope*

$$\mathrm{CUT}(G) := conv\{\chi^{\delta(W)} \in \mathbb{R}^E \mid W \subseteq V\}.$$

Cut problems have a wide range of applications and arise in various, sometimes disguised, forms. One such different looking but equivalent appearance is quadratic 0/1-programming. The polyhedron arising here is the Boolean quadratic polytope investigated, e.g., in [7].

Recall that Eulerian subgraphs and cuts are cycles of the corresponding graphic and cographic matroids, respectively; i.e., the Chinese postman and the cut polytope are special instances of a *cycle polytope*

$$P(M) := conv\{\chi^C \in \mathbb{R}^E \mid C \text{ is a cycle of } M\},$$

which is the convex hull of the incidence vectors of all cycles of a matroid M on a ground set E.

Guided by the complete characterization of the Chinese postman polytope for all graphs by Edmonds and Johnson [3] and of the cut polytope for graphs not contractible to the complete graph K_5 by Barahona [1] and based on a deep theorem of Seymour [9] characterizing matroids with the "sum of circuits property," Barahona and Grötschel [2] characterized polytopes of certain binary matroids as follows.

Let M be a matroid on E. Consider the systems of inequalities

$$0 \leq x_e \leq 1 \quad \text{for all } e \in E \tag{8.1}$$

and

$$x(F) - x(C \backslash F) \leq |F| - 1 \quad \text{for all cocircuits } C \subseteq E \text{ and all } F \subseteq C, |F| \text{ odd}, \tag{8.2}$$

and define

$$Q(M) := \{x \in \mathbb{R}^E | x \text{ satisfies (8.1) and (8.2)}\}.$$

Because of Theorem 8.1(ii), every incidence vector of a cycle of a binary matroid satisfies (8.1) and (8.2). And if $J \subseteq E$ is not a cycle, there must be, by Theorem 8.1(ii) and (iii), a cocircuit C and an odd subset F of C such that χ^J violates the corresponding inequality of (8.2). Thus, all integral points of $Q(M)$ are incidence vectors of cycles—provided M is binary. The main theorem of [2] is as follows.

Theorem 8.2. *For a binary matroid M, $P(M) = Q(M)$ if and only if M has no F_7^*, R_{10}, and $M(K_5)^*$ minor.*

Here, $M(K_5)^*$ is the cographic matroid of the complete graph on five nodes, F_7^* is the matroid dual to the Fano matroid, and R_{10} is the binary matroid associated with the 5×10 matrix whose columns are the ten 0/1-vectors with three ones and two zeros. A *minor* of a matroid $M = (E, \mathcal{I})$ is a matroid that can be obtained from M by deleting and contracting some elements of E.

A precise description of all the facets of $P(M)$ is given in [2], i.e., a complete and nonredundant characterization of $P(M)$ for this class of binary matroids M. This yields, in particular, complete and nonredundant characterizations of the Chinese postman polytope for any graph [3] and for the cut polytope of all graphs not contractible to K_5 [1].

Grötschel and Truemper [5] have shown, among other things, that one can solve the separation problem for $Q(M)$ for the class of matroids not containing F_7^*; hence by [4], for this class of matroids, one can maximize any linear function over $Q(M)$. This implies that one can maximize over $P(M)$ if M has no F_7^*, R_{10}, $M(K_5)^*$ minor; thus, for this class of binary matroids, the weighted cycle problem can be solved in polynomial time.

It turns out that knowledge about cycles in matroids and the associated polytopes is rather poor for matroids not in the class considered in Theorem 8.2. There is, e.g., a characterization of so-called master polytopes for cycles in binary matroids; see [5]. For another example, the facets of $P(F_7^*)$ are known; but—in contrast to Theorem 8.2—none of the inequalities defining $Q(F_7^*)$ defines a facet of $P(F_7^*)$; see [2]. The situation is even worse in the nonbinary case. Not even a decent integer programming formulation, such as $\max \ c^T x, x \in Q(M) \cap \{0, 1\}^E$ for binary matroids M, is known in this case.

Just as it was worthwhile to investigate a joint generalization of the Chinese postman and the max-cut problems yielding, e.g., a unified description of the associated polytopes, it may be rewarding to better understand cycles of those matroids that are more general than the matroids of Theorem 8.2, in particular, cycles of nonbinary matroids.

Strangely enough, it is not even completely obvious how to generalize the concept of cycle to the nonbinary case. Looking at the proofs, e.g., in [2], it becomes clear that, although cycles are usually defined as disjoint unions of circuits, the (in the binary case) equivalent definition that a cycle is a set that can be obtained from the set of circuits by taking symmetric differences (see Theorem 8.1) is of much greater help in proofs. It turns out that, for nonbinary matroids, this second definition does not lead to anything interesting in general. It is also worth noting that condition (ii) of Theorem 8.1 is the one that yields the so-called cocircuit inequalities (8.2), which provide an integer programming formulation and enable Theorem 8.2. This condition is not available in the nonbinary case. Is there a condition that can replace it?

To leave the class of binary matroids, there is a wonderful excluded minor theorem of Tutte [10] that, as one might hope, could lead the way.

Theorem 8.3. *A matroid is binary if and only if it has no minor isomorphic to* $U_{2,4}$.

This result shows that all uniform matroids are nonbinary except for $U_{1,n}$, $n \geq 1$, and $U_{2,3}$. It also suggests that investigating the cycles of uniform matroids may provide some polyhedral insight.

The cycles of $U_{2,4}$ are its circuits, which are the four sets of size three, and the empty set. The convex hull of the corresponding five points $(0,0,0,0)$, $(0,1,1,1)$, $(1,0,1,1)$, $(1,1,0,1)$, $(1,1,1,0)$ in $\mathbb{R}^4$ is a simplex defined by the inequalities

$$\begin{aligned}
-x_1 - x_2 - x_3 + 2x_4 &\leq 0,\\
-x_1 - x_2 + 2x_3 - x_4 &\leq 0,\\
-x_1 + 2x_2 - x_3 - x_4 &\leq 0,\\
+2x_1 - x_2 - x_3 - x_4 &\leq 0,\\
+x_1 + x_2 + x_3 + x_4 &\leq 3.
\end{aligned}$$

Unfortunately, there is not much one can learn from this observation.

8.4 Cardinality Homogeneous Set Systems

The initial proof of a linear characterization of the class of cycle polytopes of uniform matroids became easier by generalizing this result to a more abstract setting. This will be presented here.

Let $E = \{1, \dots, n\}$ be a finite set. We will assume throughout the paper that $E \neq \emptyset$, i.e., $n \geq 1$. We call a subset $\mathcal{C} \subseteq 2^E$ *cardinality homogeneous* if, whenever $\mathcal{C}$ contains some subset of cardinality k, $0 \leq k \leq n$, then $\mathcal{C}$ contains all subsets of cardinality k.

Example 8.4. The following set systems are cardinality homogeneous.

(i) $\mathcal{C} = 2^E$, the set of all subsets of E;
(ii) $\mathcal{C} = \{F \subseteq E \mid |F| \text{ is even}\}$;
(iii) $\mathcal{C} = \{F \subseteq E \mid |F| \text{ is odd}\}$;
(iv) $\mathcal{C} =$ set of circuits of $U_{k,n}$;
(v) $\mathcal{C} =$ set of cycles of $U_{k,n}$.

To simplify statements and proofs we introduce the following notation. Let $E = \{1, \dots, n\}$ be given. From now on, $a = (a_1, \dots, a_m)$ denotes a nonempty sequence of integers such that $a_i \in \{0, 1, \dots, n\}$ and $0 \le a_1 < a_2 < \dots < a_m \le n$ holds. We call such a sequence a *cardinality sequence*. We set

$$C(n; a_i) := \{C \in E \mid |C| = a_i\}, i = 1, \dots, m,$$

$$C(n; a) := C(n; a_1, \dots, a_m) := \bigcup_{i=1}^{m} C(n; a_i).$$

Clearly, each cardinality homogeneous set system $\mathcal{C}$ is of the form $C(n; a)$ for some ground set $E = \{1, \dots, n\}$ and some cardinality sequence $a = (a_1, \dots, a_m)$; thus

$$P(n; a) := P(n; a_1, \dots, a_m) := conv\{\chi^C \in \mathbb{R}^E \mid C \in C(n; a)\}$$

is a generic member of the class of polytopes associated with cardinality homogeneous set systems. We want to find a system of linear inequalities and equations describing the members of the class of polytopes $P(n; a)$ completely and nonredundantly.

There are some inequalities that are obviously valid for $P(n; a)$:
the *trivial inequalities*

$$0 \le x_j \le 1, \ j = 1, \dots, n, \tag{8.3}$$

and the *cardinality bounds*

$$a_1 \le x(E) \le a_m, \tag{8.4}$$

where $x(E)$ denotes the sum $\sum_{e \in E} x_e = x_1 + \dots + x_n$.

We introduce now a new class of inequalities that we call *cardinality-forcing inequalities* (or briefly CF-inequalities). For a given cardinality sequence $a = (a_1, \dots, a_m)$ set

$$\mathcal{F} := \mathcal{F}(a_1, \dots, a_m) := \{F \subseteq E \mid a_1 < |F| < a_m, |F| \ne a_j, j = 2, \dots, m-1\},$$

$$f := f(F) := \max\{j \in \{1, \dots, m-1\} \mid a_j < |F|\} \quad \text{for all } F \in \mathcal{F},$$

where $\mathcal{F}$ consists of all sets that are not in $C(n; a)$ and have a number of elements that is between a_1 and a_m. For $F \in \mathcal{F}$, $f(F)$ denotes the index $f \in \{1, \dots, m\}$ with $a_f < |F| < a_{f+1}$.

For each $F \in \mathcal{F}$, its corresponding CF-inequality, where $f = f(F)$, is the following:

$$\mathrm{CF}_F(x) := \sum_{j \in F} (a_{f+1} - |F|)x_j - \sum_{j \in E \setminus F} (|F| - a_f)x_j \le (a_{f+1} - |F|)a_f =: s(F). \tag{8.5}$$

Proposition 8.5.

(i) *Every CF-inequality is valid for $P(n; a)$.*

(ii) *For every 0/1-vector $y \in \mathbb{R}^E \setminus P(n; a)$ with $a_1 < y(E) < a_m$ there is at least one CF-inequality separating y from $P(n; a)$.*

(iii) *There are $\sum_{i=1}^{m-1} \sum_{k=a_i+1}^{a_{i+1}-1} \binom{n}{k}$ CF-inequalities; i.e., the number of CF-inequalities is, in general, not bounded by a polynomial in n.*

(iv) *CF-inequalities are completely dense; i.e., all coefficients are different from zero.*

Proof.

(iv) The coefficient of a variable x_j, $j \in E$, in a CF-inequality is either $a_{f+1} - |F|$ or $|F| - a_f$. These values are different from zero by definition.

(iii) This follows from simple counting.

(i) Let $F \in \mathcal{F}$, $f = f(F)$, and $S \in C(n; a)$. Substituting the incidence vector χ^S into the left-hand side of the CF-inequality $\text{CF}_F(x) \le s(F)$ results in

$$\sum_{j \in F}(a_{f+1} - |F|)\chi_j^S - \sum_{j \in E \setminus F}(|F| - a_f)\chi_j^S = (a_{f+1} - |F|)\,|F \cap S| \\ -(|F| - a_f)\,|(E \setminus F) \cap S|\,.$$

If $|S| \le a_f$, then $|F \cap S| \le a_f$ and χ^S obviously does not violate (8.5). If $|S| > a_f$, then $|S| \ge a_{f+1}$ and hence $|(E \setminus F) \cap S| = |S \setminus F| \ge a_{f+1} - |F|$. Trivially, $|F \cap S| \le |F| = a_f + |F| - a_f$ and we obtain

$$\begin{aligned}(a_{f+1} - |F|)\,|F \cap S| &- (|F| - a_f)\,|(E \setminus F) \cap S| \\ &\le (a_{f+1} - |F|)a_f + (a_{f+1} - |F|)(|F| - a_f) - (|F| - a_f)(a_{f+1} - |F|) \\ &= (a_{f+1} - |F|)a_f,\end{aligned}$$

which shows that the incidence vectors of all sets in $C(n; a)$ satisfy (8.5).

(ii) Let $y \in \{0, 1\}^E \setminus P(n; a)$, $a_1 < y(E) < a_m$, be given and let F be the subset of E with $\chi^F = y$. By our choice $F \in \mathcal{F}$. Substituting y into the CF-inequality associated with F yields the value $(a_{f+1} - |F|)\,|F|$ on the left-hand side. This is larger than the right-hand side since $|F| > a_f$; hence y violates the CF-inequality $\text{CF}_F(x) \le s(F)$ associated with F. □

Given a cardinality sequence $a = (a_1, \ldots, a_m)$, we introduce the polyhedron

$$Q(n; a) := Q(n; a_1, \ldots, a_m) := \{x \in \mathbb{R}^E | x \text{ satisfies (8.3), (8.4), (8.5)}\}.$$

Proposition 8.5(i) yields

$$P(n; a) \subseteq Q(n; a)$$

and Proposition 8.5(ii) together with the cardinality bounds yields

$$P(n; a) = conv\{x \in \{0, 1\}^E | x \in Q(n; a)\}.$$

In other words,

$$\max\ c^T x, x \in Q(n; a),$$

is a linear programming relaxation of

$$\max\ c^T x, x \in P(n; a).$$

Our main result is the following.

Theorem 8.6. *For all $E = \{1, \ldots, n\}$ and all cardinality sequences $a = (a_1, \ldots, a_m)$, $P(n; a) = Q(n; a)$.*

We will prove this in several steps and give, moreover, a characterization of all facets of $P(n; a)$.

8.5 A Primal and a Dual Greedy Algorithm

The proof of Theorem 8.6 consists of two algorithms and their analysis. We first state a greedy algorithm that finds, for every objective function c, a feasible solution for max $c^T x$, $x \in P(n; a)$. Then we describe an algorithm that produces a feasible solution of the LP dual to max $c^T x, x \in Q(n; a)$. We then show that the objective function values of the primal and the dual solution are identical. This yields, by a standard argument, that $P(n; a) = Q(n; a)$.

We are given a ground set $E = \{1, \dots, n\}$, a cardinality sequence $a = (a_1, \dots, a_m)$, and weights c_j, $j \in E$. We want to find a cardinality homogeneous set of largest weight. We do this with the following heuristic.

Algorithm 8.7 (Primal Greedy Algorithm).

1. Sort the elements of E such that $c_1 \geq c_2 \geq \dots \geq c_n$.
2. If $c_{a_m} \geq 0$, set $C_g := \{1, \dots, a_m\}$ and go to 6.
3. If $c_{a_1} \leq 0$, set $C_g := \{1, \dots, a_1\}$ and go to 6.
4. Otherwise (i.e., $c_{a_m} < 0 < c_{a_1}$), let us define the following integers:
   - p is the largest integer in $\{1, \dots, n\}$ such that $c_p > 0 \geq c_{p+1}$,
   - q is the index in $\{1, \dots, m\}$ such that $a_q \leq p < a_{q+1}$,
   - $h := \sum_{j=a_q+1}^{a_{q+1}} c_j$.
5. If $h > 0$, set $C_g := \{1, \dots, a_{q+1}\}$,
   else $C_g := \{1, \dots, a_q\}$.
6. Output C_g.

We call C_g the *greedy solution*; χ^{C_g} is a vertex of $P(n; a)$, so its objective function value $c^T \chi^{C_g}$ is a lower bound for max $c^T x, x \in P(n; a)$, which in turn is not larger than the value of its linear programming relaxation, i.e., of the corresponding LP over $Q(n; a)$:

$$\max c^T x$$

$$\begin{aligned} x_j &\leq 1, \ j = 1, \dots, n, \\ -x(E) &\leq -a_1, \\ x(E) &\leq a_m, \end{aligned}$$

$$\sum_{j \in F}(a_{f+1} - |F|)x_j - \sum_{j \in E \setminus F} (|F| - a_f)x_j \leq (a_{f+1} - |F|)a_f \quad \text{for all } F \in \mathcal{F},$$

$$x_j \geq 0, \ j = 1, \dots, n.$$

We denote this LP by $L(n; a; c)$. Let us state the LP dual to $L(n; a; c)$, for which we assume, without loss of generality, that the elements of E are ordered such that $c_1 \geq c_2 \geq \ldots \geq c_n$:

$$\min \sum_{j=1}^{n} u_j - a_1 v + a_m w + \sum_{F \in \mathcal{F}} (a_{f+1} - |F|) a_f y_F$$

$$u_j - v + w + \sum_{\substack{F \in \mathcal{F} \\ j \in F}} (a_{f+1} - |F|)\, y_F - \sum_{\substack{F \in \mathcal{F} \\ j \notin F}} (|F| - a_f)\, y_F \geq c_j, \; j = 1, \ldots, n, \tag{8.6}$$

$$v, w \geq 0,$$

$$u_j \geq 0, \; j = 1, \ldots, n,$$

$$y_F \geq 0, \; F \in \mathcal{F}.$$

We denote this dual LP by $D(n; a; c)$. We call the inequalities (8.6) above *dual CF inequalities.*

If the objective function c satisfies $c_{a_m} \geq 0$ or $c_{a_1} \leq 0$, the optimality of the greedy solution is easy to see.

Remark 8.8. If $c_{a_m} \geq 0$, set $w := c_{a_m}$, $u_j := c_j - c_{a_m}$ for $j = 1, \ldots, a_m$, and set all other variables to zero.

If $c_{a_1} \leq 0$, set $v := -c_{a_1}$, $u_j := c_j - c_{a_1}$ for $j = 1, \ldots, a_1$, and set all other variables to zero.

In both cases, the solution is feasible for $D(n; a; c)$ and the objective function value is equal to the value of the greedy solution C_g.

Let us now assume that the primal greedy algorithm has to enter step 4 and thus that the index q is defined. We will handle this case by discussing three different possibilities: $h = 0$, $h < 0$, and $h > 0$.

Before entering the case distinction, we define a set $\mathcal{F}_0$ that consists of the following subsets of $\mathcal{F}$:

$$F_k := \{1, 2, \ldots, k\}, \; k = a_q + 1, a_q + 2, \ldots, a_{q+1} - 1.$$

We claim that an optimal solution of $L(n; a; c)$ can be found by solving the relaxed LP $L_{\mathcal{F}_0}(n; a; c)$ that is obtained by dropping the cardinality constraints and all CF-inequalities but those coming from the sets $F \in \mathcal{F}_0$. This means that $L_{\mathcal{F}_0}(n; a; c)$ has the following form:

$$\max \; c^T x$$

$$x_j \leq 1, \; j = 1, \ldots, n,$$

$$\sum_{j=1}^{k} (a_{q+1} - k)\, x_j - \sum_{j=k+1}^{n} (k - a_q)\, x_j \leq (a_{q+1} - k)\, a_q, \; k = a_q + 1, \ldots, a_{q+1} - 1,$$

$$x \geq 0.$$

We point out that the incidence vector χ^{C_g} of the greedy solution C_g satisfies all CF-inequalities associated with sets $F \in \mathcal{F}_0$ with equality.

The dual to this relaxed LP, denoted by $D_{\mathcal{F}_0}(n; a; c)$, is

$$\min \sum_{j=1}^{n} u_j + \sum_{k=a_q+1}^{a_{q+1}-1} (a_{q+1} - k)\, a_q\, y_{F_k}$$

$$u_j + \sum_{\substack{k=a_q+1\\ k\geq j}}^{a_{q+1}-1} (a_{q+1} - k)\, y_{F_k} - \sum_{\substack{k=a_q+1\\ k< j}}^{a_{q+1}-1} (k - a_q)\, y_{F_k} \geq c_j, \;\; j = 1, \ldots, n,$$

$$u_j \geq 0, \;\; j = 1, \ldots, n,$$
$$y_{F_k} \geq 0, \;\; k = a_q + 1, \ldots, a_{q+1} - 1.$$

We claim that, for objective functions not covered by Remark 8.8 and for which $h = 0$, $D_{\mathcal{F}_0}(n; a; c)$ can be solved as follows.

Algorithm 8.9 (Dual Greedy Algorithm for $h = 0$).

1. For $k = a_q + 1, \ldots, a_{q+1} - 1$ set
$$y^*_{F_k} := \frac{c_k - c_{k+1}}{a_{q+1} - a_q}.$$
2. For $j = 1, \ldots, a_{q+1}$ set
$$u^*_j := c_j - \sum_{\substack{k=a_q+1\\ k\geq j}}^{a_{q+1}-1} (a_{q+1} - k)\, y^*_{F_k} + \sum_{\substack{k=a_q+1\\ k< j}}^{a_{q+1}-1} (k - a_q) y^*_{F_k}.$$
3. Set all other variables to zero.

We call the solution u^*, y^* defined in Algorithm 8.9 the *dual greedy solution*. Let us state a few observations that follow directly from the definitions.

Remark 8.10.

(a) Since $c_k \geq c_{k+1}$ and $a_{q+1} > a_q$, all values $y^*_{F_k}$ are nonnegative.

(b) Deleting all variables set in step 3 to zero, the dual CF-inequalities for $j = a_{q+1} + 1, \ldots, n$ reduce to
$$-\sum_{k=a_q+1}^{a_{q+1}-1} (k - a_q) y_{F_k} \geq c_j.$$
Since $c_j \geq c_{j+1}$, checking whether these inequalities are satisfied by the dual greedy solution, it suffices to prove that
$$-\sum_{k=a_q+1}^{a_{q+1}-1} (k - a_q) y^*_{F_k} \geq c_{a_{q+1}}.$$
This is the case if we can prove that $u^*_{a_{q+1}} = 0$.

(c) Deleting all variables set in step 3 to zero, the dual CF-inequalities for $j = 1, 2, \ldots, a_q + 1$ reduce to

$$u_j + \sum_{k=a_q+1}^{a_{q+1}-1} (a_{q+1} - k)\, y_{F_k} \geq c_j.$$

The values u_j^* are set in step 2 of Algorithm 8.9 in such a way that these inequalities are satisfied with equality by the dual greedy solution. Since $c_j \geq c_{j+1}$, to prove that $u_j^* \geq 0$ it remains to show that $u_{a_q+1}^* \geq 0$.

(d) Proving feasibility of the dual greedy solution for $D_{\mathcal{F}_0}(n; a; c)$ reduces to showing that

$$u_j^* \geq 0, \;\; j = a_q + 1, \ldots, a_{q+1}.$$

We will show that, in fact, $u_j^* = 0$, $j = a_q + 1, \ldots, a_{q+1}$.

Remark 8.11. If $h = \sum_{j=a_q+1}^{a_{q+1}} c_j = 0$, then

$$u_j^* = \frac{h}{a_{q+1} - a_q} = 0, \;\; j = a_q + 1, \ldots, a_{q+1}.$$

Proof. Let $a_q + 1 \leq j \leq a_{q+1}$.

$$\begin{aligned}
u_j^* &= c_j + \sum_{k=a_q+1}^{j-1} (k - a_q)\frac{c_k - c_{k+1}}{a_{q+1} - a_q} - \sum_{k=j}^{a_{q+1}-1} (a_{q+1} - k)\frac{c_k - c_{k+1}}{a_{q+1} - a_q} \\
&= c_j + \frac{1}{a_{q+1} - a_q}\left(\sum_{k=a_q+1}^{j-1} kc_k - \sum_{k=a_q+2}^{j} (k-1)c_k - \sum_{k=a_q+1}^{j-1} a_q c_k + \sum_{k=a_q+2}^{j} a_q c_k \right. \\
&\qquad \left. + \sum_{k=j}^{a_{q+1}-1} kc_k - \sum_{k=j+1}^{a_{q+1}} (k-1)c_k - \sum_{k=j}^{a_{q+1}-1} a_{q+1}c_k + \sum_{k=j+1}^{a_{q+1}} a_{q+1}c_k \right) \\
&= c_j + \frac{1}{a_{q+1} - a_q}\left(\sum_{k=a_q+1}^{a_{q+1}-1} kc_k - \sum_{k=a_q+2}^{a_{q+1}} (k-1)c_k - a_q c_{a_q+1} + (a_q - a_{q+1})c_j \right. \\
&\qquad \left. + a_{q+1}c_{a_{q+1}} \right) \\
&= c_j + \frac{1}{a_{q+1} - a_q}\left(\sum_{k=a_q+1}^{a_{q+1}} c_k + a_q c_{a_q+1} - a_{q+1}c_{a_{q+1}} - a_q c_{a_q+1} + a_{q+1}c_{a_{q+1}} \right. \\
&\qquad \left. - (a_{q+1} - a_q)c_j \right) \\
&= \frac{h}{a_{q+1} - a_q}.
\end{aligned}$$

□

The definitions of the values u_j^* in Algorithm 8.9 and Remark 8.11 imply immediately the following remark.

Remark 8.12. If $h = 0$, then

$$u_j^* = c_j - c_{a_q+1}, \ j = 1, \ldots, a_q.$$

Let us now determine the objective function value $\sum_{j=1}^{n} u_j^* + \sum_{k=a_q+1}^{a_{q+1}-1} s(F_k) y_{F_k}^*$ of the dual greedy solution.

By definition and Remark 8.11, $u_j^* = 0$ for $j > a_q$. Taking the values of the other variables from Remark 8.12 and recalling that $h = \sum_{j=a_q+1}^{a_{q+1}} c_j$ we obtain

$$\sum_{j=1}^{n} u_j^* = \sum_{j=1}^{a_q} u_j^* + \sum_{j=a_q+1}^{a_{q+1}} u_j^* = \sum_{j=1}^{a_q} c_j - a_q c_{a_q+1}.$$

The second term in the dual objective function yields

$$\begin{aligned}
\sum_{k=a_q+1}^{a_{q+1}-1} s(F_k) y_{F_k}^* &= \sum_{k=a_q+1}^{a_{q+1}-1} \left(a_{q+1} - k\right) a_q \frac{c_k - c_{k+1}}{a_{q+1} - a_q} \\
&= \frac{a_q}{a_{q+1} - a_q} \sum_{k=a_q+1}^{a_{q+1}-1} \left(a_{q+1} - k\right) (c_k - c_{k+1}) \\
&= \frac{a_q}{a_{q+1} - a_q} \left(\sum_{k=a_q+1}^{a_{q+1}-1} a_{q+1} c_k - \sum_{k=a_q+1}^{a_{q+1}-1} k c_k - \sum_{k=a_q+2}^{a_{q+1}} a_{q+1} c_k \right. \\
&\qquad \left. + \sum_{k=a_q+2}^{a_{q+1}} (k-1) c_k \right) \\
&= \frac{a_q}{a_{q+1} - a_q} \left(a_{q+1}(c_{a_q+1} - c_{a_{q+1}}) - (a_q + 1) c_{a_q+1} + (a_{q+1} - 1) c_{a_{q+1}} \right. \\
&\qquad \left. - \sum_{k=a_q+2}^{a_{q+1}-1} c_k \right) \\
&= \frac{a_q}{a_{q+1} - a_q} \left((a_{q+1} - a_q) c_{a_q+1} - h \right) \\
&= a_q \left(c_{a_q+1} - \frac{h}{a_{q+1} - a_q} \right) = a_q c_{a_q+1}.
\end{aligned}$$

Adding the two objective function terms we obtain

$$\sum_{j=1}^{n} u_j^* + \sum_{k=a_q+1}^{a_{q+1}-1} (a_{q+1} - k) a_q y_k^* = \sum_{j=1}^{a_q} c_j = c(C_g),$$

which is the value of the primal greedy solution. These calculations prove the following.

Remark 8.13. If $h = 0$, the dual greedy solution u^*, y^* is optimal for the LP $D(n; a; c)$ and has the same value as the primal greedy solution.

We now indicate how the solution of the case $h = 0$ can be utilized to handle the cases $h < 0$ and $h > 0$.

Remark 8.14. If $h < 0$, we increase some of the objective function coefficients c_j, $j = a_q + 1, \ldots, a_{q+1}$, such that, after the increase, the ordering of the variables is still respected and such that $h = 0$. Note that this change of the c_j values does not change the value of the primal greedy solution (in fact, now $\{1, \ldots, a_q\}$ and $\{1, \ldots, a_{q+1}\}$ are both optimal) and that any feasible solution of $D(n; a; c)$ after increase is feasible for the LP without modification. Thus applying Algorithm 8.9 to the modified dual LP $D(n; a; c)$ provides a solution u^*, y^* that is feasible and optimal for the unmodified $D(n; a; c)$ and has the same value as the primal greedy solution.

Remark 8.15. If $h > 0$, we modify the objective function vector c into a vector c' by decreasing some of the coefficients c_j, $j = a_q + 1, \ldots, a_{q+1}$, to values c'_j such that $c'_1 \geq c'_2 \geq \ldots \geq c'_n$ and $h' = \sum_{j=a_q+1}^{a_{q+1}} c'_j = 0$. If I_g and I'_g are the primal greedy solutions with respect to c and c', respectively, then clearly $c(I_g) = c'(I'_g) + h$. If we now use Algorithm 8.9 to solve $D_{\mathcal{F}_0}(n; a; c')$, we obtain an optimal solution u', y' for $D(n; a; c')$ with value $c'(I'_g)$. Setting $u^*_j := u'_j + c_j - c'_j$, $j = 1, \ldots, n$, and $y^* := y'$ yields a solution u^*, y^* with value $c'(I'_g) + h = c(I_g)$ that is feasible for $D(n; a; c)$. This implies the optimality of χ^{I_g} for $L(n; a; c)$ and of u^*, y^* for $D(n; a; c)$.

This finishes the discussion of all cases occurring in the treatment of the dual LP $D(n; a; c)$. Hence, the proof of Theorem 8.6 providing a complete linear description of all polytopes associated with cardinality homogeneous systems is also finished.

We now put together all the pieces of the dual greedy algorithm discussed above to specify the complete greedy algorithm that solves the dual LP.

Algorithm 8.16 (Complete Dual Greedy Algorithm). Let $E = \{1, \ldots, n\}$, a cardinality sequence $a = (a_1, \ldots, a_m)$, and an objective function $c = (c_1, \ldots, c_n)$ be given.

1. Set all variables v, w, u_j, y_F of $D(n; a; c)$ to zero.
2. Sort the elements of E such that $c_1 \geq c_2 \geq \ldots \geq c_n$ holds and set $c' := c$.
3. If $c_{a_m} \geq 0$, set

$$w := c_{a_m},$$
$$u_j := c_j - c_{a_m} \quad \text{for } j = 1, \ldots, a_m.$$

   Go to 11.
4. If $c_{a_1} \leq 0$, set

$$v := -c_{a_1},$$
$$u_j := c_j - c_{a_1} \quad \text{for } j = 1, \ldots, a_1.$$

   Go to 11.

5. Otherwise, let p be the largest integer in $\{1, \ldots, n\}$ such that $c_p > 0 \geq c_{p+1}$, and let q be the index in $\{1, \ldots, m\}$ such that $a_q \leq p < a_{q+1}$. Set

$$h := h' := \sum_{j=a_q+1}^{a_{q+1}} c_j .$$

6. If $h < 0$, modify the objective function values as follows.
   For $k = a_q + 1, a_q + 2, \ldots, a_{q+1}$ do

$$\begin{aligned} \Delta &:= \min\{c'_{k-1} - c'_k, -h'\}, \\ c'_k &:= c'_k + \Delta, \\ h' &:= h' + \Delta. \end{aligned}$$

7. If $h > 0$, modify the objective function values as follows.
   For $k = p, p-1, \ldots, a_q + 1$ do

$$\begin{aligned} \Delta &:= \min\{c'_k - c'_{k+1}, h'\}, \\ c'_k &:= c'_k - \Delta, \\ h' &:= h' - \Delta. \end{aligned}$$

8. For $k = a_q + 1, a_q + 2, \ldots, a_{q+1} - 1$ set

$$y^*_{F_k} := \frac{1}{a_{q+1} - a_q}(c'_k - c'_{k+1}) .$$

9. If $h \leq 0$, do the following.
   For $j = 1, 2, \ldots, a_q$ set

$$u^*_j := c'_j - c'_{a_q+1} .$$

10. If $h > 0$, do the following.
    For $j = 1, 2, \ldots, a_q$ set

$$u^*_j := c_j - c'_{a_q+1} .$$

    For $j = a_q + 1, a_q + 2, \ldots, a_{q+1}$ set

$$u^*_j := c_j - c'_j .$$

11. Output the nonzero variables.

As outlined before, the solution u^*, y^* is feasible and optimal for the dual LP $D(n; a; c)$ and has the same value as the primal greedy solution.

Let us remark that the dual solution constructed above is one of typically very many optimal solutions. For instance, any modification of the c_j's in step 6 that makes h equal to zero and maintains the ordering $c_j \geq c_{j+1}$ and that is different from the one chosen in step 6 yields a different optimal dual solution. Even if we assume that all objective function coefficients are integral, the above solution is, in general, fractional. There are cases where all or some optimal dual solutions are integral, but we know examples where, for $c \in \mathbb{Z}^n$, no optimal solution of $D(n; a; c)$ is integral; see Example 8.18 below.

Remark 8.17. If the objective function values are sorted, then the Primal Greedy Algorithm 8.7 (steps 2–6) and the Complete Dual Greedy Algorithm 8.16 (steps 3–11) perform a number of arithmetic steps that is linear in n on numbers whose size is linear in the input length. Thus, the running time of the algorithm is dominated by sorting, which requires $\mathcal{O}(n \log n)$ steps.

Recall that a system of linear equations and inequalities is called *totally dual integral* (TDI) if, for any integral objective function, the LP dual to this LP has an integral optimum solution. We now indicate that none of the three linear systems that can be naturally associated with cardinality homogeneous set systems is TDI.

Example 8.18. Consider the ground set $E = \{1, 2, 3, 4\}$, the cardinality vector $a = (a_1, a_2) = (1, 4)$, and the objective function vector $c^T = (2, 2, 1, -3)$. The linear system $Q(4; a)$ gives rise to the LP

$$\text{(Q)} \qquad \begin{array}{rl} & \max\ 2x_1 + 2x_2 + x_3 - 3x_4 \\ 0 \le x_j \le 1, & j = 1, \dots, 4, \\ 1 \le x(E) \le 4, & \\ \mathrm{CF}_F(x) \le s(F) & \text{for all } F \subseteq E \text{ with } |F| \in \{2, 3\}. \end{array}$$

The linear system consists of 20 inequalities that describe $P(4; 1, 4)$ completely. This system, however, is redundant; see Proposition 8.21. The following LP has only five inequalities, has the same solution set, and is nonredundant:

$$\text{(NRQ)} \qquad \begin{array}{l} \max\ 2x_1 + 2x_2 + x_3 - 3x_4 \\ x(E) \ge 1, \\ \mathrm{CF}_F(x) \le 1 = s(F) \ \text{ for all } F \subseteq E,\ |F| = 3. \end{array}$$

In the proof of the Dual Greedy Algorithm we showed that (for this ordered objective function) the LP $L_{\mathcal{F}_0}(4; a; c)$:

$$\text{(GQ)} \qquad \begin{array}{rl} & \max\ 2x_1 + 2x_2 + x_3 - 3x_4 \\ 2x_1 + 2x_2 - x_3 - x_4 \le 2 & (F_2 = \{1, 2\}), \\ x_1 + x_2 + x_3 - 2x_4 \le 1 & (F_3 = \{1, 2, 3\}), \\ 0 \le x_j \le 1 & j = 1, 2, 3, 4, \end{array}$$

yields an optimum solution of (Q). Note that the LPs (Q), (NRQ), and (GQ) have three optimum solutions, namely, the incidence vectors of the sets $\{1\}$, $\{2\}$, and $\{1, 2, 3, 4\}$. (Q) and (NRQ) have, as mentioned, the same solution set. However, (GQ) is a strict relaxation. The solution set of (GQ) has some fractional vertices, such as $x' = (0, 1, 1, 1/2)$.

The LP dual to the "greedy LP" (GQ) has a unique optimum solution, which is the one provided by the Dual Greedy Algorithm: $y^*_{\{1,2\}} = 1/3$, $y^*_{\{1,2,3\}} = 4/3$, and all other variables equal to zero. The dual program of (NRQ) also has a unique optimum solution: $y^*_{\{1,2,3\}} = 5/3$, $y^*_{\{1,2,4\}} = 1/3$, and all other variables equal to zero. The dual to (Q) has a face of dimension 1 as the set of optimum solutions. This face is the convex hull of the two vertices just mentioned. It contains no integral point. Thus none of the three linear systems is TDI. (These computations have been carried out by PORTA [8] and were verified by hand.)

8.6 Facets

We now address the nonredundancy issue and determine the inequalities of $Q(n; a)$ that define facets of $P(n; a)$. As before, we assume throughout this section that $E = \{1, \ldots, n\}$, $n \geq 1$, and that $a = (a_1, \ldots, a_m)$ is a cardinality vector.

We indicate only a few of the relatively simple proofs. They are all based on well-known facts about 0/1-matrices. The fact used most is that, for $0 < k < n$, the 0/1-matrix $M(n; k)$ with n columns and the $\binom{n}{k}$ rows consisting of all 0/1-vectors with k ones and $n - k$ zeros has rank n. In other words, the incidence vectors of the sets in the set system $C(n; k) = \{C \subseteq E \mid |C| = k\}$ (which form the rows of $M(n; k)$) are linearly, and thus affinely, independent. Clearly, if $k = 0$ or $k = n$, there is only one such vector, the zero vector or the all-ones vector. Proving that a certain inequality $c^T x \leq \alpha$ defines a facet of $P(n; a)$ amounts to observing that certain incidence vectors of sets in $C(n; a)$ (with additional properties) satisfy $c^T x \leq \alpha$ with equality and form a set of vectors of affine rank equal to dim $P(n; a)$.

Using the facts mentioned above we can easily determine the dimension of $P(n; a)$.

Proposition 8.19. *Let $E = \{1, \ldots, n\}$ and let $a = (a_1, \ldots, a_m)$ be a cardinality vector.*

(a) *If $m = 1$ and $a_1 = 0$ or $a_1 = n$, then* dim $P(n; a) = 0$.
(b) *If $m = 1$ and $0 < a_1 < n$, then* dim $P(n; a) = n - 1$.
(c) *If $m = 2$ and $a_1 = 0$, $a_2 = n$, then* dim $P(n; a) = 1$.
(d) *In all other cases,* dim $P(n; a) = n$.

The case $m = 1$ is very special and easy to handle.

Proposition 8.20. *Let $m = 1$; i.e., we are only interested in the system of subsets of E with cardinality a_1.*

(a) *If $a_1 = 0$, then $P(n; a_1) = \{x \in \mathbb{R}^n \mid x_1 = x_2 = \ldots = x_n = 0\}$.*
(b) *If $a_1 = n$, then $P(n; a_1) = \{x \in \mathbb{R}^n \mid x_1 = x_2 = \ldots = x_n = 1\}$.*
(c) *If $a_1 = 1$ and $n \geq 2$, then $P(n; a_1) = \{x \in \mathbb{R}^n \mid x(E) = 1,\ x_j \geq 0,\ j = 1, \ldots, n\}$.*
(d) *If $a_1 = n - 1$ and $n \geq 2$, then $P(n; a_1) = \{x \in \mathbb{R}^n \mid x(E) = n - 1,$ $x_j \leq 1,\ j = 1, \ldots, n\}$.*
(e) *If $1 < a_1 < n - 1$ and $n \geq 4$, then $P(n; a_1) = \{x \in \mathbb{R}^n \mid x(E) = a_1,$ $0 \leq x_j \leq 1,\ j = 1, \ldots, n\}$.*

The linear systems above define $P(n; a_1)$ completely and nonredundantly.

Proposition 8.20 provides a complete investigation of the nonredundancy issue for the case $m = 1$. The term *hypersimplex* is often used to name a polytope of type $P(n; a_1)$. In the terminology of this chapter, a hypersimplex is the circuit polytope of some uniform matroid $U_{k,n}$; i.e., Proposition 8.20 covers the circuit polytopes of uniform matroids.

We also refrain from providing all facet proofs in detail because many special cases have to be considered. Let us just, as one example, discuss the nonnegativity constraints thoroughly.

Given $E = \{1, \ldots, n\}$ and a cardinality vector $(a = a_1, \ldots, a_m)$, when does $x_j \geq 0$, $j = 1, \ldots, n$, define a facet of $P(n; a)$? First of all, because of symmetry, we have to consider just one of the indices, say $j = 1$.

If $m = 1$ and $a_1 = 0$ or $a_1 = n$ (see Propositions 8.19(a) and 8.20(a), (b)), then $P(n; a)$ is an affine space and has no facets at all.

Let $m = 1$ and $0 < a_1 < n$. The set of vertices of $P(n; a_1)$ satisfying $x_1 = 0$ is nothing but the set of incidence vectors of $C(n-1; a_1)$ to which a first component with value zero has been added. The matrix $M(n-1; a_1)$ has rank $n-1$ unless $a_1 = n-1$. Adding a first column of zeros to $M(n-1; a_1)$ yields a matrix of affine rank n with one exception. If $a_1 = n-1$, the affine rank is 1 only. Thus we conclude that $x_j \geq 0$ defines a facet of $P(n; a_1)$ if $m = 1$ and $1 \leq a_1 \leq n-2$ but not if $a_1 = n-1$; see Proposition 8.20(c), (d), (e).

Suppose now that $m = 2$. If $a_1 = 0$ and $a_2 = n$ (see Proposition 8.19(c)), then $P(n; a)$ is just the piece of line from the zero vector to the all-ones vector. In this case, all nonnegativity constraints $x_j \geq 0$, $j = 1, \dots, n$, define one and the same facet of $P(n; a)$, which consists of the zero vector only. If $a_1 = 0$ and $a_2 = n-1$, then $x_j \geq 0$ does not define a facet of $P(n; a)$ except when $n = 2$ (and in this case $x_j \geq 0$ appears as a degenerate case of a CF-inequality; see Proposition 8.21(c)). If $a_1 = 0$ and $1 \leq a_2 \leq n-2$, then $x_j \geq 0$ defines a facet of $P(n; a_1, a_2)$.

Because of symmetry all observations about $x_j \geq 0$ can be easily translated into corresponding observations about $x_j \leq 1$.

If $a_1 = 0$, then CF-inequalities exist for all F with $a_1 < |F| < a_2$. A moment's thought reveals that these inequalities are redundant unless $|F| = 1$. In this case the CF-inequality $(a_2 - 1)x_k - \sum_{j \neq k} x_j \leq 0$ defines a facet for all $k \in \{1, \dots, n\}$. This observation immediately translates into an equivalent observation for the case $a_2 = n$.

We summarize the situation for $m = 2$, except for the case $1 \leq a_1 < a_2 \leq n-1$, in the following.

Proposition 8.21. *Suppose* $m = 2$.

(a) *If* $a_1 = 0$ *and* $a_2 = 1$, *then*
$$P(n; 0, 1) = \{x \in \mathbb{R}^n \mid x(E) \leq 1,\ x_j \geq 0,\ j = 1, \dots, n\}.$$

(b) *If* $a_1 = 0$ *and* $1 < a_2 < n-1$, *then*
$$P(n; 0, a_2) = \{x \in \mathbb{R}^n \mid x(E) \leq a_2,\ (a_2 - 1)x_k - \textstyle\sum_{j \neq k} x_j \leq 0,\ k = 1, \dots, n,\ x_j \geq 0,\ j = 1, \dots, n\}.$$

(c) *If* $a_1 = 0$ *and* $a_2 = n-1$, *then*
$$P(n; 0, n-1) = \{x \in \mathbb{R}^n \mid x(E) \leq n-1,\ (n-2)x_k - \textstyle\sum_{j \neq k} x_j \leq 0,\ k = 1, \dots, n\}.$$

(d) *If* $a_1 = 0$ *and* $a_2 = n$, *then*
$$P(n; 0, n) = \{x \in \mathbb{R}^n \mid x_i - x_{i+1} = 0,\ i = 1, \dots, n-1,\ 0 \leq x_1 \leq 1\}.$$

(e) *If* $a_1 = 1$ *and* $a_2 = n$, *then*
$$P(n; 1, n) = \{x \in \mathbb{R}^n \mid x(E) \geq 1, \textstyle\sum_{j \neq k} x_j - (n-2)x_k \leq 1,\ k = 1, \dots, n\}.$$

(f) *If* $1 < a_1 < n-1$ *and* $a_2 = n$, *then*
$$P(n; a_1, n) = \{x \in \mathbb{R}^n \mid x(E) \geq a_1, x_j \leq 1,\ j = 1, \dots, n,\ \textstyle\sum_{j \neq k} x_j - (n-1-a_1)\, x_k \leq a_1,\ k = 1, \dots, n\}.$$

(g) *If* $a_1 = n-1$ *and* $a_2 = n$, *then*
$$P(n; n-1, n) = \{x \in \mathbb{R}^n \mid x(E) \geq n-1,\ x_j \leq 1,\ j = 1, \dots, n\}.$$

All linear systems above are complete and nonredundant.

To finish the discussion of the nonnegativity constraints we observe that, whenever there is an index i such that $0 < a_i < a_{i+1} < n$, then $x_j \geq 0$ (and for symmetry $x_j \leq 1$) defines a facet of $P(n; a)$.

The cardinality constraints are, of course, equations if $m = 1$. They define facets in the following cases.

Proposition 8.22. *Let* $m \geq 2$.

(a) *If* $a_1 \geq 1$, *then* $x(E) \geq a_1$ *defines a facet of* $P(n; a)$.

(b) *If* $a_m \leq n - 1$, *then* $x(E) \leq a_m$ *defines a facet of* $P(n; a)$.

Let us finish the discussion with the CF constraints. We already considered the special cases when $a_1 = 0$ or $a_m = n$. The general case is as follows.

Proposition 8.23. *Let* $m \geq 2$ *and* $1 < a_i < a_{i+1} < n$. *Then for all* $F \subseteq E$ *with* $a_i < |F| < a_{i+1}$ *the corresponding CF-inequality*

$$\mathrm{CF}_F(x) = \sum_{j \in F}(a_{i+1} - |F|)\, x_j - \sum_{j \in E \setminus F}(|F| - a_i)\, x_j \leq (a_{i+1} - |F|)\, a_i = s(F)$$

defines a facet of $P(n; a)$.

The proof of Proposition 8.23 is based on the fact that the incidence vectors of sets in $C(n; a)$ satisfying the CF-inequality are the subsets of F of cardinality a_i and the subsets of E of cardinality a_{i+1} containing F. A simple calculation shows that these incidence vectors form a set of affine (in fact linear) rank n.

With this observation we can finish the discussion of the case $m = 2$.

Proposition 8.24. *Let* $m = 2$ *and* $1 \leq a_1 < a_2 \leq n - 1$. *Then the inequality system defining* $Q(n; a)$ *provides a complete and nonredundant description of* $P(n; a)$.

The remarks above also immediately yield a full characterization of the case $m \geq 3$.

Theorem 8.25. *Let* $E = \{1, \ldots, n\}$, $n \geq 2$, *and let* $a = (a_1, \ldots, a_m)$, $m \geq 3$, *be a cardinality vector. Then the following system of inequalities provides a complete and nonredundant description of* $P(n; a)$.

(a) $x_j \geq 0$ *for all* $j \in E$ *unless* $m = 3$ *and* $a = (0, n - 1, n)$.

(b) $x_j \leq 1$ *for all* $j \in E$ *unless* $m = 3$ *and* $a = (0, 1, n)$.

(c) $x(E) \geq a_1$ *unless* $a_1 = 0$.

(d) $x(E) \leq a_m$ *unless* $a_m = n$.

(e) $\sum_{j \in F}(a_{f+1} - |F|)\, x_j - \sum_{j \in E \setminus F}(|F| - a_f)\, x_j \leq (a_{f+1} - |F|)\, a_f$ *for all* $F \in \mathcal{F}$ *unless* $a_1 = 0$ *and* $2 \leq |F| < a_2$ *or* $a_m = n$ *and* $a_{m-1} < |F| \leq n - 2$.

Summarizing the results above, we can state that the linear system defining $Q(n; a)$ is not only a complete description of $P(n; a)$ but also is nonredundant, with a few exceptions for $m \leq 3$ and whenever $a_1 = 0$ and $a_m = n$.

Theorem 8.25 (and the discussion of the cases $m = 2$ and $m = 3$) yields, for every uniform matroid $U_{k,n}$, a complete and nonredundant description of its cycle polytope and its circuit polytope. As a byproduct we obtain the well-known characterization of the convex hull of all 0/1-vectors with an even or odd number of ones.

A consequence of Theorem 8.25 is that, among the polytopes associated with cardinality homogeneous set systems, the polytope $P(n; 1, n-1)$, which has $2n$ vertices where any pair of vertices is adjacent, has the largest number of facets, namely, 2^n.

Example 8.26. To finish the facet discussion and give another example for the execution of the Dual Greedy Algorithm we consider the uniform matroid $U_{3,9}$. The circuits of $U_{3,9}$ are all subsets of $E = \{1, \dots, 9\}$ of cardinality 4; the cycles of $U_{3,9}$ consist of its circuits together with the empty set and all subsets of E of cardinality 8. In the notation of this chapter the set of cycles of $U_{3,9}$ is the cardinality homogeneous set system $C(9; 0, 4, 8)$. The cycle polytope $P(U_{3,9}) = P(9; 0, 4, 8)$ has $1 + \binom{9}{4} + \binom{9}{8} = 136$ vertices. The system describing the polytope $Q(n; 0, 4, 8)$ has the form

$$0 \le x_j \le 1, \qquad j = 1, \dots, 9,$$

$$0 \le x(E) \le 8,$$

$$\sum_{j \in F} (a_{f+1} - |F|)\, x_j - \sum_{j \in E \setminus F} (|F| - a_f)\, x_j \le (a_{f+1} - |F|)\, a_f$$

$$\text{for all } F \subseteq E \text{ with } |F| \in \{1, 2, 3, 5, 6, 7\}.$$

This system has 395 inequalities. By Theorem 8.25(c) the lower cardinality bound and by (e) the CF-inequalities for $|F| \in \{2, 3\}$ do not define facets. It follows that $P(9; 0, 4, 8)$ has exactly 274 facets.

Let us now maximize the objective function $c^T = (15, 12, 11, 10, 8, 6, -2, -5, -8)$ over $P(9; 0, 4, 8)$. The Primal Greedy Algorithm yields $C_g = \{1, 2, \dots, 8\}$ with $c(C_g) = 55$ and determines $p = 6$, $a_q = a_2 = 4$, $a_{q+1} = a_3 = 8$, and $h = c_5 + \dots + c_8 = 7$. The Complete Dual Greedy Algorithm 8.16 first modifies in step 7 the objective function to $c' = (15, 12, 11, 10, 8, -1, -2, -5, -8)$ so that $h' = c'_5 + \dots + c'_8 = 0$. We have shown in Section 8.5 that we can replace the LP with 274 facet-defining inequalities with the system $L_{\mathcal{F}_0}(9; a; c')$ consisting of 18 upper and lower bounds and only 3 additional CF-inequalities corresponding to the sets $F_k = \{1, \dots, k\}, k \in \{5, 6, 7\}$:

$$\max\ 15x_1 + 12x_2 + 11x_3 + 10x_4 + 8x_5 - x_6 - 2x_7 - 5x_8 - 8x_9$$

$$0 \le x_j \le 1 \quad j = 1, \dots, 9,$$

$$3x_1 + 3x_2 + 3x_3 + 3x_4 + 3x_5 - x_6 - x_7 - x_8 - x_9 \le 12 \ (F_5),$$

$$2x_1 + 2x_2 + 2x_3 + 2x_4 + 2x_5 + 2x_6 - 2x_7 - 2x_8 - 2x_9 \le 8 \ (F_6),$$

$$x_1 + x_2 + x_3 + x_4 + x_5 + x_6 + x_7 - 3x_8 - 3x_9 \le 4 \ (F_7).$$

The dual LP $D_{\mathcal{F}_0}(9; a; c')$ has the following form (where $y_k = y_{F_k}$):

$$\min u_1 + \ldots + u_9 + 12y_5 + 8y_6 + 4y_7$$

$$\begin{aligned}
u_1 + 3y_5 + 2y_6 + y_7 &\geq 15,\\
u_2 + 3y_5 + 2y_6 + y_7 &\geq 12,\\
u_3 + 3y_5 + 2y_6 + y_7 &\geq 11,\\
u_4 + 3y_5 + 2y_6 + y_7 &\geq 10,\\
u_5 + 3y_5 + 2y_6 + y_7 &\geq 8,\\
u_6 - y_5 + 2y_6 + y_7 &\geq -1,\\
u_7 - y_5 - 2y_6 + y_7 &\geq -2,\\
u_8 - y_5 - 2y_6 - 3y_7 &\geq -5,\\
u_9 - y_5 - 2y_6 - 3y_7 &\geq -8,\\
u_1, \ldots, u_9, y_5, y_6, y_7 &\geq 0.
\end{aligned}$$

The Dual Greedy Algorithm 8.9, which is step 8 of Algorithm 8.16, yields the following c'-optimal solution:

$$\begin{aligned}
y_5^* &:= \frac{1}{4}(c_5' - c_6') = \frac{9}{4},\\
y_6^* &:= \frac{1}{4}(c_6' - c_7') = \frac{1}{4},\\
y_7^* &:= \frac{1}{4}(c_7' - c_8') = \frac{3}{4},\\
u_1' &:= 7, u_2' = 4, u_3' = 3, u_4' = 2, u_5' = \ldots = u_9' = 0,
\end{aligned}$$

of $D_{\mathcal{F}_0}(9; a; c')$. To turn this solution u', y' into a solution u^*, y^* of $D_{\mathcal{F}_0}(9; a; c)$ we have to modify the values u_j' belonging to indices j where the objective function c was changed. In our case we only have to modify $u_6' = 0$ to $u_6^* = u_6' + c_j - c_j' = 7$ (this is step 10 of Algorithm 8.16), i.e., $y^* := y'$, $u_j^* := u_j'$ with the exception that $u_6^* = 7$ is an optimal solution of the dual LP $D_{\mathcal{F}_0}(9; a; c)$ and, as we have shown, also of the LP $D_{\mathcal{F}_0}(9; a; c)$. The value of this solution is $55 = c(I_g)$.

The optimum solution of $D_{\mathcal{F}_0}(9; a; c)$ is by no means unique. The face of the optimal solution of this LP has, in fact, 10 vertices.

8.7 Separation

Since we can optimize over $P(n; a)$ in polynomial time we can also solve the separation problem for $P(n; a)$ in polynomial time by the general results described in [4]. There is, however, a much simpler separation algorithm.

Let a vector $y \in Q^n$ be given. It is, of course, trivial to check the bounds $0 \leq x_j \leq 1$, $j = 1, \ldots, n$, and the cardinality constraints $a_1 \leq x(E) \leq a_m$ by substituting y into these inequalities. We may, thus, assume that y satisfies them.

Suppose now that y violates, for some $F \in \mathcal{F}$ of cardinality k, the corresponding CF-inequality, i.e.,

$$\sum_{j \in F}(a_{f+1} - k)\, y_j - \sum_{j \in E \setminus F} (k - a_f)\, y_j > (a_{f+1} - k)a_f.$$

Let F^* be a set in $\mathcal{F}$ of cardinality k such that $\sum_{j \in F^*} y_j$ is maximum. Then, clearly, y violates the corresponding CF-inequality as well. In fact, the CF-inequality associated with F^* is a "most violated" inequality among all CF-inequalities coming from sets in $\mathcal{F}$ of cardinality k. Finding such a set F^* is easy. We sort the components of y such that $y_1 \geq y_2 \geq \ldots \geq y_n$. We set $F^* := \{1, \ldots, k\}$. Then y satisfies the CF-inequality associated with F^* if and only if y satisfies all CF-inequalities associated with sets in $\mathcal{F}$ of cardinality k. This observation gives the following very simple polynomial-time separation algorithm for $P(n; a)$, which, in its major step, can also be viewed as a greedy algorithm.

Algorithm 8.27 (Greedy Separation Algorithm for $P(n; a)$). Let $E = \{1, \ldots, n\}$, a cardinality vector $a = (a_1, \ldots, a_n)$, and a vector $y \in Q^n$ be given.

1. If y has a component smaller than zero or larger than one, report that a bound is violated by y and stop.
2. If $y(E) < a_1$ or $y(E) > a_m$, report that a cardinality constraint is violated by y and stop.
3. Sort the components of y such that $y_1 \geq y_2 \geq \cdots \geq y_n$.
4. For $k = a_1 + 1$ to $a_m - 1$ and $k \neq a_i$, $i = 2, \ldots, m - 1$ do

   If $\sum_{j=1}^{k}(a_{f+1} - k)\, y_j - \sum_{j \in E \setminus F} (k - a_f)\, y_j > (a_{f+1} - k)\, a_f$, where $f = f(\{1, \ldots, k\})$,

   then output that y violates the CF-inequality corresponding to $\{1, \ldots, k\}$.

If the greedy separation algorithm produces no violated inequality, then y is in $P(n; a)$.

Bibliography

[1] F. Barahona. The max-cut problem in graphs not contractible to K_5. *Operations Research Letters*, 2:107–111, 1983.

[2] F. Barahona and M. Grötschel. On the Cycle Polytope of a Binary Matroid. *Journal of Combinatorial Theory. Series* B, 40:40–62, 1986.

[3] J. Edmonds and E.L. Johnson. Matching, Euler tours, and the Chinese postman. *Mathematical Programming*, 5:88–124, 1973.

[4] M. Grötschel, L. Lovász, and A. Schrijver. *Geometric Algorithms and Combinatorial Optimization*. Springer-Verlag, Berlin, second corrected edition, 1993.

[5] M. Grötschel and K. Truemper. Decomposition and optimization over cycles in binary matroids. *Journal of Combinatorial Theory. Series* B, 46:306–337, 1989.

[6] J. G. Oxley. *Matroid Theory*. Oxford University Press, Oxford, U.K., 1992.

[7] M. Padberg. The Boolean quadratic polytope: Some characteristics, facets and relatives. *Mathematical Programming. Series* B, 45:139–172, 1989.

[8] PORTA. http://www.zib.de/Optimization/Software/Porta/.

[9] P.D. Seymour. Matroids and multicommodity flows. *European Journal of Combinatorics*, 2:257–290, 1981.

[10] W.T. Tutte. Lectures on matroids. *Journal of Research of the National Bureau of Standands*, 69B:49–53, 1965.

[11] D.J.A. Welsh. *Matroid Theory*. Academic Press, London, 1976.

Chapter 9

(1, 2)-Survivable Networks: Facets and Branch-and-Cut

Hervé Kerivin,* Ali Ridha Mahjoub,† and Charles Nocq‡

Dedicated to Manfred Padberg
on the occasion of his 60th birthday.

Abstract. Given a graph $G = (V, E)$ with edge weights and an integer vector $r \in \mathbb{Z}_+^V$ associated with the nodes of V, the survivable network design problem is to find a minimum weighted subgraph of G such that between every pair of nodes s, t of V there are at least $\min\{r(s), r(t)\}$ edge-disjoint paths. In this chapter we consider that problem when $r \in \{1, 2\}^V$. This case is of particular interest to the telecommunication industry. We first consider the case when $r(v) = 2$ for all $v \in V$. We describe sufficient conditions for the so-called F-partition inequalities to define facets for the associated polytope. As a consequence, we show that the critical extreme points of the linear relaxation of that polytope may be separated in polynomial time using F-partition facets.

Next we consider the case where $r \in \{1, 2\}^V$. We first describe valid inequalities that generalize the F-partition inequalities. We discuss separation algorithms for these inequalities as well as for the so-called partition inequalities. Finally, we introduce a branch-and-cut algorithm based on these results and present some computational results. These show that the F-partition inequalities are very effective for the 2-connected subgraph problems.

*Institute of Mathematics and Its Applications, University of Minnesota, 357 Lind Hall, 207 Church Street S.E., Minneapolis, Minnesota 55455 (kerivin@ima.umn.edu).

†LIMOS, CNRS, Université de Clermont II, Complexe Scientifique des Cézeaux, 63177 Aubière Cedex, France (Ridha.Mahjoub@math.univ-bpclermont.fr).

‡176 avenue Adolphe Buyl, 1050 Brussels, Belgium (Charles_Nocq@hotmail.com). Currently at KPMG-consultants, Department of Planning and Simulation, Brussels, Belgium.

MSC 2000. 90C57, 90C10, 68M10

Key words. Survivable network, polytope, 2-edge connected subgraph, critical extreme point, cut, separation problem, facet, branch-and-cut

9.1 Introduction

The introduction of fiber optic technology in telecommunications has increased the need for designing survivable networks. Survivable networks must satisfy some connectivity requirements, that is, networks that are still functional after the failure of certain links. More precisely, we are given a graph $G = (V, E)$, where each edge $e \in E$ has a cost $c(e)$. For each node $v \in V$ there is a nonnegative integer $r(v)$, called the *connectivity type*, that represents the importance of communication from and to node v. The *survivable conditions* require that between every pair of nodes (s, t) there are at least

$$\min\{r(s), r(t)\}$$

edge-disjoint paths. The *survivable network design problem* (SNDP) is to determine a subgraph of G that minimizes the total cost subject to the survivable conditions.

SNDP is NP-hard in general. It includes as special cases a number of well-known NP-hard combinatorial optimization problems, such as the Steiner tree problem ($r(v) \in \{0, 1\}$ for all $v \in V$).

SNDP has been shown to be polynomially solvable in some particular cases. For instance, if $r(v) = 1$ for all $v \in V$, SNDP is nothing but the minimum spanning tree problem, which is well known to be polynomially solvable. For a complete survey of SNDP, see Grötschel, Monma, and Stoer [20] and Stoer [36].

In fiber optic networks, nodes are generally of connectivity type one or two and are called ordinary and special offices, respectively. This topology has proved to be cost effective and provides an adequate level of survivability [21, 31]. In this chapter we consider SNDP in such a case (i.e., $r(v) \in \{1, 2\}$ for all $v \in V$) and we write (1, 2)-SNDP.

Given a graph $G = (V, E)$ and an edge subset $F \subseteq E$ of G, the 0/1-vector x^F of $\mathbb{R}^E$ such that $x^F(e) = 1$ if $e \in F$ and $x^F(e) = 0$ otherwise is called the *incidence vector* of F.

Given $b : E \to \mathbb{R}$ and F a subset of E, $b(F)$ denotes $\sum_{e \in F} b(e)$. If $W \subset V$ is a node subset of G, then the set of edges that have only one node in W is called a *cut* and denoted by $\delta_G(W)$. If the context prevents any ambiguity, then we usually omit the subscript and simply write $\delta(W)$. If $W = \{v\}$, where $v \in V$, then we write $\delta(v)$ for $\delta(W)$.

For $W \subset V$ let $r(W) = \max\{r(v) \mid v \in W\}$ and $\text{con}(W) = \min\{r(W), r(V \setminus W)\}$.

If $G = (V, E)$ is a graph and (V, F) is a survivable subgraph of G, then x^F satisfies the following inequalities:

$$x(e) \geq 0 \quad \text{for all } e \in E, \tag{9.1}$$

$$x(e) \leq 1 \quad \text{for all } e \in E, \tag{9.2}$$

$$x(\delta(W)) \geq \text{con}(W) \quad \text{for all } W \subset V,\ W \neq \emptyset. \tag{9.3}$$

Inequalities (9.1) and (9.2) are called *trivial inequalities* and inequalities (9.3) are called *cut inequalities*.

A graph is called *k-edge* (resp. *k-node*) *connected* if for every pair of nodes s, t there are at least k edge-disjoint (resp. node-disjoint) (s, t)-paths. The *2-edge* (resp. *2-node*) *connected subgraph problem* TECSP (resp. TNCSP) is to find a 2-edge (resp. 2-node) connected spanning subgraph of minimum weight. TECSP corresponds to SNDP with $r(v) = 2$ for all $v \in V$. Hence inequalities (9.3) are valid for both TECSP and TNCSP and can, for these problems, be written as

$$x(\delta(W)) \geq 2 \qquad \text{for all } W \subset V, W \neq \emptyset. \tag{9.4}$$

Given a graph $G = (V, E)$, we will denote by TECSP(G) (resp. TNCSP(G)) the polytope whose extreme points are the solutions of TECSP (resp. TNCSP).

Let $P(G)$ be the polytope defined by inequalities (9.1), (9.2), and (9.4). In [13] Fonlupt and Mahjoub introduced the concept of critical extreme points of $P(G)$. They described necessary conditions for a fractional extreme point of $P(G)$ to be critical. As a consequence, they obtained a characterization of the so-called perfectly 2-edge connected graphs [28], the graphs for which $P(G)$ is integral. In this chapter we first discuss that concept. We then describe sufficient conditions for the so-called F-partition inequalities to define facets for TECSP(G). As a consequence, we show that the critical extreme points may be separated in polynomial time from TECSP(G) using F-partition facets. We also provide separation techniques. Finally, we describe a branch-and-cut algorithm for the (1, 2)-SNDP and present computational results on problems from the traveling salesman problem (TSP)-library.

SNDP has been extensively investigated in the past. Steiglitz, Weiner, and Kleitman [35] proposed a heuristic for SNDP based on local search. Monma and Shallcross [31] devised heuristics to design survivable networks with node connectivity types $r \in \{1, 2\}^V$. They used these heuristics to obtain near-optimal solutions to both real-world and randomly generated problems. Ko and Monma [26] extended these heuristics to the design of k-edge and k-node connected networks. Grötschel, Monma, and Stoer [19, 18, 21] studied a polyhedral approach to SNDP. In [19] they derived valid and facet-defining inequalities for the associated polytope. In [18, 21] they devised cutting plane algorithms and presented some experimental results. Goemans and Bertsimas [14] devised a heuristic with worst-case guarantee for SNDP when the use of multiple copies of an edge is allowed.

Much work has been done on TECSP and TNCSP. In [30] Monma, Munson, and Pulleyblank studied TECSP (resp. TNCSP) in the metric case, where the underlying graph is complete and the weight function $c(\cdot)$ satisfies the triangle inequalities (i.e., $c(e_1) \leq c(e_2) + c(e_3)$ for every three edges e_1, e_2, e_3 defining a triangle). Even in this case TECSP (resp. TNCSP) is NP-hard. They showed in this case that $\tau \leq \frac{4}{3}Q_2$, where τ is the weight of an optimal traveling salesman tour and Q_k is the weight of an optimal k-edge connected spanning subgraph of G for k fixed. This implies that $\tau' \leq \frac{4}{3}Q_2$, where τ' is the value of an optimal solution of the linear relaxation of the TSP. Cunningham (see [30]) strengthened this by showing that $\tau' \leq Q_2$. In [14] Goemans and Bertsimas extended this result to k-edge connected subgraphs by showing that $\tau' \leq \frac{2}{k}Q_k$ for every k.

The *subtour polytope* of the TSP is the set of all the solutions of the system given by inequalities (9.1), (9.2), and (9.4) together with the equations $x(\delta(v)) = 2$ for all $v \in V$. Clearly, the polytope $P(G)$ is a relaxation of both the 2-edge connected subgraph polytope and the subtour polytope. Thus minimizing cx over the polytope $P(G)$ may provide a good lower bound for both TECSP and the TSP.

In [9] Cornuéjols, Fonlupt, and Naddef studied TECSP when multiple copies of an edge may be used. They showed that, when the graph is series-parallel, the associated polytope is completely described by inequalities (9.1) and (9.4). In [8] Chopra studied that problem on directed graphs. He showed how facets of the associated polyhedron on undirected graphs can be obtained by projection. He also devised a cutting plane algorithm. Boyd and Hao [7] described a class of "comb inequalities" that are valid for TECSP(G). This class is a special case of a more general class of inequalities given by Grötschel, Monma, and Stoer [19] for the survivable network polytope. They gave necessary and sufficient conditions for these inequalities to define facets for TECSP(G) when the graph is complete. In [5] Barahona and Mahjoub characterized the polytopes TECSP(G) and TNCSP(G) for the class of Halin graphs. Baïou and Mahjoub [3] characterized the Steiner 2-edge connected subgraph polytope for series-parallel graphs. In [10, 11] Coullard et al. studied the Steiner TNCSP. In [10] they devised a linear-time algorithm for this problem on special classes of graphs. In [11] they characterized the dominant of the polytope associated with this problem on the graphs that do not have W_4 (the wheel on five nodes) as a minor.

This chapter is organized as follows. In the following section we discuss the critical extreme points of $P(G)$. In Section 9.3 we give sufficient conditions for the F-partition inequalities to be facet-defining for the TECSP(G). In Section 9.4 we discuss separation techniques and describe a branch-and-cut algorithm for the (1, 2)-SNDP and TNCSP. Our computational results are presented in Section 9.5 and finally some concluding remarks are given in Section 9.6. The rest of this section is devoted to more definitions and notations.

The graphs we consider are finite, undirected, loopless, and connected and may have parallel edges. We denote a graph by $G = (V, E)$, where V is the *nodeset* and E is the *edgeset*. If $G = (V, E)$ is a graph and $e \in E$ is the unique edge between two nodes i and j, we also write ij to denote e. For $W, W' \subseteq V$ with $W \cap W' = \emptyset$, (W, W') denotes the set of edges with one endnode in W and the other in W'. For $F \subseteq E$, $V(F)$ denotes the set of nodes of the edges of F. For $W \subseteq V$, we let $\overline{W} = V \setminus W$. We denote by $E(W)$ the set of edges having both endnodes in W and by $G(W)$ the subgraph $(W, E(W))$. $G(W)$ is called the subgraph *induced* by W. For $e \in E$, $G \setminus e$ denotes the graph obtained by deleting e. For $v \in V$, we denote by $G \setminus v$ the graph obtained by removing v and the edges incident to it.

An *edge cutset* $F \subseteq E$ of G is a set of edges such that $F = \delta(S) = \delta(V \setminus S)$ for some nonempty set $S \subset V$. We write *k-edge cutset* for an edge cutset having k edges.

9.2 Critical Extreme Points

In [13] Fonlupt and Mahjoub introduced the concept of critical extreme points of the polytope $P(G)$. In this section we discuss these extreme points.

Let $\overline{x}$ be a noninteger extreme point of $P(G)$. Let $\overline{x}'$ be a solution obtained by replacing some (but at least one) noninteger components of $\overline{x}$ with 0 or 1 (and keeping all the other components of $\overline{x}$ unchanged). If $\overline{x}'$ is a point of $P(G)$, then $\overline{x}'$ can be written as a convex combination of extreme points of $P(G)$. If $\overline{y}$ is such an extreme point, then $\overline{y}$ is said to be *dominated* by $\overline{x}$, and we write $\overline{x} \succ \overline{y}$. Note that an extreme point of $P(G)$ may dominate more than one extreme point of $P(G)$. Also note that, if $\overline{x}$ dominates $\overline{y}$, then $\{e \in E \mid 0 < \overline{y}(e) < 1\} \subset \{e \in E \mid 0 < \overline{x}(e) < 1\}$, $\{e \in E \mid \overline{x}(e) = 0\} \subseteq \{e \in E \mid \overline{y}(e) = 0\}$, and $\{e \in E \mid \overline{x}(e) = 1\} \subseteq \{e \in E \mid \overline{y}(e) = 1\}$. The relation $\succ$ defines a partial ordering on the extreme points of $P(G)$. The minimal elements of this ordering (i.e., the extreme points x for which there is no extreme point y such that $x \succ y$) correspond to the integer extreme points of $P(G)$. The minimal extreme points of $P(G)$ are called extreme points of *rank* 0. An extreme point x of $P(G)$ is said to be of *rank* k, for fixed k, if x dominates only extreme points of rank not greater than $k-1$ and if it dominates at least one extreme point of rank $k-1$. We notice that, if $\overline{x}$ is an extreme point of $P(G)$ of rank 1 and if we replace one fractional component of $\overline{x}$ with 1, keeping the other components unchanged, we obtain a feasible point $\overline{x}'$ of $P(G)$ that can be written as a convex combination of integer extreme points of $P(G)$.

Note that the extreme points of $P(G)$ may have rank at most $|V|$.

Fonlupt and Mahjoub [13] introduced the following reduction operations with respect to a solution $\overline{x}$ of $P(G)$:

θ_1: Delete an edge e with $\overline{x}(e) = 0$.

θ_2: Contract an edge e if one of its endnodes is of degree 2.

θ_3: Contract a node subset W such that $G(W)$ is 2-edge connected and $\overline{x}(e) = 1$ for all $e \in E(W)$.

Starting from a graph G and a point $\overline{x}$ of $P(G)$ and applying operations $\theta_1, \theta_2, \theta_3$, we obtain a reduced graph G' and a solution $\overline{x}' \in P(G')$. It is not hard to see that $\overline{x}$ is an extreme point of $P(G)$ if and only if $\overline{x}'$ is an extreme point of $P(G')$. Moreover, we have the following lemma.

Lemma 9.1 (Fonlupt and Mahjoub [13]). *$\overline{x}$ is an extreme point of $P(G)$ of rank 1 if and only if $\overline{x}'$ is an extreme point of $P(G')$ of rank 1.*

An extreme point of $P(G)$ is said to be *critical* [13] if it is of rank 1 and if none of the operations $\theta_1, \theta_2, \theta_3$ can be applied to it. By Lemma 9.1 the characterization of the extreme points of rank 1 reduces to those of the critical extreme points of $P(G)$. In [13] Mahjoub and Fonlupt gave the following necessary conditions for a fractional extreme point of $P(G)$ to be critical.

Theorem 9.2 (Fonlupt and Mahjoub [13]). *Let $G = (V, E)$ be a 2-edge connected graph and $\overline{x}$ a fractional extreme point of $P(G)$. If $\overline{x}$ is a critical extreme point of $P(G)$, then the following hold.*

(i) $V = V^1 \cup V^2$ *with* $V^1 \cap V^2 = \emptyset$.

$E = E^1 \cup E^2$ *with* $E^1 \cap E^2 = \emptyset$.

(V^1, E^1) *is an odd cycle.*

$(V^1 \cup V^2, E^2)$ *is a forest whose set of leaves is V^1 such that all the nodes in V^1 have degree 3.*

(ii) $\overline{x}(e) = \frac{1}{2}$ *for $e \in E^1$ and* $\overline{x}(e) = 1$ *for all* $e \in E^2$.

(iii) $\overline{x}(\delta(W)) > 2$ *for all cut $\delta(W)$ such that* $|W| \geq 2$ *and* $|\overline{W}| \geq 2$.

Remark 9.3. By (ii) and (iii) of Theorem 9.2, if G supports a critical extreme point, then G is 3-edge connected and $|\delta(S)| \geq 4$ for every cut $\delta(S)$ such that $|S| \geq 2$ and $|\overline{S}| \geq 2$.

Let $G = (V, E)$ be a graph and $\overline{x} \in \mathbb{R}^E$ a critical extreme point of $P(G)$. We may then suppose that G and $\overline{x}$ satisfy properties (i), (ii), and (iii) of Theorem 9.2. The following result has been obtained after several discussions with J. Fonlupt.

Lemma 9.4. $|\delta(S) \cap E^2| \geq 2$ *for every cut* $\delta(S)$ *such that* $|S| \geq 2$, $|\overline{S}| \geq 2$, *and* $G(S)$ *and* $G(\overline{S})$ *are both 2-edge connected.*

Proof. Assume the contrary. Then $\delta(S) \cap E^1 \neq \emptyset$. Let $E^1 = \{f_1, \dots, f_{2k+1}\}$. Let $\delta(S) \cap E^1 = \{f_{i_1}, f_{i_2}, \dots, f_{i_s}\}$ with $i_1 < i_2 < \cdots < i_s$. Since $|E^1|$ is odd, we may, without loss of generality, assume that $i_2 - i_1$ is even and $\{f_{i_1+1}, \dots, f_{i_2-1}\} \subset E(S)$.

Suppose that $|\delta(S)|$ is minimum, that is, if there is $S' \subset V$ with $|\delta(S')| < |\delta(S)|$ such that $|S'| \geq 2$, $|\overline{S}'| \geq 2$, and $G(S')$ and $G(\overline{S}')$ are both 2-edge connected, then $|\delta(S') \cap E^2| \geq 2$.

If $\delta(S) \cap E^2 = \emptyset$ (resp. $|\delta(S) \cap E^2| = 1$), then from Theorem 9.2(ii) it follows that $|\delta(S) \cap E^1| \geq 4$ (resp. $|\delta(S) \cap E^1| \geq 2$). Let $\overline{x}' \in \mathbb{R}^E$ be the solution such that

$$\overline{x}'(e) = \begin{cases} \frac{1}{2} & \text{if } e \in \{f_{i_1}, f_{i_1+1}, \dots, f_{i_2}\}, \\ 0 & \text{if } e \in \{f_{i_4}, f_{i_5}, \dots, f_{i_s}\}, \\ 1 & \text{otherwise} \end{cases}$$

(resp.

$$\overline{x}'(e) = \begin{cases} \frac{1}{2} & \text{if } e \in \{f_{i_1}, f_{i_1+1}, \dots, f_{i_2}\}, \\ 0 & \text{if } e \in \{f_{i_3}, f_{i_4}, \dots, f_{i_s}\}, \\ 1 & \text{otherwise}). \end{cases}$$

In what follows we shall show that $\overline{x}'$ is an extreme point of $P(G)$. We show this when $\delta(S) \cap E^2 = \emptyset$; the proof when $|\delta(S) \cap E^2| = 1$ is along the same lines. For this we first show the following claim.

Claim 1. $\overline{x}' \in P(G)$.

Proof. First observe that $\bar{x}'(f_{i_3}) = 1$. Let $\delta(W)$ be a cut of G. If $W = S$, then clearly $\overline{x}'(\delta(W)) = 2$. If $\overline{S} \neq W \cap \overline{S} \neq \emptyset$, as $\overline{x}'(e) = 1$ for all $e \in E(\overline{S})$ and $G(\overline{S})$ is 2-edge connected, it follows that $\overline{x}'(\delta(W)) \geq 2$. Thus we may suppose that $\overline{S} \subset W$. In consequence, $\delta(W)$ is an edge cutset of the graph G' obtained by contracting $\overline{S}$. Let $C' = \{f_{i_1}, f_{i_1+1}, \dots, f_{i_2}\}$. Note that C' is a cycle of G'. Also note that, as $i_2 - i_1$ is even, C' is odd. If $\delta(W) \cap C' = \emptyset$, as $\overline{x}'(e) = 1$ for all $e \in E(S) \setminus C'$ and $G(S)$ is 2-edge connected, we have that $\overline{x}'(\delta(W)) \geq 2$. Now suppose that $\delta(W) \cap C' \neq \emptyset$. As C' is a cycle, $|\delta(W) \cap C'|$ is even. If $|\delta(W) \cap C'| \geq 4$, as $\overline{x}'(e) = \frac{1}{2}$ for all $e \in C'$, we have that $\overline{x}'(\delta(W)) \geq 2$. In what follows we assume that $|\delta(W) \cap C'| = 2$. We consider two cases.

Case 1. $\{f_{i_1}, f_{i_2}\} \cap \delta(W) \neq \emptyset$.

If $\{f_{i_1}, f_{i_2}\} \subset \delta(W)$, then $\delta(W)$ intersects $E(S) \setminus C'$. Since $\overline{x}'(e) = 1$ for all $e \in E(S) \setminus C'$, we obtain that $\overline{x}'(\delta(W)) \geq 2$.

If only one edge among $\{f_{i_1}, f_{i_2}\}$ is in $\delta(W)$, then $\delta(W)$ contains exactly one edge from $C' \setminus \{f_{i_1}, f_{i_2}\}$. As $G(S)$ is 2-edge connected, $\delta(W)$ must contain at least one more edge, say g, from $E(S) \setminus C'$. As $\overline{x}'(g) = 1$, we have $\overline{x}'(\delta(W)) \geq 2$.

Case 2. $\{f_{i_1}, f_{i_2}\} \cap \delta(W) = \emptyset$.

Thus $\delta(W)$ contains exactly two edges, say g_1 and g_2, from $C' \setminus \{f_{i_1}, f_{i_2}\}$. Let $S_1 = S \cap W$ and $S_2 = S \setminus S_1$. Note that $S_2 = \overline{W}$. Also note that $E(S_2) \cap C'$ is a path. If either $f_{i_3} \in \delta(W)$ or $\delta(W) \cap (E(S) \setminus C') \neq \emptyset$, then clearly $\overline{x}'(\delta(W)) \geq 2$. Now suppose that $f_{i_3} \notin \delta(W)$ and $\delta(W) \cap (E(S) \setminus C') = \emptyset$. Hence $f_{i_3} \in E(W)$ and $\delta(W) \cap E(S) = (S_1, S_2) = \{g_1, g_2\}$. Since $G(S)$ is 2-edge connected, as a consequence, we have that $G(S_1)$ and $G(S_2)$ are both connected. Let L be the path given by the edges of $E(S_2) \cap C'$, and let w_1 and w_2 be the endnodes of L (w_1 and w_2 may be identical). We may assume, without loss of generality, that w_1 and w_2 are incident to g_1 and g_2, respectively. Furthermore, since $\overline{S} \subset W$, it follows that $f_{i_1}, f_{i_2} \in E(W)$ (see Figure 9.1; the edges between $\overline{S}$ and S_2 are omitted).

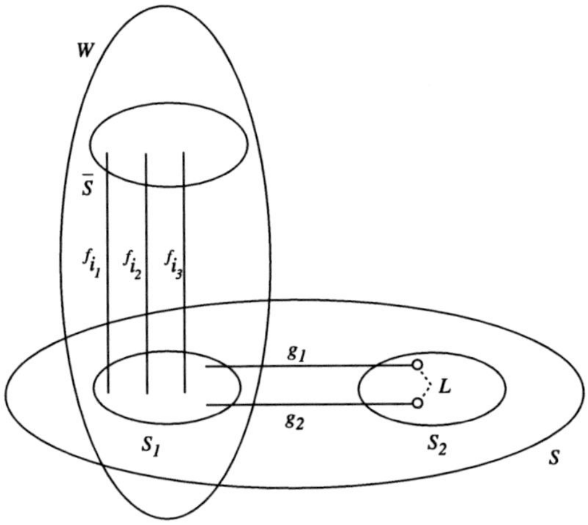

Figure 9.1. *The case where* $\{f_{i_1}, f_{i_2}\} \cap \delta(W) = \emptyset$.

We claim that $G(W)$ is 2-edge connected. Indeed, if $G(S_1)$ is 2-edge connected, as $f_{i_1}, f_{i_2} \in (\overline{S}, S_1)$ and $G(\overline{S})$ is 2-edge connected, we have that $G(W)$ is also 2-edge connected. If this is not the case, as $G(S_1)$ is connected, by contracting the 2-edge connected components of $G(S_1)$, we get a tree, say $\mathcal{T}$. As $|(S_1, S_2)| = 2$ and $G(S)$ is 2-edge connected, $\mathcal{T}$ must be a path. Let S_1^1 and S_1^2 be the leaves of $\mathcal{T}$. As, by Remark 9.3, G is 3-edge connected, both S_1^1 and S_1^2 must be linked by edges to $\overline{S}$. (Recall that $(S_1, S_2) = \{g_1, g_2\}$.) As $G(\overline{S})$ is 2-edge connected, $G(W)$ is also 2-edge connected.

Now observe that $\delta(W) = (\overline{S}, S_2) \cup \{g_1, g_2\}$. Since $(\overline{S}, S_2) \subset \delta(S)$ and $f_{i_1}, f_{i_2}, f_{i_3} \in \delta(S) \cap E(W)$, it follows that $|\delta(W)| < |\delta(S)|$. If $G(S_2)$ is 2-edge connected, since $G(W)$

is 2-edge connected, $|W| \geq 2$, and $|\overline{W}| = |S_2| \geq 2$, by the minimality hypothesis, it follows that $|\delta(W) \cap E^2| \geq 2$, a contradiction. Thus $G(S_2)$ is not 2-edge connected. Now, since $G(S_2)$ is connected, there is a partition $S_2^1, \ldots, S_2^t$ of S_2 such that $G(S_2^j)$ is 2-edge connected for $j = 1, \ldots, t$, and the graph obtained from $G(S_2)$ by contracting the sets S_2^j, $j = 1, \ldots, t$, is a tree, say $\mathcal{H}$. As $|(S_1, S_2)| = 2$ and $G(S)$ is 2-edge connected, $\mathcal{H}$ is a path. Let us suppose, without loss of generality, that S_2^1 and S_2^t are the leaves of $\mathcal{H}$. As $G(S)$ is 2-edge connected, we may also suppose, for instance, that g_1 is incident to S_2^1 and g_2 to S_2^t. Hence $w_1 \in S_2^1$ and $w_2 \in S_2^t$. Since w_1 and w_2 are joined by L in $G(S_2)$, it follows that all the edges of $\mathcal{H}$ are among the edges of L. Moreover, as, by Remark 9.3, G is 3-edge connected, S_2^t (resp. S_2^j, $j = 2, \ldots, t-1$) must be linked to W by at least two (resp. one) edges. As $G(W)$ is 2-edge connected, it thus follows that the graph induced by $W' = W \cup (\bigcup_{2 \leq j \leq t} S_2^j)$ is also 2-edge connected. Now we claim that $|S_2^1| \geq 2$. In fact, if $S_2^1 = \{w_1\}$, as w_1 is a node of $V(C')$, there is an edge, say f, of E^2, which is incident to w_1. But this implies that $f \in \delta(W)$, a contradiction. Thus $|S_2^1| \geq 2$. Since $\overline{W}' = S_2^1$, $|W'| \geq 2$, $G(\overline{W}')$ and $G(S_2^1)$ are both 2-edge connected, and $|\delta(W')| < |\delta(S)|$, this contradicts the minimality hypothesis, and our claim is proved.

In consequence, $\overline{x}' \in P(G)$. Moreover, $\overline{x}'$ is the unique solution of the system

$$\begin{cases} x(e) = 0 & \text{for all } e \in F_0, \\ x(e) = 1 & \text{for all } e \in F_1, \\ x(\delta(v)) = 2 & \text{for all } v \in V(C') \setminus \overline{S}, \\ x(\delta(S)) = 2, \end{cases}$$

where F_0 (resp. F_1) is the set of edges $e \in E$ with $\overline{x}'(e) = 0$ (resp. $\overline{x}'(e) = 1$). This implies that $\overline{x}'$ is an extreme point of $P(G)$.

Since $\overline{x}$ dominates $\overline{x}'$ and $\overline{x}'$ is fractional, this contradicts the fact that $\overline{x}$ is a critical extreme point. □

The concept of critical extreme points has also been studied by Mahjoub and Nocq [29] for TNCSP.

The following inequalities are valid for TNCSP(G):

$$x(\delta_{G\setminus v}(W)) \geq 1 \qquad \text{for all } v \in V,\ W \subset V \setminus \{v\},\ W \neq \emptyset. \tag{9.5}$$

These inequalities are a special case of the following more general valid inequalities for TNCSP(G):

$$x(\delta_{G\setminus v}(V_1, \ldots, V_p)) \geq p - 1 \text{ for all partitions } (V_1, \ldots, V_p) \text{ of } V \setminus \{v\}, \tag{9.6}$$

where $\delta_{G\setminus v}(V_1, \ldots, V_p)$ denotes the edgeset of $G \setminus v$ having nodes in different members of the partition. Inequalities (9.6) are called *node-partition inequalities*. Grötschel and Monma [17] gave necessary and sufficient conditions for inequalities (9.6) to be facet-defining.

In [29] Mahjoub and Nocq studied the polytope $Q(G)$ given by inequalities (9.1), (9.2), (9.4), and (9.5). Note that this polytope is the linear relaxation of TNCSP(G). They extended the concept of extreme points of rank 1 and critical extreme points to $Q(G)$ and gave necessary and sufficient conditions for an extreme point of $Q(G)$ to be critical.

In particular, they introduced the following operations defined with respect to a point $\overline{x}$ of $Q(G)$:

θ_1': Replace a set of parallel edges with only one edge.

θ_2': Contract $W \subset V$ such that $\overline{x}(e) = 1$ for all $e \in E(W)$ and $|\delta(W)| \leq 3$.

They then proved the following.

Lemma 9.5 (Mahjoub and Nocq [29]). *Let $\overline{x}$ be an extreme point of $Q(G)$ and $\bar{x}'$ and G' the solution and the graph obtained from $\overline{x}$ and G by repeated applications of the operations θ_1, θ_2, θ_3, θ_1', and θ_2'. Then $\overline{x}$ is an extreme point of $Q(G)$ of rank 1 if and only if $\bar{x}'$ is an extreme point of $Q(G')$ of rank 1.*

Operations $\theta_1, \theta_2, \theta_3$ (resp. θ_1, θ_2, θ_3, θ_1', θ_2') can be used in a preprocessing phase of a cutting plane algorithm for the TECSP (resp. TNCSP). As it will turn out, they are very effective for solving these problems as well as the more general (1, 2)-SNDP. In fact, they may considerably reduce the size of the graph supporting the fractional solution and then accelerate the separation process. This aspect will be discussed in Sections 9.4 and 9.5.

9.3 Facets of TECSP(G)

In this section we shall address some polyhedral consequences of the properties of the critical extreme points discussed in the previous section.

Let $G = (V, E)$ be a graph and $\bar{x}$ a critical extreme point of $P(G)$. From Theorem 9.2 it follows that there exists an odd cycle C of G such that $\bar{x}(e) = \frac{1}{2}$ for $e \in C$ and $\bar{x}(e) = 1$ for $e \in E \setminus C$. Moreover, $E \setminus C$ induces a forest whose leaves are precisely the nodes of $V(C)$. It is not hard to see that the inequality

$$\sum_{e \in C} x(e) \geq \frac{|C| + 1}{2} \tag{9.7}$$

is valid for the polytope TECSP(G) and violated by $\bar{x}$. As it will turn out, inequality (9.7) defines a facet of TECSP(G) under certain conditions. In this section we are going to prove this as a special case of a more general class of facet-defining inequalities of the polytope TECSP(G). This class generalizes the odd-wheel inequalities introduced by Mahjoub [27].

In [27] a class of valid inequalities for TECSP(G) was introduced as follows. Consider a partition $V_1, \ldots, V_p$ of V and let $F \subseteq \delta(V_1)$ with $|F|$ odd. Let $\delta(V_1, \ldots, V_p)$ be the set of edges between the elements of the partition. If we add the inequalities

$$\begin{aligned}
x(\delta(V_i)) &\geq 2, && 2 \leq i \leq p, \\
-x(e) &\geq -1, && e \in F, \\
x(e) &\geq 0, && e \in \delta(V_1) \setminus F,
\end{aligned}$$

we obtain

$$2x(\Delta) \geq 2(p-1) - |F|,$$

where $\Delta = \delta(V_1, \ldots, V_p) \setminus F$. Dividing by two and rounding up the right-hand side we obtain

$$x(\Delta) \geq p - \left\lceil \frac{|F|}{2} \right\rceil. \tag{9.8}$$

Inequalities (9.8) are called *F-partition inequalities.* Note that, if $|F|$ is even, then inequality (9.8) is implied by inequalities (9.1), (9.2), and (9.4).

The F-partition inequalities are a special case of a more general class of inequalities given by Grötschel, Monma, and Stoer [19] for the survivable network polytope.

In what follows we are going to describe a class of inequalities that is a subclass of the F-partition inequalities.

Let $G = (V, E)$ be a 2-edge connected graph. Let $k \geq 1$ and $1 \leq q \leq k$ be two integers. Let I be an odd subset of $\{1, 2, \ldots, 2k+1\}$. Suppose there exists a partition of I into $I_1, \ldots, I_q$, $q \leq k$, and a partition of V into

$$\begin{cases} U_1, \ldots, U_q, & \\ V_i^0, & i \in \{1, \ldots, 2k+1\} \setminus I, \\ V_i^j, & i \in I, j = 0, \ldots, p_i, \end{cases}$$

where p_i is a nonnegative integer, such that the following hold.

1. $|I_l| \geq 2$ for $l = 1, \ldots, q$.
2. There is an edgeset M that defines a perfect matching between the sets V_i^0, $i \in \{1, \ldots, 2k+1\} \setminus I$. (That is, every V_i^0, $i \in \{1, \ldots, 2k+1\} \setminus I$, is adjacent to exactly one edge of M.) (Note that $|\{1, \ldots, 2k+1\} \setminus I|$ is even.)
3. $(V_i^0, V_{i+1}^0) \setminus M \neq \emptyset$ for $i = 1, \ldots, 2k+1$, $(U_i, U_j) = \emptyset$ for all $i, j \in \{1, \ldots, q\}$, $i \neq j$, and $(V_i^0, V_i^j) = \emptyset$ if $i \in I$, $p_i \geq 2$, and $2 \leq j \leq p_i$. (The indices are taken modulo $2k+1$.)
4. The graphs $G(V_i^j)$ for $i = 1, \ldots, 2k+1$ and $j = 0, 1, \ldots, p_i$ are 3-edge connected and the graphs $G(U_i)$, $i = 1, \ldots, q$, are connected.
5. The edgeset (V_i^j, V_i^{j+1}) is nonempty and, if $p_i > 0$, $|(V_i^j, V_i^{j+1})| = 1$ for $i \in I$ and $j = 0, 1, \ldots, p_i$, where $V_i^{p_i+1} = U_l$ for $i \in I_l$ and $l = 1, \ldots, q$.
6. If the sets V_i^0, $i = 1, \ldots, 2k+1$, are deleted, the only edges that remain between the elements of the partition of V are among those described in 5.

Such a partition will be called a *generalized odd-wheel configuration* (see Figure 9.2). An odd-wheel configuration [27] corresponds to the case where $M = \emptyset$, $q = 1$, and $G(U_1)$ is 3-edge connected.

Let r_i, for $i \in I$, denote the largest integer such that $0 \leq r_i \leq p_i$ and $|\delta(V_i^{r_i})| \geq 3$. We denote by $e_{i,j}$ a fixed edge in (V_i^j, V_i^{j+1}) for $i \in I$ and $j = 0, 1, \ldots, p_i$. (For convenience we denote by $e_{i,0}$ the edge of M adjacent to V_i^0 for $i \in \{1, \ldots, 2k+1\} \setminus I$ and let $p_i = r_i = 0$ for $i \notin I$.) Note that the sets $V_1^0, V_2^0, \ldots, V_{2k+1}^0$ form an odd cycle and that between every

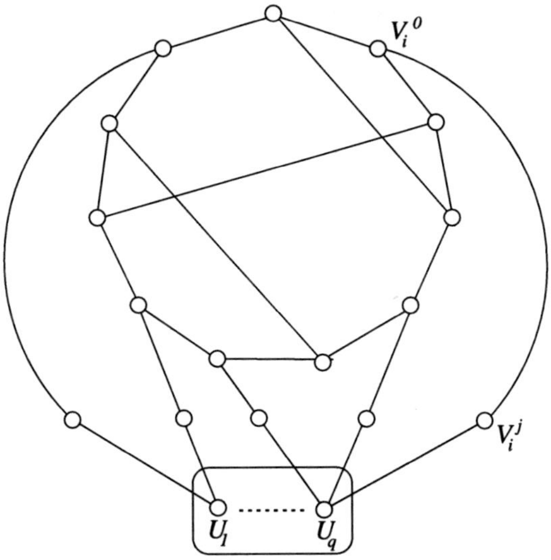

Figure 9.2. *A generalized odd-wheel configuration.*

V_i^0 and $V_i^{p_i+1}$, $i \in I$, there is a path consisting of the edges $e_{i,0}, e_{i,1}, \ldots, e_{i,p_i}$. Moreover, these paths are edge disjoint. Let

$$T = \Lambda \setminus \{e_{i,j} \mid i \in I, r_i \leq j \leq p_i\},$$

where Λ is the set of edges that are in the edge cutsets $\delta(V_i^j)$ for $i \in I, 0 \leq j \leq p_i$, and $\delta(V_i^0)$ for $i \in \{1, \ldots, 2k+1\} \setminus I$, that is,

$$\Lambda = \bigcup_{\substack{1 \leq i \leq 2k+1 \\ 0 \leq j \leq p_i}} \delta(V_i^j).$$

With a generalized odd-wheel configuration we associate the inequality

$$x(T) \geq k + 1 + |M| + \sum_{i \in I} r_i. \tag{9.9}$$

It is not hard to see that inequality (9.9) is the F-partition inequality corresponding to the partition $V_1, V_2, \ldots, V_p$ where

- $V_1 = \bigcup_{\substack{i \in I \\ j \geq r_i+1}} V_i^j$,
- $V_2, \ldots, V_p$ correspond to the sets V_i^j, $1 \leq i \leq 2k+1$, $0 \leq j \leq r_i$, and
- $F = \{e_{i,j} \mid i \in I, \ j = r_i\}$.

Hence inequalities (9.9) are valid for TECSP(G). Inequalities of type (9.9) will be called *generalized odd-wheel inequalities.*

In what follows we shall describe sufficient conditions for inequalities (9.9) to define facets for TECSP(G).

We shall denote by f_i, $f_i \notin M$, a fixed edge in (V_i^0, V_{i+1}^0) for $i = 1, \ldots, 2k+1$. Let $C = \{f_1, \ldots, f_{2k+1}\}$. Let

$$\begin{aligned}\Gamma &= E \setminus \Lambda,\\ \Gamma' &= \bigcup_{i=1,\ldots,q} E(U_i),\\ \overline{E} &= \{e_{i,j} \mid i \in I,\ 0 \le j \le p_i\} \cup M.\end{aligned}$$

Note that Γ is the set of edges having both nodes in the same V_i^j for $i = 1, \ldots, 2k+1$, $j = 0, \ldots, p_i + 1$.

Given a graph $G = (V, E)$ inducing a generalized odd-wheel configuration, we denote by $G^* = (V^*, E^*)$ the graph whose nodeset V^* is the set obtained from V by contracting the sets V_i^j, $i = 1, \ldots, 2k+1$, $j = 0, \ldots, p_i$, and $E^* = C \cup \Gamma' \cup \overline{E}$. Let $V(C) = \{v_1, v_2, \ldots, v_{2k+1}\}$, where v_i corresponds to the set V_i^0 for $i = 1, \ldots, 2k+1$.

Given two edgesets $A \subset E^* \setminus C$ and $B \subset \Lambda \setminus E^*$ (A and B may be empty sets), we denote by $G^*_{A,B}$ the graph obtained from G^* by

(i) removing the edges of A,

(ii) adding the edges of B, and

(iii) applying recursively operation θ_2 (given in Section 9.2) on the graph obtained by (i) and (ii) until no node of degree two is left in the graph.

Note that the graph $G^*_{\emptyset,\emptyset}$ is the graph obtained from G^* by repeated application of operation θ_2 until each node is of degree at least three.

Let $C_{A,B}$ stand for the restriction of C on $G^*_{A,B}$. Note that $C_{A,B}$ is a cycle, which may be either even or odd. We say that $G^*_{A,B}$ satisfies *property (P)* if $|\delta(S) \setminus C_{A,B}| \ge 2$ for every cut $\delta(S)$ of $G^*_{A,B}$ such that $|S| \ge 2$, $|\overline{S}| \ge 2$, and the subgraphs of $G^*_{A,B}$ induced by S and $\overline{S}$ are both 2-edge connected.

Observe that, by Theorem 9.2 together with Lemma 9.4, if G is a graph and $\overline{x}$ is a critical extreme point of $P(G)$, then G is a generalized odd-wheel configuration with $C = \{e \in E \mid 0 < \overline{x}(e) < 1\}$ and $G = G^* = G^*_{\emptyset,\emptyset}$. Moreover, $G^*_{\emptyset,\emptyset}$ satisfies property (P).

Now we give a technical lemma.

Lemma 9.6. *Let $A \subset E^* \setminus C$ and $B \subset \Lambda \setminus E^*$. Let $H = G^*_{A,B} = (W, D)$. Suppose that $C_{A,B} \neq \emptyset$. Let $C_{A,B} = \{g_1, \ldots, g_p\}$. Let $\{w_1, \ldots, w_p\}$ be the nodes of $C_{A,B}$ such that $g_i = w_i w_{i+1}$ for $i = 1, \ldots, p$. (The indices are modulo p.) Suppose that H is 3-edge connected and satisfies property (P). Let $t \in \{1, \ldots, p\}$. If p is odd (resp. even), then*

$$\tilde{D} = \{e_0, g_{t+1}, g_{t+3}, \ldots, g_{t+p-2}\} \cup \overline{D}$$
$$(\text{resp. } \tilde{D} = \{g_{t+1}, g_{t+3}, \ldots, g_{t+p-1}\} \cup \overline{D})$$

induces a 2-edge connected spanning subgraph of H, where $\overline{D} = D \setminus C_{A,B}$ and $e_0 \in \delta_H(w_t)$ such that $\delta_H(w_t) \setminus \{e_0, g_{t-1}, g_t\} \neq \emptyset$.

Proof. Let $\tilde{H}$ be the graph induced by $\tilde{D}$. First remark that, as H is 3-edge connected, in both cases, every node of $\tilde{H}$ is incident to at least two edges of $\tilde{D}$. Now suppose, on the contrary, that $\tilde{H}$ is not 2-edge connected and let $\delta_{\tilde{H}}(S)$ be a cut of $\tilde{H}$ such that $|\delta_{\tilde{H}}(S)| \leq 1$. As every node of $\tilde{H}$ is incident to at least two edges of $\tilde{D}$, it follows that $|S| \geq 2$ and $|\overline{S}| \geq 2$. We distinguish two cases.

Case 1. $\tilde{H}(S)$ is 2-edge connected.

Then $H(S)$ is 2-edge connected. As H satisfies property (P), it follows that $H(\overline{S})$ is not 2-edge connected. Otherwise, as $\overline{D} \subseteq \tilde{D}$, $|S| \geq 2$, and $|\overline{S}| \geq 2$, one would have $|\delta_{\tilde{H}}(S)| \geq |\delta_H(S) \setminus C_{A,B}| \geq 2$, a contradiction. In what follows we suppose that $|S|$ is maximum. That is, if $|S'| > |S|$, $|S'| \geq 2$, $|\overline{S}'| \geq 2$, and $\tilde{H}(S')$ is 2-edge connected, then $|\delta_{\tilde{H}}(S')| \geq 2$. We claim that $H(\overline{S})$ is connected. In fact, suppose this is not the case, and let W' be a node subset inducing a connected component of $H(\overline{S})$. Obviously, $|\delta_{\tilde{H}}(W')| \leq 1$. As every node of H is incident to at least two edges of $\tilde{D}$, we have that $|W'| \geq 2$. Furthermore, since H is 3-edge connected, every leaf of the graph $H(\overline{S} \setminus W')$ is linked to S by at least two edges. Thus the graph $H(S^*)$, where $S^* = S \cup (\overline{S} \setminus W')$, is 2-edge connected. Since $|\delta_{\tilde{H}}(S^*)| \leq 1$ and $|S^*| > |S|$, this contradicts the maximality of S.

Thus $H(\overline{S})$ is connected. In consequence, by contracting the 2-edge components of $H(\overline{S})$, we get a tree. Let $W_1, \ldots, W_l$, $l \geq 2$, be a partition of $\overline{S}$ such that $H(W_i)$ is 2-edge connected for $i = 1, \ldots, l$ and the graph obtained from H by contracting $W_1, \ldots, W_l$ is a tree, say $\mathcal{K}$. Without loss of generality, we may assume that W_1 and W_2 are leaves of $\mathcal{K}$. As $|\delta_{\tilde{H}}(S)| \leq 1$, we may also assume, for instance, that $\delta_H(W_1) \cap \delta_{\tilde{H}}(S) = \emptyset$. We claim that $|W_1| \geq 2$. Indeed, if $W_1 = \{w\}$, as $\delta_H(w)$ contains at least two edges of $\tilde{D}$, this implies that $\delta_H(w) \cap \delta_{\tilde{H}}(S) \neq \emptyset$, a contradiction. Furthermore, as H is 3-edge connected, every leaf (resp. node of degree two) of $\mathcal{K}$ must be linked to S by at least two (resp. one) edges. Hence the subgraph $H(W^*)$, where $W^* = S \cup (\bigcup_{i=2}^{l} W_i)$, is 2-edge connected. Moreover, we have that $|\delta_H(W^*) \setminus C_{A,B}| \leq 1$. Since $|W^*| \geq 2$ and $|\overline{W}^*| \geq 2$, this contradicts the fact that H satisfies property (P).

Case 2. $\tilde{H}(S)$ is not 2-edge connected.

Hence there is a partition $S_1, \ldots, S_h$ of S, $h \geq 2$, such that $\tilde{H}(S_i)$ is 2-edge connected for $i = 1, \ldots, h$, and the graph obtained by contracting the sets S_i, $i = 1, \ldots, h$, is a forest, say $\mathcal{K}'$. Since $|\delta_{\tilde{H}}(S)| \leq 1$, at least one of the leaves of $\mathcal{K}'$, say S_1, is not incident to any edge of $\delta_{\tilde{H}}(S)$. It thus follows that $|S_1| \geq 2$. As $|\overline{S}_1| \geq 2$ and $\tilde{H}(S_1)$ is 2-edge connected, Case 1 applies, which finishes the proof of the lemma. □

A consequence of Lemma 9.6 is the following.

Lemma 9.7. *Suppose that $G^*_{\emptyset,\emptyset}$ is 3-edge connected and satisfies property (P). Let $t \in \{1, \ldots, 2k+1\}$ and $e_0 \in \delta_G(V_t^0) \setminus \{e_{t,0}\}$. Then the edgeset*

$$\tilde{E} = \{e_0, f_{t+1}, f_{t+3}, \ldots, f_{t+2k-1}\} \cup \Gamma \cup \overline{E}$$

induces a 2-edge connected spanning subgraph of G.

Proof. Let $\tilde{G}$ be the graph induced by $\tilde{E}$ and consider a cut $\delta_{\tilde{G}}(S)$ of $\tilde{G}$. If $V_i^j \neq S \cap V_i^j \neq \emptyset$ for some $i \in \{1, \ldots, 2k+1\}$ and $j \in \{0, 1, \ldots, p_i\}$, since $G(V_i^j)$ is 3-edge connected, it then follows that $|\delta_{\tilde{G}}(S)| \geq 2$.

So suppose that either $V_i^j \subseteq S$ or $V_i^j \subseteq \overline{S}$ for $i = 1, \ldots, 2k+1$ and $j = 0, \ldots, p_i$. Let $\tilde{G}' = (\tilde{V}', \tilde{E}')$ be the graph obtained from $\tilde{G}$ by contracting

- the sets V_i^j, $i = 1, \ldots, 2k+1$, $j = 0, \ldots, p_i$, and
- the edges of $\tilde{E} \setminus (C \cup \{e_{i,0},\ i \in I\})$ having at least one node of degree two.

Clearly, $\tilde{G}$ is 2-edge connected if and only if $\tilde{G}'$ is so. Now, as $C = C_{\emptyset,\emptyset}$ and $e_0 \neq e_{t,0}$, we have that $\delta_G(V_t^0) \setminus \{e_0, f_t, f_{t-1}\} \neq \emptyset$. In addition, since $G^*_{\emptyset,\emptyset}$ is 3-edge connected and satisfies property (P), by Lemma 9.6, it follows that $\tilde{G}'$ is 2-edge connected, and hence $\tilde{G}$ is 2-edge connected. □

Let us denote by $E^0 \subseteq E$ the set of edges of E that belong to 2-edge cutsets of G and by Γ'' the set of edges of Γ' that belong to edge cutsets of $G(U_i)$, $i = 1, \ldots, q$, having no more than 2 edges. Let $\tau(G) = \{D \subseteq E \mid G(D)$ is 2-edge connected$\}$.

Theorem 9.8. *Inequality (9.9) defines a facet of TECSP(G) if the following hold.*

(i) $|\delta(S)| \geq 4$ *for every cut* $\delta(S)$ *of* $G^*_{\emptyset,\emptyset}$ *such that* $|S| \geq 2$ *and* $|\overline{S}| \geq 2$.

(ii) $G^*_{\emptyset,\emptyset}$ *satisfies property (P).*

(iii) $G^*_{\{e\},\emptyset}$ *satisfies property (P) for* $e \in M \cup \Gamma'' \cup \{e_{i,0} \mid i \in I,\ p_i = 0\}$.

(iv) (iv.1) *For every edge* $e_{i,j} \in \overline{E} \setminus E_0$ *where* $i \in I$, $1 \leq j \leq r_i - 1$, *there exist two edges* $e_1 \in \delta(V_i^j) \setminus \overline{E}$, $e_2 \in \delta(V_i^{j+1}) \setminus \overline{E}$ *such that the graph* $G^*_{\{e_{i,j}\},\{e_1,e_2\}}$ *satisfies property (P).*

(iv.2) *For every edge* $e_{i,0} \in \overline{E} \setminus E_0$ *where* $i \in I$ *there exists an edge* $f \in \delta(V_i^1) \setminus \overline{E}$ *such that the graph* $G^*_{\{e_{i,0}\},\{f\}}$ *satisfies property (P).*

(iv.3) *For every edge* $e_{i,p_i} \in \overline{E} \setminus E_0$ *where* $i \in I$ *there exists an edge* $g \in \delta(V_i^{p_i}) \setminus \overline{E}$ *such that the graph* $G^*_{\{e_{i,p_i}\},\{g\}}$ *satisfies property (P).*

Proof. Since $G = (V, E)$ is 2-edge connected and x^E does not satisfy (9.9) with equality, this implies that (9.9) defines a face of TECSP(G). Let $B = (B^1, B^2)$ be the matrix where B^1 is the identity matrix whose rows correspond to the edges of E^0 and B^2 is the $(|E^0|, |E \setminus E^0|)$ matrix formed by zeros.

Let us denote inequality (9.9) by $a^T x \geq a_0$, and let us assume that there exists a facet-defining inequality $b^T x \geq b_0$ of TECSP(G) such that $\{x \in \text{TECSP}(G) \mid a^T x = a_0\} \subseteq \{x \in \text{TECSP}(G) \mid b^T x = b_0\}$. It suffices to show that there exists $\rho > 0$ and $\lambda \in \mathbb{R}^{E^0}$ such that $a^T = \rho b^T + \lambda^T B$.

We first show that $b(e)$ has the same value for all edges of $\delta(V_i^0) \setminus \{e_{i,0}\}$, $i = 1, \ldots, 2k+1$. For this consider the following edgesets:

$$E_1 = \{f_1, f_3, \ldots, f_{2k+1}\} \cup \Gamma \cup \overline{E},$$
$$E_2 = (E_1 \setminus \{f_1\}) \cup \{e\},$$

where e is an arbitrary edge in $\delta(V_2^0) \setminus \{f_1, e_{2,0}\}$. Since every node of $G^*_{\emptyset,\emptyset}$ is of degree at least three, by (i) it follows that $G^*_{\emptyset,\emptyset}$ is 3-edge connected. Now, as by (ii), $G^*_{\emptyset,\emptyset}$ also satisfies property (P), by Lemma 9.7 we have that $E_1,\ E_2 \in \tau(G)$. Moreover, we have $a^T x^{E_j} = a_0$ for $j = 1, 2$. So

$$0 = b^T x^{E_1} - b^T x^{E_2} = b(f_1) - b(e),$$

which implies that $b(e) = \alpha$ for every edge in $\delta(V_2^0) \setminus \{e_{2,0}\}$ for some $\alpha \in \mathbb{R}$. Exchanging the roles of the nodesets V_i^0 we then obtain by symmetry that

$$b(e) = \alpha \qquad \text{for all } e \in \delta(V_i^0) \setminus \{e_{i,0}\},\ i = 1, \ldots, 2k+1. \tag{9.10}$$

Next we show that $b(e) = \alpha$ holds for all $e \in M$. Suppose that e is between V_i^0 and V_j^0, where $i, j \in \{1, \ldots, 2k+1\} \setminus I$. Without loss of generality, we may suppose that $i = 1$ and j is even. Let

$$E_3 = (E_1 \setminus \{e\}) \cup \{f_j\}.$$

Observe that $f_{j-1}, f_j \in E_3$. Let E_3' be the restriction of E_3 on $G^*_{\{e\},\emptyset}$. Note that $G^*_{\{e\},\emptyset}$ is the graph obtained from $G^*_{\emptyset,\emptyset}$ by contracting one edge among $\{f_1, f_{2k+1}\}$ and one edge among $\{f_{j-1}, f_j\}$. Therefore $E_3 \in \tau(G)$ if $E_3' \in \tau(G^*_{\{e\},\emptyset})$. Now from (i) we have that $G^*_{\{e\},\emptyset}$ is 3-edge connected. In fact, it is clear that every node of $G^*_{\{e\},\emptyset}$ is of degree at least three. Let $K = (S, \overline{S})$ be an edge cutset of $G^*_{\{e\},\emptyset}$ with $|S| \geq 2$ and $|\overline{S}| \geq 2$, and let $K' = K \cup \{e\}$. It is easy to see that K' contains an edge cutset of $G^*_{\emptyset,\emptyset}$. As, by (i), $|K'| \geq 4$, it follows that $|K| \geq 3$.

Thus $G^*_{\{e\},\emptyset}$ is 3-edge connected. As, by (iii), $G^*_{\{e\},\emptyset}$ satisfies property (P), by setting $e_0 = f_{j+1}$ (and $w_t = v_{j+2}$), it follows from Lemma 9.6 that $E_3' \in \tau(G^*_{\{e\},\emptyset})$. Hence $E_3 \in \tau(G)$. Moreover, since $a^T x^{E_1} = a^T x^{E_3} = a_0$, we obtain that

$$0 = b^T x^{E_1} - b^T x^{E_3} = b(e) - b(f_j).$$

From (9.10) we then have $b(e) = \alpha$.

Now we show that $b(e_{i,j}) = \alpha$ for every edge $e_{i,j}$ in $\overline{E} \setminus E^0$ where $r_i > 0$ and $j \leq r_i - 1$. We show this for the edges $e_{i,j}$ of $\overline{E} \setminus E^0$ where $i \in I$ and $1 \leq j \leq r_i - 1$. For the edges of $\{e_{i,0} \mid i \in I\} \setminus E^0$ the proof is similar.

Since $e_{i,j}$ is not in a 2-edge cutset (and $1 \leq j \leq r_i - 1$), by property 6 of a generalized odd wheel configuration, there must exist two integers $q_1, q_2 \in \{1, \ldots, 2k+1\}$ (q_1 and q_2 may be equal) such that $(V_i^j, V_{q_1}^0) \neq \emptyset \neq (V_i^{j+1}, V_{q_2}^0)$. Let e_1 and e_2 be two edges of $(V_i^j, V_{q_1}^0)$ and $(V_i^{j+1}, V_{q_2}^0)$, respectively. Note that $e_1 \neq e_{q_1,0}$ and $e_2 \neq e_{q_2,0}$. Without loss of generality, we may assume that $q_1 = 1$ and q_2 is odd. By (iv) we may also assume that $G^*_{A,B}$ satisfies property (P), where $A = \{e_{i,j}\}$ and $B = \{e_1, e_2\}$. Let

$$\begin{aligned} F_0 &= \{f_{q_2+1}, f_{q_2+3}, \ldots, f_{q_2+2k+1}\} \cup \Gamma \cup \overline{E}, \\ F_1 &= (F_0 \setminus \{f_{q_2+2k+1}, e_{i,j}\}) \cup \{e_1, e_2\}. \end{aligned}$$

In what follows we show that $F_1 \in \tau(G)$. To this end let us set $F_1' = (F_1 \setminus e_1) \cup \{e_{i,j}\}$ and denote by H_1 and H_1' the graphs induced by F_1 and F_1', respectively. By Lemma 9.7

we have that $F_0, F_1' \in \tau(G)$. Now consider a cut $\delta_{H_1}(S)$. If either $(V_i^j \cup V_i^{j+1}) \subseteq S$ or $(V_i^j \cup V_i^{j+1}) \subseteq \overline{S}$, then $\delta_{H_1}(S) \setminus \{e_1\}$ is an edge cutset of H_1' and, as H_1' is 2-edge connected, we have that $|\delta_{H_1}(S)| \geq 2$. If either $S = V_i^j$ or $S = V_i^{j+1}$, then clearly $|\delta_{H_1}(S)| \geq 2$. If $V_h^l \neq S \cap V_h^l \neq \emptyset$ for some $h \in \{1, 2, \ldots, 2k+1\}$ and $0 \leq l \leq p_h$, since $G(V_h^l)$ is 2-edge connected, we obtain that $|\delta_{H_1}(S)| \geq 2$. Thus we may assume that either $V_h^l \subseteq S$ or $V_h^l \subseteq \overline{S}$ for all $h \in \{1, 2, \ldots, 2k+1\}$ and $0 \leq l \leq p_h$. Consider the graph $G^*_{A,B}$, where $A = \{e_{i,j}\}$ and $B = \{e_1, e_2\}$, and let $\tilde{F}_1$ be the restriction of F_1 on $G^*_{A,B}$. Note that no edge of C has been contracted in the construction of $G^*_{A,B}$ and thus $C_{A,B} = C$. It thus follows that $F_1 \in \tau(G)$ if $\tilde{F}_1 \in \tau(G^*_{A,B})$. Now note that e_1 and e_2 belong to paths all of whose internal nodes are of degree two in the graph obtained from G^* by removing $e_{i,j}$ and adding e_1, e_2. As these paths are replaced by edges in the construction of $G^*_{A,B}$, we may suppose that e_1 and e_2 are edges of $G^*_{A,B}$. Moreover, by (i) it follows as before that $G^*_{A,B}$ is 3-edge connected. By setting $t = q_2$ and $e_0 = e_2$ Lemma 9.7 implies that $\tilde{F}_1 \in \tau(G^*_{A,B})$. Hence $F_1 \in \tau(G)$. Since $a^T x^{F_0} = a^T x^{F_1} = a_0$, we get

$$0 = b^T x^{F_0} - b^T x^{F_1} = b(f_{q_2+2k+1}) + b(e_{i,j}) - b(e_1) - b(e_2).$$

From (9.10) we then have $b(e_{i,j}) = \alpha$. Hence we have shown that $b(e) = \alpha$ for all $e \in T \setminus E^0$.

Consider an edge $e_{i,j} \in \overline{E} \setminus E^0$ with $j = p_i$. (Note that $r_i = p_i$.) By (iv) there exists an edge $g \in V_i^{p_i} \setminus \overline{E}$ such that the graph $G^*_{A,B}$, where $A = \{e_{i,j}\}$ and $B = \{g\}$, verifies property (P). Let s be such that $g \in \delta(V_s^0)$. Without loss of generality, we may suppose that $s = 1$. Let $F_2 = (E_1 \setminus \{f_1, e_{i,j}\}) \cup \{g\}$, where E_1 is as defined before. From (i) it follows that $G^*_{A,B}$ is 3-edge connected. As it also verifies property (P), we can show in a similar way as before that $F_2 \in \tau(G)$. Since $F_2' = F_2 \cup \{e_{i,j}\}$ also belongs to $\tau(G)$ and $a^T x^{F_2} = a^T x^{F_2'} = a_0$, we get

$$0 = b^T x^{F_2'} - b^T x^{F_2} = b(e_{i,j}).$$

Let e be an edge having both nodes in the same V_i^j.

As $G(V_i^j)$ is 3-edge connected, the edgeset $E_1' = E_1 \setminus \{e\}$ induces a 2-edge connected spanning subgraph of G. Since $a^T x^{E_1'} = a^T x^{E_1} = a_0$, we get $b(e) = 0$.

Now let e be an edge having both endnodes in the same U_i. If $e \notin \Gamma''$, then every cut $\delta(S)$ of G containing e contains at least three edges. Let $E_1' = E_1 \setminus \{e\}$. As $E_1 \in \tau(G)$, it thus follows that $E_1' \in \tau(G)$ and hence $b(e) = 0$. Now consider an edge $e \in \Gamma''$, and consider the graph $G^*_{\{e\},\emptyset}$. By (i) it follows that $G^*_{\{e\},\emptyset}$ is 3-edge connected. Let $\tilde{E}_1$ be the restriction of E_1 on $G^*_{\{e\},\emptyset}$. As, by (iii), $G^*_{\{e\},\emptyset}$ also satisfies property (P), by Lemma 9.6 it follows that $\tilde{E}_1 \setminus \{e\} \in \tau(G^*_{\{e\},\emptyset})$. Hence $E_1 \setminus \{e\} \in \tau(G)$, which yields $b(e) = 0$. If $e \in \{e_{i,0} \mid i \in I,\, p_i = 0\}$, we can show similarly that $b(e) = 0$.

Altogether we have now shown that

$$\begin{aligned} b(e) &= \alpha && \text{for all } e \in T \setminus E^0, \\ b(e) &= 0 && \text{for all } e \in E \setminus (T \cup E^0). \end{aligned}$$

Moreover, it is clear that for every edge $e \notin E^0$ there exists an edgeset $E' \subset E$ that induces a 2-edge connected spanning subgraph of G such that $e \notin E'$ and $a^T x^{E'} = a_0$. This implies that the face $a^T x \geq a_0$ is not contained in a trivial face $\{x \in \text{TECSP}(G) \mid x(e) = 1\}$ for some $e \in E \setminus E^0$. Hence $b^T x \geq b_0$ defines a nontrivial facet of TECSP(G), which implies that $\alpha > 0$. Now by setting $\rho = \frac{1}{\alpha}$ and

$$\lambda_e = \begin{cases} 1 - \frac{b_e}{\alpha} & \text{if } e \in T \cap E^0, \\ -\frac{b_e}{\alpha} & \text{if } e \in E^0 \setminus T, \end{cases}$$

we get $a^T = \rho b^T + \lambda^T B$ and our proof is complete. □

Inequality (9.9) may not define a facet if either of conditions (iii) and (iv) is not satisfied. In fact, consider the graph G shown in Figure 9.3. The solution given by $\overline{x}(e_i) = \frac{1}{2}$ for $i = 1, \ldots, 7$ and $\overline{x}(e) = 1$ otherwise is a critical extreme point of $P(G)$.

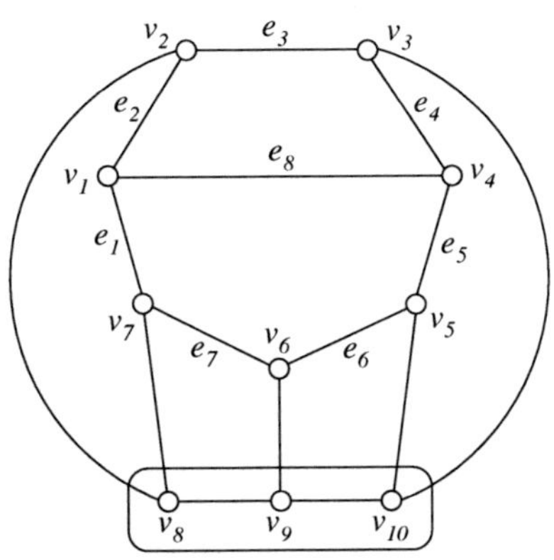

Figure 9.3. *A configuration not producing a facet.*

Moreover, the graph G induces a generalized odd wheel configuration, where $V_i^0 = \{v_i\}$ for $i = 1, \ldots, 7$, $M = \{e_8\}$, $q = 1$, and $U_1 = \{v_8, v_9, v_{10}\}$. Note that $r_i = 0$ for $i = 1, \ldots, 7$. Now note that condition (iii) is not verified for the edge e_8 of M. The corresponding generalized odd wheel inequality is given by

$$\sum_{i=1,\ldots,8} x(e_i) \geq 5. \tag{9.11}$$

It is not hard to see that every 2-edge connected spanning subgraph of G whose incident vector satisfies (9.11) with equality contains edge e_8. Hence the face defined by (9.11) is contained in the face $\{x \in \text{TECSP}(G) \mid x(e_8) = 1\}$. In consequence, inequality (9.11) is not facet-defining for TECSP(G).

This also shows that the generalized odd wheel configuration induced by a critical extreme point of $P(G)$ may not produce a facet for TECSP(G).

If $\bar{x}$ is a critical extreme point of $P(G)$, from Theorem 9.2 it follows that G is a generalized odd wheel configuration, where C is the set of edges with fractional values whose corresponding inequality is (9.9). If $M = \emptyset$ and $r_i = 0$ for $i = 1, \ldots, 2k + 1$, then the corresponding inequality is (9.7). Furthermore, as $\overline{x}$ is critical, by Remark 9.3 and Lemma 9.4, conditions (i) and (ii) of Theorem 9.8 are satisfied. If conditions (iii) and (iv) are also satisfied, the corresponding inequality (9.9) will then be facet-defining for TECSP(G).

Now let $\bar{x}$ be an extreme point of $P(G)$ of rank 1. Then $\bar{x}$ and G can be reduced by means of operations $\theta_1, \theta_2, \theta_3$ to a solution $\bar{x}'$ and a graph $G' = (V', E')$, where $\bar{x}'$ is a critical extreme point of $P(G')$. Moreover, operations $\theta_1, \theta_2, \theta_3$ can be performed in polynomial time and applied in an arbitrary order. Now by the remark above there exists an F-partition inequality $a'^T x \geq \alpha'$ of type (9.9) valid for TECSP(G'), which is violated by $\bar{x}'$. Let $U_1, \ldots, U_q$; V_i^0, $i \in \{1, \ldots, 2k+1\} \setminus I$; $V_i^j, i \in I$, $j = 0, \ldots, p_i$, be the partition of V' producing $a'^T x \geq \alpha'$. Note that the sets V_i^0, $i \in \{1, \ldots, 2k+1\} \setminus I$, and V_i^j, $i \in I$, $j = 0, \ldots, p_i$, are reduced to single nodes. Also note that the corresponding r_i are all equal to 0. Let $V_1, \ldots, V_p$ be such that

- $V_1 = \bigcup_{\substack{i \in I \\ j \geq 1}} V_i^j$,
- $V_2, \ldots, V_p$ correspond to the sets V_i^0, $1 \leq i \leq 2k+1$.

Let $a^T x \geq \alpha$, $a \in \mathbb{R}^E$, be such that

$$a(e) = \begin{cases} a'(e) & \text{if } e \in E', \\ 1 & \text{if } e \in \bigcup_{1 \leq i \leq p} \delta(V_i) \setminus E', \\ 0 & \text{otherwise,} \end{cases}$$

and $\alpha = \alpha'$. It is not hard to see that the lifted inequality $a^T x \geq \alpha$ is an F-partition inequality, valid for TECSP(G), which is violated by $\bar{x}$. Moreover, if $a'^T x \geq \alpha'$ is facet-defining for TECSP(G'), then $a^T x \geq \alpha$ is facet-defining for TECSP(G). Hence we have the following.

Corollary 9.9. *Let $\bar{x}$ be an extreme point of $P(G)$ of rank 1. Let $\bar{x}'$ and G' be the critical solution and the graph obtained, respectively, from $\bar{x}$ and G by means of the operations θ_1, θ_2, θ_3. Let $a'^T x \geq \alpha'$ be the inequality of type (9.9) induced by G' and $a^T x \geq \alpha$ the inequality obtained from $a'^T x \geq \alpha'$ by lifting. If $a'^T x \geq \alpha'$ defines a facet of TECSP(G'), then $a^T x \geq \alpha$ defines a facet of TECSP(G).*

Note that, if G is complete, then each element of the partition $(V_1, \ldots, V_p)$ given above induces a complete graph, and the coefficients $a(e)$ can then be easily computed. Also observe that, although the lifting produces an F-partition inequality, it may not give a generalized odd wheel inequality.

Corollary 9.9 has various interesting computational consequences. In particular, as will be developed in the next section, using operations $\theta_1, \theta_2, \theta_3$, one may generate cutting planes in polynomial time for extreme points that are not necessarily of rank 1.

To finish this section let us mention that there may exist F-partitions that produce facets for TECSP(G) but that are not generalized odd wheel configurations. In fact, consider the graph $G = (V, E)$ shown in Figure 9.4.

G induces the F-partition where the nodes correspond to the elements of the partition and $F = \{e_1, e_2, e_3\}$. The corresponding F-partition inequality given by

$$\sum_{e \in E \setminus F} x(e) \geq 4$$

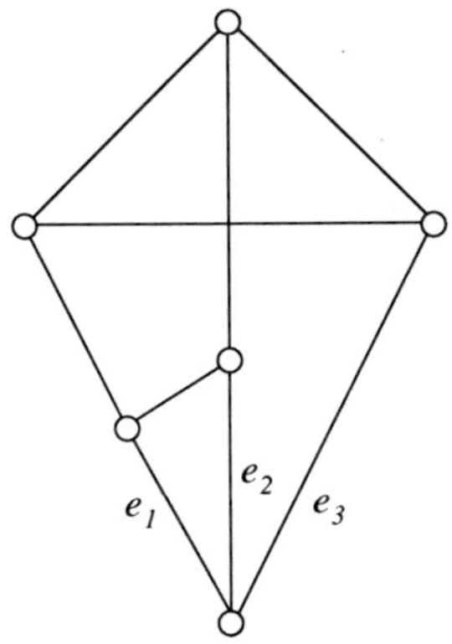

Figure 9.4. *F-partition not corresponding to a generalized odd wheel.*

defines a facet of TECSP(G). However, G is not a generalized odd wheel configuration. A natural question that may be asked is whether or not the F-partition inequalities may be extended to configurations where F is an arbitrary subset of $\delta(V_1, \ldots, V_p)$. The answer is unfortunately in the negative. In fact, if F is not contained in some $\delta(V_i)$, it is not hard to see that any inequality obtained by the Chvàtal–Gomory procedure, which has the same left-hand side as (9.8), is redundant with respect to inequalities (9.1), (9.2), and (9.4).

9.4 A Branch-and-Cut Algorithm

In this section, we describe a branch-and-cut algorithm for the SNDP when $r \in \{1, 2\}^V$. Our aim is to address the algorithmic applications of the theoretical results presented in the previous sections. So let us assume that we are given a graph $G = (V, E)$, a vector of node types $r \in \{1, 2\}^V$, and a weight vector $c \in \mathbb{R}_+^E$ associated with the edges of G.

For this kind of node type Grötschel, Monma, and Stoer [19] have introduced a class of valid inequalities for the (1, 2)-survivable network polytope, called partition inequalities, that generalize the cut inequalities. Let $V_1, \ldots, V_p$, $p \geq 3$, be a partition of V. Let $I_2 = \{i \mid \text{con}(V_i) = 2,\ i = 1, \ldots, p\}$. The *partition inequality* induced by $(V_1, \ldots, V_p)$ is given by

$$x(\delta(V_1, \ldots, V_p)) \geq \begin{cases} p - 1 & \text{if } I_2 = \emptyset, \\ p & \text{otherwise.} \end{cases} \tag{9.12}$$

Obviously, if all node types are equal to 2, the partition inequality is already implied by the cut constraints $x(\delta(V_i)) \geq 2$.

Furthermore, the F-partition inequalities (9.8) can straightforwardly be extended to the case $r \in \{1, 2\}^V$ as follows:

$$x(\delta(V_1, \ldots, V_p) \setminus F) \geq p - \left\lfloor \frac{p_1 + |F|}{2} \right\rfloor - 1, \tag{9.13}$$

where $p_1 = |\{i \mid \text{con}(V_i) = 1,\ i = 2, \ldots, p\}|$. Note that $|F|$ is not necessarily odd. In fact, inequalities (9.13) are dominated by the cut and trivial inequalities if and only if p_1 and $|F|$ have the same parity.

Let $R(G, r)$ be the polytope described by the trivial constraints (9.1), (9.2), the cut constraints (9.3), and the partition constraints (9.12). Let $\overline{x}$ be a solution of $R(G, r)$. The following operations, given with respect to $\overline{x}$, extend in a straightforward way the operation θ_2, introduced previously in Section 9.2, to the case where $r \in \{1, 2\}^V$:

θ_1'': Contract an edge uv such that $\bar{x}(uv) = 1$, $r(u) = 1$, and $\overline{x}(\delta(u)) \leq 2$.

θ_2'': Contract an edge uv such that $r(u) = 2$, $|\delta(u)| = \{uv, uw\}$, and $r(w) = 2$.

Note that these reduction operations, as those described in Section 9.2, can be realized in polynomial time. Also note that operation θ_3, given in Section 9.2, can be extended to the (1, 2)-SNDP by considering nodesets $W \subset V$ with $r(v) = 2$ for all $v \in W$.

With a graph obtained from G by contracting an edge $e = uv \in E$, we associate the node-type vector $r_e \in \{1, 2\}^{|V|-1}$ such that $r_e(w) = \text{con}(\{u, v\})$ and $r_e(i) = r(i)$ if $i \in V \setminus \{u, v\}$, where w is the node that arises from the contraction of e. Let $G' = (V', E')$ be a graph obtained by repeated applications of operations $\theta_1, \theta_2, \theta_3, \theta_1'', \theta_2''$. Let $r' \in \{1, 2\}^{V'}$ be the node-type vector corresponding to the graph G' and let $\bar{x}'$ be the restriction of $\bar{x}$ on E'. If $\bar{x}$ is an extreme point of $R(G, r)$, then $\bar{x}'$ is an extreme point of $R(G', r')$. Moreover, we have the following.

Lemma 9.10. (i) *If $a'x \geq \alpha'$ is a valid inequality of the (1, 2)-survivable network polytope on G' of type (9.3), (9.12), or (9.13), then the inequality $ax \geq \alpha$, where $a(e) = a'(e)$ if $e \in E'$, $a(e) = 1$ if e has its nodes in different sets of the partition, and $\alpha = \alpha'$, is valid for the (1, 2)-survivable network polytope on G. Moreover, if $a'x \geq \alpha'$ is violated by $\bar{x}'$, then $ax \geq \alpha$ is violated by $\bar{x}$.*

(ii) *If $ax \geq \alpha$ is a valid inequality of the (1, 2)-survivable network polytope on G of type (9.3) (resp. (9.12)) (resp. (9.13)) that is violated by $\bar{x}$, then there is an inequality valid for the (1, 2)-survivable network polytope on G' of type (9.3) (resp. (9.12)) (resp. (9.13)) that is violated by $\bar{x}'$.*

Proof. The proof is easy and therefore omitted. □

Lemma 9.10 shows that looking for inequalities of type (9.3), (9.12), or (9.13) that are violated by $\bar{x}$ reduces to looking for such inequalities that are violated by $\bar{x}'$ on G'. Note that this procedure can be applied for any solution of $R(G, r)$ and, in consequence, may permit us to separate fractional solutions that are not even extreme points of $R(G, r)$. Moreover, if $r(v) = 2$ for all $v \in V$ and $\bar{x}$ is an extreme point of $P(G)$ of rank 1, then, as mentioned above, there is an F-partition that cuts off this solution that can be found in polynomial time. In addition, this F-partition inequality defines a facet of TECSP(G) if G' induces a generalized odd wheel facet for TECSP(G').

Lemma 9.10 also holds for TNCSP when we consider the operations $\theta_1, \theta_2, \theta_3, \theta_1', \theta_2'$ and the inequalities (9.4), (9.6), (9.8). Actually, in this case, the graph G' is obtained by applications of the operations $\theta_1, \theta_2, \theta_3, \theta_1', \theta_2'$. If one of these inequalities is violated by $\bar{x}'$ in G', then one is violated by $\bar{x}$ in G. Thus, as for the (1, 2)-SNDP, the separation of $\bar{x}$ by inequalities of type (9.4), (9.6), (9.8) reduces to the separation of $\bar{x}'$ by these inequalities in G'.

In consequence and for more efficiency, our separation routines will be performed on the reduced graph G'.

We now describe the framework of our algorithm. This will be used for both the (1, 2)-SNDP and TNCSP. To start the optimization we consider the following linear program (LP):

$$\begin{aligned} \min \sum_{e\in E} c(e)x(e) \qquad & \text{subject to} \\ x(\delta(v)) \geq \text{con}(v) \qquad & \text{for all } v \in V, \\ 0 \leq x(e) \leq 1 \qquad & \text{for all } e \in E. \end{aligned}$$

The optimal solution $y \in \mathbb{R}^E$ of this relaxation of SNDP is feasible for the problem if y is an integer vector that satisfies all the cut inequalities. Usually, the solution y is not feasible, and thus, in each iteration of the branch-and-cut algorithm, it is necessary to generate further inequalities that are valid for the survivable network polytope but violated by the current solution y. The separation of valid inequalities is performed in the following order:

1. cut constraints;
2. partition constraints (only for $r \in \{1, 2\}^V$);
3. node-partition constraints (only for TNCSP);
4. F-partition constraints (inequalities (9.8) and (9.13)).

We remark that all inequalities are global (i.e., valid in all the branch-and-cut tree) and several constraints may be added at each iteration. Moreover, we go to the next class of inequalities only if we haven't found any violated inequalities. Our strategy is to try to detect some violated constraints (until no more violated inequalities are found) at each node of the branch-and-cut tree in order to obtain the best possible lower bound and thus limit the number of generated nodes.

Now we describe the separation routines used in our branch-and-cut algorithm. These may be either exact algorithms or heuristics, depending on the associated class of inequalities. All our separation algorithms are applied on G' with weights $(\overline{y}'(e);\ e \in E')$ associated with its edges, where $\overline{y}'$ is the restriction on E' of the current LP solution $\overline{y}$. Frequently, our separation algorithms are based on maximum flow computations that can be done in polynomial time using the efficient Goldberg–Tarjan algorithm [15] that runs in $O(n^3)$ time.

The separation of the cut constraints (resp. inequalities (9.5)) can be performed by computing minimum cuts in G' (resp. $G' \setminus v$). This can be done in polynomial time using the Gomory–Hu algorithm [16]. This algorithm produces the so-called Gomory–Hu tree with the property that for all pairs of nodes $s, t \in V'$ the minimum (s, t)-cut in the tree is also a minimum (s, t)-cut in the graph G'. Actually, we use the algorithm developed by Gusfield [22], which requires $|V'| - 1$ maximum flow computations. Moreover, for the SNDP, we have to distinguish between the case when all node types are equal to 2 and the one when $r \in \{1, 2\}^V$. In fact, in the second case, the right-hand side in inequalities (9.12) may be equal to 1 or 2. In consequence, our algorithm is performed in two steps. First we compute the Gomory–Hu tree involving only the nodes $v \in V'$ such that $r'(v) = 2$.

Afterward, we consider the graph $G'_1 = (V'_1, E'_1)$ obtained from G' by shrinking all the nodes having a type equal to 2. We notice that all node types are equal to 1 in G'_1. Then we calculate the Gomory–Hu tree for the graph G'_1. The time complexity to build G'_1 is $O(n^2)$ in the worst case, and this separation algorithm consists of solving $|V'| - 1$ maximum flow problems. In consequence, the exact algorithm that permits us to separate the cut inequalities is implemented to run in $O(n^4)$ time.

Now we turn our attention to the separation of partition inequalities (9.12). We recall that these constraints are exclusively considered when $r \in \{1, 2\}^V$. In [25], Kerivin and Mahjoub have shown that these inequalities can be separated in polynomial time. In fact, the separation problem for inequalities (9.12) reduces to minimizing a submodular function. This latter problem can be solved in polynomial time using either the algorithm of Schrijver [34] or the algorithm of Iwata, Fleischer, and Fujishige [23]. However, these algorithms lead to a time complexity (about $O(n^9)$) that does not permit us to implement them. Recently, Barahona and Kerivin [1] showed that the separation problem for the partition inequalities (9.12) reduces to a sequence of n submodular flow problems which leads to an $O(n^7)$ algorithm. Hence to separate partition inequalities we shall develop heuristics. We consider two cases depending on the right-hand side of inequality (9.12).

The first case deals with partitions in which all the nodes of type 2 are in the same set V_i (i.e., $I_2 = \emptyset$). By considering the graph G'_1 introduced above for the separation of cut constraints, we can write inequality (9.12) as

$$x(\delta(V_1, \dots, V_p)) \geq p - 1, \tag{9.14}$$

where $V_1, \dots, V_p$ is a partition of V'_1. In the graph G'_1, all the nodes have type 1, and thus the separation problem for inequalities (9.14) on G'_1 can be solved in polynomial time using either the algorithm of Cunningham [12] or the algorithm of Barahona [4] (see also Baïou, Barahona, and Mahjoub [2]). Cunningham's algorithm reduces the problem to $|E'_1|$ minimum cut problems, whereas Barahona's algorithm provides a reduction to $|V'_1|$ minimum cut problems. For more efficiency we have implemented Barahona's algorithm, which, in consequence, runs in $O(n^4)$ time. This algorithm is also used for separating inequalities (9.6) for TNCSP.

Now let us assume that there are at least two sets V_i and V_j with $r'(V_i) = r'(V_j) = 2$. In consequence, inequality (9.12) has the form

$$x(\delta(V_1, \dots, V_p)) \geq p. \tag{9.15}$$

In spite of having a polynomial-time algorithm based on the minimization of a submodular function to separate inequalities (9.15), we have decided to develop heuristics for a more efficient time complexity. More precisely, we consider two heuristics. The first one uses straight Barahona's algorithm, mentioned above, on the graph G'. First we may suppose that no constraints of type (9.14) are violated. If there were, it would have been detected before. Then Barahona's algorithm is applied. If there is a violated inequality of type (9.14), then we should have an inequality of type (9.15) violated by more than one. If not, then we have

$$x(\delta(V_1, \dots, V_p)) \geq p - 1 \qquad \text{for all partitions such that } I_2 \neq \emptyset. \tag{9.16}$$

In this case we use a second heuristic to find a partition that satisfies inequalities (9.16) but that induces a violated inequality of type (9.15). This second heuristic transforms cuts given by the Gomory–Hu tree into partitions. This transformation from a cut into a partition is based on the following lemma [24].

Lemma 9.11. *There exists a violated partition inequality of type (9.15) if and only if there exists* $W_1 \subset V'$ *with* $\text{con}(W_1) = 2$ *such that*

$$\overline{y}'(\delta(W_1)) - 2 + \min_{W_2,\dots,W_p} \{\overline{y}'(\delta_{G(\overline{W}_1)}(W_2, \dots, W_p)) - (p-2)\} < 0.$$

Proof. The proof is easy and therefore omitted. □

The order chosen to separate the various classes of constraints permits us to assume that all the cut inequalities are satisfied by $\overline{y}'$. In consequence, if a cut $\delta(\overline{W}_1) \subseteq E'$ can be transformed into a partition $W_1, \dots, W_p$ of V' that induces a violated inequality (9.15), we then have

$$\min_{W_2,\dots,W_p} \{\overline{y}'(\delta_{G(\overline{W}_1)}(W_2, \dots, W_p)) - (p-2)\} < 0. \tag{9.17}$$

The problem (9.17) can then be solved using Barahona's algorithm. Moreover, we would like to obtain a partition $W_1, \dots, W_p$ of V' such that $I_2 \neq \emptyset$. In consequence, we compute the Gomory–Hu tree involving only nodes having a type equal to 2. This implies that $r'(W) = r'(\overline{W}) = 2$ for all cuts $\delta(W)$ in the Gomory–Hu tree. So, our second heuristic works as follows. We try to transform each cut $\delta(W)$ generated by the Gomory–Hu tree into a partition by applying Barahona's algorithm to $G(W)$ as well as $G(\overline{W})$. The aim is to obtain partitions $W, W'_1, \dots, W'_{p_1}$ and $\overline{W}, W''_1, \dots, W''_{p_2}$ that produce violated inequalities (9.15), where $W'_1, \dots, W'_{p_1}$ and $W''_1, \dots, W''_{p_2}$ are, respectively, partitions of $\overline{W}$ and W. This heuristic requires $O(n)$ iterations, where each iteration consists of computing a minimum cut and applying Barahona's algorithm twice. Since the time complexity of each iteration is $O(n^4)$, the whole algorithm can be implemented to run in $O(n^5)$ time.

We now discuss our separation routines for the F-partition inequalities (9.8) and (9.13). The complexity of the separation problem for F-partition inequalities is still unknown. However, Baïou, Barahona, and Mahjoub [2] have given a polynomial-time algorithm that permits us to separate these inequalities if the edge subset F is fixed and $r(v) = 2$ for all $v \in V$. In order to develop more general separation routines, we have devised two heuristics that permit us to separate both of inequalities (9.8) and (9.13).

Using Theorem 9.2 for node types all equal to 2 as well as for $r \in \{1, 2\}^V$, the first heuristic works as follows. We look for cycles in G' that are formed by edges whose value is fractional in $\overline{y}'$. Thus, for each detected cycle $(v_1, \dots, v_p)$, we try to find an edge subset F among the edges having exactly one extremity in the cycle in such a way that the F-partition inequality induced by $V' \setminus \{v_1, \dots, v_p\}, \{v_1\}, \dots, \{v_p\}$ and F is violated by $\overline{y}'$. This heuristic can be implemented using a recursive algorithm that determines the 2-connected components in a graph. On the whole, the latter algorithm leads to an $O(n^2)$ time complexity.

Since the previous heuristic might fail to find violated F-partition inequalities, we have developed another heuristic that consists of transforming cuts into F-partitions. More precisely, we are interested in cuts containing the maximum number of edges $e \in E'$ such that $\overline{y}'(e) = 1$. In fact, these seem to have the larger probability of belonging to F. To determine such cuts we compute the Gomory–Hu tree for the graph G' with the weight vector $w \in \mathbb{R}^{E'}$ given by

$$w(e) = 1 - \overline{y}'(e) \qquad \text{for all } e \in E'.$$

Therefore, for each cut $\delta(W)$ in the Gomory–Hu tree, we calculate

$$p_1 = |\{v \in W \mid r'(v) = 1\}|,$$
$$p_1' = |\{v \in \overline{W} \mid r'(v) = 1\}|,$$

and then we try to select an edge subset $F_1 \subseteq \delta(W)$ (resp. $F_1' \subseteq \delta(W)$) such that $|F_1|$ and p_1 (resp. $|F_1'|$ and p_1') have different parity and

$$\overline{y}'(\delta(W) \setminus F_1) + \overline{y}'(E(W)) \geq |W| - \frac{p_1 + |F_1| - 1}{2}$$
$$\left(\text{resp. } \overline{y}'(\delta(W) \setminus F_1') + \overline{y}'(E(\overline{W})) \geq |\overline{W}| - \frac{p_1' + |F_1'| - 1}{2}\right)$$

is not satisfied. This heuristic requires $O(n)$ minimum cut problems. In consequence, the whole heuristic runs in $O(n^4)$ time.

9.5 Computational Results

The branch-and-cut algorithm described in the previous section has been implemented in C, using MINTO 3.0 [32] to manage the branch-and-bound (B&B) tree and CPLEX™ 4.0 [6] as LP solver. It was tested on a Pentium® III 450 MHz processor with 256 MB RAM, running under Linux®. We fixed the maximum CPU time to 3 hours. The test problems were obtained by taking TSP test problems from the TSPLIB library [33]. The testset consists of complete graphs whose edge costs are the rounded Euclidean distances between the edge's endnodes. Moreover, if we consider the (1, 2)-SNDP, node types are randomly generated. In all our experiments we have used the reduction operations introduced in the previous sections, unless otherwise specified. In the various tables we give, the entries are as follows:

V_1: the number of nodes with type 1 (only for $r \in \{1, 2\}^V$);

NC: the number of cut inequalities used before branching;

NP: the number of partition (resp. node-partition) inequalities used before branching (only for $r \in \{1, 2\}^V$ (resp. TNCSP));

NFP: the number of F-partition inequalities used before branching;

Gap1: the relative error between the optimal value and the lower bound achieved by the trivial and cut constraints and inequalities (9.14) (resp. (9.6)) (these latter only for $r \in \{1, 2\}^V$ (resp. TNCSP));

Gap2: the relative error between the optimal value and the lower bound achieved by the cutting plane phase;

Copt: the optimal value;

SB: the number of generated nodes in the branch-and-cut tree;

TT: the total time in seconds.

Our first series of experiments concerns TECSP (i.e., $r(v) = 2$ for all $v \in V$). We have considered graphs with 24 to 417 nodes. Table 9.1 reports results when the F-partition

Table 9.1. *Results with node types all equal to 2.*

Problem	NC	NFP	Gap1	Gap2	Copt	SB	TT
gr24*	5	2	0.00	0.00	1272	1	0.03
fri26*	10	0	0.00	0.00	937	1	0.03
bayg29*	11	4	0.12	0.00	1610	1	0.07
bays29*	9	20	0.32	0.00	2020	1	0.14
dantzig42*	19	2	0.29	0.14	699	3	0.16
swiss42*	10	1	0.08	0.00	1273	1	0.10
att48*	24	1	0.23	0.17	10628	3	0.35
gr48	21	36	1.43	0.28	5031	7	1.26
hk48*	18	3	0.14	0.00	11461	1	0.16
eil51*	22	62	0.82	0.00	426	1	2.82
berlin52*	6	0	0.00	0.00	7542	1	0.13
brazil58*	28	4	0.16	0.00	25395	1	0.27
st70*	40	78	0.59	0.22	675	3	2.70
eil76*	23	34	0.19	0.00	538	1	1.76
pr76	44	156	1.29	0.24	106492	17	14.97
rat99*	43	59	0.41	0.12	1211	4	7.04
rd100*	80	15	0.13	0.00	7910	1	2.02
kroA100	70	111	1.53	0.28	21261	44	18.05
kroB100	66	175	1.02	0.24	22059	9	79.71
kroC100*	84	143	1.33	0.26	20749	15	40.08
kroD100*	63	182	0.72	0.17	21294	7	47.66
kroE100	41	191	0.56	0.04	21923	3	14.57
eil101*	45	73	0.24	0.16	629	5	27.19
lin105*	49	7	0.06	0.00	14379	1	1.09
pr107*	86	0	0.00	0.00	44303	1	1.22
bier127*	78	245	0.72	0.17	118282	21	162.08
kroA150*	95	77	0.85	0.51	26524	182	548.29
kroB150	92	48	1.26	0.82	26060	161	206.70
u159*	59	24	0.37	0.20	42080	9	11.94
ts225	0	64	1.50	0.00	117363	1	8.74
a280*	115	182	0.50	0.05	2579	3	155.15
pr264*	90	59	0.23	0.02	49135	3	32.71
pr299	150	207	0.71	0.05	47718	5	142.68
fl417	258	64	0.20	0.07	11813	11	221.06

* The optimal solution is a tour.

inequalities (9.8) are taken into account. It appears that our separation routines for F-partition inequalities detect a large enough number of such inequalities. We can note that in about 42% of the problems, the routines that separate the cut and the F-partition constraints permit us to obtain the optimal solution in the cutting plane phase (i.e., no branching is needed). For the other problems, the F-partition inequalities improve the lower bound given at the root node of the branch-and-cut tree, and this lower bound becomes at most 0.3% over the optimum except for the instances kroA150 and kroB150. In consequence, the F-partition constraints seem to be very efficient for TECSP. In order to evaluate the effect of the reduction operations, we performed the same experiments without using them. In most cases, the algorithm spent more and more time (until 10 times over) to get the optimal solution. Moreover, we generated fewer F-partition inequalities and the number of evaluated nodes in the branch-and-cut tree increased. Thus, it seems that our heuristic separating inequalities (9.8) is less efficient without the reduction operations. On the other hand, we have also considered a branch-and-cut algorithm using only the cut constraints. In this case, we may reach the maximum CPU time (i.e., 3 hours) without having the optimal value. To illustrate this, we use ts225, which is a very significant problem. In fact, with the F-partition constraints and the reduction operations we need about 8.74 seconds to obtain the optimum, whereas 3 hours are not enough to solve the problem without both.

Moreover, we have to mention that most of the reduced graphs we obtained were generalized odd wheel configurations satisfying properties (i) and (ii) of Theorem 9.2. So the associated generalized odd wheel inequalities have been violated by the reduced solutions. In consequence, the initial fractional solutions have been cut off by (lifted) F-partition inequalities. By Corollary 9.9, it then follows that these F-partition inequalities are facet-defining on the original graph if the corresponding generalized odd wheel inequalities are facet-defining on the reduced graphs. Since the reduction operations together with the F-partition inequalities have greatly speeded up the experiments, this might imply that many of the generated fractional solutions are of rank 1 and the corresponding cuts are F-partition facets.

In addition, although the reduction operations are restricted to the extreme points of rank 1, they permitted us to generate cutting planes for extreme points of rank k, $k > 1$. These have also been cut off by F-partition constraints. The most frequent case is when all the fractional values are $\frac{1}{2}$ and the fractional edges induce edge-disjoint odd cycles. Such a point is not of rank 1. However, if we keep only one fractional cycle and give value 1 to all the remaining fractional edges, we get an extreme point (of rank 1) that is dominated by the first one. Moreover, the F-partition that cuts off the new solution also cuts off the first one.

Table 9.2 presents the results concerning TNCSP. The test problems have up to 574 nodes. The 5 last problems in the table have been solved on a SUN-Sparc5 with 128 MB RAM without time limit. We note from the table that a large number of F-partition inequalities have been generated for most of the problem instances. As for the TECSP, it seems that these inequalities are also efficient for the TNCSP. In fact, if we compare the relative errors between the optimal value and the lower bound obtained in the cutting plane phase without (Gap1) and with (Gap2) of the F-partition inequalities, we can see that the latter one has greatly decreased for most of the problems and has become 0% for some of them. This implies that these inequalities also play a central role for the TNCSP.

Finally, we notice that the solutions obtained for TECSP and TNCSP are also optimal for the TSP in about 70% of the cases. This shows that our algorithm may also be useful

Table 9.2. *Results for the TNCSP.*

Problem	NC	NP	NFP	Gap1	Gap2	Copt	SB	TT
eil51*	18	0	61	0.82	0.00	426	3	1.89
A50	33	0	50	0.86	0.06	5615	3	0.80
st70*	36	0	65	0.59	0.22	675	3	1.98
eil76*	10	0	4	0.19	0.00	538	1	0.29
pr76	51	0	78	1.29	0.24	106492	9	3.06
rat99*	31	0	61	0.41	0.13	1211	5	2.10
kroA100	62	9	154	1.53	0.10	21261	11	9.47
kroB100	75	0	186	1.02	0.21	22059	7	20.51
kroC100*	73	0	155	1.33	0.19	20749	11	7.49
kroD100*	70	0	45	0.72	0.15	21294	5	2.55
kroE100	47	0	112	0.56	0.03	21923	3	4.69
rd100*	50	0	26	0.13	0.00	7910	1	1.16
lin105*	47	0	2	0.06	0.00	14379	1	0.78
pr107*	77	0	0	0.00	0.00	44303	1	0.55
pr124*	95	11	169	1.18	1.18	59030	59	24.87
bier127*	80	0	161	0.72	0.18	118282	7	52.10
ch130*	97	16	245	0.56	0.15	6110	14	180.75
pr144*	101	0	83	0.17	0.17	58537	9	20.33
ch150*	88	0	212	0.58	0.18	6528	23	179.48
pr152*	170	74	619	0.64	0.45	73682	583	1142.27
rat195	93	578	16	0.85	0.36	2319	27	2923.05
d198*	93	7	587	0.43	0.26	15780	109	3034.25
A200	116	1	352	0.34	0.10	10947	29	165.31
kroB200*	126	0	280	0.92	0.14	29437	23	215.40
ts225	0	0	49	1.50	0.00	117363	1	6.91
tsp225	194	127	872	0.89	0.29	3913	43	19898.87
pr264*	93	64	13	0.23	0.22	49135	3	606.38
pr299	207	3	105	0.71	0.004	47718	3	518.72
lin318*	227	0	208	0.19	0.03	42029	5	4204.63
u574	426	93	840	0.41	0.04	36866	5	199424.78

* The optimal solution is a tour.

for the TSP. Now we turn our attention to the case where $r \in \{1, 2\}^V$. In this case the partition inequalities (9.12) are used in our branch-and-cut algorithm. Since the SNDP with these node types seems to be harder than the 2-edge connected spanning subgraph problem, the test problems have at most 101 nodes. We recall that the cut inequalities and the partition inequalities (9.14) can be separated in polynomial time, whereas for both partition inequalities (9.15) and F-partition inequalities (9.13) we use heuristics. In consequence, to estimate the performance of these heuristics, we compare the lower bound obtained with and without them (i.e., Gap1 and Gap2). Table 9.3 reports computational results for these problems. It appears that the relative error between the optimal value and the lower bound achieved by the cut inequalities and the partition inequalities (9.14) is often about 10%. If

Table 9.3. *Results with* $r \in \{1, 2\}^V$.

Problem	V_1	NC	NP	NFP	Gap1	Gap2	Copt	SB	TT
gr24	10	43	94	2	7.11	0.00	1265	1	1.03
gr24	15	68	135	0	13.33	0.00	1161	1	5.62
fri26	11	44	61	0	8.00	0.17	898	3	1.60
fri26	15	79	166	0	11.84	0.00	872	1	7.84
bayg29	12	61	149	0	10.66	0.00	1534	1	2.35
bayg29	18	99	67	0	13.71	0.00	1452	1	2.67
bays29	12	66	43	0	7.58	0.00	1859	1	0.60
bays29	18	77	44	0	11.41	0.00	1753	1	2.93
dantzig42	19	137	287	2	10.72	0.00	687	1	20.51
dantzig42	25	103	128	0	10.63	0.08	650	1	13.80
swiss42	19	98	314	0	10.24	0.00	1256	1	67.24
swiss42	25	130	284	0	12.80	0.00	1244	1	27.80
att48	25	120	339	4	12.01	0.44	10259	13	202.68
att48	30	172	319	0	13.20	0.00	10136	1	64.81
gr48	20	145	281	2	10.04	0.36	4860	7	755.06
gr48	21	118	421	5	9.12	0.89	4863	23	361.41
hk48	25	131	219	2	11.46	0.10	11181	9	33.83
hk48	30	156	804	0	12.99	0.00	11215	1	100.79
eil51	21	171	1143	14	11.51	0.30	413	4	345.17
eil51	23	166	1933	15	11.81	0.42	413	14	812.46
eil51	27	175	669	3	13.93	0.00	410	10	137.90
eil51	31	280	1112	2	13.76	0.23	407	14	389.44
berlin52	28	151	298	0	11.20	0.01	7435	3	115.22
berlin52	32	192	280	0	12.67	0.00	6963	1	48.76
brazil58	25	172	144	0	4.34	0.00	24524	1	28.29
brazil58	35	166	266	0	6.36	0.00	23501	1	76.60
st70	29	169	380	5	8.92	0.70	647	25	380.16
st70	30	216	1719	2	9.85	0.85	662	119	2329.06
st70	39	239	1252	12	10.79	0.16	626	3	522.32
st70	40	247	681	0	11.44	0.00	633	1	357.45
pr76	32	211	2441	13	7.25	0.16	98782	13	784.30
pr76	44	294	930	1	9.29	0.90	99680	1129	6134.49
rat99	55	390	1433	0	12.12	0.00	1164	1	2183.86
rd100	43	325	981	3	8.98	0.05	7645	3	1089.64
kroA100	46	336	1552	7	9.08	0.55	20539	93	3034.27
eil101	46	486	8396	2	12.90	0.00	598	1	8525.02

we consider the partition inequalities (9.15) and the F-partition inequalities (9.13), this error decreases considerably to become lower than 1% except for three problems. Moreover, this permits us to solve several problems without the branching phase. Thus these constraints seem to be very useful to improve the performance of our branch-and-cut algorithm. We can also notice that a large number of partition inequalities (9.12) are added in the cutting plane

Table 9.4. *Generated partition inequalities (9.12).*

Problem	PAR1	PAR2B	PAR2H
st70	16	53	311
st70	32	72	1615
st70	72	102	1078
st70	65	99	517
pr76	30	45	2366
pr76	22	50	407
pr76	57	50	823
rat99	55	101	1813
rat99	36	109	1368
rat99	70	163	1198
rd100	44	64	873
kroA100	46	49	1457
eil101	84	258	8054

phase. In Table 9.4 we report some details about the various generated partition constraints for the larger problems we solved. The entries of this table are, from left to right, the problem name, and then the following:

PAR1: the number of partition inequalities (9.14) generated by Barahona's algorithm,

PAR2B: the number of partition inequalities (9.15) generated by Barahona's algorithm, and

PAR2H: the number of partition inequalities (9.15) generated by our heuristic transforming cuts into partitions.

Usually, among the detected partition inequalities (9.12), we have at least 75% of the partition inequalities (9.15), and this percentage increases with the number of nodes in the graph. Since this kind of partition inequality is generated by a heuristic, the latter seems to be very efficient. For example, our heuristic based on transforming cuts into partitions permits us to detect 8396 violated partition inequalities of type (9.15) for the problem eil101. Together with the other separation heuristic for partition inequalities of type (9.15) and those for F-partition inequalities (9.13), the relative error between the optimal value and the lower bound obtained after the cutting plane phase decreases from 12.9% to 0%, and the problem is solved without branching.

Furthermore, the F-partition constraints (9.13) appear in a small proportion in our branch-and-cut algorithm. This does not imply that these inequalities are not necessary for this kind of node type, but just that our heuristic may be less efficient in this case. Afterward, we applied our branch-and-cut algorithm without using the reduction operations. We noticed that the solution of our problem consumes more and more CPU time and the number of generated nodes in the branch-and-cut tree grows. Moreover, the numbers of added partition inequalities are lower than those reported in Table 9.3. This implies that the relative error before branching (i.e., Gap2) increases.

9.6 Concluding Remarks

We have studied the (1, 2)-SNDP and the TNCSP. We have described sufficient conditions for the F-partition inequalities to be facet-defining for the associated polytope when $r(v) = 2$ for all $v \in V$ and shown that the fractional extreme points of rank 1 of the linear relaxation of that polytope can be separated in polynomial time using F-partition inequalities. We have provided separation algorithms for these inequalities as well as for the partition inequalities. Using these results, we have described a branch-and-cut algorithm for these problems. The algorithm uses some reduction operations that may permit us to considerably reduce the graph supporting the fractional solution and to accelerate the separation process. Our computational results have shown that the F-partition inequalities are very effective for both TECSP and TNCSP. They also show the importance of the partition inequalities for the more general model. We could also measure the performance of our separation techniques. In particular, our heuristic for separating the partition inequalities, when the node connectivity types are one and two, has shown to be very efficient. In addition, the reduction operations have been essential to good performance of our branch-and-cut algorithm.

It would be interesting to extend the ideas developed in this work to the more general survivable network design model when the node connectivity types are not restricted only to one and two. We are now making investigations in this direction.

Acknowledgements. We would like to thank the anonymous referee for constructive comments. We also wish to thank Mourad Baïou and Francisco Barahona for valuable discussions and suggestions and Pierre Fouilhoux and Pierre Pesneau for reading an earlier version of the chapter and pointing out several improvements.

Bibliography

[1] F. Barahona and H. Kerivin. *Separation of Partition Inequalities with Terminals*, preprint, 2003.

[2] M. Baïou, F. Barahona, and A.R. Mahjoub. Separation of partition inequalities. *Mathematics of Operations Research*, 25:243–254, 2000.

[3] M. Baïou and A.R. Mahjoub. Steiner 2-edge connected subgraph polytopes on series-parallel graphs. *SIAM Journal on Discrete Mathematics*, 10:505–514, 1997.

[4] F. Barahona. Separating from the dominant of the spanning tree polytope. *Operations Research Letters*, 12:201–203, 1992.

[5] F. Barahona and A.R. Mahjoub. On two-connected subgraph polytopes. *Discrete Mathematics*, 147:19–34, 1995.

[6] R.E. Bixby. *Implementing the Simplex Method: The Initial Basis*. Technical Report TR 90-32, Department of Mathematical Sciences, Rice University, Houston, TX, 1991.

[7] S.C. Boyd and T. Hao. An integer polytope related to the design of survivable communication networks. *SIAM Journal on Discrete Mathematics*, 6:612–630, 1993.

[8] S. Chopra. Polyhedra of the equivalent subgraph problem and some edge connectivity problems. *SIAM Journal on Discrete Mathematics*, 5:321–337, 1992.

[9] G. Cornuéjols, J. Fonlupt, and D. Naddef. The traveling salesman problem on a graph and some related integer polyhedra. *Mathematical Programming*, 33:1–27, 1985.

[10] R. Coullard, A. Rais, R.L. Rardin, and D.K. Wagner. Linear-time algorithm for the 2-connected Steiner subgraph problem on special classes of graphs. *Networks*, 23:195–206, 1993.

[11] R. Coullard, A. Rais, R.L. Rardin, and D.K. Wagner. The dominant of the 2-connected-Steiner subgraph polytope for w_4-free graphs. *Discrete Applied Mathematics*, 66:195–206, 1996.

[12] W.H. Cunningham. Optimal attack and reinforcement of a network. *Journal of the Association for Computing Machinery*, 32:549–561, 1985.

[13] J. Fonlupt and A.R. Mahjoub. Critical Extreme Points of the 2-Edge Connected Spanning Subgraph Polytope. *Proceedings IPCO'99*, volume 1610 of *Lecture Notes in Computer Science*, pages 166–183, Springer-Verlag, Berlin, 1999.

[14] M.X. Goemans and D.J. Bertsimas. Survivable network, linear programming and the parsimonious property. *Mathematical Programming*, 60:145–166, 1993.

[15] A.V. Goldberg and R.E. Tarjan. A new approach to the maximum-flow problem. *Journal of the Association for Computing Machinery*, 35:921–940, 1988.

[16] R.E. Gomory and T.C. Hu. Multi-terminal network flows. *Journal of the Society for Industrial and Applied Mathematics*, 9:551–570, 1961.

[17] M. Grötschel and C.L. Monma. Integer polyhedra arising from certain network design problems with connectivity constraints. *SIAM Journal on Discrete Mathematics*, 3:502–523, 1990.

[18] M. Grötschel, C.L. Monma, and M. Stoer. Computational results with a cutting plane algorithm for designing communication networks with low-connectivity constraints. *Operations Research*, 40:309–330, 1992.

[19] M. Grötschel, C.L. Monma, and M. Stoer. Facets for polyhedra arising in the design of communication with low-connectivity constraints. *SIAM Journal on Optimization*, 2:474–504, 1992.

[20] M. Grötschel, C.L. Monma, and M. Stoer. Design of survivable networks. In M.O. Ball et al., editors, *Network Models*, pages 617–671, North-Holland, Amsterdam, 1995.

[21] M. Grötschel, C.L. Monma, and M. Stoer. Polyhedral and computational investigations for designing communication networks with high survivability requirements. *Operations Research*, 43:1012–1024, 1995.

[22] D. Gusfield. *Very Simple Algorithms and Programs for All Pairs Network Flow Analysis*. Computer Science Division, University of California, Davis, 1987.

[23] S. Iwata, L. Fleischer, and S. Fujishige. *A Strongly Polynomial-Time Algorithm for Minimizing Submodular Functions*. In Algorithm Engineering on a New Paradigm, pages 11–23. Kyoto University, Japan, 1999. (In Japanese).

[24] H. Kerivin. *Réseaux fiables et polyèdres*. Ph.D. Dissertation, Université Blaise Pascal, Clermont-Ferrand, France, 2000.

[25] H. Kerivin and A.R. Mahjoub. Separation of partition inequalities for the (1,2)-survivable network design problem. *Operations Research Letters*, 30:265–268, 2002.

[26] C-W. Ko and C.L. Monma. *Heuristic Methods for Designing Highly Survivable Communication Networks*. Technical report, Bellcore, Piscataway, NJ, 1989.

[27] A.R. Mahjoub. Two-edge connected spanning subgraphs and polyhedra. *Mathematical Programming*, 64:199–208, 1994.

[28] A.R. Mahjoub. On perfectly 2-edge connected graphs. *Discrete Mathematics*, 170:153–172, 1997.

[29] A.R. Mahjoub and C. Nocq. On the linear relaxation of the 2-node connected subgraph polytope. *Discrete Applied Mathematics*, 95:389–416, 1999.

[30] C.L. Monma, B.S. Munson, and W.R. Pulleyblank. Minimum-weight two connected spanning networks. *Mathematical Programming*, 46:153–171, 1990.

[31] C.L. Monma and D.F. Shallcross. Methods for designing communication networks with certain two-connected survivability constraints. *Operations Research*, 37:531–541, 1989.

[32] G.L. Nemhauser, M.W.P. Savelsbergh, and G.C. Sigismondi. Minto, a mixed integer optimizer. *Operations Research Letters*, 15:47–58, 1994.

[33] G. Reinelt. TSPLIB—a traveling salesman problem library. *ORSA Journal on Computing*, 3:376–384, 1991.

[34] A. Schrijver. A combinatorial algorithm minimizing submodular functions in strongly polynomial time. *Journal of Combinatorial Theory*, 80:346–355, 2000.

[35] K. Steiglitz, P. Weiner, and D.J. Kleitman. The design of minimum cost survivable networks, *IEEE Transactions and Circuit Theory*, 16:455–460, 1969.

[36] M. Stoer. Design of Survivable Networks, volume 1531 of *Lecture Notes in Mathematics*. Springer-Verlag, Berlin, 1992.

Chapter 10

The Domino Inequalities for the Symmetric Traveling Salesman Problem

*Denis Naddef**

Abstract. This chapter is the first of a series of two papers that deal with the domino parity inequalities introduced by Adam Letchford [4] in an article in which he shows that such inequalities can be separated in polynomial time, provided that the support graph of the fractional point to be separated is planar. These inequalities contain the combs as a special case. In this chapter we give necessary conditions for such inequalities to be facet-inducing. This will bring a big deal of structure to these inequalities. In a second [6] paper some facet-inducing results are given for the symmetric traveling salesman polytope.

MSC 2000. 90C57, 90C27

Key words. Traveling salesman problem, traveling salesman polytope, facet, valid inequality, linear description

10.1 Introduction

Adam Letchford defines in [4] the domino parity inequalities for the symmetric traveling salesman polytope and gives a polynomial algorithm for the separation of such constraints when the support graph associated with the point to be separated is planar, generalizing a result of Fleischer and Tardos [3] for maximally violated comb inequalities. Boyd, Cockburn, and Vella [1] give a first class of facet-inducing domino parity inequalities. These domino parity inequalities, like the comb inequalities, are defined on a subset of the nodes called the *handle* and on an odd number of subsets of the nodes called teeth. Unlike the combs, teeth need not be nonintersecting and can be in quite a general position. In fact,

*Laboratoire Informatique et Distribution-IMAG, Institut National Polytechnique de Grenoble-France (Denis.Naddef@imag.fr).

these inequalities are given in [4] in a very general form and it is very difficult to understand their contribution to the linear description of the symmetric traveling salesman polytope. We redefine here these inequalities in what we hope to be a more comprehensible form.

We will first give a structural property of the teeth for the case when these inequalities are facet-inducing, showing that the teeth are nested, except possibly in one pathological case. We then give more necessary conditions for these inequalities to be facet-inducing. In the last section we explore how the basic idea behind the domino inequalities can be generalized to other inequalities by looking at the case of the path inequalities.

We end this section with some notation. We assume that we are dealing with a complete graph $K_n = (V, E)$, with $n = |V|$ the number of vertices. Let $S \subset V$. Then $\delta(S)$ (resp. $\gamma(S)$) represents the set of edges with exactly one endnode in S (resp. both endnodes in S), i.e., $\delta(S) = \{(u, v) \in E : u \in S, v \notin S\}$ (resp. $\gamma(S) = \{(u, v) \in E : u \in S, v \in S\}$). For the sake of brevity we say that an edge e is contained in S if $e \in \gamma(S)$. The edgeset $\delta(S)$ is in general called the *coboundary* of S (some authors say *cocycle* of S). We will also say that an edge of $\delta(S)$ crosses the border of S. We write $\delta(v)$ instead of $\delta(\{v\})$ for $v \in V$. For $S \subset V$ and $T \subset V \setminus S$ we denote by $(S : T) = (T : S)$ the set of edges with one endnode in S and the other in T. For a subset of edges $\tilde{E}$ we let $x(\tilde{E}) = \sum_{e \in \tilde{E}} x_e$.

Since the content of this section is not intended for a newcomer to the field of the traveling salesman polytope, we assume the reader is familiar with the basics of the traveling salesman polytope and the linear algebra it requires. The main idea is that with each Hamilton cycle Γ one associates a (0, 1)-vector indexed by the edges E, the component associated with edge e having value 1 if and only if edge $e \in \Gamma$. This vector will be referred to as the *representative* or *incidence* vector of Γ. The symmetric traveling salesman polytope is the convex hull of all the incidence vectors of Hamilton cycles.

10.2 The Domino Inequalities

A *general domino configuration* is defined by a set $H \subset V$ called the *handle* and a family of sets called teeth $\mathcal{T} = \{T_1, \dots, T_j, \dots, T_t\}$, $t \geq 3$ and odd. Moreover, each tooth is partitioned into two disjoint nonempty subsets, i.e., $T_j = A_j \cup B_j$, $A_j \neq \emptyset$, $B_j \neq \emptyset$, $A_j \cap B_j = \emptyset$. The handle H and the family of teeth $\mathcal{T}$ satisfy the following conditions:

$$H \cap T_j \neq \emptyset \quad \text{for } j = 1, \dots, t, \tag{10.1}$$

$$T_j \setminus H \neq \emptyset \quad \text{for } j = 1, \dots, t, \tag{10.2}$$

$$T_j = A_j \cup B_j \neq V \text{ for } j = 1, \dots, t. \tag{10.3}$$

Note that unlike the comb inequalities there are no other conditions on the teeth. A tooth $T_j = A_j \cup B_j$ will also be referred to as a *domino*; the sets A_j and B_j will be called its *half-dominoes*.

We call *fundamental sets* of a domino configuration the sets A_i, B_i, and T_i, $i = 1, \dots, t$.

Let $\mu_e = |\{i : e \in (A_i : B_i)\}| - \epsilon(e, H)$, where $\epsilon(e, H) = 1$ if $e \in \delta(H)$ and $\epsilon(e, H) = 0$ else, and consider the following *general domino* inequality:

$$x(\delta(H)) + \sum_{j=1}^{t} x(\delta(T_j)) + 2\sum_{e \in E} \left\lceil \frac{\mu_e}{2} \right\rceil x_e \geq 3t + 1, \tag{10.4}$$

where $\lceil a \rceil$ represents the smallest integer greater than or equal to a.

Assume one adds to the definition of the general domino configuration the following nonintersecting condition on the sets $(A_j : B_j)$:

$$(A_j : B_j) \cap (A_k : B_k) = \emptyset \text{ for } 1 \leq j < k \leq t. \tag{10.5}$$

A domino configuration satisfying this last condition is called a *noncrossing domino configuration*. Let $C = (\bigcup_{j=1}^{t}(A_j : B_j)) \setminus \delta(H)$, that is, the set of edges going from one half-domino to the other that do not cross the border of the handle H. The *noncrossing domino inequality* is

$$x(\delta(H)) + \sum_{j=1}^{t} x(\delta(T_j)) + 2x(C) \geq 3t + 1. \tag{10.6}$$

A *spanning closed walk* is a family of edges E' (the same edge may appear several times) such that $G = (V, E^\star)$ is Eulerian (i.e., spanning, connected, and all degrees even), where the set $E^\star$ is obtained from E' by creating as many copies of each edge as appear in E'. A Hamilton cycle, also called a *tour*, is a spanning closed walk of cardinality n. All that follows is true if one replaces tours with spanning closed walks, except of course for the degree of each node, which must be two for tours and may be any even number for closed walks. All the closed walks we will use are spanning; therefore we will omit from now on the term "spanning." The definition of representative vector has to be adapted to closed walks. The components are the number of times each edge appears in the closed walk. The convex hull of all these vectors is an unbounded polyhedron called the graphical traveling salesman polyhedron of K_n, denoted by $GTSP(n)$.

Theorem 10.1. *The general domino inequalities (10.4) and the noncrossing domino inequalities (10.6) are valid for the symmetric traveling salesman problem (STSP(n)).*

Proof. Call lhs the left-hand sides of any of these inequalities. It is easy to check that in both cases

$$\text{lhs} \geq \frac{1}{2}\sum_{i=1}^{t} x(\delta(A_i)) + \frac{1}{2}\sum_{i=1}^{t} x(\delta(B_i)) + \frac{1}{2}\sum_{i=1}^{t} x(\delta(T_i)) \geq 3t. \tag{10.7}$$

The lhs is even for all tours (and closed walks), since cycles and cocycles intersect in an even number of edges; therefore one can raise $3t$ to $3t + 1$. □

The reader may have noticed that this is exactly the same proof as the algebraic validity proof for comb inequalities (see [7]).

In the following we call either of the two previous inequalities the *domino inequality*.

Our first result will be that, except in a pathological case, condition (10.5) is not restrictive since no general domino inequality that is not also a noncrossing domino inequality can be facet-inducing. But we first derive a few very useful consequences of the proof of Theorem 10.1.

In what follows, a *tight tour* for an inequality is a tour whose representative vector satisfies the inequality to *equality*. When we say that a set S is *tight* for a tour, we mean that the tour intersects its coboundary in exactly two edges; that is, the subtour inequality

$x(\delta(S)) \geq 2$ is tight for that tour. We call an edge of $\bigcup_{i=1}^{t}(A_i : B_i) \setminus \delta(H)$ a *penalized edge* to refer to the fact that its coefficient is two units more than the number of coboundaries it belongs to, which would be its coefficient in "closed set form" (see [7]). The choice of this terminology will become clearer later on.

Corollary 10.2. *Let $cx \geq c_0$ be a domino inequality. Let Γ be any tight tour for this inequality. At most one of the sets A_i, B_i, and T_i, $i = 1, \ldots, t$, is not tight for Γ.*

Proof. If two or more of these sets are not tight for Γ, the proof of Theorem 10.1 shows that the value of lhs exceeds $3t + 2$ and therefore is at least $3t + 3$ by the parity argument. The proof of Theorem 10.1 also shows that the only possibly nontight fundamental set must then have exactly four edges of Γ in its coboundary. □

Corollary 10.3. *Consider a facet-inducing domino configuration. No two of the fundamental sets A_i, B_i, and T_i, $i = 1, \ldots, t$, may be identical unless they contain a unique node.*

Proof. Corollary10.2 shows that at most one term of the linear combination can contribute a value of four units. Therefore, if two fundamental sets are identical, then the face induced by the domino configuration is contained in the facet induced by the subtour elimination inequality on that set. □

Corollary 10.4. *Let $cx \geq c_0$ be a domino inequality. Unless this inequality is equivalent to a subtour inequality, there exists a tight tour Γ for this inequality for which tooth T_j is not tight and then necessarily one has $\Gamma \cap (A_j : B_j) = \emptyset$.*

Proof. The corollary holds, else one of A_j or B_j would not be tight, which is impossible. □

10.3 Minimal and Nonpathological Domino Configurations

A *minimal* tooth is a tooth that contains no other tooth. A *maximal* tooth is a tooth contained in no other tooth. A tooth can be minimal and maximal at the same time.

Combs can be seen as domino configurations with only minimal teeth. Combs can also be seen as domino configurations with a nonminimal tooth. Consider a comb with handle H and teeth $T_1, \ldots, T_t$, t odd. Consider the domino configuration made up of handle $H' = H \setminus T_1$ and teeth T'_j, $j = 1, \ldots, t$, such that $T'_1 = V \setminus (T_1 \setminus H)$, $A_1 = T_1 \cap H$, $B_1 = T'_1 \setminus A_1$, $T'_j = T_j$, $A'_j = T'_j \cap H$, $B'_j = T'_j \setminus H$, $j = 2, \ldots, t$. Figure 10.1 shows an example. A black point represents at least one node. A white point may also represent more than one node but no node needs to be in such a position. Finally, where there are no points, no node can exist. It can be checked by inspection that the comb and the domino configuration yield exactly the same inequality.

We say that a domino configuration is *minimal* if it contains a minimum number of nonminimal teeth among all domino configurations that yield the same inequality. For example, the domino configuration of the second drawing of Figure 10.1 is not minimal because, as we just mentioned, it defines the same inequality as a comb, which contains no nonminimal teeth.

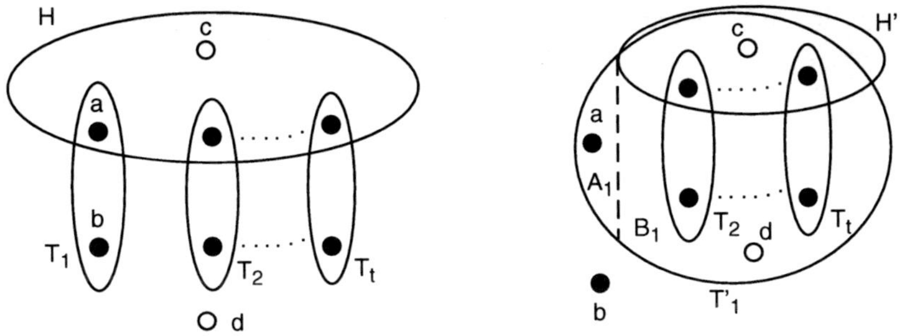

Figure 10.1. *A comb and an equivalent domino with a nonminimal tooth.*

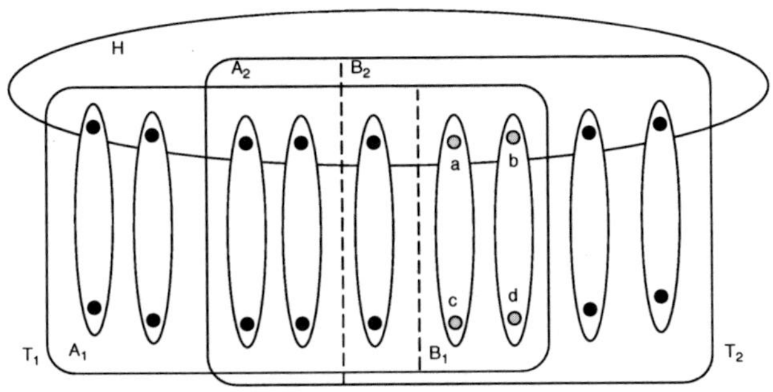

Figure 10.2. *A pathological domino configuration.*

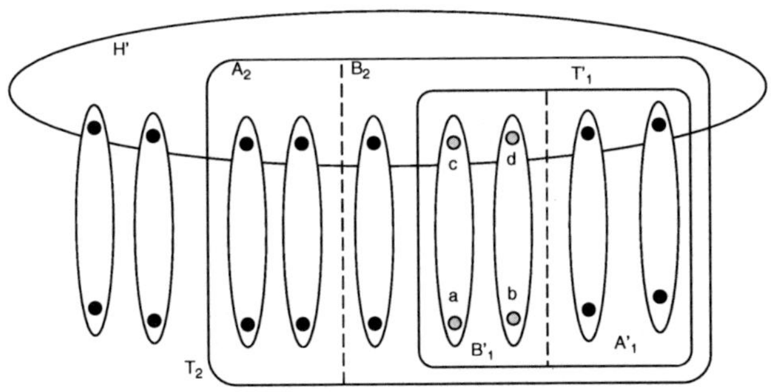

Figure 10.3. *An equivalent domino configuration.*

We say a domino configuration is *pathological* if there exist two teeth, say T_1 and T_2, such that $T_1 \cup T_2 = V$, $T_1 \cap T_2 \neq \emptyset$. Figure 10.2 shows such a pathological domino configuration that is facet-inducing. It is minimal since, if there are no nodes other than those shown, one needs at least two nonminimal teeth in order to reach the number of 11 teeth. The good news is that, in the case of this example, there is a nonpathological domino configuration shown in Figure 10.3, which yields the same inequality. The two handles H and H' differ in the two grey vertices. The two grey vertices of the second one are the grey vertices that do not belong to the first one. All the minimal teeth and tooth T_2 are the same in both cases, but tooth T_1 is now replaced by T_1', which has one of the half-dominoes of T_1, and the other half-domino consists of all the nodes outside T_1.

In the drawings, the partition for the minimal teeth is the one induced by the handle, i.e., $A = T \cap H$ and $B = T \setminus H$.

The operation used here is described in [1], and since we will use it quite often we state it now.

Theorem 10.5 (Boyd, Cockburn, and Vella [1]). *In a general domino configuration, consider a nonminimal tooth $T = (A \cup B)$, A and B being its two half-dominoes. Let $T' = ((V \setminus T) \cup B)$, with half-dominoes $V \setminus T$ and B, and $H' = H \Delta B$, the symmetric difference of B and H. Then $\mathcal{T}' = \mathcal{T} - T + T'$ and H' define the same inequality as $\mathcal{T}$ and H. In other words, they define equivalent general domino configurations.*

The following theorem shows that we do not lose any facet-inducing domino configurations by dropping the pathological ones.

Theorem 10.6. *For every pathological facet-inducing domino configuration there is a nonpathological one that induces the same facet.*

Proof. We use the transformation described in Theorem 10.5. Assume $T_1 \cup T_2 = V$. Then there is at least one of the half-dominoes of, say, T_1, which is not entirely contained in T_2. Assume it is A_1. Then the domino configuration obtained by replacing T_1 with $T_1' = (V \setminus T_1, B_1)$ and H with $H' = H \Delta B_1$ is equivalent to the pathological domino configuration but is no longer pathological. $\square$

In what follows we will only study, except when specifically mentioned, minimal and nonpathological domino configurations.

10.4 The Noncrossing Property and Nesting of Teeth

In this section we first prove the following lemma.

Lemma 10.7. *Unless the facet-inducing domino inequality is equivalent to the subtour inequality on T_j, if $T_i \subset T_j$, then either $T_i \subset A_j$ or $T_i \subset B_j$.*

Proof. Let Γ be a tight tour for which T_j is not tight, which exists since otherwise the domino inequality would be equivalent to the subtour inequality on T_j. By Corollary 10.4, no edges of Γ cross from A_j to B_j and therefore, if T_i is not entirely contained in one of these sets, it cannot be tight for Γ. $\square$

Corollary 10.8. *In a facet-inducing domino inequality that is not a subtour inequality, if $(A_i : B_i) \cap (A_j : B_j) \neq \emptyset$, $i \neq j$, then $T_i \setminus T_j \neq \emptyset$ and $T_j \setminus T_i \neq \emptyset$.*

Lemma 10.9. *Let $T_i = A_i \cup B_i$ be a tooth of a domino configuration. Any tight tour Γ that is not tight on, say, A_i, is such that $|\Gamma \cap (A_i : B_i)| = 2$ and $|\Gamma \cap (B_i : V \setminus T_i)| = 0$.*

Proof. In other words, Γ crosses the boundary between A_i and B_i twice and only crosses the border of T_i from nodes in A_i. If not, one could not have both B_i and T_i tight for Γ. $\square$

Theorem 10.10. *No general nonpathological domino inequality that is not also a noncrossing domino inequality can be facet-inducing for the symmetric traveling salesman polytope unless it is equivalent to a subtour elimination inequality.*

Proof. Consider teeth T_i and T_j, $i \neq j$, such that $(A_i : B_i) \cap (A_j : B_j) \neq \emptyset$. Let us assume $A_i \cap A_j \neq \emptyset$ and $B_i \cap B_j \neq \emptyset$. If $A_i = A_j$ and $|A_i| = |A_j| \geq 2$, we are done, by Corollary 10.3. If either $|A_i| \geq 2$ and $B_j \setminus T_i \neq \emptyset$ or $|A_j| \geq 2$ and $B_i \setminus T_j \neq \emptyset$, we are also done since, if, say, $|A_i| \geq 2$ and $B_j \setminus T_i \neq \emptyset$, then no tight tour exists for which A_i is not tight. Such a tour has to visit the nodes of $B_i \cup B_j$ consecutively, since B_i and B_j must be tight and they intersect, and therefore use an edge of the boundary of T_i with an extremity in B_i, which is not possible by Lemma 10.9.

If $|A_i| \geq 2$ and $B_j \setminus T_i = \emptyset$, then necessarily $|A_j| \geq 2$, else one would have $T_j \subset T_i$, but then one must have $B_i \setminus T_j = \emptyset$, as just shown. One cannot have $A_i \cap B_j \neq \emptyset$, since then it is impossible to have A_j not tight, since for such a tour, A_i, B_i, and B_j must be all tight, but since B_i and B_j intersect, their nodes must be visited consecutively, and it is impossible to have A_i tight. Therefore $B_j \subset B_i$. Reversing the roles of A_j and A_i one gets $B_i = B_j$, which is only possible if $|B_i| = |B_j| = 1$, by Corollary 10.3; this case can be treated just like the next case but one in which $|A_i| = |A_j| = 1$. In the case $|A_i| = 1$, if $|A_j| \geq 2$, then necessarily $B_i \setminus T_j = \emptyset$, as just noted, but then $T_i \subset T_j$, which is impossible, by Corollary 10.8.

So we are down to the case $|A_i| = |A_j| = 1$. Note that, since neither of T_i and T_j is contained in the other, one has $B_i \setminus B_j \neq \emptyset$ and $B_j \setminus B_i \neq \emptyset$. Let $u \in V \setminus T_i \cup T_j \neq \emptyset$. This is the only place where we use the nonpathological hypothesis. Let $b \in B_i \cap B_j \neq \emptyset$, and consider a tight tour Γ containing the edge (u, b). Since $B_i \setminus B_j \neq \emptyset$ (because $A_i = A_j$), if Γ is tight on B_i, it can be tight neither on T_j nor on B_j, which is impossible. The same holds when reversing the roles of i and j, and therefore neither B_i nor B_j can be tight. Therefore Γ does not exist, since at most one fundamental set can be not tight, and the domino inequality is contained in the facet induced by the nonnegativity constraint on x_{ub}. Since it is easy to build a nontight tour for the domino inequality not containing edge (u, b), this inclusion is strict. $\square$

Therefore one can from now on assume that condition (10.5) holds, we are only dealing with inequalities (10.6), and the domino configuration stands for the nonpathological and noncrossing minimal domino configuration.

We end this section with a few more properties of tight tours that will be useful.

Proposition 10.11. *Let $cx \geq c_0$ be a domino inequality. Let Γ be any tight tour for this inequality. Then Γ contains at most one penalized edge. In the case Γ contains a penalized edge, all the sets A_i, B_i, and T_i, $i = 1, \ldots, t$, are tight for Γ.*

Proof. In inequality (10.7), a penalized edge has a coefficient in lhs two units higher than in the combination of subtour elimination inequalities. Therefore, if Γ contains two penalized edges, then lhs $\geq 3t+2$ and by parity lhs $\geq 3t+3$, contradicting the fact that Γ is tight. If Γ contains a penalized edge, then, because that penalized edge has a coefficient two units higher in lhs than in the linear combination of subtour inequalities, each of these subtour inequalities must be tight. □

Proposition 10.12. *Let Γ be a tight tour such that tooth T_j is not tight. There is a one-to-one correspondence between the edges of $\delta(H) \cap \Gamma$ and the teeth other than T_j.*

Proof. Since tooth T_j is not tight, all other fundamental sets are tight for Γ, and Γ contains no penalized edge. Therefore, for each tight tooth T_i, $i \neq j$, $\Gamma \cap ((A_i : B_i) \cap \delta(H)) = \{e_i\}$ since T_i, A_i, and B_i must all be tight for Γ and $|\Gamma \cap \delta(H)| = t-1$. □

We summarize the results in the following theorem.

Theorem 10.13. *Given a domino inequality, tight tours Γ fall in one of the following categories:*

(i) *All teeth are tight and Γ contains no penalized edge. Then $|\delta(H) \cap \Gamma| = t+1$ and $|\Gamma \cap ((A_j : B_j) \cap \delta(H))| \geq 1$ for all $j = 1, \ldots, t$. Choose one edge in each set $\Gamma \cap ((A_j : B_j) \cap \delta(H))$, $j = 1, \ldots, t$. The unique nonchosen edge of $\delta(H)$ will be called the* joker. *The joker may belong to some set $(A_j : B_j)$. Edges of $\delta(H) \setminus \bigcup_{j=1}^{t} (A_j : B_j)$ can only appear in a tight tour as the joker edge.*

(ii) *Γ contains one penalized edge $e \in (A_j : B_j)$. Then all fundamental sets are tight, $|\delta(H) \cap \Gamma| = t-1$, and $|\Gamma \cap ((A_i : B_i) \cap \delta(H))| = 1$ for all $i = 1, \ldots, t$, $i \neq j$.*

(iii) *All teeth, except one T_j, are tight. Then Γ contains no penalized edge, all fundamental sets except T_j are tight, $|\delta(H) \cap \Gamma| = t-1$, and $\Gamma \cap (A_i : B_i) = \{e_i\}$ and $e_i \in \delta(H)$ for all $i = 1, \ldots, t$, $i \neq j$.*

10.5 The Structure of the Teeth in a Domino Inequality

Throughout this section we assume that the domino inequality is equivalent neither to (i.e., does not define the same face as) a subtour inequality nor to a nonnegativity constraint and that it is minimal and not pathological. In most cases we will not show the strict domination of the domino inequality by a nonnegativity or a subtour elimination inequality; it can nevertheless be easily shown.

Theorem 10.14. *For any nonpathological facet-inducing domino inequality, there is an equivalent domino configuration in which all teeth are nested.*

Proof. Consider a counterexample with a minimum number of properly intersecting teeth. Let $T_i \cap T_j \neq \emptyset$ with $T_i \setminus T_j \neq \emptyset$ and $T_j \setminus T_i \neq \emptyset$. Assume $A_i \cap A_j \neq \emptyset$. Then by Theorem 10.10 we have $B_i \cap B_j = \emptyset$. Consider a tight tour Γ not tight on B_i.

It is impossible to have both A_i and A_j tight, except if $A_j \subset A_i$, because no edge of $\delta(A_i) \cap \delta(T_i)$ may belong to Γ. But even then, one has crossed the border of T_j twice within A_i, and one should still visit the nodes of B_j, which can only be done using two more edges in the border of T_j, which is impossible since all the fundamental sets, except B_i, must be tight. Therefore no tight tour exists such that B_i is not tight, and so, if $|B_i| > 1$, the domino inequality is dominated by the subtour elimination inequality on B_i. One can do the same thing on B_j. Therefore the only case left to deal with is $|B_i| = |B_j| = 1$. If $A_i \subset A_j$, there is no tight tour Γ for which A_i is not tight, since then $\Gamma \cap (A_i : B_i) \subset \delta(T_j)$, and since T_j must be tight there is no way of leaving T_j, which has to be done since the configuration is not pathological. The same is true if $A_j \subset A_i$. Assume $A_i \setminus A_j \neq \emptyset$ and $A_j \setminus A_i \neq \emptyset$ and let u be a node of the first set and v one in the second set. Let Γ be a tight tour containing edge (u, v). If A_j is tight for Γ, then A_i cannot be tight since $A_i \cap A_j \neq \emptyset$. The path $P = (\Gamma \cap \gamma(A_j)) + \{(u, v)\}$ has its two extremities in T_i, else T_i cannot be tight. One has $P \cap \delta(T_i) \geq 2$, and since one still has the nodes of B_j to visit there is no way to have T_i tight for Γ, and therefore no tight tour can contain edge (u, v). Therefore one has, say, $A_i \subset A_j$. If $|A_i| \geq 2$, since all tight tours are tight on A_i, as just seen, the face induced by the domino inequality is dominated by the subtour elimination inequality on A_i. So we are left with the case $|A_i| = |B_i| = |B_j| = 1$. Assume $A_i \subset H$. Apply Theorem 10.5 to tooth T_i, inverting the roles of A and B. That is, the domino configuration with handle $H' = H \setminus A_i$ and tooth T_i replaced by $T_i' = (V \setminus T_i \cup A_i)$, with half-dominoes $V \setminus T_i$ and A_i, yields the same inequality, but has two fewer properly intersecting teeth since $T_i' \supset T_j$, which contradicts the minimality of the counterexample. □

A tooth is *maximal* if it is contained in no other tooth. Sometimes we will have to emphasize more and call such teeth *globally maximal*. A tooth T_j is *maximal relative to* tooth T_i (resp. to the half-domino A_i) if $T_j \subset T_i$ (resp. $T_j \subset A_i$), $T_j \neq T_i$, and maximal with that property, that is, not contained in any other tooth contained in T_i (resp. A_i).

Remark 10.15. Consider a domino configuration with nested teeth. Theorem 10.5 applied to a maximal (and nonminimal) tooth of that configuration yields another domino configuration with nested teeth. By applying that theorem iteratively to maximal teeth one can make globally maximal all maximal teeth contained in some half-domino and preserve the nesting of the teeth.

A tooth is called *odd* if either it is minimal or the number of teeth it strictly contains is *even*. That is, the number of teeth it contains, including itself, is odd. We talk here of all the teeth, not only the maximal ones contained in it. On the contrary it is called *even*. Note that a closed walk for which an odd tooth is tight, together with all the teeth contained in it, and that does not use a joker edge or a penalized edge with both extremities in that tooth, enters and exits the tooth on opposite sides of the handle, while for an even tooth one enters and exits on the same side. If a penalized edge or a joker is used inside it, then it is the reverse, that is, for an odd tooth the walk enters and exits on the same side of the handle and on opposite sides if the tooth is even.

In all the forthcoming drawings, dotted lines are used to show the partition of the dominoes.

Before proceeding, let us pause and try to understand what these inequalities mean in terms of Hamilton cycles and closed walks. Of course domino configurations with just minimal teeth are combs. Naddef and Pochet in [7] give the following intuitive insight into the valid inequalities for STSP(n), which involve handles and teeth. Look at tours for which all teeth are tight and intersect the coboundaries of the handles a minimum number of times. This will give you the right-hand side of these inequalities. The teeth in these inequalities, because of a parity argument, in some sense force edges in the coboundary of the handles.

The same argument holds here not only if we want all teeth to be tight but also if no penalized edge is to be used. What this inequality says is this: if you want a nonminimal tooth to remain tight, then at some point a tour will have to cross from one half-domino to the other. Either it does it without crossing the boundary of the handle and it pays a penalty of 2 or it crosses that boundary contributing to one edge in the coboundary of the handle. This is why we used the term "penalized edge."

Proposition 10.16. *In a facet-defining domino inequality, if T_i is a minimal tooth, then* $\{A_i, B_i\} = \{T_i \cap H, T_i \setminus H\}$.

Proof. If the condition is not satisfied, replacing A_i and B_i with $T_i \cap H$ and $T_i \setminus H$ yields a valid inequality with the same rhs but with coefficients on the lhs that are less than or equal to the former ones and that therefore dominate it. □

This proposition together with Lemma 10.7 gives the structure of the dominoes in a facet-inducing domino configuration. For the minimal teeth, the partition is given by the handle; for the others the domino separates the teeth they contain into two sets, each set contained in one half-domino. We will call *relevant* the dominoes that correspond to nonminimal teeth. In fact, from now on we never talk of the domino partition of minimal teeth.

Lemma 10.17. *A facet-inducing minimal domino inequality cannot contain a unique maximal tooth T_i such that any of $H \setminus T_i$ or $(V \setminus H) \setminus T_i$ is empty.*

Proof. If both sets are empty we are in contradiction with the definition of a domino, which assumes that $T_i \neq V$. Assume that $H \setminus T_i = \emptyset$. This implies that all other teeth are contained in T_i; thus the number of teeth strictly contained in T_i is even and each half-domino A_i and B_i contains either an even or an odd number of teeth. If, say, A_i contains no teeth but has nonempty intersection with the handle, the argument below for the even case holds. If $A_i \cap H = \emptyset$ (note that this is the case of the nonminimal domino inequality of Figure 10.1), then the domino inequality with handle $H' = H \cup A_i$, $T_j' = T_j$ for $j \neq i$ and $T_i' = A_i \cup (V \setminus T_i)$, yields the same inequality but has one fewer relevant domino partition, a contradiction with the minimality of the domino configuration. Let $e \in ((A_i : B_i) \setminus \delta(H))$ be a penalized edge with both extremities, either in H if A_i and B_i both contain an even number of teeth, or in $V \setminus H$ if A_i and B_i both contain an odd number of teeth. Let Γ be a tight tour containing e. Since e is a penalized edge, $P = \Gamma \cap \gamma(T_i)$, $P_A = \Gamma \cap \gamma(A_i)$, and $P_B = \Gamma \cap \gamma(B_i)$ are paths and $P = P_A + \{e\} + P_B$. The paths P_A and P_B have their endnodes on the same side of the handle if A_i and B_i contain an even number of teeth, and on opposite sides if these sets contain an odd number of teeth. Therefore our choice of e yields that P has its two endnodes in H. To visit the nodes of $V \setminus T_i$, which by hypothesis are outside H, one has to cross the coboundary of H at least twice, which is impossible since we have used a penalized edge and the maximum number of edges in the coboundary has been reached within the path P. Therefore the face induced by this inequality is contained in the intersection of all the nonnegativity facets for all possible choices of edge e in the proof. Reversing the roles of H and $V \setminus H$ ends the proof of the lemma. □

Proposition 10.18. *In a facet-inducing minimal domino inequality there cannot be a single maximal tooth.*

Proof. Let T_i be such that $T_j \subset T_i$ for all $j \neq i$. From Lemma 10.17 we know that $H \setminus T_i \neq \emptyset$ and $(V \setminus H) \setminus T_i = V \setminus (H \cup T_i) \neq \emptyset$; that is, there is at least one node inside and one outside the handle that does not belong to the unique maximal tooth. Then the domino inequality obtained by replacing T_i with $T_i' = V \setminus T_i$, $A_i' = T_i' \cap H$, and $B_i' = T_i' \setminus H$ has the same rhs and coefficients less than or equal to those of the first inequality. In particular, all the edges of $(A_i : B_i) \setminus \delta(H)$ have a coefficient two units lower, therefore the latter inequality strictly dominates the first one, except if $(A_i : B_i) \setminus \delta(H) = \emptyset$. In this case either $A_i = \emptyset$ or $B_i = \emptyset$, and the domino configuration obtained by replacing T_i with $V \setminus T_i$ would yield the same inequality, contradicting the minimality assumption. □

Corollary 10.19. *In a facet-inducing minimal domino inequality one can assume that no relevant half-domino can be empty of teeth.*

Proof. Say $T_i = (A_i \cup B_i)$ with B_i containing no teeth. Applying Theorem 10.5 recursively to maximal teeth containing B_i, one obtains an equivalent domino configuration with a unique maximal tooth. By the previous theorem, either the inequality is not facet-inducing or it can be transformed into a facet-inducing inequality with one fewer half-domino empty of teeth. □

We assume from now on that no relevant half-domino is empty of teeth.

Proposition 10.20. *If there is only one maximal odd tooth, that tooth cannot also be minimal, and then $H \setminus \bigcup_{i=1}^{t} T_i = \emptyset$ and $(V \setminus H) \setminus \bigcup_{i=1}^{t} T_i = \emptyset$.*

Proof. If, say, T_i is the only maximal odd tooth, then if T_i is minimal, then a tight tour such that T_i is not tight cannot exist, since all other teeth must be tight, no joker nor penalized edge can be used, and there is no way of reconnecting the nodes of $T_i \cap H$ to those of $T_i \setminus H$.

If T_i is not minimal, we show that all nodes must be contained in some tooth. Let $e \in (A_i : B_i)$ be a penalized edge. Depending on the parity of the number of teeth in A_i and B_i, a tight tour Γ containing e will be such that the two extremities of the path $\Gamma \cap \gamma(T_i)$ will be inside or outside the handle. Say they are inside the handle. There is no way of visiting the nodes of $(V \setminus H) \setminus \bigcup_{i=1}^{t} T_i$, and therefore that set must be empty in order to have tight tours containing e. Choosing e on the other side of the handle, one gets the same result for $H \setminus \bigcup_{i=1}^{t} T_i$. □

Corollary 10.21. *In a facet-inducing domino inequality there must be at least three maximal teeth.*

Proof. If there are only two teeth, one is odd and the other even, and by Proposition 10.20, their union is V, and therefore when one is not tight, neither is the other, which is impossible. □

Corollary 10.22. *Every half-domino of a facet-inducing domino configuration contains at least two maximal teeth relative to it.*

Proof. The corollary holds, else by application of Remark 10.15 one could get an equivalent domino configuration with only two maximal teeth. □

Corollary 10.23. *In a facet-inducing domino inequality, no half-domino can contain only one minimal tooth with all other maximal teeth contained in it even.*

Proof. Use Remark 10.15 to make these teeth maximal. The result follows from Proposition 10.20. □

Corollary 10.24. *In a facet-inducing domino inequality, if all maximal teeth contained in a half-domino, say A_i, are even, then $A_i \setminus \bigcup_{\{j:T_j \subset A_i\}} T_j = \emptyset$. The same is true if only one of these teeth is odd.*

Proof. Use Remark 10.15 to make these teeth maximal. The result follows from Proposition 10.20. □

At this point we may wonder whether or not we should consider as a domino inequality one in which nodes are forbidden in places where the definition of the inequality allows them to be. Let us call *unrestricted* those domino configurations for which there may be a node not contained in any tooth and any half-domino that has a node that does not belong to a tooth contained in that half-domino.

All the previous results yield the following theorem.

Theorem 10.25. *In a facet-inducing unrestricted domino configuration, there are at least three odd maximal teeth and each half-domino contains at least two odd maximal teeth.*

Proof. By Proposition 10.20, there cannot be only one maximal tooth, and therefore there must be at least three. Since, by Remark 10.15, the maximal teeth contained in a half-domino can be turned globally maximal, if there were less than two odd maximal teeth in a half-domino, there would be less than three odd maximal teeth after the transformations. □

Naddef and Wild show in [6] that all unrestricted domino configurations yield facets of the symmetric traveling salesman polytope.

Conjecture 10.26. *No domino configuration that is not unrestricted can be facet-inducing for the symmetric traveling salesman polytope.*

10.6 Extensions and Conclusion

This idea of using penalties to "force" edges in the coboundary of the handle can certainly be extended to all the inequalities defined on handles and teeth. We show an example on the path inequalities. For the sake of simplicity we restrict ourselves to a very small subfamily of those inequalities. This section assumes the reader is somewhat familiar with the path inequalities for STSP(n) defined in Cornuéjols, Fonlupt, and Naddef [2].

An *r-regular one-domino path* inequality is defined by $h = r - 1$ handles $H_1, \ldots, H_h$, an odd number $t \geq 7$ of teeth $T_1, \ldots, T_t$, together with a partition of the last tooth $T_t = A \cup B$ and integers t_A, t_B such that

$$H_i \subset H_{i+1} \quad \text{for } i = 1, \ldots, h-1, \tag{10.8}$$

$$T_i \cap T_j = \emptyset \quad \text{for } 1 \leq i < j \leq t-1, \tag{10.9}$$

$$H_1 \cap T_j \neq \emptyset \quad \text{for } j = 1, \ldots, t, \tag{10.10}$$

$$T_j \setminus H_h \neq \emptyset \quad \text{for } j = 1, \ldots, t, \tag{10.11}$$

$$(H_{i+1} \setminus H_i) \cap T_j \neq \emptyset \quad \text{for } j = 1, \ldots, t-1, \; i = 1, \ldots, h-1, \tag{10.12}$$

$$(H_{i+1} \setminus H_i) \setminus \bigcup_{j=1}^{t-1} T_j = \emptyset \quad \text{for } i, 1 \leq i < r-1, \tag{10.13}$$

$$T_i \cap T_t = \emptyset \quad \text{for } i \leq t_A - 1, \tag{10.14}$$

$$T_i \subset A \quad \text{for } t_A \leq i \leq t_B - 1, \tag{10.15}$$

$$T_i \subset B \quad \text{for } t_B \leq i \leq t - 1, \tag{10.16}$$

$$t_B - t_A \geq 2, \tag{10.17}$$

$$t - t_B \geq 2, \tag{10.18}$$

$$\text{if } t + 1 - t_A \text{ is odd, } t_A \geq 3, \text{ else } t_A \geq 4. \tag{10.19}$$

That is, we have a linearly nested set of handles and an even number of disjoint teeth such that each of these teeth has at least one node in H_{i+1} that is not in H_i for all i, at least one inside the first handle, and at least one outside the last handle. All nodes of $H_{i+1} \setminus H_i$, for all i, belong to some tooth. Finally, the last tooth has a partition into two sets each containing at least two of the other teeth. Parity being defined as in the previous sections, one requires at least three maximal odd teeth, which explains why we may require at least three of the minimal teeth to be outside the last tooth. One may wonder why we use only $r - 1$ handles in an "r-regular" definition. This is to be consistent with the path inequalities, where a comb is a 2-regular path inequality. Figure 10.4 gives two examples, in which a black point means the presence of at least one node in that position and a white one the possibility of a node in that position. No nodes are allowed in any other places.

As in the domino configurations, we let p_e be some *penalty* on the edges of $e \in (A : B) \setminus (\delta(H_1) \cap \delta(H_h))$, that is, on every edge that links a node in A to one in B but that does not cross the boundaries of all the handles. Our aim is to define those penalties in order for the following inequality to be valid and we hope facet-inducing for STSP(n). To make the following inequality more readable we set $p_e = 0$ for $e = (u, v) \in (A : B)$ crossing the

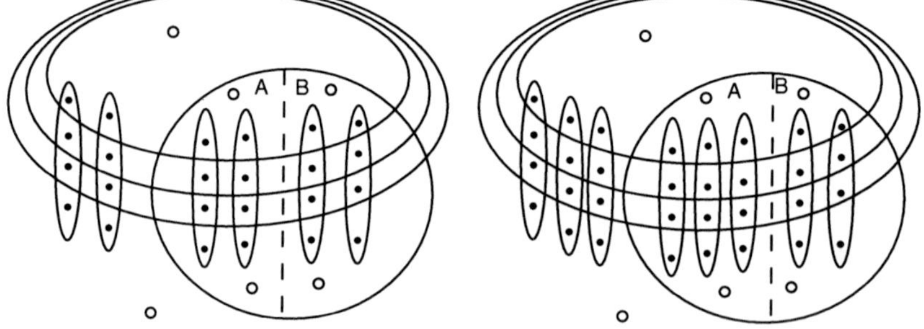

Figure 10.4. *4-regular one-domino paths.*

border of all handles, i.e., such that $u \in H_1$ and $v \notin H_h$:

$$\sum_{i=1}^{h} x(\delta(H_i)) + \sum_{j=1}^{t-1} x(\delta(T_j))$$

$$+ hx(\delta(T_t)) + \sum_{e \in (A:B)} p_e x_e \geq t(h+2) + 3h - 2. \tag{10.20}$$

We now define the *penalties* p_e for each $e \in (A : B) \setminus (\delta(H_1) \cap \delta(H_h))$, that is, for every edge that links a node in A to one in B but that does not cross the boundaries of all the handles.

All nodes in $H_i \setminus H_1$ belong to some tooth. An edge $e = (u, v) \in (A : B)$ is not penalized if and only if $u \in H_1$ and $v \notin H_h$: its coefficient in inequality (10.20) is equal to the number of boundaries of handles and teeth it crosses, that is, h or $h+1$ or $h+2$, depending on whether none, one, or two of its extremities belong to teeth.

Let $e = (u, v) \in (A : B)$ with $u \in H_i \setminus H_{i-1}$ or $u \notin H_h$ and $v \in H_j \setminus H_{j-1}$ or $v \notin H_h$. Without lost of generality, we assume $i \leq j \leq h+1$, where for simplicity one assumes that $i = h+1$ (resp. $j = h+1$) corresponds to $u \notin H_h$ (resp. $v \notin H_h$). One cannot have $i = 1$ if $j = h+1$ because it corresponds to nonpenalized edges. If $i = j = 1$ or $i = j = h+1$, the penalty is $p_e = 2h$.

In general the penalty is given by the following formula: $p_e = 2 * \max\{\min\{i-1, j-1\}, \min\{h+1-i, h+1-j\}\}$, which comes down to $p_e = 2 * \max\{i-1, h+1-j\}$ if one assumes $i \leq j$. Note that the penalty already defined for the edges inside the innermost handle or outside the outermost handle falls into this general setting.

The penalty p_e of an edge e is computed in a way that one can build a spanning closed walk Γ containing e, with all teeth tight, which has p_e fewer edges in the coboundaries of the handles relative to the case in which such an edge is not used and all teeth are tight (see Figure 10.5).

Note that we always have $p_e \geq |\{i : e \notin \delta(H_i)\}|$. Figure 10.5 shows how to build tight closed walks using exactly one penalized edge, for which all teeth are tight, and tight for the inequality. The important observation is that in each tooth involved by the penalized edge the two paths of edges not doubled must be on the same "side" of the penalized edge.

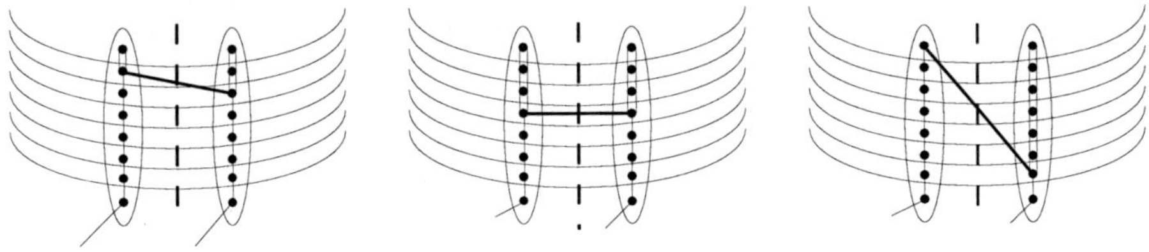

Figure 10.5. *Some tight walks with penalized edges.*

This is because we want the paths to exit tooth T_t in such a way that no more joker edges, other than those already considered, need to be used. Remember that a black point in the drawing represents all nodes in that position.

Theorem 10.27. *The r-regular one-domino path inequalities are valid for GTSP(n) and therefore for STSP(n).*

Proof. From the way all penalties have been computed, all closed walks Γ for which all teeth are tight and which contain at most one penalized edge satisfy the inequality. Let's first consider the case of closed walks Γ that are not tight on tooth T_t. Therefore $|\Gamma \cap \delta(T_t)| \geq 4$. Since the coefficient of that tooth is h, it is enough to prove that $\sum_{i=1}^{h} x^{\Gamma}(\delta(H_i)) + \sum_{j=1}^{t-1} x^{\Gamma}(\delta(T_j)) \geq h(t-1) + 2(t-1)$. If the first $t-1$ teeth are tight, it is trivially true since then $|\Gamma \cap (T_j \cap H_i : T_j \setminus H_i)| \geq 1$ for $j = 1, \ldots, t-1$. Define η_i by $x^{\Gamma}(\delta(H_i)) = t - 1 - 2\eta_i$. Note that η_i is always an integer. For each handle H_i and each tooth T_j, with $1 \leq j \leq t-1$, define $\theta_{ij} = 1$ if $x^{\Gamma}(T_j \cap H_i : T_j \setminus H_i) = 0$ and $\theta_{ij} = 0$ otherwise. That is, $\theta_{ij} = 1$ if Γ does not cross the border of H_i inside the tooth T_j. For any tooth T_j, $j \leq t-1$, one has

$$x^{\Gamma}(\delta(T_j)) \geq 2 + 2\sum_{i=1}^{h} \theta_{ij}. \qquad (10.21)$$

For each handle H_i one has $t - 1 - 2\eta_i = x^{\Gamma}(\delta(H_i)) \geq \sum_{j=1}^{t-1}(1-\theta_{ij}) = t - 1 - \sum_{j=1}^{t-1}\theta_{ij}$, which, with the fact that $\eta_i \leq 2\eta_i$, yields $\eta_i \leq 2\eta_i \leq \sum_{j=1}^{t-1}\theta_{ij}$.

Putting this all together we get

$$\sum_{i=1}^{h} x^{\Gamma}(\delta(H_i)) = (t-1)h - \sum_{i=1}^{h} 2\eta_i \geq (t-1)h - \sum_{i=1}^{h}\sum_{j=1}^{t-1} 2\theta_{ij} \geq (t-1)h - \sum_{j=1}^{t-1}(x^{\Gamma}(\delta(T_j)) - 2).$$

The last inequality is obtained by inverting the order of the summations and using equation (10.21). So if a closed walk Γ does not satisfy the inequality, it is tight on T_t.

Since T_t is tight for Γ, at least one edge of $(A : B)$ belongs to Γ. If it is an edge not crossing any handle, the penalty is at least $2h$ and the proof we just did in the case of tooth T_t not tight can be repeated here. If the edge is not penalized, i.e., it crosses all the handles, then replacing, in the definition of η_i, $t-1$ with $t+1$, yields that the inequality is valid for Γ.

Note that one can assume, although it is not used here, that a penalized edge in Γ has both extremities outside H_1 and both extremities inside H_h, i.e., it does not have an extremity inside the inner handle or outside the outmost handle. To see this, assume Γ violates the inequality, and $e = (u, v) \in \Gamma$ with $u \in H_1$ and $v \in T_s$. Then the closed walk obtained from Γ by replacing edge e with the two edges $\{(v, w), (w, u)\}$ uses no penalized edges and would violate the inequality since the coefficients of the two replacement edges equal that of e, which we already saw is impossible.

Let $e = (u, v) \in \Gamma$ be a penalized edge. Let $u \in T_r \subset A, v \in T_s \subset B, u \in H_{i^\star} \setminus H_{i^\star - 1}$, and $v \in H_{j^\star} \setminus H_{j^\star - 1}$. Moreover, assume $i^\star \leq j^\star$ and $h + 1 - j^\star > i^\star - 1$. So we have $p_e = 2 * h + 1 - j^\star$.

If all teeth are tight, then all the first $j^\star - 1$ handles have at least $t + 1$ edges in their coboundaries, and the others at least $t - 1$, as shown in Figure 10.5, if one wants to minimize $\sum_{i=1}^{h} x(\delta(H_i))$. We will show that enabling some teeth to intersect Γ in more than two edges cannot improve the quantity $\sum_{i=1}^{h} x(\delta(H_i)) + \sum_{j=1}^{t-1} x(\delta(T_j))$.

Define η_i by $x^\Gamma(\delta(H_i)) = t + 1 - 2\eta_i$ if $i < j^\star$ and by $x^\Gamma(\delta(H_i)) = t - 1 - 2\eta_i$ if $i \geq j^\star$. As before, note that η_i is always an integer. For each handle H_i and each tooth T_j, with $1 \leq j \leq t - 1$, θ_{ij} is defined as before.

For each handle $H_i, i < j^\star$, one has $t+1-2\eta_i = x^\Gamma(\delta(H_i)) \geq \sum_{j=1}^{t-1}(1-\theta_{ij}) = t-1-\sum_{j=1}^{t-1} \theta_{ij}$, which, with the fact that $\eta_i \leq 2\eta_i$ when $\eta_i \geq 0$, yields $\eta_i \leq 2\eta_i \leq \sum_{j=1}^{t-1} \theta_{ij} + 2$.

For each handle H_i, $i \geq j^\star$, one has $t - 1 - 2\eta_i = x^\Gamma(\delta(H_i)) \geq \sum_{j=1}^{t-1}(1 - \theta_{ij}) = t - 1 - \sum_{j=1}^{t-1} \theta_{ij}$, which, with the fact that $\eta_i \leq 2\eta_i$, yields $\eta_i \leq 2\eta_i \leq \sum_{j=1}^{t-1} \theta_{ij}$.

Putting this all together and using the fact that $p_e = 2(h + 1 - j^\star)$, we get

$$
\begin{aligned}
\sum_{i=1}^{h} x^\Gamma(\delta(H_i)) &= (t + 1)h - 2(h + 1 - j^\star) - \sum_{i=1}^{h} 2\eta_i \\
&\geq (t + 1)h - p_e - \sum_{i=1}^{h} \sum_{j=1}^{t-1} 2\theta_{ij} \\
&\geq (t + 1)h - p_e - \sum_{j=1}^{t-1} \left(x^\Gamma(\delta(T_j)) - 2\right),
\end{aligned}
$$

which ends the proof of validity. □

Let us assume that the configuration is node maximal, that is, that there is a node in each of the positions where such a node is possible. A configuration is *simple* if there is at most one node in each possible position. That is, in the drawings each point represents exactly one node.

In the case of a node-maximal simple configuration, let the nodes of tooth T_i, $i = 1, \ldots, t - 1$, be numbered from the inner handle outward as a_j^i, $j = 1, \ldots, h$. Tooth T_t contains four nodes not in the other teeth, which we denote $a_t, b_t, \overline{a}_t, \overline{b}_t$. Moreover, there is a node a_0 in the inner handle in no tooth and one $\overline{a}_0$ outside the larger handle and also in no tooth. It will be convenient to let $a_0^i = a_0$ and $a_{r+1}^i = \overline{a}_0$ if $1 \leq i < t_A$, $a_0^i = a_t$, and $a_{r+1}^i = \overline{a}_t$ if $t_A \leq i < t_B$, and finally $a_0^i = b_t$, and $a_{r+1}^i = \overline{b}_t$ if $t_B \leq i < t - 1$.

The skeleton of a simple configuration is the graph $\mathcal{K} = (V, \tilde{E})$ on the same nodeset V as the complete graph on which the TSP is defined. The set $\tilde{E}$ contains the following edges:

(i) $(a^i_j, a^i_{j+1}) \in \tilde{E}$ for $0 \le j \le r$ and $1 \le i \le t-1$;

(ii) $\{(a_0, a_t), (a_0, b_t), (\overline{a}_0, \overline{a}_t), (\overline{a}_0, \overline{b}_t)\} \subset \tilde{E}$;

(iii) $\{(a_t, \overline{b}_t), (\overline{a}_t, b_t)\} \subset \tilde{E}$.

The edgeset $\{(a^i_j, a^i_{j+1}) : 0 \le j \le r\}$ will be referred to as the *ith path* of the skeleton. See Figure 10.6 for an example.

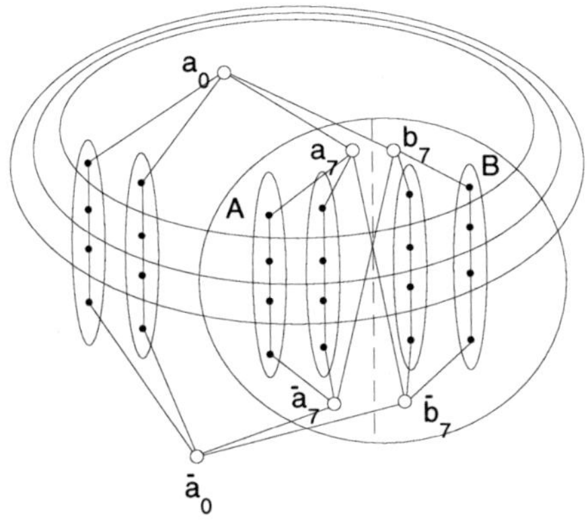

Figure 10.6. *The skeleton of a 4-regular one-domino path inequality.*

Theorem 10.28. *The node-maximal r-regular one-domino path inequalities are facet-inducing for GTSP(n).*

Proof. Since we are dealing with GTSP(n) we can assume, without loss of generality, that the r-regular one-domino path configuration is simple. We first prove that the inequality is facet-inducing when restricted to the skeleton; the result then follows easily.

Since our inequality $fx \ge f_0$ is valid, if it is not facet-inducing on GTSP(n), the face it defines is contained in a facet defined by, say, $gx \ge f_0$.

Since in Figure 10.7 the missing edge on the ith path could be any one and still yield a tight walk, we have that the g-coefficient of each edge of the same path is the same. Comparing the closed walks obtained in this way on two different paths outside T_t yields that this coefficient is the same for all edges of these paths, say γ_{out}, where the "out" is relative to tooth T_t.

Following the same argument, Figure 10.8 shows that in each half-domino all the edges of the skeleton have the same coefficient, say γ_A and γ_B, respectively.

We now consider two cases, depending on whether T_t is odd or even. We first consider the odd case. Figure 10.9 shows two tight closed walks depending on whether each domino contains an even or an odd number of teeth.

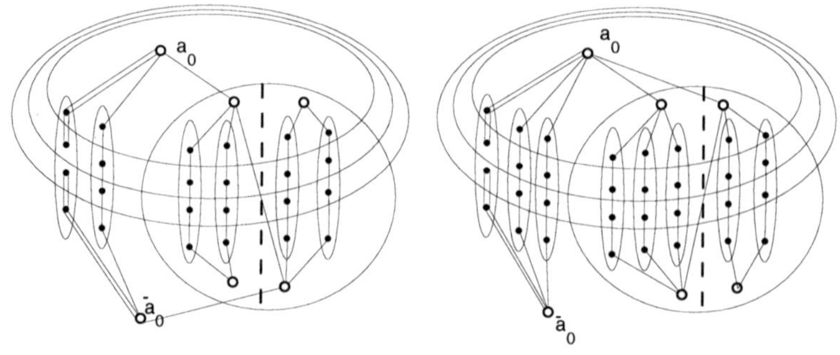

Figure 10.7. *Determining the coefficients outside T_t.*

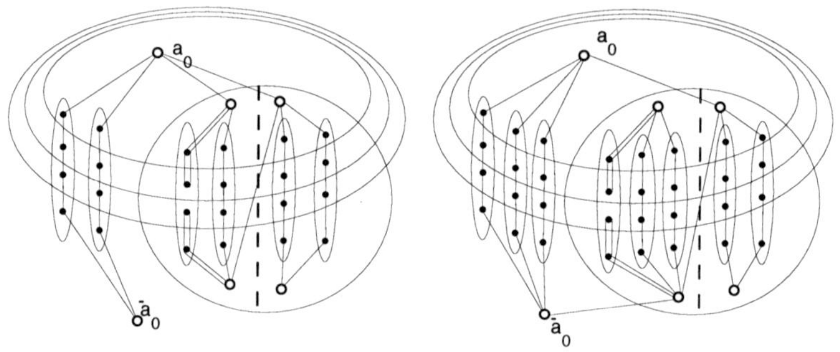

Figure 10.8. *Determining the coefficients inside T_t.*

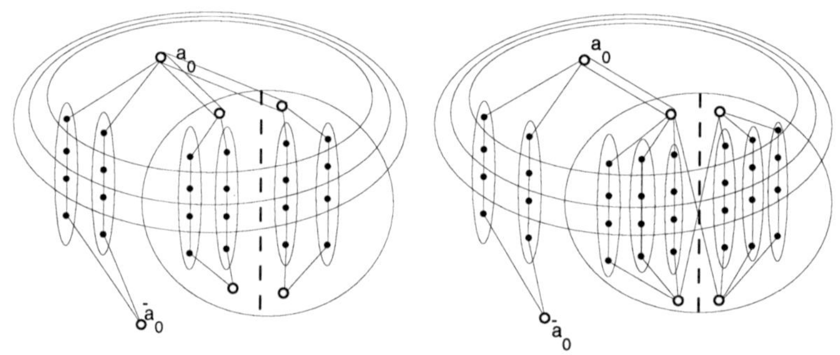

Figure 10.9. *Determining the coefficients of edges of $\delta(T_t)$, odd case.*

If both contain an even number of teeth, since one can replace the two copies of edge (a_0, a_t) with two copies of $(\overline{a}_0, \overline{a}_t)$ or two copies of $(a_t, \overline{b}_t)$ or two copies of $(\overline{a}_t, b_t)$, and can do the same with the two copies of (a_0, b_t) except for the first replacement, which is now with two copies of $(\overline{a}_0, \overline{b}_t)$, all these edges have the same coefficient, say γ_t.

If both contain an odd number of teeth, the tight walks obtained from the one in the drawing by replacing two copies of (a_0, a_t) with two copies of (a_0, b_t) or of $(\overline{a}_0, \overline{a}_t)$ or of

$(\overline{a}_0, \overline{b}_t)$ show that these four edges have the same coefficient, say γ_t. Deleting from that same walk the two edges of $(A : B)$ it contains, deleting one copy of (a_0, a_t), and adding one copy of each of (a_0, b_t), $(\overline{a}_0, \overline{a}_t)$, and $(\overline{a}_0, \overline{b}_t)$ shows that the sum of the coefficients of the two edges is $2\gamma_t$. Using two walks similar to that of the first drawing of Figure 10.7, one using edge (a_0, a_t) and the other edge (a_0, b_t), the first will contain once the edge $(\overline{a}_t, b_t)$ and not $(a_t, \overline{b}_t)$, while it will be the reverse for the other, implying that these two edges have the same coefficient, which is therefore equal to γ_t.

We now consider the case in which T_t is even. Modifying the walk of the first drawing of Figure 10.10 as we did previously proves that the edges (a_0, b_t), $(\overline{a}_0, \overline{b}_t)$, $(a_t, \overline{b}_t)$, and $(\overline{a}_t, b_t)$ have the same coefficients, say γ_t. Comparing the two walks of Figure 10.10 shows that the sum of the coefficients of the edges (a_0, a_t) and $(\overline{a}_0, \overline{a}_t)$ equals $2\gamma_t$. Comparing the walk of the second drawing of Figure 10.7 with its horizontal mirror copy yields that these two edges have equal coefficients, which are therefore γ_t.

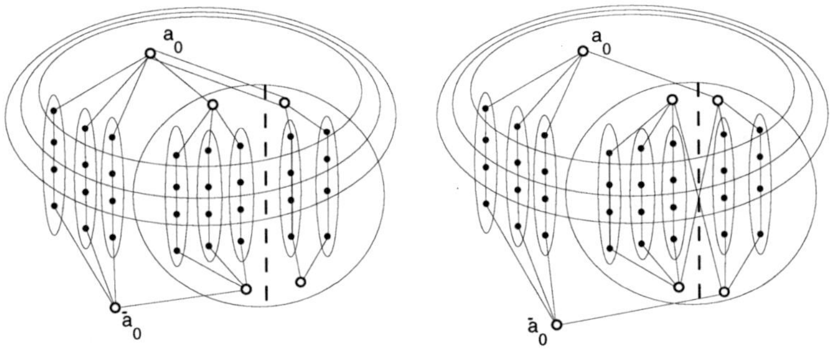

Figure 10.10. *Determining the coefficients of edges of $\delta(T_t)$, even case.*

Comparing the walks of Figures 10.7 and 10.8 yields that $\gamma_{out} = \gamma_A = \gamma_B = \gamma$. It now follows that $\gamma_t = h\gamma$, and therefore $\gamma = 1$.

To end the proof we show that the g-coefficients of the other edges are the same as in the one-domino path inequality by observing that, for any edge not in the skeleton, there is a tight walk that contains this edge and only edges of the skeleton. We leave it to the reader to check all cases. □

Remark 10.29. In the validity and facet-inducing proofs we did not use the fact that there is only one maximal tooth that is not also minimal, so these proofs carry over for the case of more than one of these teeth. Also, it is only for the sake of simplicity that we restricted ourselves to regular paths. One can rewrite the proofs with coefficients on handles and teeth as long as the interval property is satisfied and all intervals have the same weight (see, for example, [5]). The coefficient of the nonminimal teeth is twice the sum of the coefficients of the handles.

Acknowledgement. The author wants to thank Giovanni Rinaldi for the very careful reading of the manuscript and the many suggestions he made that improved the chapter.

Bibliography

[1] S. Boyd, S. Cockburn, and D. Vella. *On the Domino-Parity Inequalities for the Traveling Salesman Problem*. Technical report, University of Ottawa, 2000.

[2] G. Cornuéjols, J. Fonlupt, and D. Naddef. The traveling salesman problem on a graph and some related polyhedra. *Mathematical Programming*, 33:1–27, 1985.

[3] L. Fleischer and É. Tardos. Separating maximally violated comb inequalities in planar graphs. *Mathematics of Operations Research*, 24:130–148, 1999.

[4] A. Letchford. Separating a superclass of comb inequalities in planar graphs. *Mathematics of Operations Research*, 25:443–454, 2000.

[5] D. Naddef. Polyhedral theory and branch-and-cut algorithms for the symmetric traveling salesman problem. In G. Gutin and A. Punnen, editors, *The Traveling Salesman Problem and Its Variations*, volume 12 of Combinatorial Optimization. Kluwer Academic Publishers, Boston, Dordrecht, London, 2002.

[6] D. Naddef and E. Wild. The domino inequalities: Facets for the symmetric traveling salesman polytope. *Mathematical Programming*, 98B:223–251, 2003.

[7] D. Naddef and Y. Pochet. The traveling salesman polytope revisited. *Mathematics of Operations Research*, 26:700–722, 2001.

Chapter 11

Computing Optimal Consecutive Ones Matrices

Marcus Oswald and Gerhard Reinelt**

MSC 2000. 90C57
Key words. Polyhedral combinatorics, branch-and-bound, branch-and-cut

11.1 Introduction

A 0/1-matrix $A \in \{0, 1\}^{(m,n)}$ has the *consecutive ones property (for rows)* if the columns of A can be permuted so that the ones in each row appear consecutively. For convenience we just say that A is C1P.

Consecutive ones matrices play an important role in computational biology. However, for the purposes of this chapter, we do not discuss possible applications, but address the general problem.

The *consecutive ones problem* is the task of turning a given 0/1-matrix B into a C1P matrix by changing as few entries as possible. If individual penalties are specified for changing an entry, then we speak about the *weighted consecutive ones problem* (WC1P). This problem as well as the unweighted version is known to be NP-hard [1].

Note that, if the matrix A is C1P, then a representation of all column permutations that transform A such that all ones appear consecutively in every row can be found in time $O(mn)$ by the *PQ-tree algorithm* [2].

In the WC1P we are looking for a C1P (m, n)-matrix with 0/1-entries that minimizes a certain objective function. This matrix will be represented by variables x_{ij}, $i = 1, \ldots, m$, $j = 1, \ldots, n$, where x_{ij} represents the matrix entry of row i and column j. We will interpret $x = (x_{11}, \ldots, x_{1n}, \ldots, x_{m1}, \ldots, x_{mn})$ as a vector or as a matrix, whichever is more appropriate. In the following we deal with inequalities that have to be satisfied by x. Usually,

*Institut für Informatik, Universität Heidelberg, INF 368, 69120 Heidelberg, Germany.

the coefficients of these inequalities are specified by a matrix. Let A be an (l,k)-matrix of coefficients. For an l-tuple $I = (r_1, \ldots, r_l)$ with pairwise distinct entries $r_i \in \{1, \ldots, m\}$ and a k-tuple $J = (c_1, \ldots, c_k)$ with pairwise distinct entries $c_j \in \{1, \ldots, n\}$, we define

$$A \circ x_{IJ} = \sum_{i=1}^{l} \sum_{j=1}^{k} a_{ij} x_{r_i c_j}.$$

For simplicity we will just say, for example, "$A \circ x_{IJ} \leq a_0$ for all (l,k)-tuples (I,J)," meaning that all l-tuples $I = (r_1, \ldots, r_l)$ and all k-tuples $J = (c_1, \ldots, c_k)$ are allowed for mapping A to x. Whenever we use $A \circ x$, we assume that $I = (1, \ldots, m)$ and $J = (1, \ldots, n)$.

Now let a 0/1-matrix B be given as input. We are looking for a C1P matrix x that resembles B as closely as possible. Taking into account that a penalty c_{ij} has to be paid when switching entry B_{ij}, the following objective function value is associated with x:

$$\sum_{\substack{i,j \\ b_{ij}=0}} c_{ij} x_{ij} + \sum_{\substack{i,j \\ b_{ij}=1}} c_{ij}(1 - x_{ij}) = \sum_{\substack{i,j \\ b_{ij}=0}} c_{ij} x_{ij} - \sum_{\substack{i,j \\ b_{ij}=1}} c_{ij} x_{ij} + \sum_{\substack{i,j \\ b_{ij}=1}} c_{ij}. \tag{11.1}$$

If we set

$$\overline{c}_{ij} = \begin{cases} -c_{ij} & \text{if } b_{ij} = 1, \\ c_{ij} & \text{if } b_{ij} = 0, \end{cases}$$

then WC1P amounts to minimizing the linear objective function $\overline{c}$ over the set of C1P matrices.

11.2 The Consecutive Ones Polytope

Tucker [6] has shown that a 0/1-matrix A is C1P if and only if it is not possible to permute the rows and columns of A such that any of five forbidden matrices occurs as a submatrix. Based on this characterization it is easy to give an integer programming formulation of WC1P consisting of inequalities that forbid exactly these Tucker matrices.

We define the *consecutive ones polytope* as

$$P_{\text{C1}}^{m,n} = \text{conv}\{M \mid M \text{ is an } (m,n)\text{-matrix with C1P}\}.$$

The WC1P then is the problem of minimizing the linear function c defined above over the polytope $P_{\text{C1}}^{m,n}$ for a given (m,n)-matrix B.

It is easy to see that $P_{\text{C1}}^{m,n}$ has full dimension $m \cdot n$. Namely, the zero matrix is C1P and, for every $1 \leq i \leq m$ and $1 \leq j \leq n$, the matrix consisting of zeros only except for a one in position ij is C1P. This gives a set of $n \cdot m + 1$ affinely independent C1P matrices.

Let $a^T x \leq a_0$ be a valid inequality for $P_{\text{C1}}^{m,n}$ and let $m' \geq m$ and $n' \geq n$. We say that the inequality $\overline{a}^T x \leq a_0$ for $P_{\text{C1}}^{m',n'}$ is obtained from $a^T x \leq a_0$ by *trivial lifting* if

$$\overline{a}_{ij} = \begin{cases} a_{ij} & \text{if } i \leq m \text{ and } j \leq n, \\ 0 & \text{otherwise.} \end{cases}$$

Theorem 11.1. *Let $a^T x \leq a_0$ be a facet-defining inequality for $P_{\text{C1}}^{m,n}$ and let $m' \geq m$ and $n' \geq n$. If $a^T x \leq a_0$ is trivially lifted, then the resulting inequality defines a facet of $P_{\text{C1}}^{m',n'}$.*

Proofs of this theorem as well as of the following ones are given in [4].

Since trivial lifting is possible, larger polytopes inherit all facets of smaller polytopes.

Inequalities that are obviously valid for $P_{\mathrm{C1}}^{m,n}$ are the *trivial inequalities* $x_{ij} \geq 0$ and $x_{ij} \leq 1$ for all $1 \leq i \leq m$, $1 \leq j \leq n$. It is easily seen that they also define facets.

Theorem 11.2. *For all $m \geq 1$, $n \geq 1$, $1 \leq i \leq m$, $1 \leq j \leq n$, the inequalities $x_{ij} \geq 0$ and $x_{ij} \leq 1$ define facets of $P_{\mathrm{C1}}^{m,n}$.*

We are interested in getting more insight into the facet structure of $P_{\mathrm{C1}}^{m,n}$. In this section we will describe four types of facet-defining inequalities that can then be used to replace the integer programming formulation based on Tucker matrices with a stronger formulation.

Note that, except for a few cases, the inequalities forbidding single Tucker matrices are not facet-defining for $P_{\mathrm{C1}}^{m,n}$.

Our first two inequality classes are based on two matrices $\mathcal{F}_{1_k}$ and $\mathcal{F}_{2_k}$, $k \geq 1$. These matrices are $(k+2, k+2)$- and $(k+2, k+3)$-matrices, respectively. They have entries -1, 0, and $+1$, where for convenience we write "$-$" ("$+$") instead of "-1" ("$+1$"). The matrices are shown in Figure 11.1.

$$
\begin{pmatrix}
+ & + & & & & - \\
 & + & + & & & - \\
 & & \ddots & \ddots & & \vdots \\
 & & & + & + & - \\
- & 0 & \cdots & 0 & + & + \\
+ & 0 & \cdots & 0 & - & +
\end{pmatrix}
\qquad
\begin{pmatrix}
+ & + & & & & - & - \\
 & + & + & & & - & - \\
 & & \ddots & \ddots & & \vdots & \vdots \\
 & & & + & + & - & - \\
- & 0 & \cdots & 0 & + & + & - \\
- & 0 & \cdots & 0 & + & - & +
\end{pmatrix}
$$

Figure 11.1. *Matrices $\mathcal{F}_{1_k}$ and $\mathcal{F}_{2_k}$.*

These two matrix classes lead to facet-defining inequalities.

Theorem 11.3.

(i) *The inequalities $\mathcal{F}_{1_k} \circ x_{IJ} \leq 2k+3$, $k \geq 1$, for all $(k+2, k+2)$-index sets, are facet-defining for $P_{\mathrm{C1}}^{m,n}$ for all $m \geq k+2$ and $n \geq k+2$.*

(ii) *The inequalities $\mathcal{F}_{2_k} \circ x_{IJ} \leq 2k+3$, $k \geq 1$, for all $(k+2, k+3)$-index sets, are facet-defining for $P_{\mathrm{C1}}^{m,n}$ for all $m \geq k+2$ and $n \geq k+3$.*

In addition to these general classes of inequalities, there are two further ones induced by two special single matrices, namely, the (4, 6)-matrix denoted by $\mathcal{F}_3$ and the (4, 5)-matrix denoted by $\mathcal{F}_4$, shown in Figure 11.2.

$$
\begin{pmatrix}
+ & + & - & 0 & - & 0 \\
- & 0 & + & + & - & 0 \\
- & 0 & - & 0 & + & + \\
- & + & - & + & - & +
\end{pmatrix}
\qquad
\begin{pmatrix}
+ & + & 0 & + & - \\
+ & 0 & 0 & + & - \\
- & 0 & + & + & - \\
+ & - & - & + & +
\end{pmatrix}
$$

Figure 11.2. *Matrices $\mathcal{F}_3$ and $\mathcal{F}_4$.*

Theorem 11.4.

(i) *The inequalities* $\mathcal{F}_3 \circ x_{IJ} \leq 8$, *for all* $(4,6)$*-index sets, define facets of* $P_{\mathrm{C1}}^{m,n}$ *for all* $m \geq 4$ *and* $n \geq 6$.

(ii) *The inequalities* $\mathcal{F}_4 \circ x_{IJ} \leq 8$, *for all* $(4,5)$*-index sets, define facets of* $P_{\mathrm{C1}}^{m,n}$ *for all* $m \geq 4$ *and* $n \geq 5$.

It can be shown that these four classes of inequalities are satisfied by all C1P matrices and that every Tucker matrix violates at least one of them. The system of all of these inequalities therefore gives an integer programming formulation of WC1P consisting of facet-defining inequalities only.

If $m \leq 2$ or $n \leq 2$, all (m,n)-matrices are C1P and therefore the trivial inequalities completely describe $P_{\mathrm{C1}}^{m,n}$. The facets discussed above give a complete description if $n = 3$ and $m = 3, 4, 5$, or 6 or if $m = 3$ and $n = 4$ or 5. For $m = 3$ and $n > 5$ or for $m > 3$ and $n > 3$ further facet-defining inequalities are needed. With PORTA [3] we computed the complete description of $P_{\mathrm{C1}}^{4,4}$. It requires nine additional classes of facet-defining inequalities, and the total number of facets of $P_{\mathrm{C1}}^{4,4}$ is 1880.

11.3 Separation

In this section we address how to solve the linear programming relaxation obtained from the integer programming formulation by dropping the integrality requirement. To this end we have to give separation algorithms for the given classes of inequalities.

11.3.1 Integer vector separation

Usually, in the first phases of the cutting plane procedure, the solution x^* of the linear programming relaxation is integral but infeasible. Feasibility can easily be tested by the PQ-tree algorithm. If x^* is infeasible, let $P = \{i \mid x_i^* = 1\}$ and $Z = \{i \mid x_i^* = 0\}$. Then

$$\sum_{i \in P} x_i - \sum_{i \in Z} x_i \leq |P| - 1$$

is a cutting plane that cuts off x^* but is satisfied by all other 0/1-vectors.

One can strengthen this cutting plane by removing rows and columns of the linear programming relaxation as long as the remaining matrix stays infeasible.

A heuristic version of this separation idea can even be performed if the linear programming relaxation x^* is fractional. First we have to make it integral. One possibility is to round entry x_{ij}^* to 1 with probability x_{ij}^* and to 0 otherwise. Now for each row and for each column the total sum of differences between the linear program (LP) values and the rounded values is calculated. Now rows and columns with high value of total rounding are chosen earlier to be removed from the matrix. Again this is repeated as long as the remainder stays infeasible. And, if the total amount of rounding in the remaining matrix is less than 1, we can construct a cutting plane.

The same method of shrinking the linear programming relaxation is also used in the following subsection, where it is explained in more detail.

11.3.2 Separating inequalities from small instances

To be able to detect cuts from small instances we try to identify a small submatrix of the fractional LP solution x^* violated by such an inequality.

Given a fractional LP solution x^* we create a 0/1-matrix $\tilde{x}$ by setting $\tilde{x}_{ij}$ to 1 with probability x^*_{ij} and to 0 otherwise. For every row i we compute the coefficient

$$r_i = \sum_j |\tilde{x}_{ij} - x^*_{ij}|$$

and for every column j the coefficient

$$c_j = \sum_i |\tilde{x}_{ij} - x^*_{ij}|.$$

Then we sort all coefficients (together) in nonincreasing order. Rows and columns are now deleted from $\tilde{x}$ in this order. After every deletion, we check whether the resulting matrix is C1P. If it is C1P, then we undo the last row or column deletion. We proceed this way until all rows and columns are tested. The idea of this heuristic is to find the core of infeasibility of $\tilde{x}$.

Usually we end up with a submatrix of $\tilde{x}$ that is not C1P and has about three to five rows and three to seven columns. For the corresponding submatrix of x^* we test all known inequalities from small instances as well as the $\mathcal{F}_{1_k}$- and $\mathcal{F}_{2_k}$-inequalities.

We illustrate this separation heuristic by an example. Let the following fractional solution and rounded version of it be given:

$$x^* = \begin{pmatrix} 1 & 1 & 1 & 0 & 0 \\ 0 & 0 & 1 & 1 & 1 \\ 0 & 1 & 0.5 & 1 & 1 \\ 0 & 0 & 0.2 & 1 & 0.5 \\ 0 & 0 & 0 & 0.8 & 0.4 \end{pmatrix}, \qquad \tilde{x} = \begin{pmatrix} 1 & 1 & 1 & 0 & 0 \\ 0 & 0 & 1 & 1 & 1 \\ 0 & 1 & 0 & 1 & 1 \\ 0 & 0 & 0 & 1 & 1 \\ 0 & 0 & 0 & 1 & 1 \end{pmatrix}.$$

The sequence of ordered coefficients is, for example, $c_5, r_5, r_4, c_3, r_3, c_4, c_1, r_2, c_2$, and r_1.

We end up with the submatrix of $\tilde{x}$ consisting of rows 1, 2, 3 and columns 2, 3, 4 and thus the following corresponding submatrix of x^*:

$$\begin{pmatrix} 1 & 1 & 0 \\ 0 & 1 & 1 \\ 1 & 0.5 & 1 \end{pmatrix}.$$

This submatrix violates the facet-defining inequality

$$\begin{pmatrix} 1 & 1 & -1 \\ -1 & 1 & 1 \\ 1 & -1 & 1 \end{pmatrix} \circ x \leq 5,$$

which is actually an F_{1_1}-inequality. Note that we do not use true rounding to obtain $\tilde{x}$ from x^*. Our experiments show that this is not sufficient to have a good chance of finding violated inequalities. Random rounding performs much better.

11.3.3 Separating F_{1_k}- and F_{2_k}-inequalities

Actually, in the case of F_{1_k}-inequalities, we will separate a more general class of inequalities. These inequalities can be obtained by observing that the "-1"-entry in the last row can be moved to any position by changing the first and last columns in an appropriate way. The corresponding $\overline{F}_{1_k}$-inequalities can also be shown to be facet-defining for $P_{\text{CI}}^{m,n}$ and are visualized in Figure 11.3 (where the left-hand side matrix $\overline{\mathcal{F}}_{1_k}$ is a $(k+2, k+2)$-matrix).

$$
\begin{array}{cc}
 & \begin{array}{cccccccccc} i & & & & & j & c_2 & \dots & c_d & h \end{array} \\
\begin{array}{c} \\ \\ \\ \\ r_1 \\ \vdots \\ r_{d-1} \\ r_d \\ l \end{array} &
\left(\begin{array}{cccccccccc}
+ & + & & & & & & & & - \\
0 & + & + & & & & & & & - \\
\vdots & & \ddots & \ddots & & & & & & \vdots \\
0 & & & & + & + & & & & - \\
- & & & & & + & + & & & 0 \\
\vdots & & & & & & \ddots & \ddots & & \vdots \\
- & & & & & & & + & + & 0 \\
- & & & & & & & & + & + \\
+ & 0 & \dots & & 0 & - & 0 & \dots & 0 & +
\end{array}\right) \circ x \le 2k+3
\end{array}
$$

Figure 11.3. $\overline{F}_{1_k}$*-inequality.*

We obtain the original F_{1_k}-inequalities when there are no columns $c_2, \dots, c_d$, i.e., when $d = 1$.

The main task of the separation algorithm is to identify the row l and the columns i, j, and h and to sum up appropriate coefficients for rows and columns in between.

We proceed as follows. For every column $i = 1, \dots, n$ we create the complete undirected bipartite graph G^i with the n columns and m rows as the two nodesets. With every edge cr we associate the weight $w^i_{cr} = 1 - x^*_{rc} + \frac{1}{2}x^*_{ri}$, where x^* is the given LP solution to be cut off.

In every weighted graph G^i we now compute for every pair j, $j \neq i$, and $h, h \neq i, h \neq j$, of columns a shortest path between j and h with respect to the assigned edge weights. This way we obtain shortest lengths p^i_{jh} (see Figure 11.4).

For every quadruple i, j, h, l of columns i, j, h and row l we evaluate

$$p^i_{jh} + p^h_{ji} - x_{li} + x_{lj} - x_{lh} = 2k + 2 - \overline{\mathcal{F}}_{1_k} \circ x^*.$$

For every expression that has value less than -1 we can construct a violated $\overline{F}_{1_k}$-inequality using the shortest paths computed above to include columns $c_2, \dots, c_d$ and rows $r_1, \dots, r_d$. If none of these values is less than -1, then no violated $\overline{F}_{1_k}$-inequality and thus no violated F_{1_k}-inequality exists.

The running time is dominated by the all-pairs shortest path computation for every column, and we obtain time complexity $O(n^3(n+m))$. Therefore the F_{1_k}-inequalities can be separated in polynomial time.

For the separation of the F_{2_k}-inequalities (see Figure 11.5) we proceed as follows. For each pair of columns g and h we compute a weighted graph G^{gh} with the $n-2$ remaining

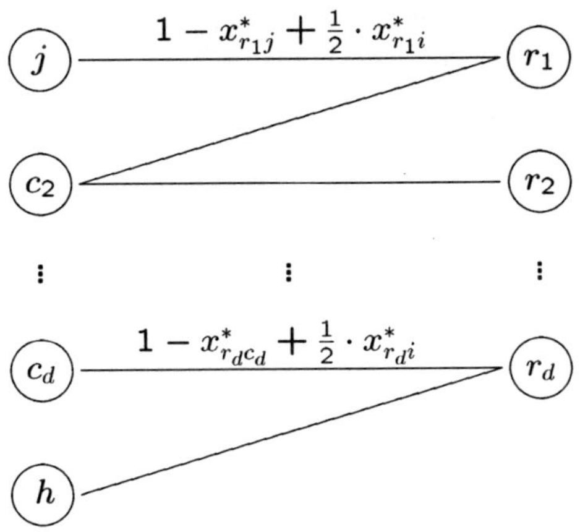

Figure 11.4. *Path between columns j and h.*

$$
\begin{array}{c}
 \\ r_1 \\ \vdots \\ r_{k-1} \\ r_k \\ l \\ m
\end{array}
\begin{array}{c}
\begin{array}{ccccccc} i & c_2 & \dots & c_k & j & g & h \end{array} \\
\begin{pmatrix}
+ & + & & & & - & - \\
 & + & + & & & - & - \\
 & & \ddots & \ddots & & \vdots & \vdots \\
 & & & + & + & - & - \\
- & 0 & \dots & 0 & + & + & - \\
- & 0 & \dots & 0 & + & - & +
\end{pmatrix}
\end{array}
\circ x \leq 2k + 3
$$

Figure 11.5. *F_{2_k}-inequality.*

columns as nodesets. The weight of the edge cd is the minimum over all rows r of the expression $2 - x^*_{cr} - x^*_{dr} + x^*_{gr} + x^*_{hr}$. The time needed to create these graphs is $O(n^4 m)$. Now we compute all the shortest paths in the graphs. This can be done in $O(n^3)$ for each graph with the algorithm of Floyd–Warshall. In our implementation we use Dijkstra's algorithm implemented with heaps. Though the worst-case running time for dense graphs is worse, it is faster in practice since computations can be stopped early as soon as cuts are found.

The weight of the shortest path from column i to column j in the graph G^{gh} is denoted by p^{gh}_{ij}.

Now for each quadruple g, h, i, j of columns we compute the row l (m) to be that row minimizing the expression $w^{gh}_{ij,l} := 2 + x^*_{il} - x^*_{jl} - x^*_{gl} + x^*_{hl}$ ($w^{gh}_{ij,m} := 2 + x^*_{im} - x^*_{jm} + x^*_{gm} - x^*_{hm}$). This also takes time $O(n^4 m)$. Finally, we only have to check if there is a quadruple g, h, i, j and corresponding rows l and m with the property

$$w^{gh}_{ij,l} + w^{gh}_{ij,m} + p^{gh}_{ij} < 1.$$

If we find columns and rows with this property we can construct a violated F_{2_k}-inequality.

The overall worst-case running time is $O(n^4(n + m))$. In practice, however, our separation routine works even for problems with $n \leq 70$ and $m \leq 1000$ in reasonable time.

11.4 Primal Heuristic

To find good feasible solutions we use a heuristic that is driven by the fractional LP solution. The idea is to generate a big C1P matrix "close" to the LP solution.

Let x^* be the current fractional LP solution. From this solution we generate a 0/1-matrix $\tilde{x}$ as above by setting $\tilde{x}_{ij}$ to 1 with probablity x^*_{ij} and to 0 otherwise and compute the coefficients r_i for every row i.

The rows are sorted with respect to nondecreasing coefficients, so we give priority to rows that resemble the LP solution.

The rows are input in this order to the PQ-tree algorithm as long as the corresponding submatrices are C1P. When the algorithm stops (either because all rows are taken into account or because the next row would lead to an infeasible matrix), we generate all possible permutations from the final PQ-tree.

Let π_k be one of these permutations. We permute the columns of the objective function accordingly and use a scan-line algorithm to find the best sequence of consecutive ones in every row. So every π_k yields a feasible solution, and the best one among them is selected.

To illustrate this approach we give a small example. Consider a fractional $(5, 5)$ solution x^* and a possible $\tilde{x}$:

$$x^* = \begin{pmatrix} 1 & 0.4 & 1 & 0 & 0.2 \\ 0 & 0 & 1 & 1 & 1 \\ 0.8 & 1 & 0.5 & 1 & 1 \\ 0 & 0.6 & 0 & 1 & 0.7 \\ 0 & 0.9 & 0 & 1 & 1 \end{pmatrix}, \qquad \tilde{x} = \begin{pmatrix} 1 & 1 & 1 & 0 & 0 \\ 0 & 0 & 1 & 1 & 1 \\ 1 & 1 & 0 & 1 & 1 \\ 0 & 0 & 0 & 1 & 1 \\ 0 & 1 & 0 & 1 & 1 \end{pmatrix}.$$

According to their coefficients r_i, the rows are sorted in the order 2, 5, 3, 1, 4. The first three rows yield feasible permutations $(1, 2, 4, 5, 3)$, $(1, 2, 5, 4, 3)$, and their reversals. Adding also the fourth row would lead to a matrix that is not C1P. Suppose the objective function permuted according to the first permutation is

$$\begin{pmatrix} +3 & -2 & +3 & -2 & -2 \\ +3 & -2 & -2 & +3 & +3 \\ -2 & -2 & +3 & +3 & -2 \\ +3 & +3 & +3 & -2 & -2 \\ -2 & +3 & -2 & +3 & +3 \end{pmatrix}.$$

Then the best C1P matrix for this objective function is

$$\begin{pmatrix} 1 & 1 & 1 & 0 & 0 \\ 0 & 0 & 0 & 1 & 1 \\ 0 & 0 & 1 & 1 & 0 \\ 1 & 1 & 1 & 0 & 0 \\ 0 & 1 & 1 & 1 & 1 \end{pmatrix}$$

with objective function value 32.

As for the separation of small instance cuts, random rounding turned out to be superior to true rounding.

11.5 Computational Results

We implemented a branch-and-cut code for the WC1P based on the framework ABACUS [5] with CPLEX 6.5.3 as LP solver. The runs were performed on a Sun Ultra™ 10 with UltraSPARC™ IIi 440 CPU.

All separation procedures and the primal heuristic described above were used. They were called in a hierarchical fashion, i.e., first the simple ones were executed and, if these did not find any cuts, then the F_{1_k}- and F_{2_k}-inequalities were checked using exact separation. The primal heuristic was performed after every solution of an LP.

For our computational tests we generated random (n, n)-matrices with 0/1-entries for $n = 5, \ldots, 19$ with densities of the 1-entries varying between 10% and 90%. For every problem we computed the minimal number of entries to be switched to obtain a C1P matrix.

For every size and density we created 10 instances. Up to $n = 12$ all instances could be solved in reasonable time. Larger problems turned out to be more difficult depending on their density. Table 11.1 displays the average percentage of entries that have to be switched for the various problem sizes. The most entries apparently have to be switched for densities around 60%. These problems are also the most difficult ones for our branch-and-cut algorithm.

Table 11.1. *Percentage of entries to be switched.*

	10%	20%	30%	40%	50%	60%	70%	80%	90%
(5, 5)	0.0	0.0	1.2	2.0	0.8	0.8	0.4	0.8	0.0
(6, 6)	0.0	0.3	2.5	2.2	1.9	3.3	2.5	1.9	0.6
(7, 7)	0.2	0.6	3.1	3.7	4.9	5.9	4.9	2.9	0.6
(8, 8)	0.0	0.3	1.6	3.9	5.2	7.5	6.3	4.4	2.7
(9, 9)	0.1	1.1	3.6	5.4	6.3	7.4	5.9	5.2	2.2
(10, 10)	0.3	1.9	4.3	6.2	7.4	8.4	8.0	6.6	2.7
(11, 11)	0.2	2.5	5.1	6.9	9.6	10.1	9.3	6.9	3.7
(12, 12)	0.4	3.2	6.0	8.5	10.1	11.7	10.7	8.8	5.0
(13, 13)	0.4	3.6	6.3	11.1		11.1	10.5	8.2	4.4
(14, 14)	0.6	4.3					11.3	8.8	4.8
(15, 15)	0.8	4.1						9.6	5.3
(16, 16)	0.9	4.6						9.3	5.3
(17, 17)	0.9								5.6
(18, 18)	1.4								5.3
(19, 19)	1.5								6.0

Table 11.2. *Average execution time (minutes:seconds) over 10 instances.*

	10%	20%	30%	40%	50%	60%	70%	80%	90%
(5, 5)	0:01	0:01	0:01	0:01	0:01	0:01	0:01	0:01	0:01
(6, 6)	0:01	0:01	0:01	0:01	0:01	0:01	0:01	0:01	0:01
(7, 7)	0:01	0:01	0:01	0:01	0:02	0:02	0:01	0:01	0:01
(8, 8)	0:01	0:01	0:01	0:01	0:02	0:04	0:02	0:02	0:01
(9, 9)	0:01	0:01	0:05	0:05	0:06	0:07	0:03	0:02	0:01
(10, 10)	0:01	0:01	0:21	0:21	0:23	0:39	0:16	0:16	0:01
(11, 11)	0:02	0:07	2:26	9:14	8:01	7:10	3:07	0:15	0:02
(12, 12)	0:02	0:09	4:21	25:26	28:16	240:58	12:29	4:23	0:13

Table 11.3. *Number of generated small instance cuts of size (m, n).*

(m, n)	3	4	5	6
3	2	29	0	14
4	0	8	15	176
5	0	11	0	0

Table 11.4. *Number of generated F_{1_k}- and F_{2_k}-cuts.*

k	3	4	5	6	7	8	9	10
F_{1_k}	13	62	77	62	34	16	3	2
F_{2_k}	93	253	103	21	2	0	0	0

Table 11.2 shows the average computation time for $n \leq 12$. The variance of CPU times for problems of the same size and density is very high. For example, for the problems of size (12, 12) and density 50%, the fastest computation took 34 seconds and the slowest 76 minutes. We have not displayed the CPU times for the bigger problems. For $n = 14$ and density 50% it can even take several days to solve a problem.

Tables 11.3 and 11.4 display some statistics about the number and type of generated cuts of a branch-and-cut run for a (19, 19)-matrix with density 10%. The running time was about 2 minutes. 113 LPs had to be solved in 45 branch-and-bound (B&B) nodes.

It is typical for all problems that the separation of inequalities from small instances proved to be very helpful. Many of them are identified quickly by our heuristic.

Furthermore, the primal heuristic described in Section 11.4 turned out to be very good. It took only about 3 minutes average time to find the optimal solutions for the (12, 12)-matrices with density 50%. But proving optimality took an additional 25 minutes. For 4 of the 10 instances the optimal primal bound was found within 2 seconds, in 8 cases within 1 minute.

11.6 Conclusion

We have described a first branch-and-cut algorithm for solving the WC1P. The approach is very promising. In fact, the problem instances discussed here (with cardinality objective function) belong to the most difficult ones. Problems obtained from real-world data could be solved easily up to a range (320, 60).

There are several possible ways to improve the algorithm for solving larger problems. Besides finding new classes of cutting planes, there is need of a good criterion for selecting violated inequalities to be added to the current LP, since already in our first version many of them are found. Branching strategies are completely unexplored; up to now we have only branched on variables. Furthermore, there are many options for parallelization.

Bibliography

[1] K.S. Booth. *PQ-Tree Algorithms*. Ph.D. thesis, University of California, Berkeley, 1975.

[2] K.S. Booth and G. Lueker. Testing for the consecutive ones property, interval graphs, and graph planarity using PQ-tree algorithms. *Journal of Computer and System Sciences*, 13:335–379, 1976.

[3] T. Christof and A. Loebel. PORTA—A polyhedron representation algorithm, 1998. www.informatik.uni-heidelberg.de/groups/comopt/software/PORTA/.

[4] M. Oswald and G. Reinelt. Polyhedral aspects of the consecutive ones problem. In *Computing and Combinatorics*, pages 373–382. Proceedings of the 5th Conference on Computing and Combinatorics (COCOON), Sydney, 2000.

[5] S. Thienel. *ABACUS: A Branch-And-CUt System*. Ph.D. thesis, Universität zu Köln, 1995.

[6] A. Tucker. A structure theorem for the consecutive 1's property. *Journal of Combinatorial Theory*, 12:153–162, 1972.

Chapter 12

Protein Folding on Lattices: An Integer Programming Approach

Vijay Chandru, M. Rammohan Rao,† and Ganesh Swaminathan‡*

Abstract. In this chapter, we initiate the study of the protein folding problem from an integer linear programming perspective. The particular variant of protein folding that we examine is known as the hydrophobic–hydrophilic (HP) model of protein folding on the integer lattice. This problem is known to be NP-hard and also maxSNP-hard. We examine various alternate formulations for the planar version of this problem and present some preliminary computational results. We hope that this will set the stage for a polyhedral combinatorics assault on this important problem.

MSC 2000. 90C10

Key words. Protein folding, lattice models, integer programming

12.1 Introduction

Proteins are biological molecules that are responsible for implementing various functions in all living organisms. Each protein has well-defined functions, which range from building up DNA and RNA molecules to controlling different parameters in living cells. It is amazing that proteins are built of very simple building blocks, known as amino acids [4]. There are 20 different amino acids. Amino acids are linked to each other by means of peptide bonds.

*Indian Institute of Science, Strand Genomics, Bangalore (chandru@csa.iisc.ernet.in).

†Indian Institute of Management, Bangalore (mrao@iimb.ernet.in).

‡Strand Genomics, Bangalore (gans@strandgenomics.com).

Determining the structure of proteins is a very important problem. The three-dimensional structure of a protein is believed to be a very important determinant of the properties of the protein. This becomes crucial in drug design, where the aim is to obtain proteins with specific functionalities. A remarkable discovery was made by Christian Anfinsen and his colleagues in the 1950s when they found that many simple proteins had a unique native structure, which just seems to depend on the sequence. This has been subsequently verified for a large number of proteins and it is now believed that the native structure is a minimum energy configuration (the Thermodynamic Hypothesis). This has led to an enormous interest in the development of methods to predict the three-dimensional structure from the sequence information via optimization techniques. Determining a protein sequence has become feasible with current technology, but determining the exact three-dimensional structure is still a very slow and expensive process that requires crystallization of the protein, and a majority of proteins cannot be crystallized.

In principle, it should be possible to predict the fold of a protein into its native conformation once we are given the sequence of the constituent amino acids. This is known as the protein structure prediction problem and is sometimes referred to as deciphering the second half of the genetic code. While large proteins fold in nature in seconds, computational chemists and biologists have found it to be a huge challenge to compute the minimum energy conformations using various formulations of this optimization problem. Recent work by theoretical computer scientists on this problem [8] has shown that the problems are NP-hard (cf. [13]), and even the very simple lattice model examined in this chapter is known to be max-SNP hard and therefore unlikely to admit polynomial-time approximation schemes as well.

The difficulty of working with the detailed atomic level model has motivated biologists to work on simple discrete models. One way to discretize this problem is to only consider embeddings on a lattice. The energy function also has to be defined appropriately in this new setting. The resulting minimum energy conformation problems are essentially combinatorial optimization problems.

Broadly, three optimization modeling strategies have been proposed for protein folding.

The Protein Structure Prediction Model (PSP model). This model is a general nondiscrete model defined formally by Ngo and Marks [19], who also give an NP-hardness result for this model. In this model, the protein is described by the complete list of the atoms in the molecules; their connectivities, bond lengths, and angles; and force constants between all pairs of atoms. The energy of a conformation is a nonconvex function obtained by summing the contributions of different kinds of interactions. The NP-hardness is shown by a reduction from the partition problem.

The Lattice Polymer Embedding Model (LPE model). The LPE model was studied by Unger and Moult [27]. The protein is modeled as a chain of beads. The space is the collection of embeddings in the three-dimensional cubic lattice. An embedding means that each bead must be placed at some lattice site, and successive beads must be adjacent on the lattice. In addition, the embedding must not be self-intersecting. The energy is defined as a weighted sum of pairwise interaction energies (functions that depend on the lattice distance between pairs of beads). The objective is to find the conformation that minimizes this energy. Unger and Moult show that this problem is NP-hard by a reduction from the optimal linear arrangement. The hydrophobic–hydrophilic (HP) model that we discuss later is a special case of this model.

The Charged Graph Embedding Model (CGE model). This model also describes the protein as a sequence of beads. A charge of -1, 0, or $+1$ is associated with each bead. For each pair of beads, the interaction energy is defined to be the product of the charges divided by the distance separating the beads (provided the distance is within a cutoff). The total energy is simply the sum of pairwise energies. One important condition is that bonds are allowed to cross as long as there is at most one bead per site. Fraenkel [12] showed that this problem is also NP-hard by reduction from three-dimensional matching. The CGE model incorporates charges on the residues, which is a realistic feature, but the bonds permitted are too general.

We now consider a popular model of protein folding called the Hydrophobic–Hydrophilic model.

The Hydrophobic–Hydrophilic Model (HP model). This model was introduced by Dill [10] as a special case of the LPE model and has been studied extensively in [11, 16, 17, 18, 28, 29] and is the simplest possible abstraction of the folding problem, which is still nontrivial and retains the hardness features of the original problem.

The model starts with classifying the 20 amino acids as H (hydrophobic or nonpolar) and P (hydrophilic or polar). This classification is known from experimental results. A protein is modeled as a sequence of H's and P's. The conformations allowed are not self-intersecting embeddings on a two- or three-dimensional cubic lattice. A pair of amino acids that occur in successive positions in the chain are called *connected neighbors*, while a pair of nonsuccessive amino acids that are adjacent in the embedding are called *topological neighbors*. The energy of any folding is proportional to the negative of the number of pairs of H's that are topological neighbors. Therefore the aim is to maximize the number of topological neighbors.

Even this simple model is NP-hard to solve, and proving this was an open question for a long time (see [1, 2, 3, 8, 9, 15, 20, 22, 26]). Even before hardness results were known, Hart and Istrail [14] gave a simple approximation algorithm, which achieves a worst-case ratio of 1/4 for two-dimensional lattices and 3/8 for the three-dimensional case. Very recently, Newman [21] improved the 1/4 bound for the two-dimensional case to a 1/3 bound with a linear-time approximation algorithm. A lot of empirical work has also been done on this model. Dill et al. [11] have extensively studied the biological properties of this model by actual enumeration of all conformations for small length sequences. Unger and Moult [28, 29] looked at this problem from a genetic algorithm viewpoint and they were able to obtain compact foldings of fairly long sequences, but they were not able to give any guaranteed bounds on their algorithms.

In this chapter we focus on the HP model and in particular on the two-dimensional lattice embedding of the main chain. The next five sections describe integer programming formulations of this problem. We report some very preliminary computational experiments carried out on these formulations in Section 12.7 and conclude with a brief agenda for research on folding proteins using integer programming.

12.2 Formulation

The two-dimensional HP protein-folding model on a rectangular lattice is formulated as an integer linear programming problem. A protein is a chain of amino acid residues. The sequence of amino acids in the chain to be folded on the two-dimensional grid is denoted as s_k, $k = 1, 2, \dots, n$.

Each amino acid s_k is either hydrophobic or hydrophilic. The set of amino acids that are hydrophobic is denoted as H. Amino acids s_t and s_{t+1}, $1 \leq t \leq n-1$, are adjacent on the chain.

In this formulation a $(2n-1) \times (2n-1)$ grid is used. Each lattice point or vertex is denoted as (i, j), $1 \leq i, j \leq 2n-1$. Two vertices (i, j) and (u, v) are said to be neighbors on the grid if one of the following holds:

- $u = i$ and $v = j + 1$ or $v = j - 1$;
- $v = j$ and $u = i + 1$ or $u = i - 1$.

The set of vertices adjacent to vertex (i, j) is denoted as N_{ij}. Note that, if $(u, v) \in N_{ij}$, then $(i, j) \in N_{uv}$. We define the *grid graph* $G = (V, E)$, where every edge e is of the form $((i, j), (u, v))$, where $(u, v) \in N_{ij}$ and $1 \leq i, j \leq 2n-1$. The first amino acid, s_1, is assumed to be anchored at the center of the grid, i.e., at the lattice point (n, n). In Section 12.4, it is shown that the size of the grid can be reduced considerably, thereby eliminating a large number of variables.

The protein-folding problem on a two-dimensional grid involves placing the amino acids s_k, $1 \leq k \leq n$, at the vertices (i, j), $1 \leq i, j \leq 2n-1$, such that the following constraints are satisfied.

(i) Each amino acid is placed at precisely one vertex.

(ii) Each vertex has at most one amino acid.

(iii) Amino acids that are adjacent on the chain must be placed at adjacent vertices.

The objective is to place the amino acids on the vertices so that a maximum number of amino acids in the set H that are nonadjacent on the chain are adjacent on the grid, i.e., are topologically adjacent.

The variables are defined as follows: For $1 \leq i, j \leq 2n-1$, and $1 \leq k \leq n$,

- x_{ij}^k is 1 if amino acid s_k is placed at the grid point (i, j) and 0 otherwise.
- y_{ij}^{uv} is 1 if some $s_a \in H$ and $s_b \in H$ are placed at the vertices (i, j) and (u, v) that are neighbors, i.e., $(u, v) \in N_{ij}$, and 0 otherwise.

The integer programming formulation is as follows:

$$\text{(P)} \qquad \max \sum_{((i,j),(u,v)) \in E} y_{ij}^{uv} \text{ subject to}$$

$$x_{nn}^1 = 1, \tag{12.1}$$

$$\sum_{k=1}^{n} x_{ij}^{k} \leq 1 \text{ for } 1 \leq i, j \leq 2n-1, \tag{12.2}$$

$$\sum_{i=1}^{2n-1} \sum_{j=1}^{2n-1} x_{ij}^{k} = 1 \text{ for } 1 \leq k \leq n, \tag{12.3}$$

$$x_{ij}^{k} \leq \sum_{(uv)\in N_{ij}} x_{uv}^{k+1} \text{ for } 1 \leq i, j \leq 2n-1, \ 1 \leq k \leq n-1, \tag{12.4}$$

$$y_{ij}^{uv} \leq \sum_{k\in H} x_{ij}^{k} \text{ for all edges } ((i, j), (u, v)), \tag{12.5}$$

$$y_{ij}^{uv} \leq \sum_{k\in H} x_{uv}^{k} \text{ for all edges } ((i, j), (u, v)), \tag{12.6}$$

$$x_{ij}^{k} = 0 \text{ or } 1 \text{ for } 1 \leq i, j \leq 2n-1, \text{ and } 1 \leq k \leq n, \tag{12.7}$$

$$y_{ij}^{uv} \geq 0 \text{ for all edges } ((i, j), (u, v)). \tag{12.8}$$

The first constraint anchors the first amino acid, s_1, in the chain. The second set of constraints ensures that at most one amino acid is placed at any vertex. The requirement that each amino acid be placed at some vertex is ensured by the constraint set (12.3).

The first amino acid is anchored at the vertex (n, n), and the constraint in (12.4) for $i = n$, $j = n$, and $k = 1$ ensures that the second amino acid is placed at a vertex, say $(a, b) \in N_{nn}$. Next, the constraint in (12.4) for $i = a$, $j = b$, and $k = 2$ ensures that the third amino acid is placed at a vertex in N_{ab}. Repeating this argument, it follows that all amino acids are placed at some vertex. However, constraint set (12.3) ensures that no amino acid is placed at more than one vertex.

The constraint sets (12.5) and (12.6) imply that y_{ij}^{uv} may be set to 1 only if an amino acid $s_a \in H$ is placed at vertex (i, j) and another amino acid $s_b \in H$ is placed at the neighboring vertex (u, v). Because of the objective function, it follows that y_{ij}^{uv} is set to 1 if and only if two hydrophobic amino acids are placed at neighboring vertices (i, j) and (u, v). The constraint sets (12.5) and (12.6) are written in a convenient form. Clearly, some constraints are duplicated since $(u, v) \in N_{ij}$ implies $(i, j) \in N_{uv}$. It is understood that such duplicates are eliminated; i.e., for any pair (i, j) and (u, v) of neighboring vertices only one constraint in (12.5) and one constraint in (12.6) is required.

Constraint set (12.7) ensures integrality of the variables x_{ij}^{k}. The variables y_{ij}^{uv} are restricted to be nonnegative variables in (12.8). It is not necessary to require that the variables y_{ij}^{uv} be 0 or 1. This is because of the integrality restriction on x_{ij}^{k} and because the objective function involves maximizing the sum of the variables y_{ij}^{uv}.

The objective function maximizes the number of hydrophobic amino acids that are placed at adjacent vertices. The optimal objective function value includes a constant that is the number of hydrophobic amino acids that are adjacent in the chain. This is because the true objective is to maximize the number of hydrophobic topological neighbors while the objective in the formulation counts the number of hydrophobic neighbors both topological and adjacent on the chain.

Remark 12.1. It is straightforward to extend the formulation to other lattices such as triangular lattices or three-dimensional lattices. The size of the lattice and the set N_{ij} change accordingly and additional restrictions, if required, can be easily imposed.

It is also possible to include interactions between amino acids that are placed on non-adjacent vertices on the lattice but within some specified distance. This requires additional variables y_{ij}^{uv}, where vertex (u, v) is within the specified distance from vertex (i, j). The formulation can easily be extended to the case of generalized hydrophobicity discussed in [2]. In this case, the amino acids s_k, $1 \leq k \leq n$, in the chain are not restricted to be only hydrophobic or hydrophilic but can be any one of the 20 possible amino acids. The energy between two topological neighbors can be an arbitrary function that depends on the type of amino acids. This extension of the problem requires that the variables y_{ij}^{uv} must now have two additional parameters, say a and b, denoting the amino acids that are not adjacent on the chain. Then the constraints (12.5) and (12.6) are to be replaced by

$$y_{ij}^{uv}(a, b) \leq x_{ij}^{a}, \quad 1 \leq i, j \leq 2n - 1, \quad 1 \leq a, b \leq n, \tag{12.9}$$

$$y_{ij}^{uv}(a, b) \leq x_{uv}^{b}, \quad 1 \leq i, j \leq 2n - 1, \quad 1 \leq a, b \leq n, \tag{12.10}$$

where a and b are nonadjacent in the chain and $(u, v) \in N_{ij}$.

The objective function coefficient for $y_{ij}^{uv}(a, b)$ would be the energy between amino acids s_a and s_b that are not adjacent on the chain but adjacent on the grid. It is understood that, if the energy between topologically adjacent amino acids s_a and s_b is zero, the corresponding variables and constraints in (12.9) and (12.10) are eliminated from the formulation.

12.3 Additional Inequalities

Instead of merely restricting the position of the neighboring amino acids s_k and s_{k+1}, we can write constraints that restrict the position of amino acids s_k and s_{k+t}, where $t \geq 1$.

A path (or a simple path) between two vertices $((i, j), (u, v))$ in the graph G is defined in the usual manner as a sequence of edges connecting vertices (i, j) and (u, v) with no intermediate vertex repeated. It follows that the shortest distance between any pair of nodes (i, j) and (u, v) is $d_{ij}^{uv} = |u - i| + |v - j|$. Then we have the constraints

$$x_{ij}^{k} \leq \sum_{(u,v):d_{ij}^{uv} \leq t} x_{uv}^{k+t} \text{ for } 1 \leq i, j \leq 2n - 1, \quad 1 \leq k \leq n - t. \tag{12.11}$$

Viewing the chain of amino acids in the reverse direction, analogous to constraint set (12.11), we have for $t \geq 1$ the constraints

$$x_{ij}^{k} \leq \sum_{(u,v):d_{ij}^{uv} \leq t} x_{uv}^{k-t} \text{ for } 1 \leq i, j \leq 2n - 1, \quad t + 1 \leq k \leq n. \tag{12.12}$$

These constraints ensure that, if amino acid s_k is placed at vertex (i, j), then amino acid s_{k-t} must be placed at a vertex at a distance less than or equal to t from vertex (i, j).

Integrality of the variables x_{ij}^k and the constraint set (12.4) imply the constraints in (12.11) and (12.12). However, some fractional solutions to (P) are cut off by (12.11) and (12.12). The usefulness of these constraints from a computational point of view needs to be explored. The inequalities

$$\sum_{k \in H} x_{ij}^k + \sum_{k \in H} x_{uv}^k \leq 1 + y_{ij}^{uv} \text{ for all edges } ((i, j), (u, v)) \in E \tag{12.13}$$

are easily verified to be valid. The formulation (P) needs to be studied from a polyhedral combinatorics perspective. For example, it would be interesting to identify the facet-defining inequalities, if any, in the above set of valid inequalities.

12.4 Grid Size and Elimination of Variables

In the formulation (P) in Section 12.2, it is possible to set a number of variables to zero, i.e., to eliminate several variables. This results in a more compact formulation and, more importantly, can help partial enumeration–based optimization algorithms by pruning off the search space.

The shortest distance between any pair of vertices is either even or odd. It is easy to verify that, if the shortest distance between any pair of vertices (i, j) and (u, v) is even (odd), then all paths between the same two vertices are of even (odd) length.

The distance between two amino acids s_a and s_b in the chain is defined as $w_{ab} = |a - b|$. Clearly, the distance between two amino acids is either even or odd. It follows that two amino acids at even (odd) distance in the chain must be at even (odd) distance on the grid. Moreover, two amino acids s_a and s_b with distance w_{ab} can never be placed at vertices (i, j) and (u, v), where $d_{ij}^{uv} > w_{ab}$. Note, however, that it is possible to place s_a and s_b at vertices (i, j) and (u, v), where $d_{ij}^{uv} < w_{ab}$.

Now, noting that the first amino acid s_1 is anchored at vertex (n, n), it follows that the second amino acid s_2 can be placed only at a vertex at a shortest distance of one from vertex (n, n), i.e., at vertex $(n - 1, n)$, $(n + 1, n)$, $(n, n - 1)$, or $(n, n + 1)$. Similarly, the third amino acid s_3 can be placed only at a vertex at a shortest distance of two. Moreover, since w_{13} is even, s_3 cannot be placed at a vertex (u, v) whose shortest distance from vertex (n, n) is odd. Continuing, the fourth amino acid s_4 can be placed only at a vertex at a shortest distance of one or three but not more, i.e., since w_{14} is odd, s_4 can be placed at a vertex (u, v) such that d_{nn}^{uv} is odd and less than or equal to three.

By repeating the above argument for amino acids s_i, $5 \leq i \leq n$, a large number of variables can be eliminated from the formulation. It should be noted that some vertices (u, v) in the grid such that $d_{nn}^{uv} > n$ can be eliminated from the grid itself. Eliminating all such variables, it follows that, when n is odd, the number of x_{ij}^k variables is $\frac{2}{3}p(p+1)(4p+5)$, where $p = \frac{n-1}{2}$. In this case, the number of y_{ij}^{uv} variables is $4(n - 1)^2$.

It is also possible to reduce the size of the grid itself by noting that any folding of the protein can be rotated as necessary. Suppose, as before, we anchor the first amino acid at vertex (n, n). Let $p = \lfloor \frac{n}{2} \rfloor$, where $\lfloor y \rfloor$ denotes the largest integer less than or equal to y. It suffices to consider the rectangular lattice with vertices (i, j), $n - p \leq i \leq 2n - 1$, $n - p \leq j \leq n + p$. In this rectangular grid as well, there are some vertices (u, v) such that $d_{nn}^{uv} > n$. Such vertices can be eliminated from the grid.

12.5 Alternative Formulation

Instead of anchoring the first amino acids s_1 at vertex (n, n), the first amino acid may be placed anywhere. In this case we require only an $n \times n$-grid to begin with. Now we need the constraint

$$\sum_i \sum_j x_{ij}^1 = 1 \tag{12.14}$$

to ensure that the first amino acid is placed at some vertex. Moreover, in this formulation, we cannot eliminate the variables as in Section 12.4. However, by increasing the grid size to $(n + 1) \times (n + 1)$, it is possible to restrict the placement of the first amino acid s_1 to a vertex (i, j), where i and j are odd. Given this restriction, it is easily verified that amino acid s_2 can be placed only at a vertex (u, v) such that one of the following holds:

$$u \text{ is odd and } v \text{ is even,} \tag{12.15}$$

$$u \text{ is even and } v \text{ is odd.} \tag{12.16}$$

Continuing, it follows that amino acid s_3 can be placed only at a vertex (a, b) such that one of the following holds:

$$a \text{ is odd and } b \text{ is odd,} \tag{12.17}$$

$$a \text{ is even and } b \text{ is even.} \tag{12.18}$$

It is easily verified that amino acids at an odd distance from s_1 in the chain may be placed at a vertex (u, v) that satisfies either (12.15) or (12.16), while amino acids at an even distance from s_1 in the chain may be placed at a vertex (a, b) that satisfies either (12.17) or (12.18).

By this argument, a large number of variables can be eliminated in this alternative formulation. Eliminating all such variables, it follows that, when n is odd, the number of x_{ij}^k variables is $\frac{1}{4}(n + 1)^2(2n - 1)$, while the number of y_{ij}^{uv} variables is $2n(n + 1)$.

The alternative formulation is analogous to the formulation in Section 12.2 except that the grid size is different and constraint (12.1) is now replaced by constraint (12.14).

12.6 Row and Column Generation

A feasible solution to (P) has exactly n of the variables x_{ij}^k equal to 1. Typically, in a feasible solution only a few (perhaps of order n) of the variables y_{ij}^{uv} are equal to 1. However, in spite of eliminating several variables, as indicated in Sections 12.4 and 12.5, the formulation in Section 12.2 and the alternative formulation in Section 12.5 have a large number of variables and constraints. The number of variables and constraints is of the order n^3. Hence, feasible solutions to (P) are highly degenerate.

In order to speed up the computations, it might be desirable to start with a small number of constraints and variables. Given an optimal solution to the linear programming relaxation of the smaller problem, it is straightforward to generate, if there is one, a violated constraint from among those not written down explicitly. Similarly, it is straightforward to generate from among those variables that have not been written down explicitly a variable, if there is one, to enter the basis. Thus, to keep the size of the working basis small and thereby speed up the computations, it is possible to resort to row and column generation.

The process of row and column generation is akin (though not identical) to starting with a thin rectangular lattice and solving the problem repeatedly by increasing the width and length of the lattice.

12.7 Computational Results

To determine the folding of a protein consisting of n amino acids the current (alternative) formulation uses a grid of n^2 lattice points. Since any of the n amino acids can occupy any of the n^2 lattice points, the number of variables x_{ij}^k is n^3. The number of y_{ij}^{uv} variables is $2n(n-1)$. However, the maximum size of the lattice spanned by a protein of length n consists of p rows and k columns such that $p + k \leq n + 1$. This can be utilized to reduce the size of the problem. The following outlines the approach that has been used for computational purposes.

(i) Initialize $p = 2$.
(ii) Use a grid of size $p \times k$ such that $p + k = n + 1$.
(iii) Calculate the optimal objective function value for the associated integer programming problem.
(iv) Increment p by 1. If $p = n$, go to step (v); else go to step (ii).
(v) Calculate the maximum of the optimal objective function values obtained in the above $n - 2$ iterations.

If $n - 2$ iterations (the value of p varies from 2 to $n - 1$) are done, then one of the solutions will be an optimal solution to the given protein-folding problem. Further, by considerations of symmetry, one can vary p from 2 to $\lfloor \frac{n}{2} \rfloor$, leading to $\lfloor \frac{n}{2} \rfloor - 1$ iterations. For a $p \times k$ lattice the number of x_{ij}^k variables is pkn and the number of y_{ij}^{uv} variables is $2pk - p - k$. The total time taken for $\lfloor \frac{n}{2} \rfloor - 1$ iterations with reduced lattice points can be expected to be less than the time taken for one iteration with n^2 lattice points. That is what we observed in our limited computational results.

The following is an example illustrating the results of the approach outlined above. The protein has 10 amino acids with 3 hydrophobic amino acids at positions 1, 4, and 7. For different lattice sizes the optimal objective function value for the integer programming problem, the number of simplex iterations that were required, and the number of nodes generated in the branch-and-bound (B&B) scheme are summarized in Table 12.1.

Table 12.1. *Partial enumeration.*

Lattice size	Opt. int. obj. value	Total number of simplex iterations	Number of nodes in B&B tree
2×9	1	13921	222
3×8	2	2805	39
4×7	2	5028	49
5×6	2	16401	131

Four problems were solved using the integer programming (IP) approach and the Hart–Istrail (HI) heuristic. The number of amino acids (n) in the chain, the positions of the hydrophobic amino acids, the optimal objective function value obtained by integer programming, and the objective function value obtained by the Hart–Istrail heuristic are given in Table 12.1. The results clearly show that the Hart–Istrail heuristic produces solutions that are far from being optimal in relative terms.

The integer programming approach uses the $n \times n$-grid. The optimal objective function values for the linear programming (LP) relaxation and the integer programming approach that are given in Table 12.2 are those obtained after subtracting the constant given by the number of hydrophobic amino acids that are adjacent in the chain.

In all the above problems, none of the variables, as suggested in Section 12.5, were eliminated. It is planned to implement these improvements in the near future.

Table 12.2. *Optimality gaps.*

Problem number	n	Position of H amino acids	Opt. value LP relax	Opt. value IP	Obj. val. HI-heuristic
1	10	1, 4, 7	5.4	2	1
2	10	1, 4, 7, 8	6.2	3	2
3	11	1, 4, 7, 8, 11	8.09	4	2
4	11	1, 3, 7, 8	6.27	2	1

12.8 Conclusion

The lattice models of protein folding have certain limitations [11].

- The resolution of the original problem is lost. Bond angles actually lie in some restricted regions, as indicated by the Ramachandran plot, rather than being right angles.
- Details of protein structure, bond energies, and charges cannot be represented accurately in such models.
- Bond lengths are not captured well enough.

An attempt to address the first limitation explored alternative lattice structures, such as triangular lattices [2]. In spite of the above limitations, discrete lattice models continue to be useful, particularly for simulations and for enumerative techniques for deducing statistical properties of small proteins and peptides [5, 6, 7, 23, 24, 25]. An important challenge is therefore to improve the integrity of lattice models.

The computational results reported in Section 12.7 are clearly at a very preliminary stage. We intend to continue with the experimentation and will report our findings in a subsequent paper.

Acknowledgements. The authors would like to thank Prof. Ramesh Hariharan for some early discussions on the integer programming approach and in particular for the idea of incrementally expanding lattices, which led us to the interpretation in Section 12.6. They would also like to thank an anonymous referee for suggesting the valid inequalities (12.13).

Bibliography

[1] J. Atkins and W.E. Hart. On the intractability of protein folding with a finite alphabet of amino acids. *Algorithmica*, 25:279–294, 1999.

[2] R. Agarwala, S. Batzogloa, V. Dancik, S.E. Decatur, S. Hannenhalli, M. Farach, S. Muthukrishnan, and S. Skiena. Local rules for protein folding on a triangular lattice and generalized hydrophobicity. In *Proceedings of the First Annual International Conference on Research in Computational Molecular Biology*, pages 1–2, ACM, New York, 1997.

[3] B. Berger and T. Leighton. Protein folding in the hydrophobic-hydrophilic model is NP complete. *Journal of Compututational Biology*, 5:27–40, 1998.

[4] C. Brandon and J. Tooze. *Introduction to Protein Structure*, second edition. Garland, New York, 1999.

[5] H.S. Chan and K.A. Dill. Origins of structure in globular proteins. *Proceedings of the National Academy of Science USA*, 87:6388–6392, 1990.

[6] H.S. Chan and K.A. Dill. Polymer principles in protein structure and stability. *Annual Review of Biophysics and Biophysical Chemistry*, 20:447–490, 1991.

[7] P. Clote. Protein folding, the Levinthal paradox and rapidly mixing Markov chains. In *Proceedings of the 26th International Colloquium on Automata, Languages, and Programming*, pages 240–249, Springer, Berlin, 1999.

[8] V. Chandru, A. Dattasharma, V.S. Kumar. The algorithmics of folding proteins on lattices. *Discrete Applied Mathematics*, 127:145–161, 2003.

[9] P. Crescenzi, D. Goldman, C. Papadimitrou, A. Piccolboni, and M. Yannakakis. On the complexity of protein folding, *Journal of Compututational Biology*, 5:423–446, 1998.

[10] K.A. Dill. *Biochemistry*, 24:1501, 1985.

[11] K.A. Dill, S. Bromberg, K. Yue, K.M. Fiebig, D.P. Yee, P.D. Thomas, and H.S. Chan. Principles of protein folding: A perspective from simple exact models. *Protein Science*, 4:561–602, 1995.

[12] A.S. Fraenkel. Complexity of protein folding. *Bulletin of Mathematical Biology*, 55:1199–1210, 1993.

[13] M.R. Garey and D.S. Johnson. *Computers and Intractability—A Guide to the Theory of NP-completeness*. Freeman, San Francisco, 1979.

[14] W.E. Hart and S. Istrail. Fast protein folding in the hydrophobic-hydrophilic model within three-eighths of optimal. *Journal of Computational Biology*, 3:53–96, 1996.

[15] V. Heun. Folding. In *Proceedings of the European Symposium on Algebra*, Springer, Berlin, 1999.

[16] K.F. Lau and K.A. Dill. A lattice statistical mechanics model of the conformation and sequence spaces of proteins. *Macromolecules*, 22:3986–3997, 1989.

[17] K.F. Lau and K.A. Dill. Theory for protein mutability and biogenesis. *Proceedings of the National Academy of Science USA*, 87:638–642, 1990.

[18] D. Lipman and J. Wilber. Modelling neutral and selective evolution of protein folding. *Proceedings of the Royal Society of London, Series* B, 245(1312):7–11, 1991.

[19] J.T. Ngo and J. Marks. Computational complexity of a problem in molecular-structure prediction. *Protein Engineering*, 5(4):313–321, 1992.

[20] A. Nayak, A. Sinclair, and U. Zwick. Spatial codes and the hardness of string folding problems. In *Proceedings of the 9th Annual ACM-SIAM Symposium on Discrete Algorithms*, pages 639–648, SIAM, Philadelphia, ACM, New York, 1998.

[21] A. Newman. A new algorithm for protein folding in the HP model. In *Proceedings of the 13th Annual ACM-SIAM Symposium on Discrete Algorithms*, pages 876–884, SIAM, Philadelphia, ACM, New York, 2002.

[22] M. Paterson and T. Przytycka. On the complexity of string folding. *Discrete Applied Mathematics*, 71:217–230, 1996.

[23] A. Sinclair. *Algorithms for Random Generation and Counting: A Markov Chain Approach*. Birkhäuser, Boston, 1993.

[24] A. Sali, E. Shakhnovich, and M. Karplus. How does a protein fold?. *Nature*, 369:248–251, 1994.

[25] A. Sali, E. Shakhnovich, and M. Karplus. Kinetics of protein folding: A lattice model study of the requirements for folding to the native state. *Journal of Molecular Biology*, 235:1614–1636, 1994.

[26] L. Trevisan. When Hamming meets Euclid: The approximability of geometric TSP and Steiner tree. *SIAM Journal on Computing*, 30:475–485, 2000.

[27] R. Unger and J. Moult. Finding the lowest free energy conformation of a protein is an NP-hard problem: Proof and implications. *Bulletin of Mathematical Biology*, 55:1183–1198, 1993.

[28] R. Unger and J. Moult. Genetic algorithms for protein folding simulations. *Journal of Molecular Biology*, 231:75–81, 1993.

[29] R. Unger and J. Moult. A genetic algorithm for three dimensional protein folding simulations. In *Proceedings of the 5th International Conference on Genetic Algorithms*, pages 581–588, 1993.

Part IV

General Polytopes

Chapter 13

On the Expansion of Graphs of 0/1-Polytopes

*Volker Kaibel**

Abstract. The edge expansion of a graph is the minimum quotient of the number of edges in a cut and the size of the smaller one among the two node sets separated by the cut. Bounding the edge expansion from below is important for bounding the "mixing time" of a random walk on the graph from above. It has been conjectured by Mihail and Vazirani (see [9]) that the graph of every 0/1-polytope has edge expansion at least one. A proof of this (or even a weaker) conjecture would imply solutions of several long-standing open problems in the theory of randomized approximate counting. We present different techniques for bounding the edge expansion of a 0/1-polytope from below. By means of these tools we show that several classes of 0/1-polytopes indeed have graphs with edge expansion at least one. These classes include all 0/1-polytopes of dimension at most five, all simple 0/1-polytopes, all hypersimplices, all stable set polytopes, and all (perfect) matching polytopes.

MSC 2000. 52B12, 52B11, 52B05, 68W20, 60G50
Key words. 0/1-polytope, graph, expansion, Mihail–Vazirani conjecture, random walk, rapid mixing, stable set polytope, matching polytope

13.1 Introduction

In the early days of polyhedral combinatorics there was some hope that investigations of the graphs of 0/1-polytopes that are associated with certain sets of combinatorial objects might yield insights that could be exploited in designing algorithms for related combinatorial optimization problems. Certainly this hope was inspired by the success of Dantzig's simplex

*Sekr. MA 6–2, Institut für Mathematik, Fakultät II, TU Berlin, Straße des 17. Juni 136, 10623 Berlin, Germany (kaibel@math.tu-berlin.de). Supported by the Deutsche Forschungsgemeinschaft, FOR 413/1–1 (Zi 475/3–1).

algorithm for linear programming. Quite soon, people came across astonishing facts like the one that the diameter of the asymmetric traveling salesman polytope equals one for at most five cities and two for more than five cities (Padberg and Rao [22]; apparently already discovered, but not published, in the early 1950s by Kuhn [15]). This was even outperformed by the cut polytope of the complete graph on n nodes that has diameter one for all $n \geq 2$ (Barahona and Mahjoub [6]). Other polytopes turned out to have more complicated graphs, e.g., the stable set polytopes, for which two vertices are adjacent if and only if the symmetric difference of the corresponding stable sets induces a connected graph [8]. Another interesting example is the basis polytope of a matroid (i.e., the convex hull of the characteristic vectors of its bases), where two vertices are adjacent if and only if the corresponding bases have a symmetric difference of cardinality two (observed by Edmonds in the early 1970s). All in all lots of interesting results on the graphs of *special* 0/1-polytopes have been obtained, however, usually without much impact on algorithms for related optimization problems.

Maybe the best known result on graphs of *general* 0/1-polytopes is due to Naddef. He proved [20] that the graph of any d-dimensional 0/1-polytope has diameter at most d, and thus 0/1-polytopes satisfy the Hirsch conjecture (claiming that the graph of any d-dimensional polytope with n facets has diameter at most $n - d$). Some results on cycles of the graphs of general 0/1-polytopes were proved as well by Naddef and Pulleyblank in the 1980s [19, 21]. Nevertheless, the graphs of (general) 0/1-polytopes did not receive too much attention. Probably this was due to the fact that people did not see how to exploit potential knowledge on this topic with respect to algorithms for combinatorial optimization problems, where the interest in 0/1-polytopes originally came from. As a source of general results on 0/1-polytopes, we refer to [25].

The question on graphs of 0/1-polytopes treated in this chapter is mainly motivated by the goal of designing algorithms that generate random elements in classes of combinatorial objects, which often translates into the task of generating random vertices of 0/1-polytopes. Of course, in general, this includes combinatorial optimization problems via appropriate choices of random distributions, but here we will be more concerned with the task of drawing a vertex according to the uniform distribution. Maybe the most important motivation for generating (uniformly distributed) random elements from a set of combinatorial objects is the fact that in many cases this allows us to count the number of objects approximately by a randomized algorithm. The first spectacular success of this method was Jerrum and Sinclair's randomized approximation algorithm for computing the permanent in a certain large class of 0/1-matrices [11] (extended to arbitrary matrices with nonnegative integer entries by Jerrum, Sinclair, and Vigoda [13]).

For an introduction to the topic of randomized approximate counting and random generation see [12] or [7]. Here we briefly sketch the ideas on the example of the spanning trees of a given graph, although the exact number of spanning trees can be computed efficiently by Kirchhoff's matrix tree theorem (see, e.g., [2, Chap. 24]).

Let $\mathcal{T}(G)$ be the set of spanning trees of a graph G. The basic idea for counting spanning trees via generating them randomly is the following. Suppose G' is the graph G plus an additional edge e', and assume that we already know a number τ' approximating $|\mathcal{T}(G')|$. If we generate a large set $\mathcal{T}'$ of spanning trees in G' uniformly at random, and if α is the fraction of those trees in $\mathcal{T}'$ that do not contain e', then we might hope that $|\mathcal{T}(G)|$ approximately equals $\alpha \cdot \tau'$. Since the number of spanning trees of the complete graph

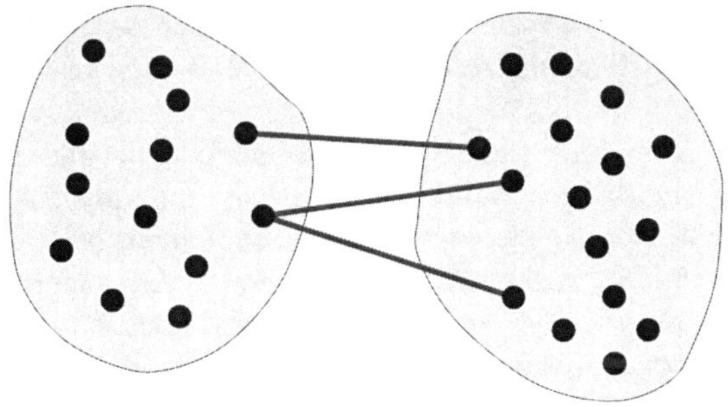

Figure 13.1. *If the neighborhood structure on which a random walk is performed allows us to partition the objects into two large parts with only a few connections between them, then the random walk cannot converge quickly. Fortunately, the converse of this statement is true as well.*

on n nodes is well known to be n^{n-2}, this suggests an iterative method to approximately compute $|\mathcal{T}(G)|$ by a randomized algorithm.

We do not go into the details of this algorithm and its analysis but rather turn to the question of how to generate a spanning tree in a graph G uniformly at random, where our exposition here is meant only to give an idea of the method as far as it is useful for understanding the motivation of the questions on 0/1-polytopes that we consider in this chapter. The strategy is to perform a (finite) random walk on the set $\mathcal{T}(G)$, meaning that one starts with an arbitrary spanning tree $T_0 \in \mathcal{T}(G)$, slightly modifies T_0 randomly to a spanning tree T_1, slightly modifies T_1 randomly to T_2, and so on. After a certain number of steps, one stops and takes the current tree as the desired random object. The passage from T_i to T_{i+1} could be performed in the following way. For technical reasons, we first flip an unbiased coin in order to decide if we "do nothing" and stay at $T_{i+1} := T_i$, or if we try to get to a modified tree, as described subsequently. We first choose a pair (e, f) of edges of G uniformly at random. If it happens that $e \notin T_i$ and f lies on the cycle in $T_i \cup \{e\}$, then we proceed to $T_{i+1} := T \setminus \{f\} \cup \{e\}$. Otherwise, we stay at $T_{i+1} := T_i$.

Thus, we perform a *random walk* in the graph $\mathcal{G}(\mathcal{T}(G))$ that has the spanning trees of G as its nodes, where two trees are connected if and only if their symmetric difference consists of two edges. All *transition probabilities* (i.e., for each ordered pair T and T' of adjacent nodes in $\mathcal{G}(\mathcal{T}(G))$ the probability that we proceed to T' if we currently are at T) equal $\frac{1}{2 \cdot m^2}$, where m is the number of edges of G. By standard arguments (see Section 13.2) one can prove that, if i tends to infinity, at step i the random walk will be at each spanning tree with the same probability, no matter at which spanning tree we started. However, for algorithmic purposes, it is of course important that this convergence not be too slow. Responsible for the speed of convergence is the *edge expansion* of $\mathcal{G}(\mathcal{T}(G))$ (see Figure 13.1), where the edge expansion of a graph $H = (V, E)$ is the number

$$\begin{aligned}\mathcal{X}(H) &:= \min\left\{\frac{|\delta(S)|}{\min\{|S|, |V \setminus S|\}} \ : \ S \subset V,\ S \neq \emptyset, V\right\}\\ &= \min\left\{\frac{|\delta(S)|}{|S|} \ : \ S \subset V,\ 0 < |S| \leq \frac{|V|}{2}\right\}\end{aligned}$$

(with $\delta(S)$ denoting the set of all edges with one endnode in S and the other one in $V \setminus S$). If $\mathcal{X}(\mathcal{G}(\mathcal{T}(G)))$ is bounded by the reciprocal of a polynomial in the size of G, then the random walk described above converges "sufficiently fast." Actually, it is well known that in our case even $\mathcal{X}(\mathcal{G}(\mathcal{T}(G))) \geq 1$ holds (see the remarks at the end of Subsection 13.4.4).

Viewing this example of generating spanning trees randomly as a prototype, one might formulate a strategy for random generation of certain combinatorial objects as follows. First, one has to choose a neighborhood structure on the objects and then transition probabilities have to be assigned appropriately. Here, "appropriately" means (a) that the random walk should asymptotically behave according to the desired probability distribution, and (b) that it should do so approximately after only a small number of steps. Let us assume that the distribution we aim at is the uniform distribution. Then, provided that the neighborhood structure is (as in the example) symmetric and connected, we can always achieve goal (a) by choosing the same probability for all proper transitions. In this case, goal (b) is equivalent to choosing a neighborhood structure with a "not too small" edge expansion. Of course, in order to be able to efficiently simulate the random walk, it should also be possible to draw for each object uniformly at random one of its neighboring objects. However, this will not be our focus here.

Thus we are faced with the task of coming up with good candidates for neighborhood structures. Suppose that the set of objects we are interested in is a family of subsets of a finite set (as in the example of spanning trees). Then the graph of the associated polytope (the convex hull of the characteristic vectors of the subsets in the family) is a natural candidate, where the graph is defined by the 1-skeleton, i.e., the zero- and the one-dimensional faces. In fact, the neighborhood structure we considered in the example is given by the graph of the spanning-tree polytope. Two vertices of that polytope are adjacent if and only if the symmetric difference of the corresponding spanning trees consists of two edges (since the spanning trees of some graph are the bases of a matroid—the *graphic* matroid defined by that graph).

As mentioned above, two vertices of a stable set polytope are adjacent if and only if the symmetric difference of the corresponding stable sets induces a connected subgraph [8]. Since matchings correspond to stable sets in the line graph, two vertices of a matching polytope are thus adjacent if and only if the symmetric difference of the corresponding matchings is connected. The same is true for perfect matching polytopes, since they are faces of matching polytopes. Two of the most prominent random walks in combinatorics are the ones designed and analyzed by Jerrum and Sinclair [11] on the set of (near-)perfect matchings of a bipartite graph and on the set of all matchings of an arbitrary graph. While the first one led to a randomized approximation algorithm for the permanent (for a certain class of 0/1-matrices), the second one yielded a randomized approximation algorithm for evaluating the partition function of a monomer-dimer system in statistical physics, which is the same as the generating function of the matchings in an arbitrary graph. In both cases the random walk was performed on a subgraph of the graph of the associated 0/1-polytope, and the crucial step was to prove that this subgraph has a large edge expansion. Another example is the random generation of 0/1-knapsack solutions (leading to a randomized approximation algorithm for counting as well) due to Morris and Sinclair [18]. Again, the key step for their result was to show that a certain subgraph of the graph of the 0/1-knapsack polytope has a large edge expansion.

It seems clear from these examples that it is important to investigate the question for the edge expansion of general 0/1-polytopes (i.e., the convex hulls of arbitrary sets of points with coordinates from $\{0, 1\}$). Actually, it appears from a citation in a paper of Feder and Mihail [9] (which we will be concerned with in Section 13.4) that Mihail and Vazirani considered this question some time ago. Feder and Mihail (and also Mihail [17]) quote their conjecture that the graph of every 0/1-polytope has edge expansion at least one. Of course, even a proof showing that the edge expansion of the graph of any d-dimensional 0/1-polytope is bounded by one over a polynomial in d would be very important (see also Section 13.5).

While this extensive introduction was intended to shed some light on the relevance of the question for expansion properties of graphs of 0/1-polytopes, the rest of the chapter is meant to support the conjecture of Mihail and Vazirani with some partial results. In Section 13.3 we show that the conjecture is indeed true for every 0/1-polytope whose dimension does not exceed five. In Section 13.2 we list a few well-known facts on random walks; our main goal is to provide some background that is relevant to Section 13.3. As a consequence, the concepts treated in this introduction may become a bit more clear. In Section 13.4 we present some methods for bounding the edge expansion that are especially suited for graphs of (certain) 0/1-polytopes. In particular, it will turn out that simple 0/1-polytopes, hypersimplices, and stable set polytopes satisfy Mihail and Vazirani's conjecture. We conclude with some remarks in Section 13.5.

The results presented in Section 13.3 were obtained jointly with Werner [24].

13.2 Expansion and Eigenvalues

The aim of the present section is to explain the connection between the edge expansion of a graph and the second-largest eigenvalue of a certain matrix, which will be relevant in Section 13.3. This connection originates in the work of Alon and Milman [4, 5] and was specifically adapted for our context by Aldous [3]. Our treatment closely follows Behrends's book [7].

Let $G = (V, E)$ be a connected graph (without loops or multiple edges) on $n := |V|$ nodes. We define a random walk (i.e., transition probabilities for all edges—in both directions) on G in a canonical way. Let $\Delta_{\max}$ be the maximum degree of a vertex in G. Each pair (v, w) of vertices such that $\{v, w\} \in E$ is an edge of G receives a constant transition probability $p_{vw} := \tau := \frac{1}{2\cdot\Delta_{\max}}$. If $v \in V$ is a node of degree Δ_v, then we set $p_{vv} := \frac{1}{2} + (\Delta_{\max} - \Delta_v) \cdot \tau$. Let $P \in \mathbb{R}^{V\times V}$ be the matrix with entries p_{vw} $(v, w \in V)$. As defined here, P is a symmetric doubly stochastic matrix with real spectrum

$$\lambda_1 = 1 > \lambda_2 \geq \lambda_3 \geq \ldots \geq \lambda_n > 0.$$

Let $M \in \mathbb{R}^V$ be a matrix whose columns are eigenvectors of P that form an orthonormal basis of $\mathbb{R}^V$ such that the ith column is an eigenvector for the eigenvalue λ_i. In particular, the first column of M is $(\frac{1}{\sqrt{n}}, \ldots, \frac{1}{\sqrt{n}})$. Then we have

$$P = M \cdot \begin{pmatrix} 1 & 0 & \ldots & 0 \\ 0 & \lambda_2 & \ldots & 0 \\ \vdots & \vdots & \ddots & \vdots \\ 0 & 0 & \ldots & \lambda_n \end{pmatrix} \cdot M^T$$

(after suitably numbering the vertices of G).

If the row vector $\pi \in \mathbb{R}^V$ describes the probability distribution for the start vertex of the random walk, then the distribution after performing i steps of our random walk is given by $\pi \cdot P^i$, i.e., by

$$\pi \cdot M \cdot \begin{pmatrix} 1 & 0 & \dots & 0 \\ 0 & \lambda_2^i & \dots & 0 \\ \vdots & \vdots & \ddots & \vdots \\ 0 & 0 & \dots & \lambda_n^i \end{pmatrix} \cdot M^T. \tag{13.1}$$

For $i \longrightarrow \infty$ this converges to

$$\pi \cdot M \cdot \begin{pmatrix} 1 & 0 & \dots & 0 \\ 0 & 0 & \dots & 0 \\ \vdots & \vdots & \ddots & \vdots \\ 0 & 0 & \dots & 0 \end{pmatrix} \cdot M^T = \pi \cdot \begin{pmatrix} \frac{1}{n} & \frac{1}{n} & \dots & \frac{1}{n} \\ \frac{1}{n} & \frac{1}{n} & \dots & \frac{1}{n} \\ \vdots & \vdots & \ddots & \vdots \\ \frac{1}{n} & \frac{1}{n} & \dots & \frac{1}{n} \end{pmatrix} = \left(\frac{1}{n}, \dots, \frac{1}{n}\right).$$

Thus, as was intended, asymptotically the random walk will give convergence to the uniform distribution over V, independently of the start distribution (e.g., independently of the start vertex). Moreover, it follows from (13.1) that the speed of convergence is determined by the second-largest eigenvalue λ_2. Intuitively, it seems to be clear that the edge expansion of G determines how fast the convergence occurs. And, in fact, there exists the following strong connection between the edge expansion and λ_2 (see [7, Theorem 11.3]).

Theorem 13.1. *Let G be a graph with maximum degree $\Delta_{\max}$, and let $0 \le \lambda_2 < 1$ be the second-largest eigenvalue of the matrix P defined as above. Then we have*

$$(1 - \lambda_2) \cdot \Delta_{\max} \;\le\; \mathcal{X}(G) \;\le\; \sqrt{8 \cdot (1 - \lambda_2)} \cdot \Delta_{\max}.$$

The original application of this theorem was, of course, to derive upper bounds on the size of λ_2 by the edge expansion, since the latter seems to be easier to access in structural analyses than the former. However, with respect to algorithmic issues, the situation is somehow the other way around. While computing the edge expansion is NP-hard (see Theorem 13.2), the second-largest eigenvalue can be calculated efficiently. We will exploit this fact in the next section.

13.3 Small Dimensions

Aichholzer classified all 0/1-polytopes of dimension less than or equal to five up to isometries of the cube, i.e., up to flipping and permuting the coordinates [1]. The following table shows the number of classes for each dimension.

Dimension	1	2	3	4	5
# Classes	1	2	12	349	1226525

Thus, in principle one can compute the edge expansion of the graph of each 0/1-polytope up to dimension five by computer. Unfortunately, the following result shows that, in general, computing the edge expansion is difficult. This has been well known for some time (see, e.g., [16]). However, since we could not find an explicit proof in the literature, we include one here.

Theorem 13.2. *The problem of computing $\mathcal{X}(G)$ for arbitrary graphs G is NP-hard.*

Proof. We reduce the problem of finding a maximum (unweighted) cut in a graph (which was proved to be NP-hard by Karp [14]) to the problem of computing the edge expansion of some related graph. The proof is an extension of the proof of the NP-hardness of the equicut problem given by Garey, Johnson, and Stockmeyer [10].

Let $G = (V, E)$ be a graph with $n := |V|$ nodes. We construct a graph $G' = (V', E')$, where $V' = V \uplus W$ for some set W, disjoint from V, with $|W| = n$, and with E' containing all possible edges except the ones in E. Thus G' has $n' = 2n$ nodes. We denote by $\delta_G(S)$ and $\delta_{G'}(S')$ the set of all edges of G (resp. G') having precisely one endnode in S (resp. S') and define

$$\eta_{G'}(S') := \frac{|\delta_{G'}(S')|}{\min\{|S'|, |V' \setminus S'|\}}.$$

We first show that it suffices to consider node subsets of cardinality $\frac{n'}{2} = n$ in order to compute the edge expansion of G'. Let $S \subseteq V$ and $T \subseteq W$ be two sets of nodes of G' with $k := |S| + |T| \leq n$. We have $|\delta_{G'}(S \cup T)| = k \cdot (2n - k) - |\delta_G(S)|$ and

$$\eta_{G'}(S \cup T) = 2n - k - \frac{|\delta_G(S)|}{k}. \tag{13.2}$$

In particular, if $k = n$, then

$$\eta_{G'}(S \cup T) = n - \frac{|\delta_G(S)|}{n}. \tag{13.3}$$

We claim that the right-hand side of (13.3) is less than or equal to the right-hand side of (13.2) for each $1 \leq k \leq n$. Indeed, this claim is equivalent to

$$n - k + \left(\frac{1}{n} - \frac{1}{k}\right) \cdot |\delta_G(S)| \geq 0,$$

which follows from

$$|\delta_G(S)| \leq |S| \cdot n \leq k \cdot n.$$

Thus we have (where the second equation follows from (13.3))

$$\begin{aligned}
\mathcal{X}(G') &= \min\{\eta_{G'}(S \cup T) \ :\ S \subseteq V, W \subseteq W, |S| + |T| = n\} \\
&= \min\left\{n - \frac{\delta_G(S)}{n} \ :\ S \subseteq V, W \subseteq W, |S| + |T| = n\right\} \\
&= n - \frac{\max\{|\delta_G(S)| \ :\ S \subseteq V\}}{n}. \qquad \square
\end{aligned}$$

In view of Theorem 13.2 we decided first to calculate the lower bounds on the edge expansion provided by Theorem 13.1 for each 0/1-polytope of dimensions four and five.

And, somewhat surprisingly, it turned out that for none of the polytopes was this bound less than one. Thus, the conjecture of Mihail and Vazirani is true for 0/1-polytopes up to dimension five.

Theorem 13.3. *The graph of each 0/1-polytope of dimension less than or equal to five has edge expansion at least one.*

Figure 13.2 shows that in many cases the lower bound given by the second-largest eigenvalue was significantly larger than one.

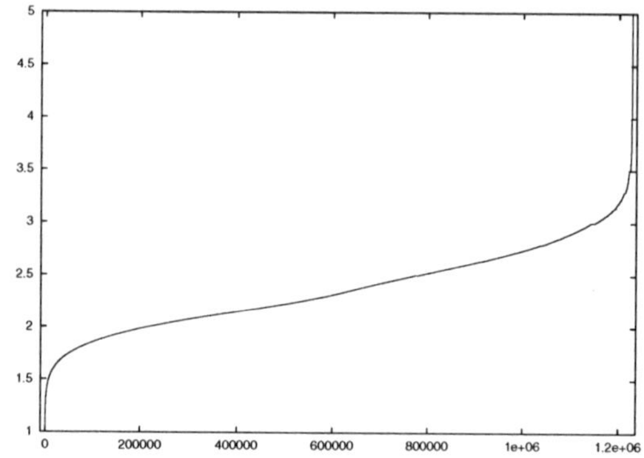

Figure 13.2. *The (lower) eigenvalue bound on the edge expansion for all 1,226,525 five-dimensional 0/1-polytopes.*

13.4 Flow Methods

In this section, we describe methods for proving that a graph has good edge expansion properties that are specifically suited for graphs of 0/1-polytopes. Applying these methods, we will show that the conjecture of Mihail and Vazirani is true for well-known classes of 0/1-polytopes (see Corollaries 13.8 and 13.14). On the other hand, it will be quite obvious that the methods are not sufficient to prove the conjecture in its whole generality.

13.4.1 Expansion and flows

In order to bound the edge expansion of a graph $G = (V, E)$ from below, we will construct certain flows in the (uncapacitated) network $\mathcal{N}(G) = (V, A)$, where A contains for each edge $\{u, v\} \in E$ both of arcs (u, v) and (v, u). This strategy dates back to the method of "canonical paths" developed by Sinclair (see [23]). The extension to flows was explicitly exploited by Morris and Sinclair [18]. Feder and Mihail [9] use random canonical paths, which can equivalently be formulated in terms of flows.

The crucial idea is to construct for each ordered pair $(s, t) \in V \times V$ a flow $\phi_{(s,t)} : A \longrightarrow \mathbb{Q}^{\geq 0}$ in the network $\mathcal{N}(G)$ sending one unit of some commodity from s to t. Let

$\phi := \sum_{(s,t)\in V\times V} \phi_{(s,t)}$ be the sum of all these flows. By

$$\phi_{\max} := \max\{\phi(a) \ : \ a \in A\}$$

we denote the maximal amount of ϕ-flow on any arc. By construction of ϕ, the total amount $\phi(S : V\backslash S)$ of ϕ-flow leaving S is at least $|S| \cdot (n - |S|)$, where $n = |V|$. On the other hand, we have $\phi(S : V\backslash S) \leq \phi_{\max} \cdot |\delta(S)|$. This implies $|S| \cdot (n - |S|) \leq \phi_{\max} \cdot |\delta(S)|$, and hence, if $|S| \leq \frac{n}{2}$ holds,

$$\frac{|\delta(S)|}{|S|} \geq \frac{n}{2 \cdot \phi_{\max}}.$$

Thus, we have proved

$$\mathcal{X}(G) \geq \frac{n}{2 \cdot \phi_{\max}}. \tag{13.4}$$

In light of inequality (13.4), it is clear that the task is to construct a flow ϕ as above with $\phi_{\max}$ as small as possible in order to prove a strong lower bound on the edge expansion of G.

13.4.2 Fractional wall matchings

While the setting presented so far applies to general graphs, we now derive a method to construct ϕ in the special situation where G is the graph of a 0/1-polytope. The method generalizes ideas for analyzing random walks on the bases-exchange graph of matroids due to Feder and Mihail [9].

Let $P \subset \mathbb{R}^d$ be a 0/1-polytope. A *wall* of P is the intersection of P with any face of the cube $C_d := \left\{x \in \mathbb{R}^d \ : \ 0 \leq x_i \leq 1 \text{ for all } i\right\} \supseteq P$. Thus, the walls of P are special faces of P. Usually, we will identify a wall of P with its vertices. The faces F of C_d are in one-to-one correspondence with the vectors $\sigma(F) \in \{0, 1, \star\}^d$ (and vice versa) via

$$F = \{x \in C_d \ : \ x_i = \sigma(F)_i \text{ for all } i \text{ with } \sigma(F)_i \neq \star\}.$$

For a face $F \neq C_d$ of C_d let $\mu(F) := \min\{i \ : \ \sigma(F)_i = \star\}$ be the "smallest direction" of F. Let W be a wall of P and let F be the inclusion minimal face of C_d with $W = P \cap F$. The vector $\sigma(W) := \sigma(F)$ indicates the components in which all vertices of W agree, and $\mu(W) := \mu(F)$ is the smallest coordinate direction of any edge of W. We define $W_0 := \left\{w \in W \ : \ w_{\mu(W)} = 0\right\}$ and $W_1 := \left\{w \in W \ : \ w_{\mu(W)} = 1\right\}$ and denote by $\mathcal{B}(W)$ the bipartite subgraph of $\mathcal{G}(P)$ induced by the two disjoint subsets W_0 and W_1 of nodes of $\mathcal{G}(P)$.

A wall W of P is called *initial* if there is some $i \in \{0, 1, \ldots, d\}$ such that $\sigma(W)_j \in \{0, 1\}$ for $1 \leq j \leq i$ and $\sigma(W)_j = \star$ for all $j > i$. The following fact follows immediately from the definitions.

Lemma 13.4. *For every edge e of a 0/1-polytope P there is a unique initial wall W of P such that e is an edge of $\mathcal{B}(W)$.*

Thus the bipartite graphs associated with the initial walls of P induce a partition of the edges of P.

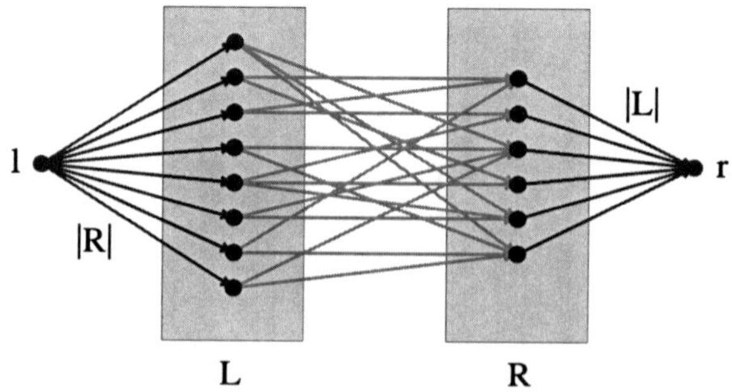

Figure 13.3. *The bipartite graph on $L \uplus R$ has a fractional matching if and only if in the network indicated in the figure there is a (nonnegative) flow sending $|L| \cdot |R|$ units of some commodity from l to r. The arcs leaving l have capacities $|R|$, the arcs entering r have capacities $|L|$, and the arcs connecting L to R have infinite capacities.*

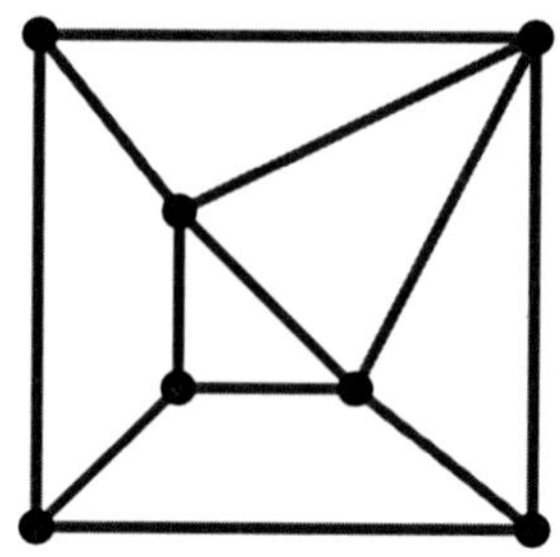

Figure 13.4. *The graph of the 0/1-polytope P arising from C_3 by removing one vertex. Independently of the numbering of the coordinate directions, $\mathcal{B}(P)$ has no fractional matching.*

A bipartite graph with bipartition $L \uplus R$ has a *fractional matching* if one can assign nonnegative weights to its edges such that all nodes in L have the same weighted degree, and the same holds for all nodes in R as well (see Figure 13.3).

Observation 13.5. If a bipartite graph B with bipartition $L \uplus R$ has a fractional matching and there is a constant amount of some commodity located in each node in L (or R, respectively), then one can distribute the entire amount of the commodity from L to R through the edges of B such that each node in R (or L, respectively) receives the same amount of the commodity.

A 0/1-polytope P *has fractional wall matchings* if $\mathcal{B}(W)$ has a fractional matching for every wall W of P. In general, the bipartite graph $\mathcal{B}(W)$ associated with a wall W of a 0/1-polytope P does not necessarily have a fractional matching (see Figure 13.4). However, several interesting classes of 0/1-polytopes have fractional wall matchings, as we will show below. The method we will describe to construct suitable flows ϕ only works for such 0/1-polytopes. Thus from now on we assume that $P \subset \mathbb{R}^d$ is a 0/1-polytope that has fractional wall matchings.

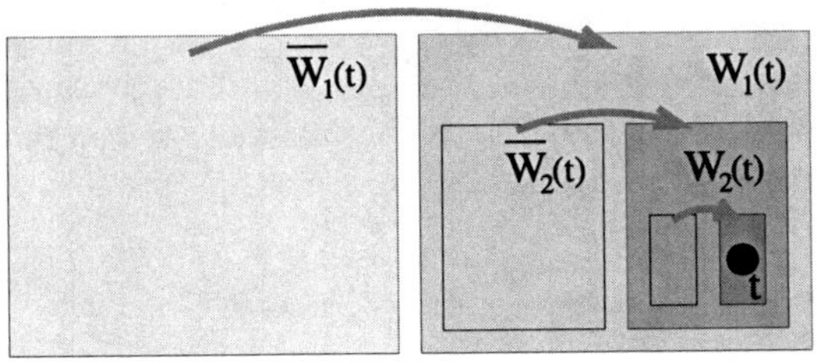

Figure 13.5. *Simultaneous construction of the flows $\phi_{(s,t)}$ for all $s \in V$.*

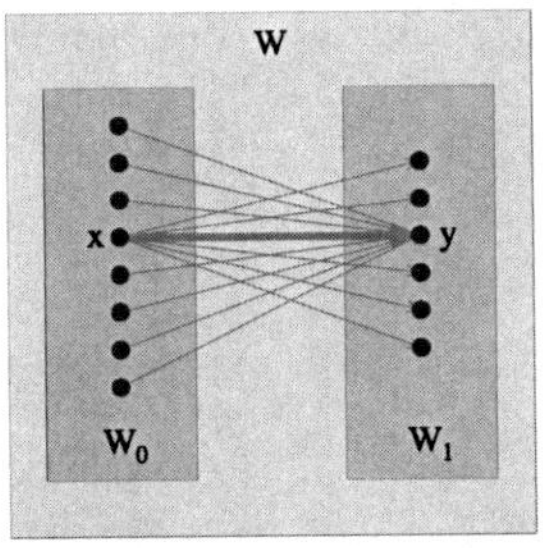

Figure 13.6. *The arc sets $A^-(x)$ and $A^+(y)$.*

Let $t \in \text{vert}(P)$ be a vertex of P. We will particularly be concerned with the initial walls

$$W_i(t) := \{w \in \text{vert}(P) \ : \ w_1 = t_1, \ldots, w_i = t_i\} \qquad (i = 0, 1, \ldots, d).$$

These walls form a *flag* of P, i.e., we have

$$\{t\} = W_d(t) \subseteq W_{d-1}(t) \subseteq \ldots \subseteq W_1(t) \subseteq W_0(t) = P.$$

For each $i \in \{1, \ldots, d\}$ we define $\overline{W}_i(t) := W_{i-1}(t) \setminus W_i(t)$. Now we are ready to construct all flows $\phi_{(s,t)}$, $s \in V$, simultaneously in d steps. Imagine a single unit of some commodity initially placed at each node. Suppose that, before we perform step $i \in \{1, \ldots, d\}$, the n units of the commodity are distributed uniformly among the nodes in $W_{i-1}(t)$ (as is the case before the first step). Since we have assumed that P has fractional wall matchings, we can route (see Observation 13.5) the amount of commodity distributed at the nodes in $\overline{W}_i(t)$ through the arcs corresponding to the edges of $\mathcal{B}(W_{i-1}(t))$ such that afterward the n units of our commodity are uniformly distributed among the nodes in $W_i(t)$. Figure 13.5 illustrates the construction.

For each pair $(s, t) \in V \times V$ we have thus defined a flow in the network $\mathcal{N}(\mathcal{G}(P))$ sending one unit of some commodity from s to t. It remains to bound the maximal flow $\phi_{\max}$ produced by $\phi := \sum_{(s,t) \in V \times V} \phi_{(s,t)}$ at any arc. Therefore, let (x, y) be any arc of $\mathcal{N}(\mathcal{G}(P))$. By Lemma 13.4 there is a unique initial wall W of P such that $\mathcal{B}(W)$ contains the edge $\{x, y\}$ (see Figure 13.6). Due to reasons of symmetry we might assume $x \in W_0$ and $y \in W_1$. Let $A^-(x)$ and $A^+(y)$ be the sets of out-arcs (resp. in-arcs) incident to x (resp. y) corresponding to edges of $\mathcal{B}(W)$. In particular, we have $(x, y) \in A^-(x)$ and $(x, y) \in A^+(y)$. The arcs going from W_0 to W_1 are used only by the flows $\phi_{(s,t)}$ with $s \notin W_1$ and $t \in W_1$. Thus, the total amount of flow carried by these arcs is $\frac{|W_0|\cdot|W_1|}{|W|}\cdot n$. Consequently, precisely $\frac{|W_1|}{|W|}\cdot n$ units of flow are sent through $A^-(x)$ and $\frac{|W_0|}{|W|}\cdot n$ units of flow are sent through $A^+(y)$. Hence, (x, y) carries at most

$$\min\left\{\frac{|W_1|}{|W|}, \frac{|W_0|}{|W|}\right\}\cdot n \leq \frac{n}{2}$$

units of flow. Since this holds for every arc of $\mathcal{N}(\mathcal{G}(P))$, we have $\phi_{\max} \leq \frac{n}{2}$. By (13.4) this proves the following result.

Theorem 13.6. *If P is a 0/1-polytope that has fractional wall matchings, then* $\mathcal{X}(\mathcal{G}(P)) \geq 1$ *holds.*

Thus, 0/1-polytopes that have fractional wall matchings satisfy the conjecture of Mihail and Vazirani.

13.4.3 Walls with regular graphs

Let us say that a 0/1-polytope P has *regular walls* if the graph of every wall of P is regular, i.e., all its vertices have the same degree. It is obvious that every 0/1-polytope with regular walls has fractional wall matchings (see Figure 13.7). This proves the following consequence of Theorem 13.6.

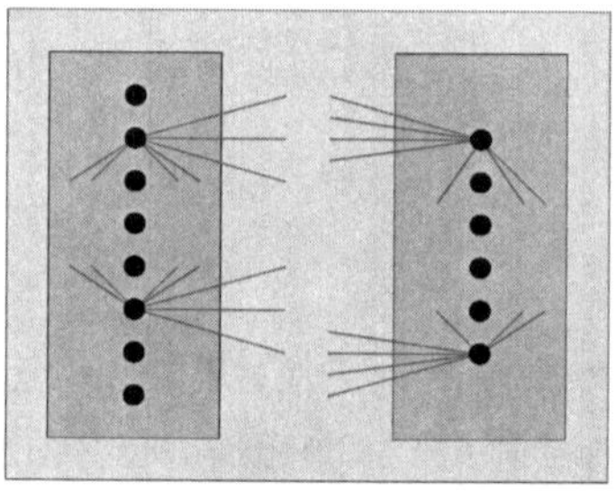

Figure 13.7. *Regular walls yield fractional wall matchings, since in each of the relevant bipartite graphs all vertices in the left shore have the same degree, and the same is true for all vertices in the right shore.*

Corollary 13.7. *If a 0/1-polytope P has regular walls, then* $\mathcal{X}(\mathcal{G}(P)) \geq 1$ *holds.*

A d-dimensional polytope P is *simple* if every vertex lies in precisely d facets or, equivalently, if $\mathcal{G}(P)$ is d-regular. The polytopes

$$\operatorname{conv}\left\{v \in \{0,1\}^d : \sum_{i=1}^{d} v_i = \varrho\right\} \qquad (\varrho \in \{0, 1, \ldots, d\})$$

are called *hypersimplices* (they are special knapsack polytopes).

Corollary 13.8. *If a 0/1-polytope P is simple or a hypersimplex, then $\mathcal{X}(\mathcal{G}(P)) \geq 1$ holds.*

Proof. Every face of a simple polytope is simple and thus has a regular graph. Every wall of a hypersimplex is a hypersimplex, again. Since hypersimplices obviously have a transitive automorphism group, they have regular graphs. Thus, in any of the two cases of the claim, $\mathcal{X}(\mathcal{G}(P)) \geq 1$ holds by Corollary 13.7. □

13.4.4 Balanced uniform 0/1-polytopes

A 0/1-polytope $P \subset \mathbb{R}^d$ is called *ϱ-uniform* ($\varrho \in \{0, 1, \dots, d\}$) if it is contained in the hyperplane $\{x \in \mathbb{R}^d : \sum_{i=1}^d x_i = \varrho\}$, i.e., if all vertices of P have precisely ϱ ones. For instance, hypersimplices and basis polytopes of matroids are uniform. Obviously, every wall of a uniform 0/1-polytope is uniform as well.

A 0/1-polytope $P \subset \mathbb{R}^d$ is *balanced* if, for every $\sigma \in \{0, 1, \star\}$ and for each pair $i, j \in \{1, \dots, d\}$ with $i \neq j$ and $\sigma_i = \sigma_j = \star$, the relation

$$|W_{0,0}| \cdot |W_{1,1}| \leq |W_{1,0}| \cdot |W_{0,1}| \tag{13.5}$$

holds, where W is the wall of P defined by σ and $W_{\alpha,\beta} := \{w \in W : w_i = \alpha, w_j = \beta\}$. If $W_{0,1} \cup W_{1,1} \neq \emptyset$ (i.e., there is some $w \in W$ with $w_j = 1$), then (13.5) is equivalent to

$$\frac{|W_{1,1}|}{|W_{0,1}| + |W_{1,1}|} \leq \frac{|W_{1,1}| + |W_{1,0}|}{|W_{0,1}| + |W_{1,1}| + |W_{0,0}| + |W_{1,0}|}. \tag{13.6}$$

This means that for a vertex w chosen uniformly at random from W the probability of the event $w_i = 1$ does not increase by conditioning on the event $w_j = 1$. Similarly, (13.5) is equivalent to the fact that for a vertex w chosen uniformly at random from W the probability of the event $w_i = 0$ does not increase by conditioning on the event $w_j = 0$.

The property of being balanced is not invariant under arbitrary symmetries of the cube. However, it is invariant under simultaneous "flipping" of all coordinates (and under arbitrary permutations of the coordinates).

Proposition 13.9. *Balanced uniform 0/1-polytopes have fractional wall matchings.*

We omit the proof, which closely follows the corresponding proof on the bases-exchange graph of balanced matroids due to Feder and Mihail [9].

Proposition 13.9 and Theorem 13.6 imply the following.

Theorem 13.10. *Every balanced uniform 0/1-polytope P satisfies $\mathcal{X}(\mathcal{G}(P)) \geq 1$.*

A matroid $\mathcal{M}$ on the ground set E has the *negative correlation property* if, for a basis B chosen uniformly at random from the set of bases of $\mathcal{M}$ and for every pair of elements $e, f \in E$,

$$prob[e \in B] \geq prob[e \in B \mid f \in B]$$

holds; i.e., the probability of the event $e \in B$ does not increase by conditioning on the event $f \in B$. A matroid $\mathcal{M}$ is *balanced* if every minor of $\mathcal{M}$ has the negative correlation property. Regular (in particular, graphic) matroids are known to be balanced.

It is obvious that the basis polytope $P(\mathcal{M}) := \text{conv}\{\chi(B) : B \text{ basis of } \mathcal{M}\}$ (where $\chi(B)$ is the characteristic vector of $B \subseteq E$) of a balanced matroid $\mathcal{M}$ is uniform and balanced. Thus, Theorem 13.10 immediately yields $\mathcal{X}(\mathcal{G}(P(\mathcal{M}))) \geq 1$. Notice that the actual adjacency structure on $P(\mathcal{M})$ is irrelevant for this.

Hence, Theorem 13.10 generalizes the result of Feder and Mihail [9], saying that the bases-exchange graphs (where two bases are adjacent if and only if their symmetric difference has two elements) of balanced matroids have edge expansion at least one.

13.4.5 Cube-spanned walls

The technique described in this subsection is particularly suited for proving that 0/1-polytopes coming from certain combinatorial problems have graphs with large edge expansion (see Corollary 13.14). It relies on the high symmetry of the graph $\mathcal{G}(Q)$ of a cube Q, from which one easily derives the following fact (where the *antipodal vertex* of some vertex x of Q is the vertex with maximum distance from x in $\mathcal{G}(Q)$).

Observation 13.11. For a cube Q it is possible to define for each pair (s, t) of *antipodal* vertices a flow $\psi_{(s,t)}$ in $\mathcal{N}(\mathcal{G}(Q))$ sending one unit of some commodity from s to t, such that for the total flow $\psi := \sum_{s,t} \psi_{(s,t)}$, one has $\psi(a) = 1$ for each arc a in $\mathcal{N}(\mathcal{G}(Q))$.

Let $P \subset \mathbb{R}^d$ be any 0/1-polytope. A subset $C \subseteq \text{vert}(P)$ of vertices of P is called an *affine cube in* P if C is affinely isomorphic to $\{0, 1\}^k$ for some k or, equivalently, if there is a subset $I \subseteq \{1, \ldots, d\}$ (with $|I| = k$) such that the orthogonal projection of $\mathbb{R}^d$ onto $\mathbb{R}^I$ induces a bijection between C and $\{0, 1\}^I$. It is not hard to see that $C \subseteq \text{vert}(P)$ is an affine cube (in P) if and only if there are 0/1-vectors $z^{(1)}, \ldots, z^{(k)} \in \{0, 1\}^d$, pairwise orthogonal to each other, such that, for every $x \in C$,

$$C = \left\{x \oplus \epsilon_1 z^{(1)} \oplus \ldots \oplus \epsilon_k z^{(k)} \;:\; \epsilon \in \{0, 1\}^k\right\}$$

(where $\oplus$ denotes addition modulo two). The vertex $x \oplus z^{(1)} \oplus \ldots \oplus z^{(k)}$ is the *antipodal vertex of* x *in* C. In particular, we will be interested in *affine edge-cubes* in P, i.e., affine cubes in P on which $\mathcal{G}(P)$ induces the graph of a cube.

For a subset $A \subseteq \text{vert}(P)$ let us call the intersection of all walls of P that contain A the *wall spanned by* A, denoted by $W(A)$. A wall W of P is *edge-cube spanned* if there is an affine edge-cube C with $W(C) = W$ (which is equivalent to the fact that each pair of antipodal vertices in C spans W). A wall W of P is *uniquely edge-cube spanned* if it is spanned by an affine edge-cube C in P and if it is not spanned by any other affine edge-cube $C' \neq C$ in P. In this case, we call the vertices in C the *cube vertices of* W. See Figure 13.8 for examples.

For a vertex $x \in W$ in a wall W of P the vertex $x^{(W)} := x \oplus t^{(W)}$ is the *mirror image* of x with respect to W, where $t^{(W)}$ is the 0/1-vector having ones precisely in those components where $\sigma(W)$ has stars. In general, $x^{(W)}$ need not be contained in W. If, however, a wall W is spanned by an affine cube C, then $x^{(W)}$ is the antipodal vertex of x in C for every $x \in C$; in particular, $x^{(W)} \in W$.

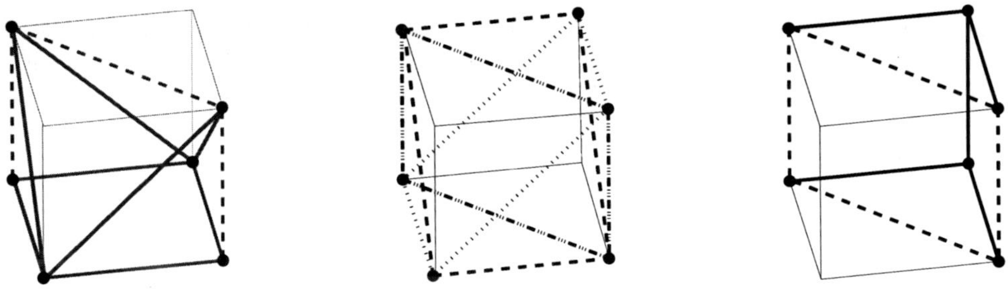

Figure 13.8. *Three three-dimensional walls. The first one is spanned by a cube (but is not edge-cube spanned), the second one is edge-cube spanned (but not uniquely edge-cube spanned), and the third one is uniquely edge-cube spanned.*

Lemma 13.12. *Let P be a 0/1-polytope, and let $u, v \in \text{vert}(P)$, $u \neq v$, be two distinct vertices of P. There are at most $\frac{1}{2}|\text{vert}(P)|$ walls W of P such that u, v, and their mirror images with respect to W are contained in W.*

Proof. Let $\mathcal{W}(u, v)$ be the set of all walls W of P such that u, v, and their mirror images $u^{(W)}$, $v^{(W)}$ are contained in W. We have

$$u \oplus v^{(W)}, v \oplus u^{(W)} \leq t^{(W)}, \qquad \text{and} \qquad u \oplus u^{(W)} = v \oplus v^{(W)} = t^{(W)} \tag{13.7}$$

(where $\leq$ is meant to hold componentwise). Since $u \neq v$ we have $u^{(W)} \neq v^{(W)}$. Thus, we can define a map ω assigning to each $W \in \mathcal{W}(u, v)$ the two-element subset $\omega(W) := \{u^{(W)}, v^{(W)}\}$ of $\text{vert}(P)$.

Suppose that for $W, W' \in \mathcal{W}(u, v)$ we have $x \in \omega(W) \cap \omega(W')$. After possibly interchanging the roles of u and v, by (13.7) we have $v \oplus x \leq u \oplus x$ and thus

$$t^{(W)} = u \oplus x = t^{(W')},$$

yielding $W = W'$. Hence, the images of ω have pairwise empty intersections, which implies the lemma. □

Theorem 13.13. *Let P be a 0/1-polytope such that each pair $s, t \in \text{vert}(P)$, $s \neq t$, of distinct vertices s and t is a pair of antipodal cube vertices in a uniquely edge-cube spanned wall of P. Then $\mathcal{X}(\mathcal{G}(P)) \geq 1$ holds.*

Proof. In each affine edge-cube spanning a uniquely edge-cube spanned wall of P we construct a flow as described in Observation 13.11. Let ϕ be the sum of all these flows. Since each pair $s, t \in \text{vert}(P)$, $s \neq t$, is a pair of antipodal cube vertices in a uniquely edge-cube spanned wall of P, the flow ϕ has the properties required in Subsection 13.4.1. Lemma 13.12 ensures that each arc (u, v) in the network $\mathcal{N}(\mathcal{G}(P))$ is a cube-arc in at most $\frac{n}{2}$ uniquely edge-cube spanned walls if n is the number of vertices of P. Thus we have $\phi_{\max} \leq \frac{n}{2}$, and by (13.4) we obtain the claim of the theorem. □

Theorem 13.13 in particular yields a unified proof for the following results, which appeared in [17] (where only a proof for the statement concerning the perfect matching polytope is given).

Corollary 13.14. *The graphs of the stable set polytope, the matching polytope, and the perfect matching polytope associated with an arbitrary graph have edge expansion at least one.*

Proof. Let $G = (V, E)$ be a graph and let P be its stable set polytope. For two vertices s and t of P let $A_s, A_t \subseteq V$ be the corresponding stable sets in G, and denote by $A^{(1)}, \dots, A^{(k)} \subseteq V$ the nodesets of the connected components of the subgraph of G induced by the symmetric difference of A_s and A_t. Define $A_s^{(i)} := A^{(i)} \cap A_s$ and $A_t^{(i)} := A^{(i)} \cap A_t$. For each $\epsilon \in \{s, t\}^k$ the set

$$S_\epsilon := (A_s \cap A_t) \cup A_{\epsilon_1}^{(1)} \cup \dots \cup A_{\epsilon_k}^{(k)}$$

is stable in G. By Chvátal's result [8] two vertices of P are adjacent if (and only if) the symmetric difference of the corresponding stable sets induces a connected subgraph of G. Thus, the set C of vertices of P corresponding to $\{S_\epsilon : \epsilon \in \{s, t\}^k\}$ is an affine edge-cube in P, spanning the wall W that is defined by the equations $x_v = 0$, $v \in V \setminus (A_s \cup A_t)$, and $x_v = 1$, $v \in A_s \cap A_t$. Clearly, s and t are antipodal vertices of C. Since all pairs of mirror images in W belong to C (these pairs correspond to bipartitions of the subgraph of G that is induced by $A^{(1)} \cup \dots \cup A^{(k)}$), W is uniquely edge-cube spanned by C. Thus, $\mathcal{X}(P) \geq 1$ by Theorem 13.13.

Since the matching polytope of a graph G is the stable set polytope of the line graph of G (having the edges of G as vertices, which are adjacent if and only if the corresponding edges of G have a common endnode), the claim on matching polytopes follows.

Perfect matching polytopes satisfy the requirements of Theorem 13.13 as well; they even have the property that each pair of vertices spans a wall that *is* an affine edge-cube. □

13.5 Some Remarks

The results presented in this paper support the conjecture that graphs of 0/1-polytopes inherently have good expansion properties and therefore may in principle be good candidates for defining neighborhood structures in the context of random walks. In fact, we have proved for some classes of 0/1-polytopes, including simple 0/1-polytopes, stable set polytopes, and all 0/1-polytopes up to dimension five, that their graphs have edge expansion at least one.

A proof of the conjecture that the edge expansion of the graph of any d-dimensional 0/1-polytopes is bounded by the reciprocal of a polynomial in d would have important consequences, even if this was proved only for uniform 0/1-polytopes. For instance, such a result would imply that indeed the bases-exchange graphs of arbitrary matroids have sufficiently large edge expansion in order to construct a randomized approximate counting algorithm. In particular, this would solve the open questions for randomized approximation algorithms for counting connected spanning subgraphs of a graph, forests of a prescribed size in a graph, or maximal independent subsets in a given set of vectors over **GF(2)** (see [12]).

Therefore, one might hope that, while the concept of the graph of a 0/1-polytope has not proved to be very useful in the context of combinatorial optimization, it might have a successful revival in the context of random generation and counting of certain combinatorial objects.

Bibliography

[1] O. Aichholzer. Extremal properties of 0/1-polytopes of dimension 5. In G. Kalai and G.M. Ziegler, editors, *Polytopes—Combinatorics and Computation*, volume 29 of *DMV Seminar Band*, pages 111–130. Birkhäuser, Basel, Boston, 2000.

[2] M. Aigner and G.M. Ziegler. *Proofs from THE BOOK*, second edition. Springer, Berlin, 2001.

[3] D. Aldous. On the Markov chain simulation method for uniform combinatorial distributions and simulated annealing. *Probability in the Engineering and Informational Sciences*, 1:33–46, 1987.

[4] N. Alon. Eigenvalues and expanders. *Combinatorica*, 6:83–96, 1986.

[5] N. Alon and V.D. Milman. λ_1, isoperimetric inequalities for graphs, and superconcentrators. *Journal of Combinatorial Theory. Series* B, 38:73–88, 1985.

[6] F. Barahona and A.R. Mahjoub. On the cut polytope. *Mathematical Programming*, 36:157–173, 1986.

[7] E. Behrends. *Introduction to Markov Chains. With Special Emphasis on Rapid Mixing.* Advanced Lectures in Mathematics. Vieweg, Braunschweig, 1999.

[8] V. Chvátal. On certain polytopes associated with graphs. *Journal of Combinatorial Theory. Series* B, 18:138–154, 1975.

[9] T. Feder and M. Mihail. Balanced matroids. In *Proceedings of the 24th Annual ACM Symposium on the Theory of Computing (STOC)*, pages 26–38, Victoria, British Columbia, 1992. ACM Press, New York.

[10] M.R. Garey, D.S. Johnson, and L. Stockmeyer. Some simplified NP-complete graph problems. *Theoretical Computer Science*, 1:237–267, 1976.

[11] M. Jerrum and A. Sinclair. Approximating the permanent. *SIAM Journal of Computing*, 18:1149–1178, 1989.

[12] M. Jerrum and A. Sinclair. The Markov chain Monte Carlo method: An approach to approximate counting and integration. In D. Hochbaum, editor, *Approximation Algorithms*, pages 482–520. PWS Publishing Company, Boston, 1997.

[13] M. Jerrum, A. Sinclair, and E. Vigoda. A Polynomial-Time Approximation Algorithm for the Permanent of a Matrix with Non-negative Entries. Technical report, ECCC Report, TR00-079, September 2000.

ftp://ftp.eccc.uni-trier.de/pub/eccc/reports/2000/TR00-079/index.html.

[14] R.M. Karp. Reducibility among combinatorial problems. In R.E. Miller and J.W. Thatcher, editors, *Complexity of Computer Computations*, pages 85–103. Plenum Press, New York, 1972.

[15] H.W. Kuhn. Talk given at the workshop "The Sharpest Cut" in honor of Manfred Padberg's 60th anniversary, Berlin, October 11–13, 2001.

[16] T. Leighton and S. Rao. An approximate max-flow min-cut theorem for uniform multicommodity flow problems with applications to approximation algorithms. In *Proceedings of the 29th Annual Symposium on Foundations of Computer Science*, pages 422–431, White Plains, New York, 1988.

[17] M. Mihail. On the expansion of combinatorial polytopes. In I.M. Havel and V. Koubek, editors, *Proceedings of the 17th International Symposium on Mathematical Foundations of Computer Science*, volume 629 of *Lecture Notes in Computer Science*, pages 37–49. Springer-Verlag, Berlin, 1992.

[18] B. Morris and A. Sinclair. Random walks on truncated cubes and sampling 0-1 knapsack problem. In *Proceedings of the 40th IEEE Symposium on Foundations of Computer Science*, pages 230–240, IEEE, New York, 1999.

[19] D.J. Naddef. Pancyclic properties of the graph of some 0-1 polyhedra. *Journal of Combinatorial Theory. Series* B, 37:10–26, 1984.

[20] D.J. Naddef. The Hirsch conjecture is true for (0,1)-polytopes. *Mathematical Programming. Series* B, 45:109–110, 1989.

[21] D.J. Naddef and W.R. Pulleyblank. Hamiltonicity in (0-1)-polyhedra. *Journal of Combinatorial Theory. Series* B, 37:41–52, 1984.

[22] M.W. Padberg and M.R. Rao. The travelling salesman problem and a class of polyhedra of diameter two. *Mathematical Programming*, 7:32–45, 1974.

[23] A. Sinclair. *Algorithms for Random Generation and Counting: A Markov Chain Approach. Progress in Theoretical Computer Science*. Birkhäuser, Boston, 1993.

[24] J. Werner. Berechnung der Quotientenschnittzahl von Graphen von 0/1-Polytopen. Diplomarbeit, Technische Universität Berlin, 2001.

[25] G.M. Ziegler. Lectures on 0/1-polytopes. In G. Kalai and G.M. Ziegler, editors, *Polytopes—Combinatorics and Computation*, volume 29 of *DMV Seminar Band*, pages 1–41. Birkhäuser, Basel, Boston, 2000.

Chapter 14
Typical and Extremal Linear Programs

Günter M. Ziegler *

Abstract. Viewed geometrically, the simplex algorithm on a (primally and dually non-degenerate) linear program traces a monotone edge path from the starting vertex to the (unique) optimum. *Which* path it takes depends on the pivot rule. In this chapter we survey geometric and combinatorial aspects of the following situation: What do "real" linear programs and their polyhedra look like? How long can simplex paths be in the worst case? Do short paths always exist? Can we expect randomized pivot rules (such as `Random_Edge`) or deterministic rules (such as Zadeh's rule) to find short paths?

MSC 2000. 90C05, 52B11
Key words. Geometry of linear programs, polytopes and polyhedra, degeneracy, 2-dimensional shadows, long paths, deformed products, Hirsch conjecture, short paths, deterministic and randomized pivot rules

14.1 Introduction

What can geometry contribute to the study and understanding of linear programs (LPs) and the simplex algorithm? This short chapter attempts to sketch a variety of answers to this question. We want to show that geometry and geometric insights can contribute both to the understanding of "real" LPs and to the construction of "extremal" examples on which (certain variants of) the simplex algorithm would or should "behave badly." For this, we will not be so naive as to assume that the nice, symmetrical, and "interesting" examples of polytopes that lie at the core of modern polytope theory, such as the permuto-associahedron

*Inst. Mathematics, MA 6-2, TU Berlin, 10623 Berlin, Germany (ziegler@math.tu-berlin.de). This work was partially supported by the DFG.

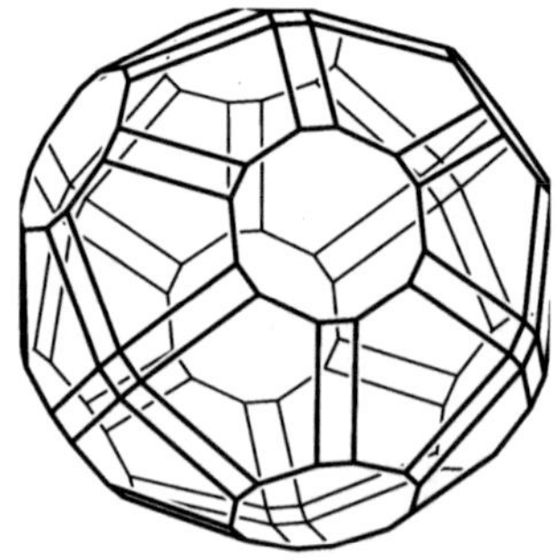

Figure 14.1. *A nice polytope [30].*

displayed in Figure 14.1, have much to do in their combinatorics and geometry with the polytopes and polyhedra that arise as the feasible regions of LPs "in practice." Indeed, we have to admit that of course we do not know what a "real" LP in, say, $d = 1000$ variables and with $n = 5000$ constraints "looks like" (that is, if you ask for a geometric impression rather than just looking at the input file).

But this still may be a reasonable question. Take for an example the smallest entry of the netlib collection, the `afiro` LP, as displayed in Figure 14.2. What can we say about the geometry of such a program? Well, we can say a lot. Before we start the discussion, let us just fix a bit of notation, and—more importantly—match a few terms that have become standard in linear programming and in polytope theory, respectively. So, for the following, d will be used for the dimension of a problem; in geometry this will be the dimension of a polytope; in linear programming this will be the number of (free) variables. Similarly, n will denote the number of facets of a polytope, corresponding to the number of (essential) inequalities for an LP. Standard arguments (linear algebra, perturbation) allow one—at least for the analysis of extremal examples—to assume that the LP under consideration is *primally nondegenerate* (that is, the polyhedron is *simple*), that the *objective function* is given by the last variable x_d (and thus we are trying to maximize the *height* of a point in the polyhedron), that the program is *dually nondegenerate* (that is, the polyhedron has *no horizontal edges*), and finally that the feasible region is *bounded* (and hence we are looking at a *polytope* rather than a *polyhedron*).

14.2 Real LPs

What do "real" LPs "look like," and how can we analyze and picture them? In this section, we will sketch two approaches to this question. The first one is based on the fact that computational polytope theory (cf. [16]) has made enormous progress since the 1960s, so that now we can compute and analyze the full polyhedron, at least for small LPs, and try to understand it. The second approach is based on the shadow boundary algorithm, which can be (ab)used to compute two-dimensional pictures (projections) of LPs.

The shadow boundary algorithm for linear programming starts at a feasible vertex of the polyhedron, for which it is easy to construct a linear function $c^t x$ that is optimal at the given vertex, and then traces the sequence of optimal vertices that appear while the

```
/* objective function: */
min: + 10 X39 + -0.48 X36 + -0.6 X23 + -0.32 X14 + -0.4 X02 ;

/* constraints */
X51 : + 1 X38 + 1 X16 <= 300;
X50 : + 1 X26 + 1 X04 <= 310;
X49 : + -1 X37 + 0.326 X09 + 0.313 X08 + 0.313 X07 + 0.301 X06 <= 0;
X48 : + -1 X24 + 0.301 X01 <= 0;
X47 : + 0.107 X31 + 0.108 X30 + 0.108 X29 + 0.109 X28 + -1 X15 <= 0;
X46 : + 0.109 X22 + -1 X03 <= 0;
X45 : + 2.279 X35 + 2.249 X34 + 2.219 X33 + 2.191 X32 + -1 X25 + 2.429 X13\\
                                + 2.408 X12 + 2.386 X11 + 2.364 X10 <= 0;
X43 : + -1 X35 + 1 X31 <= 0;
X42 : + -1 X34 + 1 X30 <= 0;
X41 : + -1 X33 + 1 X29 <= 0;
X40 : + -1 X32 + 1 X28 <= 500;
R23 : + 1 X39 + 1 X37 + -1 X36 + 1 X31 + 1 X30 + 1 X29 + 1 X28 = 44;
R22 : + 1 X38 + -0.37 X31 + -0.39 X30 + -0.43 X29 + -0.43 X28 = 0;
X44 : + 1.4 X36 + -1 X23 <= 0;
X27 : + 1 X22 <= 500;
R20 : + 1 X26 + -0.43 X22 = 0;
R19 : + 1 X25 + 1 X24 + 1 X23 + -1 X22 = 0;
X20 : + -1 X13 + 1 X09 <= 0;
X19 : + -1 X12 + 1 X08 <= 0;
X18 : + -1 X11 + 1 X07 <= 0;
X17 : + -1 X10 + 1 X06 <= 80;
R13 : + 1 X16 + -0.86 X09 + -0.96 X08 + -1.06 X07 + -1.06 X06 = 0;
R12 : + 1 X15 + 1 X14 + -1 X09 + -1 X08 + -1 X07 + -1 X06 = 0;
X21 : + 1.4 X14 + -1 X02 <= 0;
X05 : + 1 X01 <= 80;
R10 : + 1 X04 + -1.06 X01 = 0;
R09 : + 1 X03 + 1 X02 + -1 X01 = 0;%
```

Figure 14.2. *The* `afiro` *LP.*

linear objective function is linearly interpolated between the starting objective function $c^t x$ and the final objective function $d^t x$. Geometrically, this procedure computes a part of the boundary of a two-dimensional projection of the feasible region of the LP. If we continue the procedure, by next interpolating between $d^t x$ and $-c^t x$, then between $-c^t x$ and $-d^t x$, and finally between $-d^t x$ and $c^t x$, we compute the full two-dimensional projection. Thus we obtain *pictures*. This idea was developed, implemented, and tested in the Diplomarbeit of S. Fischer [4], who produced interesting and somewhat surprising pictures of the polyhedra of LPs in the netlib library. The picture gallery of Figure 14.3 of rather typical examples is supposed to illustrate that some LPs have rather sharp angles (in their two-dimensional projections) while others appear to be rather round (with many vertices very close to each other in a two-dimensional projection). Here the two directions of projections are typically taken to be the given objective function and the first or last variable.

On the other hand, for "small" LPs such as `afiro`, a complete analysis of the LP is possible. The `Polymake` system of Gawrilow and Joswig [8, 9] is a tool that, for a LP/polyhedron given by a list of inequalities, would (attempt to) run a convex hull algorithm to determine the list of vertices, then produce all kinds of combinatorial information "asked for," such as the number of vertices, the graph, the graph diameter, etc. This was carried out in the Diplomarbeit of D. Weber [34, pp. 20, 21].

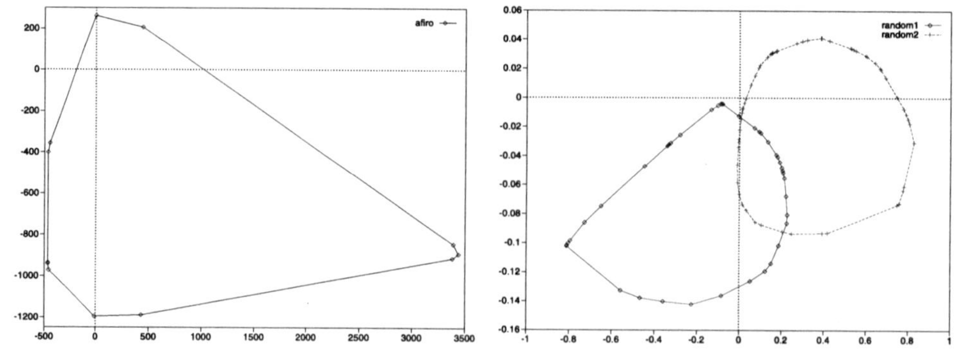

The afiro problem ... and two random projections of the afiro LP

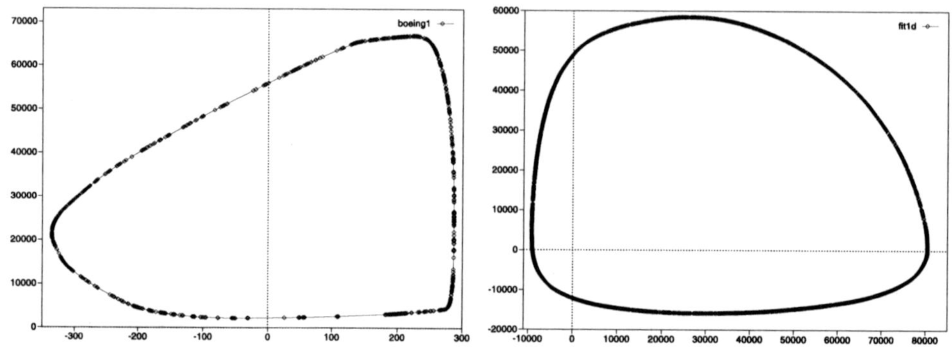

The boeing1 problem and the fit1d problem have rather "round" projections

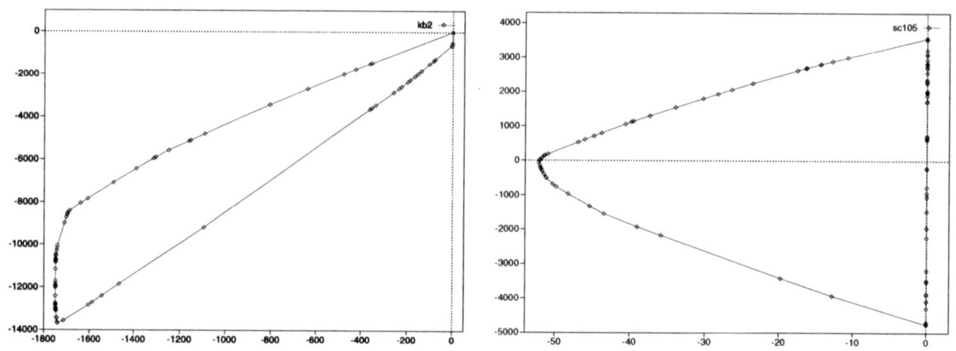

while kb2 and ... sc105 display distinctive "corners."

Figure 14.3. *Six shadows of LPs (from [4]).*

For example, the afiro problem has 32 variables, but it includes 8 equations, so the dimension of the polytope is $d = 24$. It has 19 explicit inequality constraints, plus the 32 nonnegativities, but many constraints are redundant. Indeed, the polytope has only $n = 29$ facets. It has 1654 vertices, of which only 78 are degenerate. Thus, the minimal degree is 24 and the maximal degree is 39. (These degeneracy data are, of course, only observed if we interpret the coefficients of the `afiro` problem as rational numbers.) The average vertex degree is 24.71, not much more than the minimal degree. The LP has 20,433 edges, of which 11,718 are horizontal: This is more than half of them, perhaps the first big surprise! With respect to the given objective function, the LP has a unique maximal vertex, which has the maximal degree 39; on the other side, there are 4 minimal vertices for the given objective function, all of which are simple (of degree 24). The minimal vertices describe a face of dimension 2 (a quadrilateral). The (graph-theoretic) distance from the minimal vertices to the maximal one is just 2, while the diameter of the polyhedron is 5. (Thus the problem satisfies the Hirsch bound, as discussed below, with equality!) What else do you want to know about this LP? Chances are that your question can easily be answered by the `Polymake` system.

14.3 Long Paths

A lot of effort has been put into the goal of understanding the "worst case" of the simplex algorithm. In particular, we are trying to resolve the following central question:

The complexity of linear programming. *Is there a strongly polynomial (simplex) algorithm for linear programming?*

A natural approach to resolve this question is to construct and understand "bad examples" of LPs for (selected) variants of the simplex algorithm. Thus the development of bad examples and the understanding of pivot rules should, ideally, be closely connected. We would expect bad examples to show that certain pivot rules are not good; on the other hand, they should tell us how pivot rules have to be designed in order to escape the "bad examples." This program has been worked out only partially up to now. It has produced the "deformed product" examples of bad LPs, which managed to fool all the *classical deterministic* pivot rules for linear programming into exponential behavior.

The first and classical example of bad LPs is given by the Klee–Minty cubes [21]. These are deformed d-dimensional cubes for which there is a monotone path (that is, a path on which the objective function increases strictly) through all the 2^d vertices. The classical Dantzig pivot rule as well as various lexicographic rules can be made to be exponential on these examples.

In his linear programming book [29, p. 76] Manfred Padberg talks about what he calls "worstcasitis": Following Klee and Minty's initial breakthrough there appeared a flood of papers that produced bad/exponential examples for all kinds of pivot rules in linear programming. Although some of these constructions are quite ingenious, one gets the feeling that they all more or less work along the same lines and produce the same type of bad examples, namely, "deformed products." And, indeed, a precise and systematic concept of "deformed products" was formalized by Amenta and Ziegler [1], for which the following is essentially true.

"Theorem." *The Klee–Minty cubes and all other published bad examples of LPs (that is, exponential examples for the simplex algorithm with various pivot rules) can be constructed and analyzed as iterated deformed products.*

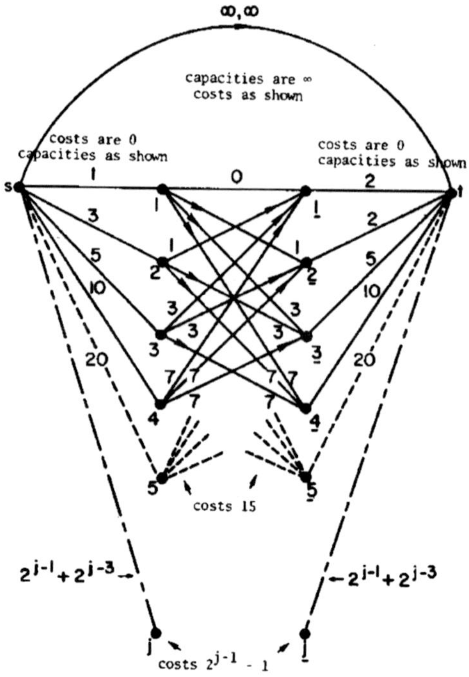

Figure 14.4. *The bad network problem from [35].*

We refer to Amenta and Ziegler [1] for details of this construction, including many examples and pictures. Here we want to present just one additional example not shown in [1]. Namely, Zadeh [35] presented a sequence of min-cost flow problems N_k for which the network simplex algorithm produces an exponential number of steps (see Figure 14.4). It may seem surprising that even these examples are iterated deformed products.

Theorem [35], [27]. *The polytope P_k of Zadeh's network N_k satisfies the following conditions:*

- *The network N_k has $2k + 2$ nodes and $k^2 + k + 2$ edges.*
- *The corresponding polytope P_k has dimension $k^2 - k + 1$ and $k^2 + k - 1$ facets.*
- *The polytope P_k is an iterated deformed product of $2k - 2$ simplices:*

$$P_k \cong \Delta_k \times \Delta_{k-1} \times (\Delta_{k-2})^2 \times \ldots \times (\Delta_1)^2.$$

  Thus P_k has $(k + 1)!(k - 1)!$ vertices.
- *On this iterated deformed product the network simplex algorithm (with the common "path," "M-path," "primal dual," and "cycle" pivot rules) will trace a monotone path through $2^k + 2^{k-2} - 1$ vertices.*

14.4 Longest Paths

How long can monotone paths be on an LP of dimension d with n constraints? This is an extremal problem that was perhaps first asked by Klee in 1965 [19], for which the complete answer is not really known (although the general impression about this may be different).

Indeed, let us assume here and in the following that we are dealing with d-dimensional polytopes with n facets that are simple, with an objective function that is nondegenerate. Then we can consider the following three quantities:

- $M(d, n)$ is defined as the maximal number of vertices on a monotone path on a simple d-polytope with n facets. This is the quantity we are after: it represents the worst case for the simplex algorithm with the stupidest choice of pivots, in the worst possible example.
- $M_{ubt}(d, n)$ is defined as the maximal number of vertices for a d-dimensional polytope with n facets. Clearly, this represents an upper bound for $M(d, n)$, and a claim by Motzkin [28] led to the "upper bound conjecture" that the maximum is given by the dual of a cyclic polytope $C_d(n)$. The upper bound theorem was proved by McMullen in 1970 [26], so we now know that

$$M_{ubt}(d, n) \quad = \quad f_{d-1}(C_d(n)) \quad = \quad \binom{n - \lceil \frac{d}{2} \rceil}{\lfloor \frac{d}{2} \rfloor} + \binom{n - 1 - \lceil \frac{d-1}{2} \rceil}{\lfloor \frac{d-1}{2} \rfloor}.$$

- $M_{sh}(d, n)$ is the maximal number of any two-dimensional projection of a simple d-polytope with n facets. This quantity is of interest since it represents the worst case for the simplex algorithm with the shadow vertex rule on a d-dimensional problem with n constraints. This is also the rule for which Borgwardt [2] has shown that the simplex algorithm is polynomial (essentially linear) on average for a reasonable model of "random LP." It is also the Gass–Saaty rule for parametric linear programming.

In summary, we now have a chain of inequalities

$$M_{sh}(d, n) \quad \leq \quad M(d, n) \quad \leq \quad M_{ubt}(d, n),$$

where we know the exact value of the right-hand side, we have exponential lower bounds for the left-hand side, but the quantity in the middle is the one that "we are after." But how tight are these bounds? Do we always have equality?

To illustrate the gap in our knowledge, let us first discuss the "diagonal" case of $n = 2d$. In this case the maximal number of vertices is roughly given by $M_{ubt} \approx \left(\frac{27}{4}\right)^{d/2} \approx 2.6^d$, while the lower bound is at least $2^d \leq M_{sh}(d, 2d)$, as is shown by the deformed cubes of Goldfarb [10]; also see [1, Section 4.3]. Of course there is a huge gap between 2^d and 2.6^d, and we do not know which of the two bounds is closer to the truth.

For a second example, let us just consider the case $d = 4$. In this case we have $M_{ubt}(4, n) = \frac{1}{2}n(n-3)$, and this value is achieved by the duals of cyclic 4-polytopes. On the other hand, the two-dimensional projections of duals of cyclic polytopes have at most $3n$ vertices, and these do not represent the best lower bound for $M_{sh}(d, n)$ [1, Section 5.4].

It is clear (see [1]) that $M_{sh}(d, n) = M(d, n) = M_{ubt}(d, n)$ holds for $d \leq 3$ as well as for $n \leq d + 2$. A current study of Kaibel, Pfeifle, and Ziegler [15] is resolving at least the first few nontrivial precise values. In particular, for $d = 4$ and $n = 7, 8$, we have

$$12 = M_{sh}(4,7) < M(4,7) = M_{ubt}(4,7) = 14$$

and

$$16 \le M_{sh}(4,8) \le M(4,8) = M_{ubt}(4,8) = 20.$$

In particular, the lower bound of $M(4, 8) \ge 17$, first achieved by C. Schultz [31] on the dual of a cyclic polytope, is better than the lower bound of 16 that one gets from the Klee–Minty cubes.

On the other hand, we know from an enumeration of "Hamiltonian abstract objective functions with the Holt–Klee property" that the value of 20 is *not* achieved for duals of cyclic polytopes $C_4(8)$: these are customarily taken as the canonical examples of polytopes that yield equality in the upper bound theorem, but they are not the only ones. For $d = 4$ and $n = 8$ there are exactly two other types of neighborly 4-polytopes with 8 vertices, called N_8 and N_8^* by Grünbaum [11, p. 125], and on both of these, equality $M(4, 8) = 20$ is achieved.

In summary, the available data do not contradict a conjecture that

$$M(d,n) = M_{ubt}(d,n)$$

holds for all $n > d > 1$. The author believes that this conjecture is indeed quite plausible. In any case, it is interesting that neither deformed products nor dual-to-cyclic polytopes give worst-possible results.

14.5 Short Paths

Let us now reverse the question: We are no longer asking for bad examples for a given pivot rule, but we rather ask for a pivot rule that is good on all examples. Thus we are trying to answer the tandem question:

- Is there always a short path "to the top,"
- and can one find one?

Our geometric model/interpretation is again that we are studying a simple d-dimensional polytope for which the last coordinate x_d is a linear objective function that is not constant on any edge. A *short path* is any path whose length (number of vertices) is polynomial in the number n of facets (and in the dimension $d < n$).

A natural approach to provide a positive answer to the questions above is to construct and analyze (new) pivot rules that have a chance to be (at least) polynomial. Thus we are looking for short monotone paths from any given vertex of the polytope to the (unique) top vertex. This is doomed to fail if there are no such short paths at all, perhaps not even nonmonotone short paths. A key question, first apparently posed by W. Hirsch in 1957, is the following.

Conjecture 14.1 (The Hirsch Conjecture [3, pp. 160, 168]). *Does every d-dimensional polytope with n facets have graph-theoretic diameter at most $n - d$?*

This is a famous/notorious question, and there have been many diverse attempts to provide an answer. We refer to the extensive survey by Klee and Kleinschmidt [20] for information, as well as to [36, Lecture 3] for more recent updates. In particular, the following two conjectures are known to be equivalent to the Hirsch Conjecture.

Conjecture 14.2 (The d-Step Conjecture). *Is it true that, for all $d > 1$ and for all d-polytopes with $2d$ facets, and for any two vertices u and v that do not lie in a common facet, there is a path from u to v of length d?*

Conjecture 14.3 (The Nonrevisiting Path Conjecture). *For any two vertices on a simple polytope, is there always a path between them that does not leave and then revisit any one of the facets?*

The Hirsch Conjecture is old, classical, interesting, important, and still unsolved. More concretely, the status of the conjecture may be summarized as follows:

- The Hirsch Conjecture is true for $d \leq 3$, but not proved for any $d > 3$ [22].
- The d-Step Conjecture is known to be true for $d \leq 5$ [22].
- The Hirsch Conjecture is tight for all $n > d \geq 8$: for any parameters in this range there is a d-polytope with n facets that has graph-theoretic diameter exactly $n - d$ [12, 5].
- No polynomial upper bound is known for the diameter of a simple d-polytope with n facets; the best upper bound of $n^{\log d+1}$ is due to Kalai and Kleitman [18].

In an attempt to provoke the construction of interesting (counter)examples for the Hirsch Conjecture, the following "rather daring" conjecture was published in 1995.

Conjecture 14.4 (The "Strong Monotone" Hirsch Conjecture [36]). *For any simple d-polytope with n faces and for any generic linear objective function, is there always a* monotone *path from the (unique) minimal vertex to the (unique) maximal vertex of length at most $n - d$?*

This conjecture escapes the counterexamples to the "Monotone Hirsch Conjecture" by Todd [33], for which the starting vertex is not the minimal one. It also escaped an attack by Holt and Klee [13]. Thus this conjecture is still open.

How about pivot rules that have a chance to find polynomial paths on LPs? In the following, we discuss three such rules.

Pivot Rule I ("Random_Edge"). Given any vertex that is not the top vertex, choose one of the outgoing improving edges, uniformly with equal probability.

Warning: This pivot rule sounds simple, but it seems to be awful to analyze in any nontrivial example. Its status may be summarized as follows:

- On the Klee–Minty cubes this pivot rule has essentially quadratic running times:

$$\frac{d^2}{8\log d} \leq E_d \leq \binom{d+1}{2};$$

see Gärtner, Henk, and Ziegler [7].

- The running time of this pivot rule is at most quadratic on all iterated deformed product examples.

- Thus this rule might as well be quadratic in expected running time on every example, but no subexponential upper bound on the expected running time has been proved.

As a challenge (and a nice example to illustrate the complexity and behavior of the RANDOM_EDGE rule), we ask for the maximal expected running time on a three-dimensional simple polytope with n facets and $2n-4$ vertices. For this Figure 14.5 indicates a class of examples for which the expected running time from the "worst" vertex is roughly $\frac{4}{3}n$. The expected running time for every starting vertex can be computed recursively for the RANDOM_EDGE rule: It is 0 for the top vertex, and for every other vertex it is 1 plus the average of the expected running times when starting at its upper neighbors. The resulting values are also indicated for the example in Figure 14.5. We refer to Kaibel et al. [14] for a detailed discussion of the RANDOM_EDGE rule on three-dimensional polytopes, whose worst-case behavior poses surprisingly tricky problems (for example, the factor $\frac{4}{3}$ is not worst possible).

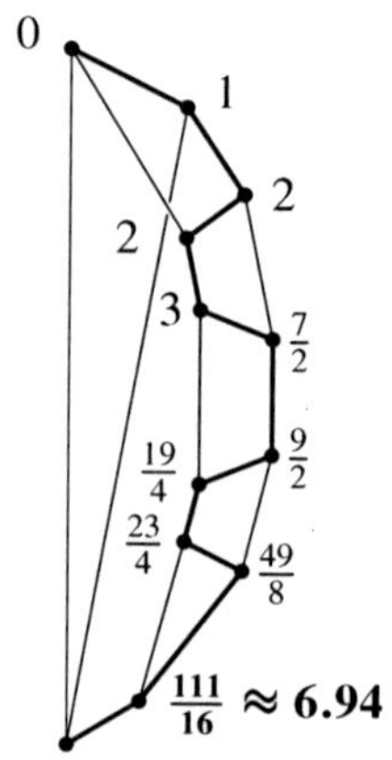

Figure 14.5. *A bad three-dimensional example for* RANDOM_EDGE *(from [32]).*

Pivot Rule II ("RANDOM_FACET"). At any given starting vertex that is not the top vertex, move up if there is a unique edge on which the objective function is increasing. If there is more than one such edge, choose a random facet that contains the given vertex, and solve the LP restricted to this facet recursively (that is, by calling RANDOM_FACET).

This rule may seem a bit more contrived than RANDOM_EDGE, but it turns out to be sometimes much more accessible to analysis. It was introduced by Kalai [17], and simultaneously (in a dual simplex algorithm setting) by Matoušek, Sharir, and Welzl [25]. Its status may be summarized as follows:

- The running time of RANDOM_FACET is at most quadratic (in d) on the Klee–Minty cubes. In fact, one can come up with an exact formula that yields the exact expected running time for every single starting vertex on the Klee–Minty cubes [7].

- There is a subexponential (but not polynomial) upper bound for the expected running time of the RANDOM_FACET simplex algorithm of

$$O\left(d^2 n + c^{\sqrt{d \log d}}\right),$$

due to Kalai [17] and Matoušek, Sharir, and Welzl [25].

- RANDOM_FACET is *slow* on the Matoušek-cubes [24]: these are edge orientations of the d-dimensional cubes that are not, in general, geometrically realizable. Combinatorially, they may be described recursively as follows: In the bottom facet of the d-cube take any Matoušek-orientation; all the vertical edges are directed upward; on the top facet, we copy the directions from the bottom facet, except that all the edges of any given parallel class may be reversed (simultaneously).
- But, amazingly, there is again a *quadratic* upper bound for the running time of RANDOM_FACET on any Matoušek-cube that is geometrically realizable, as was shown by Gärtner [6].

Finally, we want to discuss a deterministic rule that still (as far as we know) has a chance to be polynomial in the worst case.

Pivot Rule III ("LEAST_ENTERED"). At any given vertex that is not the top vertex, among the increasing edges that leave the vertex, choose any edge that leaves a facet that has been left least often on the previous moves.

Thus this rule is deterministic, but its choices depend heavily on previous choices. Also it cannot be purely implemented just on the graph: It is essential that we know for each edge the (unique!) facet of the polytope that it leaves. The formulation given here depends heavily on the fact that we are dealing with a simple polytope: Given any edge incident to a vertex v of a simple polytope, there is always a *unique* facet that contains v but not the edge.

The LEAST_ENTERED rule was proposed by Norman Zadeh around 1980, and he offered \$1000 to anyone who could prove or disprove that this rule is polynomial in the worst case; see the text of Figure 14.6 in Zadeh's handwriting (from a letter to Victor Klee,

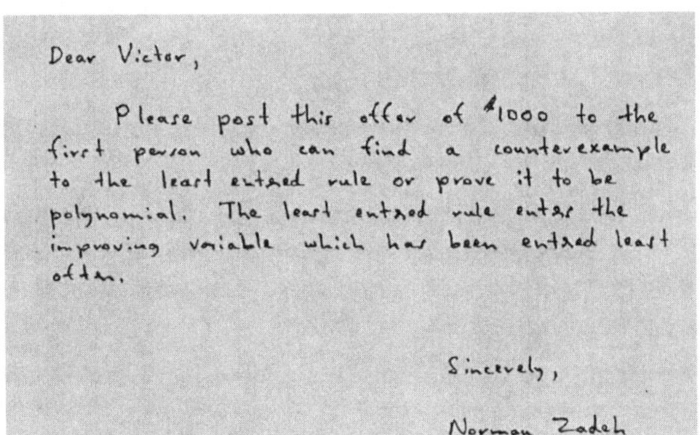
Dear Victor,

Please post this offer of \$1000 to the first person who can find a counterexample to the least entered rule or prove it to be polynomial. The least entered rule enters the improving variable which has been entered least often.

Sincerely,

Norman Zadeh

Figure 14.6. *Zadeh's offer.*

reproduced with his kind permission). Just to encourage the readers to try their luck on this problem, we want to mention that, according to a recent magazine report [23], Norman Zadeh is now a successful businessman for whom it should be no problem to pay for the prize once you have solved the problem. Good luck!

Note added in proof. There is substantial recent progress on the "monotone upper bound problem" discussed in Section 14.4: The inequality $M_{ubt}(d, n) \leq M(d, n)$ is tight (holds with equality) for the cases

$n \leq d + 2$: see B. Gärtner, J. Solymosi, F. Tschirschnitz, P. Valtr, and E. Welzl, One line and n points. In *Proceedings of the 33rd Annual ACM Symposium on Theory of Computing (STOC),* ACM, New York, 2001, pp. 306-315; and

$d \leq 4$: see J. Pfeifle, *Extremal Constructions for Polytopes and Spheres*, Dissertation, TU Berlin, 2003.

However, the inequality is not tight in general:

$$M(6, 9) < M_{ubt}(6, 9) = 30$$

was established by J. Pfeifle and G. M. Ziegler, On the Monotone Upper Bound Problem, Preprint math.CO/0308186, TU Berlin, 2003; *Experimental Mathematics*, to appear.

Bibliography

[1] N. Amenta and G.M. Ziegler. Deformed products and maximal shadows. In B. Chazelle, J.E. Goodman, and R. Pollack, editors, *Advances in Discrete and Computational Geometry*, volume 223 of *Contemporary Mathematics*, pages 57–90. American Mathematical Society, Providence, RI, 1998.

[2] K.H. Borgwardt. The Simplex Method. A Probabilistic Analysis, volume 1 of *Algorithms and Combinatorics*. Springer-Verlag, Berlin, Heidelberg, 1987.

[3] G.B. Dantzig. *Linear Programming and Extensions*. Princeton University Press, Princeton, NJ, 1963. Reprint 1998.

[4] S. Fischer. *Zweidimensionale Projektionen von linearen Programmen*, Diplomarbeit, Technische Universität Berlin, 1998.

[5] K. Fritzsche and F.B. Holt. More polytopes meeting the conjectured Hirsch bound. *Discrete Mathematics*, 205:77–84, 1999.

[6] B. Gärtner. Combinatorial linear programming: Geometry can help. In *Proceedings of the 2nd Workshop "Randomization and Approximation Techniques in Computer Science (RANDOM)" (Heidelberg)*, volume 1518 of *Lecture Notes in Computer Science*, pages 82–96, Springer-Verlag, Berlin, 1998.

[7] B. Gärtner, M. Henk, and G. M. Ziegler. Randomized simplex algorithms on Klee-Minty cubes. *Combinatorica*, 18:349–372, 1998.

[8] E. Gawrilow and M. Joswig. Polymake: A software package for analyzing convex polytopes. http://www.math.tu-berlin.de/diskregeom/polymake/.

[9] E. Gawrilow and M. Joswig. Polymake: A framework for analyzing convex polytopes. In G. Kalai and G.M. Ziegler, editors, *Polytopes—Combinatorics and Computation*, volume 29 of *DMV Seminar*, pages 43–73. Birkhäuser-Verlag, Basel, 2000.

[10] D. Goldfarb. On the complexity of the simplex algorithm. In S. Gomez, editor, *Advances in Optimization and Numerical Analysis*, pages 25–38. Kluwer, Dordrecht, The Netherlands, 1994.

[11] B. Grünbaum. *Convex Polytopes.* 2nd edition. Prepared and with a Preface by V. Kaibel, V. Klee, G.M. Ziegler, Graduate Texts in Math. 221, Springer-Verlag New York, 2003.

[12] F.B. Holt and V. Klee. Counterexamples to the strong d-step conjecture for $d \geq 5$. *Discrete Computational Geometry*, 19:33–46, 1998.

[13] F.B. Holt and V. Klee. Many polytopes meeting the conjectured Hirsch bound. *Discrete Computational Geometry*, 20:1–17, 1998.

[14] V. Kaibel, R. Mechtel, M. Sharir, and G. M. Ziegler. *The Simplex Algorithm in Dimension Three*, Preprint math.CO/0309351, September 2003.

[15] V. Kaibel, J. Pfeifle, and G.M. Ziegler. On the Monotone Upper Bound Problem, in preparation, 2002.

[16] G. Kalai and G.M. Ziegler (eds.). *Polytopes—Combinatorics and Computation*, volume 29 of *DMV Seminars*. Birkhäuser-Verlag, Basel, 2000.

[17] G. Kalai. A subexponential randomized simplex algorithm. In *Proceedings of the 24th ACM Symposium on the Theory of Computing (STOC)*, pages 475–482. ACM Press, New York, 1992.

[18] G. Kalai and D.J. Kleitman. A quasi-polynomial bound for the diameter of graphs of polyhedra. *Bulletin of the American Mathematical Society*, 26:315–316, 1992.

[19] V. Klee. Heights of convex polytopes. *Journal of Mathematical Analysis and Applications*, 11:176–190, 1965.

[20] V. Klee and P. Kleinschmidt. The d-step conjecture and its relatives. *Mathematics of Operations Research*, 12:718–755, 1987.

[21] V. Klee and G. J. Minty. How good is the simplex algorithm?. In O. Shisha, editor, *Inequalitites*, III, pages 159–175. Academic Press, New York, 1972.

[22] V. Klee and D.W. Walkup. The d-step conjecture for polyhedra of dimension $d < 6$. *Acta Mathematica*, 117:53–78, 1967.

[23] S. Lubove. See no evil. Much of the raunchy porn on the internet wouldn't exist were it not for the help of a handful of legitimate companies operating quietly in the background. *Forbes* (September 17, 2001), 68–70.

[24] J. Matoušek. Lower bounds for a subexponential optimization algorithm. *Random Structures and Algorithms*, 5:591–607, 1994.

[25] J. Matoušek, M. Sharir, and E. Welzl. A subexponential bound for linear programming. In *Proceedings of the Eighth Annual ACM Symposium on Computational Geometry (Berlin 1992)*, pages 1–8. ACM Press, New York, 1992.

[26] P. McMullen. *The maximum numbers of faces of a convex polytope*, Mathematika 17:179–184, 1970.

[27] T. Morstein. *Die Probleme von Zadeh zum Netzwerk-Simplex sind deformierte Produkte von Polytopen*, Diplomarbeit, Technische Universität Berlin, 1999.

[28] T. S. Motzkin. Comonotone curves and polyhedra. *Bulletin of the American Mathematical Society*, 63:35, 1957. Abstract.

[29] M. Padberg. *Linear Programming*, second edition, volume 12 of *Algorithms and Combinatorics*. Springer-Verlag, Heidelberg, 1999.

[30] V. Reiner and G.M. Ziegler. Coxeter-associahedra. *Mathematika*, 41:364–393, 1994.

[31] C. Schultz. *Schwierige lineare Programme für den Simplex-Algorithmus*, Diplomarbeit, Technische Universität Berlin, 2001.

[32] M. Sharir and G.M. Ziegler. The Random-Edge Simplex Algorithm on Three-Dimensional Polytopes, in preparation, 2002.

[33] M.J. Todd. The monotonic bounded Hirsch conjecture is false for dimension at least 4. *Mathematics of Operations Research*, 5:599–601, 1980.

[34] D. Weber. *Kombinatorische Analyse einiger linearer Programme*, Diplomarbeit, Technische Universität Berlin, 1999.

[35] N. Zadeh. A bad network problem for the simplex method and other minimum cost flow algorithms. *Mathematical Programming*, 5:255–266, 1973.

[36] G.M. Ziegler. *Lectures on Polytopes*, volume 152 of *Graduate Texts in Mathematics*. Springer-Verlag, New York, 1995. Revised edition, 1998.

Part V

Semidefinite Programming

Chapter 15

A Cutting Plane Algorithm for Large Scale Semidefinite Relaxations

*Christoph Helmberg**

Abstract. The recent spectral bundle method allows one to compute, within reasonable time, approximate dual solutions of large scale semidefinite quadratic 0/1-programming relaxations. We show that it also generates a sequence of primal approximations that converge to a primal optimal solution. Separating with respect to these approximations gives rise to a cutting plane algorithm that converges to the optimal solution under reasonable assumptions on the separation oracle and the feasible set. We have implemented a practical variant of the cutting plane algorithm for improving semidefinite relaxations of constrained quadratic 0/1-programming problems by odd-cycle inequalities. We also consider separating odd-cycle inequalities with respect to a larger support than given by the cost matrix and present a heuristic for selecting this support. Our preliminary computational results for max-cut instances on toroidal grid graphs and balanced bisection instances indicate that warm start is highly efficient and that enlarging the support may sometimes improve the quality of relaxations considerably.

MSC 2000. 90C22, 90C25, 90C27, 90C09, 90C20, 90C06
Key words. Bisection, equicut, max-cut, semidefinite programming, spectral bundle method, subgradient method, quadratic 0/1-programming

15.1 Introduction

Crowder, Johnson, and Padberg [7] initiated the rise of general mixed-integer programming by solving several large scale, unstructured 0/1-linear programming problems via a unified

*Konrad-Zuse-Zentrum für Informationstechnik Berlin, Takustraße 7, D-14158 Berlin (helmberg@zib.de), http://www.zib.de/helmberg.

cutting plane framework. Can we set up a similar framework for large scale quadratic 0/1-programming problems?

It seems likely that this question motivated much of the work on the Boolean quadric polytope and the max-cut polytope; see [28, 9] and references therein. In the late 1980s basic techniques for lifting linear inequalities into quadratic space were developed [33, 26, 1]; Lovász and Schrijver [26] linked this to a semidefinite relaxation of quadratic 0/1-programming [34] and demonstrated by means of the stable set problem that much can be gained by doing so. Further evidence on the effectiveness of the semidefinite approach was provided by Goemans and Williamson [12] via their approximation algorithm for max-cut. These works provide a clear guideline on how one should set up and improve relaxations of constrained quadratic 0/1-programming problems. Unfortunately, the final ingredient, an efficient algorithm that solves large scale semidefinite relaxations in acceptable time and allows for the addition of cutting planes on the fly, is still missing. In this work we would like to convince the reader that the spectral bundle method [18, 17] provides all that is needed—certainly not for all applications, but for many cases of relevance.

The dual of a semidefinite program (SDP) with bounded feasible set can be transformed into a problem of minimizing the maximum eigenvalue of an affine matrix function (see, e.g., [15]). The latter is a nonsmooth convex optimization problem that may be tackled by subgradient and bundle methods; see [20] and references therein. The spectral bundle method [18, 17] is tuned to eigenvalue optimization problems and their associated SDPs in that it uses a quadratic semidefinite subproblem. It was already pointed out in [18] that the solutions of these quadratic semidefinite subproblems may be interpreted as (infeasible) approximate solutions to the primal SDP. Here, we prove rigorously that these approximations converge to a primal optimal solution within the setting of the spectral bundle method with bounds [17]. Feltenmark and Kiwiel [10] proved a related result for a classical proximal bundle method; see also [4] and references therein for other approaches for generating primal solutions from subgradient methods.

The primal approximations will serve as input for a separation oracle. Since these approximations are infeasible in general, precautions have to be taken against the separation oracle returning the same inequality again and again. A realistic assumption that is often fulfilled in practice is that the oracle returns a maximally violated inequality from a finite representation $Ax \leq b$ of the polyhedron. The combination of the spectral bundle method with such an oracle yields a cutting plane algorithm that generates a sequence of iterates converging to primal and dual optimal solutions whenever a strictly feasible primal solution exists. To the best of our knowledge this is the first provably convergent cutting plane approach based on bundle methods. Since semidefinite programming includes linear and second order cone programming, the algorithm easily extends to cutting plane algorithms over products of these cones as long as the primal feasible set is bounded.

In our implementation we concentrate on the quadratic 0/1-programming setting and do not obey all requirements for theoretic convergence in favor of computational efficiency. For various reasons we prefer to work with the equivalent semidefinite relaxation of quadratic ± 1-programming, which is better known as the max-cut relaxation. We improve the basic relaxation by adding odd-cycle inequalities as cutting planes. In contrast to linear programming it may also help to separate these inequalities with respect to support not contained in the cost function. We present a simple heuristic for enlarging the support that turned out to considerably improve the quality of the bound for several bisection instances. In order to illustrate this and the behavior of the cutting plane approach in general, we present preliminary numerical results for max-cut instances on toroidal grid graphs and bisection instances from numerical linear algebra.

Here is an outline of the chapter. In Section 15.2 we review the equivalence between 0/1- and ± 1-formulations and list some important properties of the odd-cycle inequalities. Next, in Section 15.3, we explain the basic steps of the spectral bundle method with bounds and prove primal convergence of the iterates. This part relies heavily on [17], which should be at hand. Based on these convergence properties we develop a conceptual cutting plane algorithm with convergence guarantee in Section 15.4. For efficiency we employ a slightly different approach in practice. The implementational choices are described in Section 15.5. Finally, we present preliminary computational results in Section 15.6.

Our notation is quite standard. The set of symmetric matrices of order n will be denoted by S_n. $A \succeq 0$, $A \in S_n^+$ refers to positive semidefinite matrices; $A \succ 0$ is used for positive definiteness. The trace $\operatorname{tr} A$ is the sum of the diagonal elements; $\operatorname{diag}(A)$ denotes the vector of diagonal elements. For $A, B \in S_n$ or $A, B \in \mathbb{R}^{m \times n}$, we employ the inner product $\langle A, B \rangle = \operatorname{tr} B^T A$. When minimizing some function $f(y)$, $\operatorname{argmin} f$ refers to a unique minimizer of f and $\operatorname{Argmin} f$ to the set of minimizers. An (undirected) graph $G = (V, E)$ consists of a finite set of nodes $V \subset \mathbb{N}$ and a set of edges $E \subset \{\{i, j\} : i < j,\ i, j \in V\}$. We only consider graphs without loops or multiple edges. For an edge $\{i, j\}$ we will also write ij, because we typically associate edges with matrix elements a_{ij}. A set of edges $C \subset E$ is called a cycle (of length k) if $C = \{v_1v_2, v_2v_3, \ldots, v_kv_1\}$ for pairwise distinct $v_i \in V$, $i = 1, \ldots, k$. For a matrix $A \in S_n$, the support graph refers to $G = (V, E)$ with $V = \{1, \ldots, n\}$ and $E = \{ij : i < j, a_{ij} \neq 0, i, j \in V\}$.

15.2 Semidefinite Programming Relaxations for Quadratic 0/1- and ±1-Programming

We first review the process by which the semidefinite Lovász–Schrijver relaxation for constrained quadratic 0/1-programming [26] can be transformed into an equivalent semidefinite relaxation for quadratic ± 1-programming (we refer to the latter as the max-cut setting). For a cost matrix $\bar{C} \in S_n$, constraints matrices $\bar{A}_i \in S_n$ for $i \in M := \{1, \ldots, m\}$, and right-hand side b an SDP arising from a Lovász–Schrijver relaxation over $n-1$ binary (0/1) variables, may read

$$\text{(SQP)} \qquad \begin{array}{ll} \max & \langle \bar{C}, \bar{Y} \rangle \text{ subject to} \\ & \langle \bar{A}_i, \bar{Y} \rangle + s_i = b_i \qquad \text{for } i \in M, \\ & \bar{Y} = \begin{bmatrix} 1 & \operatorname{diag}(Y)^T, \\ \operatorname{diag}(Y) & Y \end{bmatrix} \succeq 0,\ s \geq 0. \end{array}$$

By employing the scaling

$$X = Q^{-1} \bar{Y} Q^{-T} \qquad \text{with} \quad Q = \begin{bmatrix} 1 & 0 \\ \frac{1}{2}e & \frac{1}{2}I_n \end{bmatrix} \quad \text{and} \quad Q^{-1} = \begin{bmatrix} 1 & 0 \\ -e & 2I_n \end{bmatrix}$$

and by transforming the coefficient matrices according to the identity $\langle \bar{A}, \bar{Y} \rangle = \langle Q\bar{A}Q^T, X \rangle$,

$$C := Q\bar{C}Q^T, \qquad A_i := Q\bar{A}_iQ^T \quad \text{for } i \in M, \tag{15.1}$$

we obtain an equivalent semidefinite relaxation within the max-cut setting [24, 14] (e denotes the vector of all ones)

$$
\text{(SMC)} \qquad \begin{array}{lll} \max & \langle C, X\rangle \text{ subject to} & \\ & \langle A_i, \bar{X}\rangle + s_i = b_i & \text{for } i \in M, \\ & \operatorname{diag}(X) = e, & \\ & X \succeq 0, s \geq 0. & \end{array}
$$

Indeed, if the equality constraints describing the structure of $\bar{Y}$ are chosen appropriately, then (SQP) and (SMC) share the same slack and dual variables. Furthermore, the transformation (15.1) preserves sparsity and low rank structure of the constraints [14]. Thus, we may switch between both formulations without loss, in theory and in practice.

Likewise, the Boolean quadric polytope and the max-cut polytope are isomorphic; this has already been proved in [8]. Both polytopes have been studied extensively (see [28] for the Boolean quadric polytope and [9] for the max-cut polytope).

Our goal is to devise a cutting plane approach for generically improving semidefinite relaxations of type (SMC) or (SQP) by exploiting polyhedral knowledge about the underlying polytopes. Within our computational framework, the semidefinite relaxation of max-cut (SMC) offers significant advantages over (SQP) and therefore we will concentrate on the max-cut setting. In particular, we are interested in cutting planes that are suitable for large sparse cost matrices C.

Numerous classes of facet-defining inequalities of the cut polytope appear in the literature, but for most of them no efficient separation algorithm or heuristic is available. In the case of large unstructured support graphs $G = (V, E)$, the only class that has proved to be of practical value is, so far, the class of odd-cycle inequalities. Formulated within the ± 1-setting of (SMC) they read

$$
\sum_{ij \in \mathcal{C} \setminus F} x_{ij} - \sum_{ij \in F} x_{ij} \leq |\mathcal{C}| - 2 \qquad \text{for } \mathcal{C} \subset E \text{ a cycle, } F \subseteq \mathcal{C}, |F| \text{ odd.} \tag{15.2}
$$

Odd-cycle inequalities can be separated in polynomial time [5] by solving shortest path problems in an auxiliary graph with twice the number of nodes and four times the number of edges. They provide a complete description of the cut polytope for graphs not contractible to K_5 (the complete graph on five nodes) [32, 2].

In a pure polyhedral setting, enlarging the number of cycles in a graph (and therefore the number of separable odd-cycle inequalities) by adding edges with weight 0 does not improve the relaxation [3]. This is not true in combination with the semidefinite relaxation (SMC), as has been observed in many examples.

15.3 Primal Convergence of the Spectral Bundle Method

We first introduce some basic objects and concepts that we will need throughout the next two sections. For $a > 0$ let

$$
\mathcal{W} = \{X \succeq 0 : \langle I, X\rangle = a\} \tag{15.3}
$$

denote the set of positive semidefinite matrices of order n with constant trace a. Consider the SDP

$$\text{(PSDP)} \qquad \begin{array}{ll} \max & \langle C, X\rangle \text{ subject to} \\ & \mathcal{A}X + s = b, \\ & X \in \mathcal{W}, s \geq 0, \end{array}$$

with variables $X \in S_n$, $s \in \mathbb{R}^m_+$, cost matrix $C \in S_n$, right-hand-side vector $b \in \mathbb{R}^m$, and a constraint matrix (or linear map) $\mathcal{A} : S_n \to \mathbb{R}^m$. This covers all forms of linear programs (LPs) over symmetric cones with bounded feasible sets.

Introducing Lagrange multipliers $y \in \mathbb{R}^m$ and dualizing we arrive at a dual problem

$$\min_{y\in\mathbb{R}^m} f(y) := \sup_{(X,s)\in\mathcal{W}\times\mathbb{R}^m_+} \langle C, X\rangle + \langle b - s - \mathcal{A}X, y\rangle . \tag{15.4}$$

This problem is tightly related to the standard dual SDP of (PSDP). By making use of (15.3), it can be verified that strong duality holds for (15.4) and (PSDP) [18]. We denote the joint optimal value of (15.4) and (PSDP) (it may be $-\infty$) by

$$f^* = \inf_{y\in\mathbb{R}^m} f(y).$$

The function f is the supremum over the family of linear functions

$$f_{W,\eta}(y) := \langle C, W\rangle + \langle b - \eta - \mathcal{A}W, y\rangle , \qquad (W, \eta) \in \mathcal{W} \times \mathbb{R}^m_+, \tag{15.5}$$

and therefore convex. Denoting by $\mathcal{A}^T$ the adjoint of $\mathcal{A}$ (by definition it satisfies $\langle \mathcal{A}X, y\rangle = \langle X, \mathcal{A}^T y\rangle$ for all $(X, y) \in S_n \times \mathbb{R}^m$) we may express f as a sum of two well-known supremums:

$$\begin{aligned} f(y) &= \langle b, y\rangle + \sup_{W\in\mathcal{W}} \langle C - \mathcal{A}^T y, W\rangle + \sup_{\eta\in\mathbb{R}^m_+} \langle -\eta, y\rangle \\ &= \langle b, y\rangle + a\lambda_{\max}(C - \mathcal{A}^T y) + \iota_{\mathbb{R}^m_+}(y). \end{aligned} \tag{15.6}$$

The first supremum is a "max" and is related to the maximum eigenvalue function $\lambda_{\max}(\cdot)$ by the fact that $\lambda_{\max}(A) = \max\{\langle A, X\rangle : X \succeq 0, \langle I, X\rangle = 1\}$ (see, e.g., [25]). The second supremum yields the indicator function ι_Y for $Y := \mathbb{R}^m_+$ ($\iota_Y(y) = 0$ for $y \in Y$ and ∞ otherwise). Thus, the effective domain of f is Y. For a feasible $\bar{y} \in Y$, the function value and a subgradient may be determined by computing $\lambda_{\max}(C - \mathcal{A}^T\bar{y})$ and a corresponding eigenvector v. With $W_S := avv^T$ a subgradient of f in $\bar{y}$ is, e.g., $\nabla f_{W_S,0} = b - \mathcal{A}W_s$. By (15.4), the subdifferential of f at $\bar{y} \in Y$ is

$$\partial f(\bar{y}) = \left\{\nabla f_{W,\eta} : f_{W,\eta}(\bar{y}) = f(\bar{y}), (W, \eta) \in \mathcal{W} \times Y\right\}.$$

In order to solve (15.4), we employ the spectral bundle method with bounds [17], whose main steps we summarize next. Using subgradient information, it forms a model $\hat{f}$ minorizing f and determines a new *candidate* y^+ as the minimizer of the augmented model $\hat{f} + \frac{u}{2}\|\cdot - \hat{y}\|^2$, where $\hat{y}$ is the *center of stability* and the *weight* $u > 0$ provides indirect control on the distance of y^+ to $\hat{y}$. At this candidate, f is evaluated and a subgradient is computed. If progress is good in comparison to the progress predicted by the model, the algorithm moves its center to the new candidate (a *descent step*; $\hat{y}$ is set to y^+). Otherwise the center is left unchanged (a *null step*) but the subgradient is used to improve the model.

The model $\hat{f}$ is formed as follows. For arbitrary subsets $\widehat{\mathcal{W}} \subseteq \mathcal{W}$ and $\widehat{Y} \subseteq Y$ define

$$f_{\widehat{\mathcal{W}},\widehat{Y}}(y) := \sup_{(W,\eta)\in\widehat{\mathcal{W}}\times\widehat{Y}} f_{W,\eta}(y) \qquad \leq f(y) \qquad \text{for all } y \in \mathbb{R}^m. \tag{15.7}$$

For example, $f_{\mathcal{W},Y} = f$ and $f_{\mathcal{W},0}(y) = f(y)$ for all $y \in Y$. Instead of $f_{\{W\},\widehat{Y}}$ or $f_{\widehat{\mathcal{W}},\{\eta\}}$ we will also write $f_{W,\widehat{Y}}$ and $f_{\widehat{\mathcal{W}},\eta}$. In our algorithm we choose $\widehat{Y} = Y$ and

$$\widehat{\mathcal{W}} = \left\{PVP^T + \alpha\overline{W} : \langle I, V\rangle + \alpha a = a,\, V \in S_r^+,\, \alpha \geq 0\right\} \tag{15.8}$$

for a given *bundle* $P \in \mathbb{R}^{n\times r}$, $P^TP = I_r$, and *aggregate matrix* $\overline{W} \in \mathcal{W}$. The matrices P and $\overline{W}$ will be updated at the end of each iteration.

Given a center of stability $\hat{y}$ and a weight u, the candidate is now determined by computing argmin $f_{\widehat{\mathcal{W}},Y} + \frac{u}{2}\|\cdot - \hat{y}\|^2$. Using standard saddle-point arguments from convex analysis [31] one can show that

$$\min_{y\in\mathbb{R}^m} f_{\widehat{\mathcal{W}},Y}(y) + \frac{u}{2}\|y - \hat{y}\|^2 = \max_{(W,\eta)\in\widehat{\mathcal{W}}\times Y} \min_{y\in\mathbb{R}^m} f_{W,\eta}(y) + \frac{u}{2}\|y - \hat{y}\|^2 \tag{15.9}$$

and that solving the right-hand side yields the left-hand side minimizer as well. The minimizer of the right-hand-side inner minimization is (using (15.5))

$$y_{W,\eta} = \hat{y} - \frac{1}{u}\nabla f_{W,\eta} = \hat{y} - \frac{1}{u}(b - \eta - \mathcal{A}W). \tag{15.10}$$

Substituting this into the right-hand side of (15.9) and using the definition (15.5) of $f_{W,\eta}$ we obtain the dual function to the augmented model:

$$\psi(W,\eta) := \langle C, W\rangle + \langle b - \eta - \mathcal{A}W, \hat{y}\rangle - \frac{1}{2u}\|b - \eta - \mathcal{A}W\|^2. \tag{15.11}$$

An exact maximizing pair of ψ would yield the exact candidate via (15.10). For efficiency, however, we prefer to compute a rough approximation by a coordinatewise approach in Gauss–Seidel fashion. In particular, we first fix an $\hat{\eta}$ and compute a

$$W^+ \in \operatorname{Argmax}\left\{\psi(W,\hat{\eta}) : W \in \widehat{\mathcal{W}}\right\} \tag{15.12}$$

by solving, by means of an interior-point algorithm, the quadratic semidefinite subproblem

$$\begin{array}{ll} \max & \langle C, W\rangle + \langle b - \hat{\eta} - \mathcal{A}W, \hat{y}\rangle - \frac{1}{2u}\|b - \hat{\eta} - \mathcal{A}W\|^2 \text{ subject to} \\ & W = PVP^T + \alpha\overline{W}, \\ & \operatorname{tr} V + a\alpha = a, \\ & V \succeq 0, \alpha \geq 0. \end{array} \tag{15.13}$$

(Observe that in solving (15.13) we need only $\mathcal{A}\overline{W}$ and $\langle C, \overline{W}\rangle$ and not $\overline{W}$ itself.) Then we determine the next η^+ as the maximizer for this fixed W^+:

$$\eta^+ := \operatorname{argmax}\left\{\psi(W^+,\eta) : \eta \in Y\right\} = \max\{0, -u\hat{y} + b - \mathcal{A}W^+\}. \tag{15.14}$$

The corresponding approximate candidate is feasible (i.e., in the effective domain Y of f) and satisfies complementarity:

$$y^+ := y_{W^+,\eta^+} = \max\left\{\hat{y} - \frac{1}{u}(b - \mathcal{A}W^+), 0\right\}, \qquad \langle \eta^+, y^+ \rangle = 0. \tag{15.15}$$

Even though we allow for several repetitions of these coordinatewise steps in Algorithm 15.1 (we call these *inner iterations*), it is shown in [17] that one inner iteration suffices to ensure convergence.

As pointed out in the text following (15.6), evaluating $f(y^+)$ can be done by computing the maximum eigenvalue $\lambda_{\max}(C - \mathcal{A}^T y^+)$. To exploit available structure in the matrix $C - \mathcal{A}^T y^+$ it is advantageous to employ iterative methods like the Lanczos method (see, e.g., [13]) that rely on matrix-vector multiplications only. These methods produce successively better lower estimates and approximate eigenvectors of $\lambda_{\max}(C - \mathcal{A}^T y^+)$. As soon as this lower estimate indicates a null step, the $W_S = avv^T$ corresponding to the approximate eigenvector v guarantees sufficient improvement of the model for convergence [17]. This is the rationale for combining the descent test and evaluation in Step 2 of Algorithm 15.1 below.

In order to guarantee progress of the algorithm after a null step, the new model $\widehat{\mathcal{W}}^+$ has to contain W^+ and W_S. The minimal choice is $P^+ = v$ and $\overline{W}^+ = W^+$. A better strategy, which successively adapts the subspace spanned by the columns of P, is described in [17].

Algorithm 15.1 (Spectral Bundle Method with Bounds [17]).
Input: $y^0 \in \mathbb{R}^m_+$, $\varepsilon_{\text{opt}} \geq 0$, $\kappa_{\text{M}} \in (0, \infty]$, $\kappa \in (0, 1)$, $\bar{\kappa} \in [\kappa, 1)$, a weight $u > 0$.

Step 0 (Initialization.) Set $k = 0$, $\hat{y}^0 = y^0$, $\eta^0 = 0$, $f(\hat{y}^0)$, and $\widehat{\mathcal{W}}^0$.

Step 1 (Trial point finding.) Set $\hat{y} = \hat{y}^k$, $\widehat{\mathcal{W}} = \widehat{\mathcal{W}}^k$, $\hat{\eta} = \eta^k$.

(a) Find $W^+ \in \text{Argmax}_{W \in \widehat{\mathcal{W}}} \psi(W, \hat{\eta})$ (see (15.11)).

(b) Set $\eta^+ = \text{argmax}_{\eta \geq 0} \psi(W^+, \eta)$ (see (15.14)) and $y^+ = y_{W^+,\eta^+}$ (feasible by (15.15)).

(c) (stopping criterion.) If $f(\hat{y}) - f_{W^+,\eta^+}(y^+) \leq \varepsilon_{\text{opt}}(|f(\hat{y})| + 1)$, then STOP.

(d) If $f_{\widehat{\mathcal{W}},0}(y^+) - f_{W^+,\eta^+}(y^+) > \kappa_{\text{M}}[f(\hat{y}) - f_{W^+,\eta^+}(y^+)]$, then set $\hat{\eta} = \eta^+$ and go to (a).

(e) Set $y^{k+1} = y^+$, $W^{k+1} = W^+$, and $\eta^{k+1} = \eta^+$.

Step 2 (Descent test.) Find $W_S^{k+1} \in \mathcal{W}$ such that either

(a) $f(\hat{y}^k) - f_{W_S^{k+1},0}(y^{k+1}) \leq \bar{\kappa}[f(\hat{y}^k) - f_{W^{k+1},\eta^{k+1}}(y^{k+1})]$ or

(b) $f_{W_S^{k+1},0}(y^{k+1}) = f(y^{k+1})$ and $f(\hat{y}^k) - f(y^{k+1}) \geq \kappa[f(\hat{y}^k) - f_{W^{k+1},\eta^{k+1}}(y^{k+1})]$.

In case (a), set $\hat{y}^{k+1} = \hat{y}^k$ (null step); otherwise set $\hat{y}^{k+1} = y^{k+1}$ (descent step).

Step 3 (Model updating.) Choose a closed convex $\widehat{\mathcal{W}}^{k+1} \supset \{W^{k+1}, W_S^{k+1}\}$.

Step 4. Increase k by 1 and go to Step 1.

The following theorem is proved in [17].

Theorem 15.2. *Either* $\hat{y}^k \to \bar{y} \in \text{Argmin}\, f$ *or* $\text{Argmin}\, f = \emptyset$ *and* $\|\hat{y}^k\| \to \infty$. *In both cases* $f(\hat{y}^k) \downarrow f^*$.

A close inspection of the proof yields an important observation.

Lemma 15.3. *If* $\text{Argmin}\, f \neq \emptyset$ *and Algorithm 15.1 does not stop, it generates a subsequence* $K \subseteq \mathbb{N}$ *satisfying* $\nabla f_{W^k,\eta^k} \xrightarrow{K} 0$ *and* $f_{W^k,\eta^k}(y^k) \xrightarrow{K} f^*$.

Proof. First consider the case of a finite number of descent steps. Then an infinite number of inner iterations or null steps occurs starting with some iteration $\bar{k}$. Let the final stability center be $\hat{y} = \hat{y}^{\bar{k}}$. Then it is shown in [17, Lemma 3.2(c)][1] and [17, Lemma 3.4] that $f_{W^k,\eta^k}(y^k) \to f(\hat{y})$ and $y^k \to \hat{y} \in \text{Argmin}\, f$. By $y^k = y_{W^k,\eta^k}$ (see (15.15)) and (15.10) we conclude that $\nabla f_{W^k,\eta^k} = b - \eta^k - \mathcal{A}W^k = u(\hat{y} - y^k) \to 0$.

In the case of an infinite number of descent steps, assumption $\text{Argmin}\, f \neq \emptyset$ ensures condition [17, (3.17)] holds. Then the last paragraph of the proof of [17, Lemma 3.5] establishes that the subsequence $K := \{k : \hat{y}^k = y^k\}$ of candidates yielding descent steps satisfies the desired properties. □

This motivates the following lemma.

Lemma 15.4. *Assume* $\text{Argmin}\, f \neq \emptyset$. *Let* $K \subseteq \mathbb{N}$ *be a subsequence of iterates satisfying* $\nabla f_{W^k,\eta^k} \xrightarrow{K} 0$ *and* $f_{W^k,\eta^k}(y^k) \xrightarrow{K} f^*$. *Then all cluster points of* $(W^k, \eta^k)_{k\in K}$ *are optimal solutions of (PSDP).*

Proof. By construction $W^k \in \widehat{\mathcal{W}}^{k-1} \subseteq \mathcal{W}$ for all $k \geq 1$, and $\mathcal{W}$ is compact. Theorem 15.2 and $\text{Argmin}\, f \neq \emptyset$ imply that the $\hat{y}^k$ remain bounded. Therefore the vectors $-u\hat{y}^k + b - \mathcal{A}W^k$ remain bounded. By (15.14) and (15.15) the same is true for the η^k and the y^k for all k. Furthermore, the η^k are nonnegative by (15.14). Compactness ensures that there is at least one cluster point for $(W^k, \eta^k)_{k\in K}$. Now let $(\bar{W}, \bar{\eta})$ be such a cluster point and $\bar{K} \subseteq K$ a corresponding subsequence with $(W^k, \eta^k) \xrightarrow{\bar{K}} (\bar{W}, \bar{\eta})$. Then $\bar{W} \in \mathcal{W}$, $\bar{\eta} \geq 0$, and (see (15.5)) $\nabla f_{W^k,\eta^k} = b - \eta^k - \mathcal{A}W^k \xrightarrow{\bar{K}} b - \bar{\eta} - \mathcal{A}\bar{W} = 0$. Thus, $(\bar{W}, \bar{\eta})$ is feasible for (PSDP). Furthermore, $f_{W^k,\eta^k}(y^k) = \langle C, W^k \rangle + \langle b - \eta^k - \mathcal{A}W^k, y^k \rangle \xrightarrow{\bar{K}} \langle C, \bar{W} \rangle = f^*$. Since f^* is an upper bound on the objective value of (PSDP), this implies optimality. □

Before stating the main theorem of this section we need one more result.

Lemma 15.5. *If Algorithm 15.1 terminates for* $\varepsilon_{\text{opt}} = 0$, *then the terminating* (W^+, η^+) *is an optimal solution of (PSDP).*

[1] Strictly speaking, the fact that [17, Lemma 3.2(c)] holds for inexact evaluation in Step 2 requires a slightly different proof; see [15]. Alternatively, one may avoid additional inner iterations by setting $\kappa_M = \infty$.

Proof. If the algorithm terminates, then $f(\hat{y}) = f_{W^+,\eta^+}(y^+)$ and [17, (2.22)] yields $\hat{y} = y^+$. So by (15.15) and (15.10) we conclude that $b - \eta^+ - \mathcal{A}W^+ = 0$. Together with $W^+ \in \widehat{\mathcal{W}} \subseteq \mathcal{W}$ (15.12), $\eta^+ \geq 0$ (15.14), and $f(\hat{y}) = f_{W^+,\eta^+}(\hat{y}) = \langle C, W^+ \rangle$ (see (15.5)), we obtain feasibility and optimality of (W^+, η^+) for (PSDP). $\square$

Theorem 15.6. *Assume* Argmin $f \neq \emptyset$ *and let* $\varepsilon_{\text{opt}} = 0$. *If Algorithm 15.1 terminates, then the terminating* (W^+, η^+) *is an optimal solution of (PSDP). If the algorithm does not terminate, there is a subsequence* $K \subseteq \mathbb{N}$ *so that all cluster points of* $(W^k, \eta^k)_{k \in K}$ *are optimal solutions of (PSDP).*

Proof. If the algorithm terminates, then Lemma 15.5 applies. Otherwise the result follows from Lemmas 15.3 and 15.4. $\square$

15.4 Extension to a Cutting Plane Algorithm

In this section we extend Algorithm 15.1 to a cutting plane algorithm for optimizing over the intersection of $\mathcal{W}$ with a polyhedron $\{X : \mathcal{A}X \leq b\}$ that is given by a special type of separation oracle. Note that within our setting there is no hope for polynomiality; our aim is to establish convergence. Encouraged by Theorem 15.6 we would like to separate with respect to W^+. Unfortunately, W^+ is never feasible unless it is optimal. Thus, one is faced with the problem that a separation oracle may return the same cut over and over again without disclosing any further information. In order to avoid this, we require the oracle to return a maximally violated inequality out of a finite set of inequalities describing the polyhedron (many actual separation routines satisfy this requirement).

Definition 15.7. A separation oracle for a polyhedron $P = \{x : Ax \leq b\}$ with $A \in \mathbb{R}^{m \times n}$ and $b \in \mathbb{R}^m$ is called a *maximum violation oracle with respect to* (A, b) if, for a given point $\bar{x} \in \mathbb{R}^n$, it either asserts that $\bar{x} \in P$ or returns an inequality $A_{\bar{\imath},\cdot}x \leq b_{\bar{\imath}}$ with $b_{\bar{\imath}} - A_{\bar{\imath},\cdot}\bar{x} \leq \min_{i \in \{1,\ldots,m\}} b_i - A_{i,\cdot}\bar{x} < 0$ ($A_{i,\cdot}$ refers to the ith row of matrix A).

In the following we assume that the polyhedron is given by a maximum violation oracle with respect to $(\mathcal{A}, b)$. A call to this oracle for a given point W^+ will be denoted by $\mathcal{O}(W^+)$. For convenience we will refer to the inequality returned by the oracle by its index in $M = \{1, \ldots, m\}$.

At any point in time the algorithm will work with a subset of the constraints of $\mathcal{A}X \leq b$. We will call this the *active index set* $J \subseteq M$ and denote the corresponding subsystem by $\mathcal{A}_J X \leq b_J$. In particular, the feasible set corresponding to J is

$$\mathcal{P}_J = \{X \in \mathcal{W} : \mathcal{A}_J X \leq b_J\} \tag{15.16}$$

and the corresponding optimization problem reads

$$(\text{PSDP}_J) \qquad \max \langle C, X \rangle \text{ s.t. } X \in \mathcal{P}_J. \tag{15.17}$$

In the analysis of the algorithm we need to investigate the convergence of the slack variables η and dual variables y; so in choosing our notation we must take care that modifications of J do not affect their dimension. Therefore we regard them as elements of $\mathbb{R}^m$ with

$y_j = \eta_j = 0$ for $j \notin J$. In particular, we will use $\mathbb{R}^m_J = \{y \in \mathbb{R}^m : y_j = 0\ \forall j \in M \setminus J\}$ and $Y_J = \{y \in \mathbb{R}^m_J : y \geq 0\}$. We define $[\cdot]_J : \mathbb{R}^m \to \mathbb{R}^m_J$ to extract the support on J, i.e., for $y \in \mathbb{R}^m$ the vector $\bar{y} = [y]_J$ is defined by $\bar{y}_j = y_j$ for $j \in J$ and $\bar{y}_j = 0$ for $j \in M \setminus J$. Modifications of J also affect all formulas and functions involving $\mathcal{A}$ and b of the previous section. We will indicate this by a superscript J:

$$f^J_{W,\eta}(y) := \langle C, W\rangle + \langle [b - \eta - \mathcal{A}W]_J, y\rangle, \qquad (W,\eta) \in \mathcal{W} \times \mathbb{R}^m_+, \tag{15.18}$$

$$f^J_{\widehat{\mathcal{W}},\widehat{Y}}(y) := \sup_{(W,\eta)\in\widehat{\mathcal{W}}\times\widehat{Y}_J} f^J_{W,\eta}(y), \qquad \widehat{\mathcal{W}} \subseteq \mathcal{W}, \widehat{Y}_J \subseteq Y_J, \tag{15.19}$$

$$f^J(y) := f^J_{\mathcal{W},Y_J}(y), \tag{15.20}$$

$$y^J_{W,\eta} := \hat{y} - \frac{1}{u}\nabla f^J_{W,\eta} = \hat{y} - \frac{1}{u}[b - \eta - \mathcal{A}W]_J \qquad (\hat{y} \in Y_J), \tag{15.21}$$

$$\psi^J(W,\eta) := \langle C, W\rangle + \big\langle [b - \eta - \mathcal{A}W]_J, \hat{y}\big\rangle - \frac{1}{2u}\|[b - \eta - \mathcal{A}W]_J\|^2, \tag{15.22}$$

$$\begin{aligned} \eta^+ &:= \operatorname{argmax}\left\{\psi^J(W^+,\eta) : \eta \in Y_J\right\} \\ &= \max\{0, [-u\hat{y} + b - \mathcal{A}W^+]_J\}, \end{aligned} \tag{15.23}$$

$$y^+ := y^J_{W^+,\eta^+} = \max\left\{\left[\hat{y} - \frac{1}{u}(b - \mathcal{A}W^+)\right]_J, 0\right\}, \qquad \left\langle \eta^+, y^+\right\rangle = 0. \tag{15.24}$$

Now consider the following modification of Algorithm 15.1.

Algorithm 15.8 (Cutting Plane Algorithm).
Input: $J^0 \subseteq M$, $y^0 \in Y_{J^0}$, $\varepsilon_{\text{opt}} \geq 0$, $\kappa_{\text{M}} \in (0,\infty]$, $\kappa \in (0,1)$, $\bar{\kappa} \in [\kappa, 1)$, a weight $u > 0$.

Step 0 (Initialization.) Set $k = 0$, $\hat{y}^0 = y^0$, $\eta^0 = 0$, $\widehat{J}^0 = J^0$, $f^{\widehat{J}^0}(\hat{y}^0)$, and $\widehat{\mathcal{W}}^0$.

Step 1 (Trial point finding.) Set $\hat{y} = \hat{y}^k$, $\widehat{\mathcal{W}} = \widehat{\mathcal{W}}^k$, $\hat{\eta} = \eta^k$, $\widehat{J} = \widehat{J}^k$.

(a) Find $W^+ \in \operatorname{Argmax}_{W\in\widehat{\mathcal{W}}} \psi^{\widehat{J}}(W,\hat{\eta})$ (see (15.22)).

(b) Call $\mathcal{O}(W^+)$. If it returns inequality $j \notin \widehat{J}$, set $J^+ = \widehat{J} \cup \{j\}$, $\widehat{J} = J^+$ and go to (a).

(c) Set $\eta^+ = \operatorname{argmax}_{\eta\geq 0} \psi^{\widehat{J}}(W^+,\eta)$ (see (15.23)) and $y^+ = y^{\widehat{J}}_{W^+,\eta^+}$ (feasible by (15.24)).

(d) (Stopping criterion.) If $f^{\widehat{J}}(\hat{y}) - f^{\widehat{J}}_{W^+,\eta^+}(y^+) \leq \varepsilon_{\text{opt}}(|f^{\widehat{J}}(\hat{y})| + 1)$, then STOP.

(e) If $f^{\widehat{J}}_{\widehat{\mathcal{W}},0}(y^+) - f^{\widehat{J}}_{W^+,\eta^+}(y^+) > \kappa_{\text{M}}[f^{\widehat{J}}(\hat{y}) - f^{\widehat{J}}_{W^+,\eta^+}(y^+)]$, then set $\hat{\eta} = \eta^+$ and go to (a).

(f) Set $y^{k+1} = y^+$, $W^{k+1} = W^+$, $\eta^{k+1} = \eta^+$, and $J^{k+1} = \widehat{J}$.

Step 2 (Descent test.) Find $W^{k+1}_S \in \mathcal{W}$ such that either

(a) $f^{J^{k+1}}(\hat{y}^k) - f^{J^{k+1}}_{W^{k+1}_S,0}(y^{k+1}) \leq \bar{\kappa}[f^{J^{k+1}}(\hat{y}^k) - f^{J^{k+1}}_{W^{k+1},\eta^{k+1}}(y^{k+1})]$ or

(b) $f^{J^{k+1}}_{W^{k+1}_S,0}(y^{k+1}) = f^{J^{k+1}}(y^{k+1})$ and
$f^{J^{k+1}}(\hat{y}^k) - f^{J^{k+1}}(y^{k+1}) \geq \kappa[f^{J^{k+1}}(\hat{y}^k) - f^{J^{k+1}}_{W^{k+1},\eta^{k+1}}(y^{k+1})]$.

In case (a), set $\hat{y}^{k+1} = \hat{y}^k$, $\widehat{J}^{k+1} = J^{k+1}$ (null step);
otherwise set $\hat{y}^{k+1} = y^{k+1}$, $\widehat{J}^{k+1} = \{j \in J^{k+1} : \eta_j = 0\}$ (descent step).

Step 3 (Model updating.) Choose a closed convex $\widehat{\mathcal{W}}^{k+1} \supset \{W^{k+1}, W_S^{k+1}\}$.

Step 4. Increase k by 1 and go to Step 1.

Remark 15.9. (i) In each execution of Step 1, the active index set $\widehat{J}$ is enlarged in substep (b) at most $|M|$ times, so there is no danger that an infinite loop is caused by Step 1(b). In addition, if execution passes on to Step 1(c), then the maximum violation oracle ensures

$$\min_{j \in M \setminus \widehat{J}} b_j - \mathcal{A}_j W^+ \geq \min_{j \in \widehat{J}} b_j - \mathcal{A}_j W^+. \tag{15.25}$$

(ii) Throughout the inner loop in Step 1 the active index set $\widehat{J}$ satisfies $\widehat{J}^k \subseteq \widehat{J} \subseteq J^{k+1}$. Since $\hat{y}^k \in Y_{\widehat{J}^k} \subseteq Y_{\widehat{J}} \subseteq Y_{J^{k+1}}$ with $\hat{y}^k_j = 0$ for all $j \in J^{k+1} \setminus \widehat{J}^k$, we obtain

$$f^{\widehat{J}^k}(\hat{y}^k) = f^{\widehat{J}}(\hat{y}) = f^{J^{k+1}}(\hat{y}) \tag{15.26}$$

and $y^+ \in Y_{\widehat{J}}$ (see (15.24) with $J = \widehat{J}$).

(iii) The reduction of J^{k+1} in the descent Step 2(b) amounts to deleting all constraints $j \in J^{k+1}$ with $\eta_j^{k+1} > 0$ (the inactive constraints with positive slack). By complementarity (15.15), the corresponding coordinates of y^{k+1} are zero; $y_j^{k+1} = 0$ for $j \in \widehat{J}^{k+1} \setminus J^{k+1}$. Therefore $\hat{y}^{k+1} = y^{k+1} \in \widehat{J}^{k+1}$, and by (15.20) and (15.18)

$$f^{\widehat{J}^{k+1}}(\hat{y}^{k+1}) = f^{J^{k+1}}(y^{k+1}) \quad \text{and} \quad f^{\widehat{J}^{k+1}}_{W^{k+1},\eta^{k+1}}(\hat{y}^{k+1}) = f^{J^{k+1}}_{W^{k+1},\eta^{k+1}}(y^{k+1}). \tag{15.27}$$

(iv) Updating and solving the model (see (15.8) and (15.22)) after increasing $\widehat{J}$ in Step 1(b) is not an issue as long as the information stored about $\overline{W}$ allows us to compute $\langle A_j, \overline{W}\rangle$ for all $j \in \widehat{J}$. If this is not possible, the information of the aggregate is lost for the model, and $\overline{W}$ or $\mathcal{A}\overline{W}$ has to be rebuilt from scratch in the following iterations. This, however, is no obstacle to convergence.

For the proof of convergence we assume $\varepsilon_{\text{opt}} = 0$ for the rest of this section. Some steps of the proof rely directly on [17].

Lemma 15.10. *If Algorithm 15.8 stops for some finite $\bar{k}$, then $\hat{y}^{\bar{k}} \in \operatorname{Argmin} f^M$ and W^+ is an optimal solution of (PSDP$_M$).*

Proof. If Algorithm 15.8 stops, then Lemma 15.5 and its proof apply for $f^{\widehat{J}}$, so W^+ is a (feasible) optimal solution of (PSDP$_{\widehat{J}}$) and

$$\langle C, W^+ \rangle = f^{\widehat{J}}(\hat{y}^{\bar{k}}) = \min_y f^{\widehat{J}}. \tag{15.28}$$

Because the last call to the oracle in Step 1(b) for this W^+ did not yield new violated inequalities, we conclude from (15.25) that W^+ is feasible for $\mathcal{P}_M$ of (15.16). By (15.28), $\langle C, W^+\rangle = \min_y f^{\widehat{J}} = \min_{y \in Y_{\widehat{J}}} f^M \geq \min_y f^M \geq \langle C, W^+\rangle$, where the last inequality follows from the feasibility of W^+ for $\mathcal{P}_M$ and the strong duality theorem for semidefinite programming. □

We may now concentrate on the case that the algorithm does not stop.

Lemma 15.11. *Suppose that at iteration $\bar{k}$ an infinite loop occurs in Step 1 of Algorithm 15.8. Then $\hat{y}^{\bar{k}} \in \operatorname{Argmin} f^M$ and all cluster points of the W^+ are optimal solutions of (PSDP$_M$).*

Proof. In an infinite loop in Step 1, the active index set $\widehat{J}$ must reach its maximal size in Step 1(c) after finitely many subiterations. From then on [17, Lemma 3.2(c)] applies for $f^{\widehat{J}}$ yielding $\hat{y}^{\bar{k}} \in \operatorname{Argmin} f^{\widehat{J}}$. By Theorem 15.6 all cluster points of W^+ converge to optimal solutions of (PSDP$_{\widehat{J}}$). Let $\bar{W}$ be such a cluster point (existence follows from the compactness of $\mathcal{W}$). Then, for arbitrary $\varepsilon > 0$, there is a W^+ satisfying $\|W^+ - \bar{W}\| < \varepsilon$ and $b_j - \mathcal{A}_j W^+ > -\varepsilon$ for all $j \in \widehat{J}$. Therefore, by (15.25), $b_j - \mathcal{A}_j W^+ > -\varepsilon$ for all $j \in M$; so $\bar{W}$ is feasible for $\mathcal{P}_M$ of (15.16). Optimality of $\bar{W}$ for (PSDP$_M$) may now be shown as in the proof of Lemma 15.10. □

Lemma 15.12. *Suppose that, after iteration $\bar{k}$, Algorithm 15.8 produces an infinite number of null steps. Then $\hat{y}^{\bar{k}} \in \operatorname{Argmin} f^M$ and all cluster points of the W^k are optimal solutions of (PSDP$_M$).*

Proof. Arguing as in the proof of Lemma 15.11, the maximal $\widehat{J}^k$ must be reached for some $\hat{k} \geq \bar{k}$. From then on [17, Lemma 3.4] applies for $f^{\widehat{J}^{\hat{k}}}$ and the proof is completed as for Lemma 15.11. □

The proof for an infinite sequence of descent steps would be equally direct if we did not allow for the deletion of inequalities. The removal of inactive inequalities, however, is indispensable in practical applications. Unfortunately, the proof of [17, Lemma 3.5] breaks down in this setting, because the linear model $f^J_{W,\eta}$ is not necessarily a minorant of f^M for a proper subset $J \subset M$. To guarantee the boundedness of the $\hat{y}^k$ our proof needs the additional assumption that f^M is 0-coercive, i.e., $f(y) \to \infty$ whenever $\|y\| \to \infty$. This is, e.g., the case if there exists a strictly feasible X for (PSDP$_M$), i.e., an $\bar{X} \succ 0$ satisfying $\bar{X} \in \mathcal{W}$ and $\mathcal{A}\bar{X} < b$.

Lemma 15.13. *If f^M is 0-coercive, then the $\hat{y}^k$ remain bounded and all cluster points are in $\operatorname{Argmin} f^M$. Furthermore, if the set $D := \{k : \hat{y}^{k+1} = y^{k+1}\}$ of descent iterations is infinite, all cluster points of the W^{k+1} for $k \in D$ are optimal solutions of (PSDP$_M$).*

Proof. By Lemma 15.12 we may concentrate on the case of an infinite number of descent iterations D. Each iteration $k \in D$ satisfies the descent step criterion of Step 2(a) and, therefore, by (15.26) and (15.27),

$$0 \leq f^{\widehat{J}^k}(\hat{y}^k) - f^{\widehat{J}^{k+1}}_{W^{k+1},\eta^{k+1}}(\hat{y}^{k+1}) \leq \frac{1}{\kappa}\left[f^{\widehat{J}^k}(\hat{y}^k) - f^{\widehat{J}^{k+1}}(\hat{y}^{k+1})\right]. \tag{15.29}$$

Using the assumption that f^M is 0-coercive (as are then all f^J with $J \subseteq M$), there exists a minimizer $\tilde{y} \in \operatorname{Argmin} f^M$ and the $\hat{y}^k$ remain bounded, because the $f^{\widehat{J}^k}(\hat{y}^k)$ are monotonically decreasing (see (15.29)). Furthermore,

$$\sum_{k\in D}\Big[f^{\widehat{J}^k}(\hat{y}^k) - f^{\widehat{J}^{k+1}}_{W^{k+1},\eta^{k+1}}(\hat{y}^{k+1})\Big] \le \frac{1}{\kappa}\sum_{k\in D}\Big[f^{\widehat{J}^k}(\hat{y}^k) - f^{\widehat{J}^{k+1}}(\hat{y}^{k+1})\Big]$$
$$\le \frac{1}{\kappa}\Big[f^{J^0}(\hat{y}^0) - f^M(\bar{y})\Big] < \infty$$

and (15.29) imply

$$f^{\widehat{J}^k}(\hat{y}^k) - f^{\widehat{J}^{k+1}}_{W^{k+1},\eta^{k+1}}(\hat{y}^{k+1}) \xrightarrow{D} 0. \tag{15.30}$$

For $k \in D$, the gradient of $f^{J^{k+1}}_{W^{k+1},\eta^{k+1}}$ may be expressed by (15.21) (use $\hat{y}^{k+1} = y^{J^{k+1}}_{W^{k+1},\eta^{k+1}}$) as

$$\nabla f^{J^{k+1}}_{W^{k+1},\eta^{k+1}} = u(\hat{y}^k - \hat{y}^{k+1}). \tag{15.31}$$

This and $f^{J^{k+1}}_{W^{k+1},\eta^{k+1}}(\hat{y}^k) \le f^{J^{k+1}}(\hat{y}^k) = f^{\widehat{J}^k}(\hat{y}^k)$ (use (15.20) and (15.26)) yields

$$f^{\widehat{J}^k}(\hat{y}^k) \ge f^{J^{k+1}}_{W^{k+1},\eta^{k+1}}(\hat{y}^{k+1}) + \left\langle \nabla f^{J^{k+1}}_{W^{k+1},\eta^{k+1}}, \hat{y}^k - \hat{y}^{k+1}\right\rangle$$
$$= f^{J^{k+1}}_{W^{k+1},\eta^{k+1}}(\hat{y}^{k+1}) + u\|\hat{y}^{k+1} - \hat{y}^k\|^2,$$

so by (15.27)

$$u\|\hat{y}^{k+1} - \hat{y}^k\|^2 \le f^{\widehat{J}^k}(\hat{y}^k) - f^{\widehat{J}^{k+1}}_{W^{k+1},\eta^{k+1}}(\hat{y}^{k+1}). \tag{15.32}$$

Combining (15.30) to (15.32) we obtain (see (15.18))

$$\nabla f^{J^{k+1}}_{W^{k+1},\eta^{k+1}} = [b - \eta^{k+1} - AW^{k+1}]_{J^{k+1}} \xrightarrow{D} 0. \tag{15.33}$$

Now let $\bar{W}$ be a cluster point of the W^{k+1} for $k \in D$ (existence follows from the compactness of $\mathcal{W}$). Then there is a subsequence $K \subseteq D$ with $W^{k+1} \xrightarrow{K} \bar{W}$ and (15.33) and (15.25) ensure the feasibility of $\bar{W}$ for $\mathcal{P}_M$ of (15.16). Furthermore, the boundedness of the $\hat{y}^k$, (15.30), and $f^{J^{k+1}}_{W^{k+1},\eta^{k+1}}(\hat{y}^{k+1}) = \langle C, W^{k+1}\rangle + \langle \nabla f^{J^{k+1}}_{W^{k+1},\eta^{k+1}}, \hat{y}^{k+1}\rangle \xrightarrow{K} \langle C, \bar{W}\rangle = \inf_k f^{\widehat{J}^k}(\hat{y}^k) \ge \inf f^M$ imply that $\bar{W}$ is an optimal solution of (PSDP$_M$) and that all cluster points $\bar{y}$ of the $\hat{y}^k$ satisfy $\bar{y} \in \operatorname{Argmin} f^M$. □

We may now state our main result.

Theorem 15.14. *Let $\mathcal{P}_M$ of* (15.16) *have a strictly feasible point. Then Algorithm 15.8 solves (PSDP$_M$).*

Proof. In the case of finitely many descent steps the proof follows from Lemmas 15.10, 15.11, and 15.12. For infinitely many descent steps, the strict feasibility of $\mathcal{P}_M$ implies the 0-coercivity of f^M and so Lemma 15.13 completes the proof. □

Remark 15.15. (i) The strictly feasible point assumption could be dropped if the $\hat{y}^k$ remain bounded whenever there is a $\tilde{y} \in Y$ with $f^{\widehat{J}^k}(\hat{y}^k) \ge f^M(\tilde{y})$ for all k (this assumption corresponds to [17, (3.16)]). We do not know whether this is true. Primal feasibility alone is not sufficient for boundedness, as can be seen from the well-known example

$$\max x_{12} \text{ subject to } \begin{bmatrix} 1 & x_{12} \\ x_{12} & 0 \end{bmatrix} \succeq 0.$$

(ii) For efficiency our implementation of Algorithm 15.8 will call the separation oracle only after descent steps; see Section 15.5. In order to extend our proof of convergence for $\varepsilon_{\text{opt}} = 0$ to this variant, some additional safeguards are required. If $\hat{y}^k$ happens to be an optimal solution of $f^{\widehat{J}^k}$ but not of f^M, the code may either stop prematurely or produce an infinite number of inner iterations or null steps without ever enlarging $\widehat{J}^k$. These two cases could easily be handled by calling the oracle before terminating and by executing Step 1(b) every thousand iterations, say. There is another difficulty of more practical relevance. By skipping Step 1(b) we lose the property (15.25) that the active index set J^{k+1} includes an inequality that is maximally violated for W^{k+1}. The current proof of Lemma 15.13 needs this property to establish primal feasibility and, based on this, convergence to optimality of the function values. Convergence to a primal feasible solution is at stake if inequalities enter and leave the active index set infinitely often while maximal violation shows insufficient decrease (we do not know whether this can actually happen). A possible safeguard would be to eliminate inactive inequalities after a descent step only if the maximum violation oracle returns an inequality with violation at most $(\frac{1}{2})^h$, where h is the number of previously executed elimination steps (in fact, any sequence of $\varepsilon_h > 0$, $h \in \mathbb{N}$, with $\varepsilon_h \to 0$, would do, as well).

15.5 Implementation

Algorithm 15.8 is convenient for theoretical investigations but has significant drawbacks in practice. The oracle has to be called for each subproblem solution, which is often computationally too expensive; the number of inequalities may grow enormously before the next descent step occurs; finally, the frequent changes in the model may slow down convergence.

Our implementation is tuned for semidefinite relaxations of quadratic ± 1-programming problems in the style of (SMC); we separate odd-cycle inequalities (15.2) exclusively. The code employs the C++ class library `SBmethod` [16], which implements Algorithm 15.1. We delete and add inequalities only after descent steps of Algorithm 15.1 (see Remark 15.15(ii) for theoretical pitfalls and safeguards). In particular, the routines for deletion and separation are called, in this sequence, whenever the first 10 descent steps have been completed and the condition $f(\hat{y}^k) - f_{W^{k+1},\eta^{k+1}}(y^{k+1}) \leq 5 \times 10^{-2}(|f(\hat{y}^k)| + 1)$ holds (here and in the following f refers to the relaxation currently in use). The code stops when the stopping criterion of Step 1(c) of Algorithm 15.1 is satisfied for the current relaxation with $\varepsilon_{\text{opt}} = 10^{-5}$ or when a given time limit is reached.

The deletion routine runs as follows. Let $\bar{m}$ denote the number of constraints before deletion. We first determine the set $D \subset \{1, \ldots, \bar{m}\}$ of cutting planes whose slack values η_j satisfy $\eta_j \geq 10^{-5}$. Then we delete the $|D| - \max\{|D|/4, \bar{m}/100\}$ constraints of D with largest slack. By this rather cautious strategy we hope to keep constraints that alternate between being active and inactive; we do not bother about deleting inactive inequalities if their number is small in comparison to $\bar{m}$.

In the separation routine we will separate with respect to W^+, which in the following refers to the W^{k+1} that gave rise to the descent step preceding separation. Although W^+ has in theory the representation $W^+ = PV^+P^T + \alpha^+\overline{W}$ (see (15.13)), `SBmethod` does not store $\overline{W}$ by default but only $\mathcal{A}\overline{W}$ and $\langle C, \overline{W}\rangle$. However, `SBmethod` supports updating $\overline{W}$ in dense form or on sparse support. Separating with respect to a given support of W^+ requires updating $\overline{W}$ on this support in each execution of Step 3 of Algorithm 15.1. For large support this may cause a significant increase in computation time and memory consumption. In particular, for large n, say $n > 1000$, updating $\overline{W}$ in dense form is computationally too expensive. Therefore we concentrate on the sparse case.

For the separation routine we assume that $X = W^+$ is a sparse matrix given by a weighted undirected graph $G = (V, E)$ with nodeset $V = \{1, \dots, n\}$, edgeset $E \subset \{ij : i < j,\ i, j \in V\}$, and edge weights x_{ij} for $ij \in E$.

The separation routine for odd-cycle inequalities (15.2) employs the exact separation routine of [5] that uses shortest path computation in a graph having twice the number of nodes and four times the number of edges. In order to speed up this computation, we make use of the well-known trick of starting the shortest path tree from both endpoints. In fact, due to symmetries in the graph, both shortest path trees are identical and only one has to be computed. We first fix a random order of the nodes and then compute the shortest path for each node in this order. If a node is already covered by a newly separated violated odd-cycle inequality, it is ignored in all subsequent shortest path computations of this call. For each starting node the shortest path computation is stopped if an odd-cycle inequality is found that is violated by at least 10^{-6} or if there is no such inequality containing this node.

The separation routine expects $-1 \le x_{ij} \le 1$. This would be guaranteed if W^+ satisfied $\operatorname{diag}(W^+) = e$ and $W^+ \succeq 0$. Unfortunately, only the latter is guaranteed by Algorithm 15.1; the deviation $\|e - \operatorname{diag}(W^+)\|$ may still be quite considerable. We enforce the box constraints by three different approaches:

$$\bar{x}_{ij} = w^+_{ij}/\max\{w^+_{ii} : i \in V\},\ \tilde{x}_{ij} = w^+_{ij}/\sqrt{w^+_{ii}w^+_{jj}},\ \hat{x}_{ij} = \max\{-1, \min\{w^+_{ij}, 1\}\}$$

for $ij \in E$, and call the separation routine for all three: $\bar{X}$, $\tilde{X}$, and $\hat{X}$. We then normalize each new inequality $\langle A, X\rangle \le b$ to $\|A\| = 1$ (A is a sparse symmetric matrix with Frobenius norm one). We add those that are not yet contained in the current description and are violated by at least 10^{-6} with respect to the original W^+.

It remains to specify the set E. `SBmethod` offers several types of coefficient matrices; one of them is `SYMMETRIC_SPARSE`. The cost matrix C is expected to be of this type. In the standard setting, E is set to contain the union of the support of the cost matrix and all other constraint matrices of type `SYMMETRIC_SPARSE`. Since newly separated inequalities have their support within E, this choice ensures that the cost of one matrix-vector multiplication within the eigenvalue computation does not increase. Furthermore, because $\overline{W}$ is available on E, the inner product $\langle A, \overline{W}\rangle$ can be computed for all new inequalities $\langle A_i, \cdot\rangle X \le b_i$. Consequently, $\mathcal{A}\widehat{W}$ is still available after the insertion of new constraints and there is no loss in the quality of the model. As starting values for the new y_i variables we choose $y_i = 0$ and, as in (15.26), the algorithm can continue without the need to recompute any function values.

We will also consider a second setting where we modify E in the course of the algorithm. If $\overline{W}$ is updated on E, then $W^+ = PV^+P^T + \alpha^+\overline{W}$ is not available in full; but, if α^+ is small, then $W^+_{ij} \approx [PV^+P^T]_{ij}$ for $ij \notin E$. These approximations may be used in the search for edges that may be worth adding to E. Employing the separation procedure on the complete graph by including all approximations is computationally too expensive. As usual, one has to resort to heuristics. After experimenting with a few, we have settled, for the time being, on the following routine, which we call directly after the first separation step and then after every 10th separation step.

Start by setting E to the union of the support of all SYMMETRIC_SPARSE cost and constraint matrices currently contained in the description. For each node $\bar{\imath} \in V$ (in increasing order) we compute a shortest path tree on $G = (V, E)$ with respect to edge weights $x_{ij} = 1 - |w^+_{ij}/\sqrt{w^+_{ii}w^+_{jj}}|$. Each edge $\bar{\imath}j \notin E$ induces a cycle C_j with respect to this shortest path tree. For each such cycle C_j we find, with respect to the weight $x_{\bar{\imath}j} = [PV^+P^T]_{\bar{\imath}j}$, the "best" odd set $F \subseteq C_j$ (see (15.2)) and add the edge $\bar{\imath}j$ to E that gives rise to the "most violated" odd cycle (even if the inequality is not violated). We also add the edge $\bar{\imath}j$ with $j \in \operatorname{argmin}\{|[PV^+P^T]_{\bar{\imath}j}| : \bar{\imath}j \notin E\}$ to E and continue with the next node. The idea is to add exactly two edges per node, one of them offering good possibilities for separation, the other providing support for difficult decisions. Edges that do not give rise to new violated inequalities in the following 10 separation steps will be removed in reinitializing E in the next call to this routine. After adding edges to E, the old $\overline{W}$ must be reinitialized and rebuilt from scratch. This leads to a slight loss in the quality of the model but does not affect the warm start otherwise.

15.6 Computational Results

In order to illustrate the numerical behavior of the cutting plane approach we present preliminary results for a few max-cut instances on toroidal grid graphs and some bisection instances of sparse KKT system matrices provided by Boeing. The experiments seem to indicate that the proposed cutting plane approach works well. Another independent issue of interest is the quality of the relaxations. To our astonishment, the results are rather discouraging for the pure max-cut examples, but appear to be quite promising for our bisection instances.

The numerical results were computed on a Linux PC with two Intel Pentium III 800 MHz processors (256 KB cache) and 1 GB of memory, but the code makes use of one processor only. CPU time refers to the user time returned by the system routine `getrusage()`; measurement is started after completion of the input. We usually needed to run more than one process on the machine, which probably caused time measurement to be unreliable. Indeed, we observed significant deviations (up to 20% and more) for identical runs. Our computation times may therefore only serve as a rough guideline.

The performance of `SBmethod` strongly depends on numerous parameters (see the manual [16]). In order to make the runs more comparable for the various settings with and without cutting planes, we have fixed these as far as possible to the same constant values for all instances. In particular, we use a constant weight $u = 1$ and parameters $\kappa = 0.1$, $\bar{\kappa} = 0.1$, $\kappa_M = 0.6$, $\varepsilon_{\text{opt}} = 10^{-5}$; we set the maximum number of columns to keep in the bundle P (see (15.8)) to $n_K = 15$, the maximum number of columns to add to P to $n_A = 5$, and a time limit of 20 hours.

The tables contain the following columns. *Problem* gives the name of the problem; n is the order of the matrix (the number of ±1-variables in the quadratic ±1-program); $\bar{m}$ gives the number of constraints at termination; *nz* displays the maximum number of elements in which $\overline{W}$ is updated during the run ($|E|$ of Section 15.5 plus diagonal elements); in *feas.* we report the best lower bound that we know of[2] or that we could generate by Goemans–Williamson rounding based on PV_+P^T combined with simple exchange heuristics; f_{term} is the value of the upper bound at termination; *time* lists the CPU time in hh:mm:ss; k is the value of the iteration counter of Algorithm 15.1 at termination; *inner* gives the number of executions of Step 1(a) of Algorithm 15.1 ($= k + 1+$ inner iterations); *desc.* displays the number of descent steps.

15.6.1 Max-cut on toroidal grid graphs

Large scale and sparse max-cut instances in pure form do not seem to appear frequently in practice. The only application we are aware of is in the computation of ground states for Ising spin glasses; see [22] and references therein. Experimentally it was observed that on instances over toroidal grid graphs the relaxation by odd-cycle inequalities yields bounds of very good quality. Linear programming approaches proved very successful on this class of problems [22, 27]. We present results for four instances of three-dimensional toroidal grid graphs. Instances *toruspm3-8-50* and *toruspm3-15-50* have edge weights chosen uniformly from ± 1, whereas the edge weights for *torusg3-8* and *torusg3-15* are taken according to the standard Gaussian distribution. They are part of the 7th DIMACS challenge testset and turned out to be rather difficult to solve by the linear programming techniques.

Let A be the weighted adjacency matrix of the respective graph. Then we set $C = \frac{1}{4}(\frac{e^T Ae}{n} I - A)$ and use the elliptope [9] as initial semidefinite relaxation:

$$\max \langle C, X \rangle \text{ subject to } \operatorname{diag}(X) = e, \ X \succeq 0. \tag{15.34}$$

In Table 15.1 we first give the results for (15.34) (computed by the same code with identical parameter settings but without cutting planes), then for separating odd-cycle inequalities with respect to the support of C as described in Section 15.5, and, finally, the results when including the heuristic for enlarging the support on every 10th call to the separation procedure (see Section 15.5).

The improvement of the bound when including separation is considerable, but computation time reaches the limit of 20 hours in all cases. Listing the final values provides little insight into the development of the bound over time; therefore we also present plots in Figure 15.1 that show this development with respect to a logarithmic time scale. It turns out that the cutting plane approaches improve the bound considerably even before the bound on the elliptope converges. This shows that the warm start technique is very efficient. Enlarging the support does not seem to help at all for these instances, yet performance does not deteriorate much when employing it.

Even though the cutting plane approach is very successful in improving the basic semidefinite relaxation of these instances, the results are surprisingly poor in comparison to the linear programming bound that is based on separating only odd-cycle inequalities. Indeed, the values obtained by optimizing over the odd-cycle polytope alone are 464.7035 for *toruspm3-8-50*, < 3063.757 for *toruspm3-15-50*, 417.6645 for *torusg3-8*, and < 2873.773 for *torusg3-15*. What is more, the linear programming approach needs considerably less

[2]The values for *toruspm3-8-50*, *torusg3-8*, and *torusg3-15* were reported by Frauke Liers, personal communication, and 416.84814 was reported optimal for *torusg3-8*.

Table 15.1. *Max-cut on toroidal grid graphs.*

Problem	n	$\bar{m}$	nz	feas.	f_{term}	time	k	inner	desc.
elliptope without cuts									
toruspm3-8-50	512	512	0	456	527.8127	14	139	140	30
toruspm3-15-50	3375	3375	0	2944	3475.151	31:38	855	856	38
torusg3-8	512	512	0	416.84814	457.3618	19	201	202	39
torusg3-15	3375	3375	0	2841.96	3134.591	38:55	992	993	46
elliptope with odd cycles on support									
toruspm3-8-50	512	2790	2048	456	464.7413	20:00:01	73645	73675	67
toruspm3-15-50	3375	13822	13500	2944	3071.477	20:00:34	15932	15933	47
torusg3-8	512	2877	2048	416.84814	417.6862	20:00:01	50839	50870	64
torusg3-15	3375	13218	13500	2841.96	2879.676	20:00:36	15547	15549	46
elliptope with odd cycles on extended support									
toruspm3-8-50	512	3342	3910	456	464.8202	20:00:01	50303	51754	61
toruspm3-15-50	3375	15205	22323	2944	3073.534	20:01:00	16594	16866	44
torusg3-8	512	4113	4593	416.84814	417.7017	20:00:01	46787	48894	71
torusg3-15	3375	14496	22559	2841.96	2882.073	20:00:47	15057	15279	46

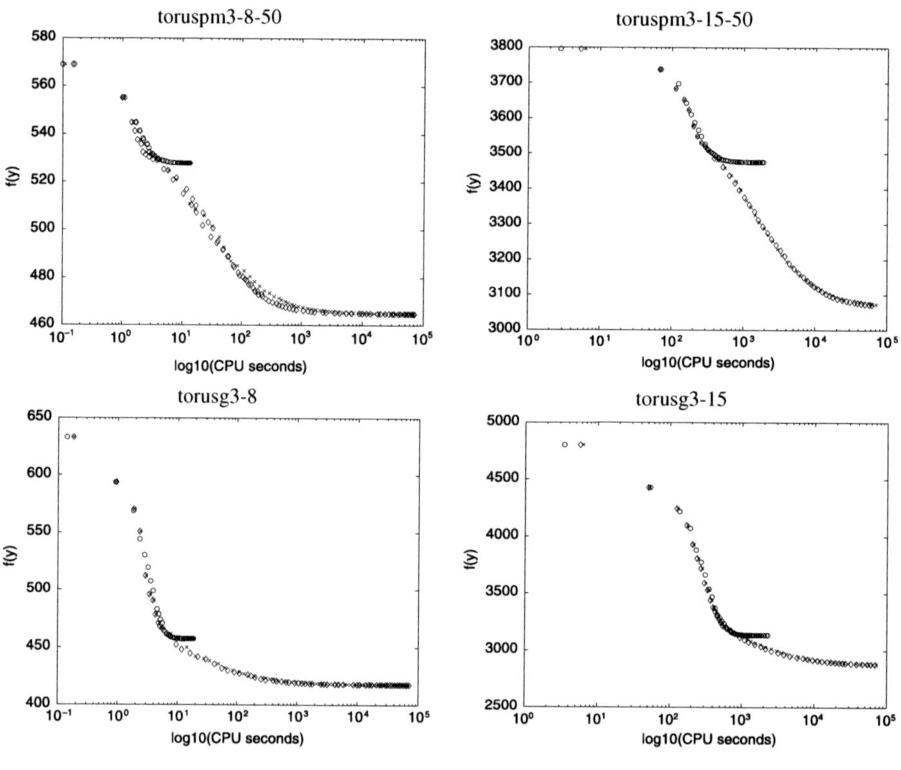

Figure 15.1. *Max-cut on toroidal grid graphs. The horizontal axis gives CPU time in seconds in logarithmic scale, the vertical axis displays the upper bound;* $\circ$ *refers to the elliptope,* $\diamond$ *to separation on the support,* $\times$ *to separation on extended support.*

computation time. Theoretically, the bounds obtained by optimizing over the odd-cycle polytope cannot be better than the bounds obtained by intersecting the odd-cycle polytope with the elliptope. Therefore we have to blame this effect on the poor final convergence rate of the spectral bundle method. The experiments suggest that for toroidal grid graphs the improvement obtained by intersecting with the elliptope is only marginal (for smaller instances we have observed at least some improvement).

15.6.2 Graph bisection

The bisection instances were communicated to us by Sharon Filipowski from Boeing and stem from nested dissection approaches for solving sparse symmetric linear systems; standard bisection heuristics seem not to work well on these instances, but no bounding method was available to judge the quality of the solutions produced. Each instance consists of a graph $G = (V, E)$ that represents the support structure of a sparse symmetric linear system. The task is to partition V into two sets $(S, V \setminus S)$ that differ in cardinality by at most $0.05 \cdot n$ so that the number of edges that have one endpoint in S and the other in $V \setminus S$ is minimized.

Let A denote the 0/1-adjacency matrix of G. Then with $C = \frac{1}{4}(A - \frac{e^T Ae}{n} I)$ (note the change in sign to obtain a maximization problem) a canonical semidefinite relaxation [29, 19, 23, 6] reads

$$\max \ \langle C, X \rangle \ \text{subject to} \ \operatorname{diag}(X) = e, \ \left\langle ee^T, X \right\rangle \leq \lfloor 0.05 \cdot n \rfloor^2, \ X \succeq 0. \qquad (15.35)$$

In `SBmethod` the constraint matrix $A_{n+1} = ee^T$ can be represented in this structured form using the constraint class `GRAM_DENSE`. Therefore there is no need to work with a dense dual matrix, and the matrix-vector multiplication is still efficient. On these instances the aggregation mechanism often reduced the bundle size too much so that we forced the minimum number of columns to be kept in P to 10 by setting $n_{\min} = 10$ (see the manual [16]).

Table 15.2 lists our numerical results for the basic semidefinite relaxation, for the semidefinite relaxation combined with odd-cycle cutting planes on the support of C, and for separating odd-cycle inequalities on the dynamically enlarged support (see Section 15.5). Plots of the progress of the algorithm with respect to a logarithmic time scale are displayed in Figures 15.2 and 15.3. In order to obtain a better resolution of the relevant part, we only show descent steps with function value below zero.

Quite contrary to the max-cut instances on toroidal grid graphs, the separation of odd-cycle inequalities on the support of C rarely yields significant improvement for our bisection instances, whereas enlarging the support seems to help a lot in most cases. We add some comments on the single instances. Using enlarged support allowed us to prove optimality of the feasible ± 1-solutions of *lowt01* and *putt01*. Extending the support also led to significant improvements of the bounds for *skwz02*, *orbe11_hc100*, and *heat02*. For *capt09* and, in particular, *plnt01* the gap between the best feasible solution known and the upper bound is still large. For the three large examples *tra27*, *lnts02*, and *traj33* the time limit was too short to reach definitive conclusions on the improvement; even the basic relaxation could not be solved to sufficient precision within this time span. Furthermore, for these three instances the computation time spent in the exact separation algorithm significantly exceeds the time spent in the spectral bundle code (e.g., for *traj33* on extended support, 17 hours are due

Table 15.2. *Bisection instances.*

Problem	n	$\bar{m}$	nz	feas.	f_{term}	time	k	inner	desc.
				relaxation (15.35) without cuts					
lowt01	82	83	0	−13	−4.541642	1	14	15	13
putt01	115	116	0	−28	−18.94562	1	17	18	15
capt09	2063	2064	0	−6	−0.6457718	34:37	360	361	199
skwz02	2117	2118	0	−567	−493.8798	1:50	73	74	37
orbe11_hc100	2186	2187	0	−2087	−1839.536	4:51	76	77	41
plnt01	2817	2818	0	−74	−4.334505	38:16	472	622	446
heat02	5150	5151	0	−150	−9.940345	16:43	379	388	63
traj27	17148	17149	0	−8174	−8140.877	20:01:13	1213	1214	96
lnts02	17990	17991	0	−6589	−6063.601	20:01:35	2181	2182	71
traj33	20006	20007	0	−9593	−9496.117	20:00:24	719	719	94
				relaxation (15.35) with odd cycles on support					
lowt01	82	179	342	−13	−4.542344	3	25	46	24
putt01	115	521	548	−28	−21.40906	16	106	188	84
capt09	2063	15613	12999	−6	−0.6586966	13:03:43	3999	11689	1369
skwz02	2117	46188	16118	−567	−506.2629	15:13:26	9430	9502	595
orbe11_hc100	2186	15860	40057	−2087	−1840.485	35:21	182	311	72
plnt01	2817	43907	27816	−74	−4.541399	20:00:56	3067	5765	793
heat02	5150	16796	25056	−150	−9.940555	1:22:02	552	990	64
traj27	17148	48005	129781	−8174	−8118.609	20:24:18	1190	1191	43
lnts02	17990	18900	63873	−6589	−6029.048	20:36:48	1384	1394	50
traj33	20006	64164	261953	−9593	−9460.441	20:40:23	529	530	40
				relaxation (15.35) with odd cycles on extended support					
lowt01	82	1230	1060	−13	−12.96362	25:17	6588	10505	101
putt01	115	1510	1482	−28	−27.99941	6:58	1887	3177	91
capt09	2063	14059	19527	−6	−4.142882	20:00:15	15332	23194	222
skwz02	2117	20347	28714	−567	−558.1742	20:00:31	19703	19718	191
orbe11_hc100	2186	20107	50048	−2087	−2033.389	20:00:55	10726	10753	168
plnt01	2817	30619	35477	−74	−8.023619	20:01:22	5011	25567	335
heat02	5150	18009	43739	−150	−144.1614	20:01:57	7964	8007	167
traj27	17148	26396	170827	−8174	−8105.315	20:33:10	1116	1116	32
lnts02	17990	21985	104148	−6589	−5961.739	20:55:52	831	833	36
traj33	20006	30269	306952	−9593	−9454.076	21:31:32	364	365	35

to separation!); the decrease in number of iterations and descent steps for the separation versions is mainly due to this effect and not to the increased work by updating a larger aggregate matrix $\overline{W}$. It should be noted that the gap between the feasible solution and the bound of the basic relaxation is relatively small for *traj27*, *lnts02*, and *traj33*, so there is not much room for improvement.

The semidefinite relaxation for balanced bisection does not lend itself to comparison with a pure linear relaxation. Although there is a wealth of heuristics for graph bisection (see [30] and references therein), we are not aware of any recent computational studies on bounds for this particular problem. In [11] a polyhedral approach is discussed for the more general node capacitated graph partitioning problem; computational results also include equipartition problems for graphs with up to 300 nodes and 500 edges. The original version of the code of [11] could not be run any longer, so we updated the code to work with CPLEX 7.1 [21] and tested it on the two small examples *lowt01* and *putt01* (the separation routines were obviously not designed for larger instances and far too slow for even *capt09*). We only computed the root node with all separation procedures switched on. The linear

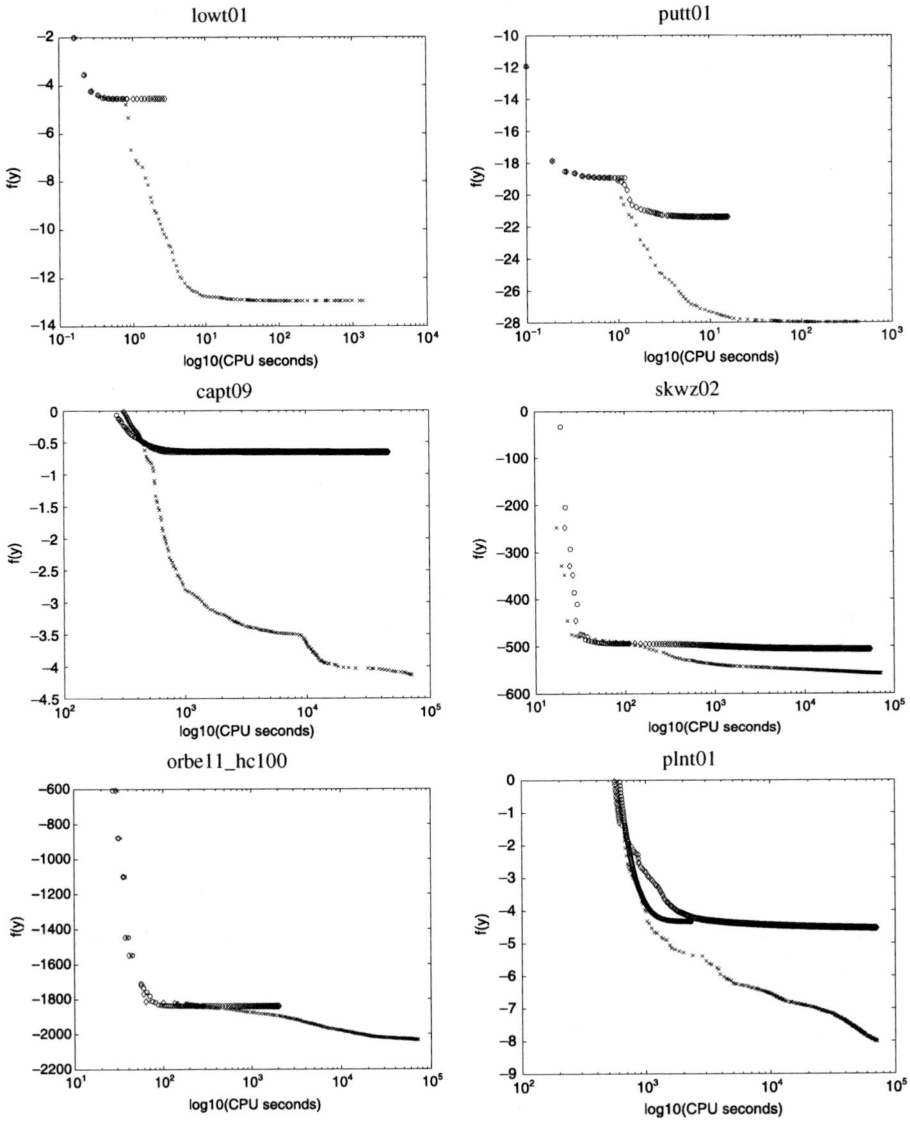

Figure 15.2. *Bisection instances. The horizontal axis gives CPU time in seconds in logarithmic scale, the vertical axis displays the upper bound; ∘ refers to relaxation* (15.35), *◇ to separation on the support, × to separation on extended support.*

programming approach needed about twice the time to arrive at the same bound; about 95% of the total time was spent in the linear programming solver CPLEX 7.1. It is quite likely that the performance of the linear programming approach can be improved by fine tuning it with respect to this class of instances. Yet we believe that this is sufficient evidence that the semidefinite cutting plane approach is competitive for these bisection problems.

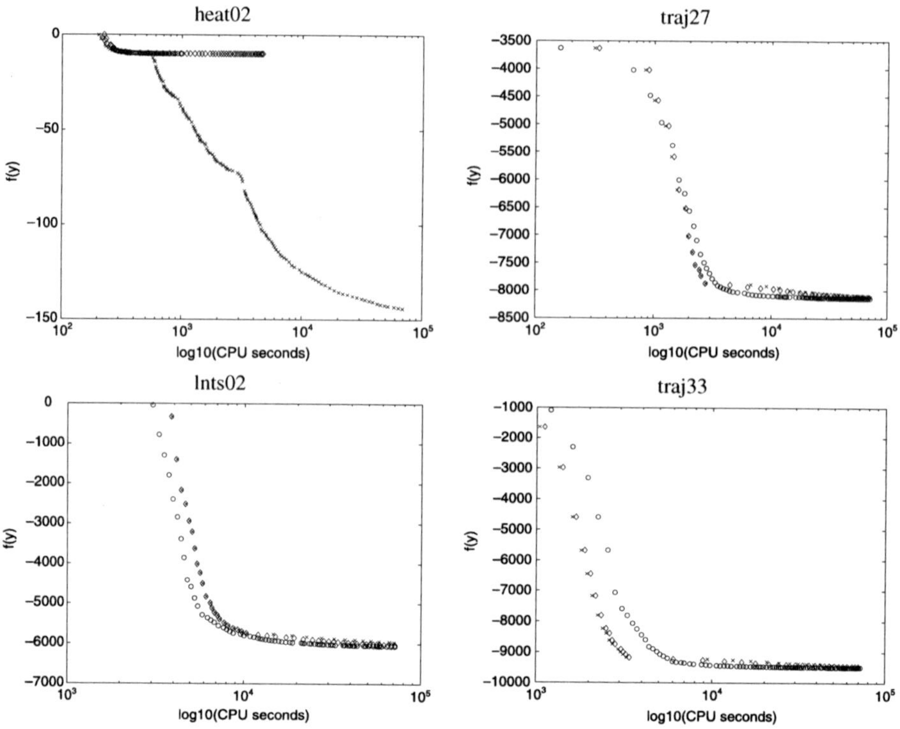

Figure 15.3. *Bisection instances. The horizontal axis gives CPU time in seconds in logarithmic scale, the vertical axis displays the upper bound;* $\circ$ *refers to relaxation* (15.35), $\diamond$ *to separation on the support,* $\times$ *to separation on extended support.*

Acknowledgements. I thank Mike Jünger and Frauke Liers for extensive discussions and comparative runs, and for making available their separation routines. I thank Sharon Filipowski for providing the bisection instances and Kurt Anstreicher for his careful reading and thoughtful comments that helped to improve the presentation.

Bibliography

[1] E. Balas, S. Ceria, and G. Cornuéjols. A lift-and-project cutting plane algorithm for mixed 0/1 programs. *Mathematical Programming*, 58:295–324, 1993.

[2] F. Barahona. The max-cut problem in graphs not contractible to K_5. *Operations Research Letters*, 2:107–111, 1983.

[3] F. Barahona. On cuts and matchings in planar graphs. *Mathematical Programming*, 60:53–68, 1993.

[4] F. Barahona and R. Anbil. The volume algorithm: Producing primal solutions with a subgradient method. *Mathematical Programming*, 87 A(3):385–399, 2000.

[5] F. Barahona and A.R. Mahjoub. On the cut polytope. *Mathematical Programming*, 36:157–173, 1986.

[6] S. Benson, Y. Ye, and X. Zhang. Mixed linear and semidefinite programming for combinatorial and quadratic optimization. *Optimization Methods and Software*, 11&12:515–544, 1999.

[7] H. Crowder, E. Johnson, and M. Padberg. Solving large-scale zero-one linear programming problems. *Operations Research*, 31:803–834, 1983.

[8] C. De Simone. The cut polytope and the Boolean quadric polytope. *Discrete Mathematics*, 79:71–75, 1989.

[9] M. Deza and M. Laurent. *Geometry of Cuts and Metrics*, volume 15 of *Algorithms and Combinatorics*. Springer, New York, 1997.

[10] S. Feltenmark and K.C. Kiwiel. Dual applications of proximal bundle methods, including Lagrangian relaxation of nonconvex problems. *SIAM Journal on Optimization*, 10:697–721, 2000.

[11] C.E. Ferreira, A. Martin, C.C. de Souza, R. Weismantel, and L.A. Wolsey. The node capacitated graph partitioning problem: A computational study. *Mathematical Programming*, 81:229–256, 1998.

[12] M.X. Goemans and D.P. Williamson. Improved approximation algorithms for maximum cut and satisfiability problems using semidefinite programming. *Journal of the Association for Computing Mechanics*, 42:1115–1145, 1995.

[13] G.H. Golub and C.F. van Loan. *Matrix Computations*, second edition. The Johns Hopkins University Press, Baltimore, MD, 1989.

[14] C. Helmberg. Fixing variables in semidefinite relaxations. *SIAM Journal on Matrix Analysis and Applications*, 21:952–969, 2000.

[15] C. Helmberg. *Semidefinite Programming for Combinatorial Optimization*. Habilitationsschrift Technische Universität Berlin, January 2000; ZIB-Report ZR 00-34, Konrad-Zuse-Zentrum für Informationstechnik Berlin, 14195 Berlin, Germany, October 2000.

[16] C. Helmberg. `SBmethod`—*A C++ Implementation of the Spectral Bundle Method. Manual to Version 1.1*, ZIB-Report ZR 00-35, Konrad-Zuse-Zentrum für Informationstechnik Berlin, Takustraße 7, 14195 Berlin, Germany, October 2000. http://www.tu-chemnitz.de/~helmberg/SBmethod.

[17] C. Helmberg and K.C. Kiwiel. A spectral Bundle Method with Bounds. *Mathematical Programming*, 93:173–194, 2002.

[18] C. Helmberg and F. Rendl. A spectral bundle method for semidefinite programming. *SIAM Journal on Optimization*, 10:673–696, 2000.

[19] C. Helmberg, F. Rendl, R.J. Vanderbei, and H. Wolkowicz. An interior-point method for semidefinite programming. *SIAM Journal on Optimization*, 6:342–361, 1996.

[20] J.-B. Hiriart-Urruty and C. Lemaréchal. *Convex Analysis and Minimization Algorithms II*, volume 306 of *Grundlehren der mathematischen Wissenschaften*. Springer, Berlin, Heidelberg, 1993.

[21] ILOG S.A., Gentilly, France. *ILOG CPLEX 7.1, User's Manual*, March 2001. Information available at http://www.ilog.com/products/CPLEX.

[22] M. Jünger and G. Rinaldi. Relaxations of the max cut problem and computation of spin glass ground states. In P. Kischka, editor, *Proceedings of Operations Research Proceedings 1997 (Jena)*, Springer, Berlin, pages 74–83, 1998.

[23] S.E. Karisch, F. Rendl, and J. Clausen. Solving graph bisection problems with semidefinite programming. *INFORMS Journal on Computation*, 12:177–191, 2000.

[24] M. Laurent, S. Poljak, and F. Rendl. Connections between semidefinite relaxations of the max-cut and stable set problems. *Mathematical Programming*, 77:225–246, 1997.

[25] A.S. Lewis and M.L. Overton. Eigenvalue optimization. *Acta Numerica*, 5:149–190, 1996.

[26] L. Lovász and A. Schrijver. Cones of matrices and set-functions and 0–1 optimization. *SIAM Journal on Optimization*, 1:166–190, 1991.

[27] J.E. Mitchell. Computational experience with an interior point cutting plane algorithm. *SIAM Journal on Optimization*, 10:1212–1227, 2000.

[28] M. Padberg. The Boolean quadric polytope: Some characteristics, facets and relatives. *Mathematical Programming*, 45:139–172, 1989.

[29] S. Poljak and F. Rendl. Nonpolyhedral relaxations of graph-bisection problems. *SIAM Journal on Optimization*, 5:467–487, 1995.

[30] R. Preis. *Analyses and Design of Efficient Graph Partitioning Methods*. Ph.D. thesis, Fachbereich Mathematik/Informatik, Universität Paderborn, Germany, July 2000.

[31] R.T. Rockafellar. *Convex Analysis*. Princeton University Press, Princeton, NJ, 1970.

[32] P.D. Seymour. Matroids and multicommodity flows. *European Journal of Combinatorics*, 2:257–290, 1981.

[33] H.D. Sherali and W.P. Adams. A hierarchy of relaxations between the continuous and convex hull representations for zero-one programming problems. *SIAM Journal on Discrete Mathematics*, 3:411–430, 1990.

[34] N.Z. Shor. Quadratic optimization problems. *Soviet Journal of Computer and Systems Sciences*, 25:1–11, 1987. Originally published in *Tekhnicheskaya Kibernetika*, No. 1, pages 128–139, 1987.

Chapter 16

Semidefinite Relaxations for Max-Cut

*Monique Laurent**

Abstract. We compare several semidefinite relaxations for the cut polytope obtained by applying the lift and project methods of Lovász and Schrijver and of Lasserre. We show that the tightest relaxation is obtained when applying the Lasserre construction to the node formulation of the max-cut problem. This relaxation $Q_t(G)$ can be defined as the projection on the edge subspace of the set $\mathcal{F}_t(n)$, which consists of the matrices indexed by all subsets of $\{1, \ldots, n\}$ of cardinality $\leq t+1$ with the same parity as $t+1$ and having the property that their (I, J)th entry depends only on the symmetric difference of the sets I and J. The set $\mathcal{F}_0(n)$ is the basic semidefinite relaxation of max-cut consisting of the semidefinite matrices of order n with an all ones diagonal, while $\mathcal{F}_{n-2}(n)$ is the (2^{n-1})-dimensional simplex with the cut matrices as vertices. We show the following geometric properties. Let $Y \in \mathcal{F}_t(n)$ and let X be its principal submatrix indexed by the first n rows and columns; if rank $X \leq t+1$, then Y can be written as a convex combination of at most 2^t cut matrices; this extends a result of Anjos and Wolkowicz for the case $t = 1$. Any 2^{t+1} cut matrices form a face of $\mathcal{F}_t(n)$ for $t = 0, 1, n-2$. The class $\mathcal{L}_t$ of the graphs G for which $Q_t(G)$ is equal to the cut polytope of G is shown to be closed under taking minors. The graph K_7 is a forbidden minor for membership in $\mathcal{L}_2$, while K_3 and K_5 are the only minimal forbidden minors for the classes $\mathcal{L}_0$ and $\mathcal{L}_1$, respectively.

MSC 2000. 05C50, 15A57, 52B12, 90C22, 90C27

Key words. 0/1-polytope, semidefinite relaxation, max-cut, moment matrix

*CWI, Kruislaan 413, 1098 SJ Amsterdam, Netherlands (monique@cwi.nl).

16.1 Introduction

16.1.1 Preamble

A basic problem in integer programming is to find the linear description of the convex hull P of the 0/1-valued points lying in a polytope $K \subseteq \mathbb{R}^d$ defined by an explicitly given linear system $Ax \leq b$, or at least to find linear relaxations of P that are good and efficient, meaning that they approximate P well and that one can optimize a linear objective function over them in polynomial time. If all the vertices of K are 0/1-valued, then $P = K$ and we are done. Otherwise we have to find "cutting planes" permitting us to strengthen the relaxation K by cutting off its fractional vertices. Extensive research has been done to find (partial) linear relaxations for many 0/1-polytopes arising from specific combinatorial optimization problems by exploiting the combinatorial structure of the problem at hand. Research has also focused on developing general purpose methods applying to arbitrary 0/1 (or integer) problems.

One of the first such methods (which applies to general integer polyhedra) is the method of Gomory, which constructs cutting planes from the linear system $Ax \leq b$ defining K using integer rounding. Namely, it constructs the *Chvátal closure*

$$K' := \{x \in \mathbb{R}^d \mid u^T Ax \leq \lfloor u^T b \rfloor \text{ for all } u \geq 0 \text{ such that } u^T A \text{ is integer}\},$$

which satisfies $P \subseteq K' \subseteq K$. Define iteratively $K^{(1)} := K'$ and $K^{(t+1)} := (K^{(t)})'$ for $t \geq 1$. Then, $K^{(t)} = P$ for some integer t [7]; the smallest such t is the *Chvátal rank* of K. Although the Chvátal rank can be very large in general (as it depends on the dimension d and on the coefficients of A), it is bounded by $O(d^2 \log d)$ when K is contained in the cube $[0, 1]^d$ [13]. From an algorithmic point of view, the first Chvátal closure does not yield an efficient relaxation in general, since optimizing a linear objective function over K' is a co-NP-hard problem [12].

Another idea has been investigated for constructing cutting planes in an implicit way, which consists of trying to represent P as the projection of another polytope Q lying in a higher dimensional space. The rationale behind this is that the projection of a polytope Q may have a much more complicated facial structure than Q itself. As a matter of fact, any d-dimensional 0/1-polytope can be realized as the projection of a $(2^d - 1)$-dimensional simplex! Several general purpose methods have been proposed for constructing projection representations for 0/1-polytopes; in particular, Balas, Ceria, and Cornuéjols [3], Sherali and Adams [30], Lovász and Schrijver [28], and, recently, Lasserre [19, 20]. A common feature of these methods is the construction of a hierarchy

$$K \supseteq K_1 \supseteq \ldots \supseteq K_d \supseteq P$$

of relaxations of P obtained as projections of higher dimensional polyhedra that finds P in d steps, that is, $K_d = P$. An important algorithmic property is that, if one can optimize a linear objective function over K in polynomial time, then the same holds for K_t for any *fixed* t (assuming that the number of rows of A is part of the input data in the case of Lasserre). The relaxations are linear or, in the case of Lovász–Schrijver and of Lasserre, semidefinite.

This idea of using semidefinite relaxations for a combinatorial 0/1-problem goes back to the seminal work of Lovász [27], who introduced the theta function $\vartheta(G)$ as bound for the stability number of a graph G obtained by optimizing over the semidefinite relaxation

TH(G) of the stable set polytope. This idea was again used successfully by Goemans and Williamson [16], who proved the first nontrivial approximation algorithm for max-cut using a semidefinite relaxation of the cut polytope. Since then, semidefinite relaxations have been widely used for approximating combinatorial problems (see, e.g., [26] for a survey).

A comparison of the various lift-and-project methods can be found in [23]. In particular, if we denote the tth iterate in the Lovász–Schrijver hierarchy by $N_+^t(K)$ and the tth iterate in the Lasserre hierarchy by $Q_t(K)$, it is shown there that

$$Q_t(K) \subseteq N_+^t(K). \tag{16.1}$$

In this chapter we study how the Lovász–Schrijver and Lasserre procedures apply to the max-cut problem. There are, in fact, two possible ways in which they can be applied, either to the edge formulation of the problem or to its node formulation. We examine the relationships existing between the various semidefinite relaxations obtained for the cut polytope. It turns out that the best relaxation is obtained when using the Lasserre construction applied to the node model. Its definition involves an interesting set of matrices (*moment matrices*) having nice geometric properties, in particular, about adjacencies of cuts and matrices of small rank versus exact resolution of max-cut.

16.1.2 The max-cut problem

Let $G = (V, E)$ be a graph with nodeset $V = \{1, \ldots, n\}$; its edgeset E is viewed as a set of unordered pairs of distinct elements of V. Given a subset $S \subseteq V$, the *cut* $\delta(S)$ defined by S is the set of edges $ij \in E$ with $|S \cap \{i, j\}| = 1$. Given edge weights $w \in \mathbb{Q}^E$, the *max-cut problem* consists of finding a cut $\delta(S)$ whose weight $\sum_{ij \in \delta(S)} w_{ij}$ is maximum. It can be formulated as the problem

$$\max \frac{1}{2} \sum_{ij \in E} w_{ij}(1 - x_i x_j) \text{ subject to } x \in \{\pm 1\}^n. \tag{16.2}$$

For a cut $\delta(S)$, we use the same symbol $\delta(S)$ for denoting its ± 1-incidence vector in $\mathbb{R}^E$ with ijth entry -1 if $ij \in \delta(S)$ and 1 otherwise. The *cut polytope* CUT(G) is the polytope in $\mathbb{R}^E$ defined as the convex hull of all cuts $\delta(S)$ ($S \subseteq V$). Thus the max-cut problem can be expressed as a linear programming problem over the polytope CUT(G):

$$\max \frac{1}{2} w(E) - \frac{1}{2} w^T z \text{ subject to } z \in \mathrm{CUT}(G). \tag{16.3}$$

In view of (16.2) and (16.3), there are two possible ways in which the lift-and-project methods can be applied to the max-cut problem.

The edge model. A first possibility is to work in the edge space and to apply the constructions to a linear relaxation K of the cut polytope CUT(G). As linear relaxation one can choose the *metric polytope* MET(G), which is the polytope in $\mathbb{R}^E$ defined by the bound constraints $-1 \leq x_{ij} \leq 1$ ($ij \in E$) and the *cycle inequalities*

$$\sum_{ij \in F} x_{ij} - \sum_{ij \in E(C) \setminus F} x_{ij} \geq 2 - |C| \tag{16.4}$$

(for C a cycle in G and $F \subseteq E(C)$ with an odd cardinality). Applying the Lovász–Schrijver and Lasserre constructions to the pair $K = \text{MET}(G)$, $P = \text{CUT}(G)$, one obtains, respectively, the semidefinite relaxations $N_+^t(\text{MET}(G))$ and $Q_t(\text{MET}(G))$ satisfying

$$\text{CUT}(G) \subseteq Q_t(\text{MET}(G)) \subseteq N_+^t(\text{MET}(G))$$

(recall (16.1)). We will consider mainly the relaxation $N_+^t(\text{MET}(G))$ obtained using the Lovász–Schrijver N_+ operator, since the definition of the relaxation $Q_t(\text{MET}(G))$ involves a large number of semidefinite constraints; precise definitions are given in Section 16.2.

Let $E_n := \{ij \mid 1 \leq i < j \leq n\}$ denote the edgeset of the complete graph K_n and let π_E denote the projection from $\mathbb{R}^{E_n}$ onto $\mathbb{R}^E$. Obviously, $\text{CUT}(G) = \pi_E(\text{CUT}(K_n))$, and Barahona [4] shows that $\text{MET}(G) = \pi_E(\text{MET}(K_n))$. In the linear description of $\text{MET}(G)$, it suffices to consider the cycle inequalities (16.4) for chordless circuits [6]; therefore, $\text{MET}(K_n)$ is defined by the $4\binom{n}{3}$ *triangle inequalities*

$$x_{ij} + x_{ik} + x_{jk} \geq -1, \; x_{ij} - x_{ik} - x_{jk} \geq -1 \tag{16.5}$$

for all distinct $i, j, k \in V$. As a consequence, one can alternatively obtain a semidefinite relaxation of $\text{CUT}(G)$ by first applying the N_+ operator to $\text{MET}(K_n)$ and then projecting onto $\mathbb{R}^E$, namely, define,

$$N_+^t(G) := \pi_E(N_+^t(\text{MET}(K_n))). \tag{16.6}$$

It can be verified (see [22]) that

$$N_+^t(G) \subseteq N_+^t(\text{MET}(G)); \tag{16.7}$$

it is not known whether equality holds, i.e., whether the two operators N_+ and π_E commute.

The node model. A second possibility is to apply the lift-and-project constructions to the set $K = [-1, 1]^n$ (lying in the node space) and to take projections onto the edge space $\mathbb{R}^E$ (instead of projections onto the node space $\mathbb{R}^n$). When applying the Lasserre construction in this framework, we obtain the semidefinite relaxation $Q_t(G)$ of $\text{CUT}(G)$, which is defined as the projection on $\mathbb{R}^E$ of the set of vectors $y = (y_I)_{\substack{I \subseteq V \\ |I| \leq 2t+2}}$ for which $y_\emptyset = 1$ and the matrix

$$M_{t+1}(y) := (y_{I \Delta J})_{\substack{I,J \subseteq V \\ |I|,|J| \leq t+1}} \tag{16.8}$$

is positive semidefinite (see Subsection 16.2.3). The first member $Q_0(K_n)$ in this hierarchy corresponds to the basic semidefinite relaxation of $\text{CUT}(K_n)$ considered in [16], while the second member $Q_1(K_n)$ tightens the semidefinite relaxation F_n introduced by Anjos and Wolkowicz [1].

16.1.3 Contents of the chapter

The chapter is organized as follows. In Section 16.2 we present the Lovász–Schrijver and Lasserre constructions and indicate how they apply to the max-cut problem. Our main result there is that $Q_t(G)$ (the relaxation obtained by applying the Lasserre construction to the node model) is contained in $N_+^{t-1}(G)$ (the relaxation obtained using the Lovász–Schrijver procedure in the edge model). In Section 16.3 we consider the class $\mathcal{L}_t$ consisting of the graphs G for which $\mathrm{CUT}(G) = Q_t(G)$. We show that $\mathcal{L}_t$ is closed under taking minors and that a graph G belongs to $\mathcal{L}_t$ if it contains an edge whose contraction produces a graph in $\mathcal{L}_{t-1}$. Section 16.5 contains a numerical comparison of the relaxations $Q_t(K_n)$ for small $n \leq 7$ and $t \leq 2$.

Section 16.4 is devoted to the study of some geometric properties of the matrix set $\mathcal{F}_t(n)$ underlying the relaxation $Q_t(K_n)$, which consists of the matrices of the form (16.8) (or rather of their principal submatrices indexed by the sets having the same parity as $t+1$; cf. (16.22)). Thus $\mathcal{F}_0(n)$ is the basic semidefinite programming relaxation of $\mathrm{CUT}(K_n)$ consisting of the semidefinite matrices of order n with an all-ones diagonal, while $\mathcal{F}_{n-2}(n)$ is the 2^{n-1}-dimensional simplex with the cut matrices as vertices. We study adjacency properties of cuts on $\mathcal{F}_t(n)$. Padberg [29] showed that any two cuts form a face of the metric polytope and Laurent and Poljak [24] showed the analogous result for $\mathcal{F}_0(n)$. We address here the question of whether, more generally, any 2^{t+1} cuts form a face of $\mathcal{F}_t(n)$; we show that this property holds for $t = 1$ and $n - 2$.

The matrix set $\mathcal{F}_t(n)$ permits us to formulate the following upper bound for the max-cut problem:

$$\max \langle Q_w, Y \rangle \text{ subject to } Y \in \mathcal{F}_t(n)$$

(defining Q_w suitably). An interesting question is to find some conditions on the rank of an optimal solution Y, ensuring that the above program solves the max-cut problem exactly. For $Y \in \mathcal{F}_t(n)$ let X be its principal submatrix indexed by the first n rows and columns. We show the following result: If rank $X \leq t+1$, then Y can be written as a convex combination of 2^t cut matrices (and thus the above relaxed program solves the max-cut problem exactly). The case $t = 0$ is obvious and the case $t = 1$ has been settled earlier by Anjos and Wolkowicz [2].

Notation. The notation $X \succeq 0$ means that X is a (symmetric) positive semidefinite matrix and PSD_n denotes the set of positive semidefinite matrices of order n. For a matrix X, $\ker X := \{u \mid Xu = 0\}$. We let $e_1, \ldots, e_n$ denote the standard unit vectors in $\mathbb{R}^n$. The following (easy to verify) properties of positive semidefinite matrices will be frequently used throughout the chapter.

Lemma 16.1. *Let X be a positive semidefinite matrix of order n.*

(i) *Write X as $X = \begin{pmatrix} B & C \\ C^T & D \end{pmatrix}$. Then $u \in \ker B$ if and only if $(u\ 0\ \ldots\ 0)^T \in \ker X$.*

(ii) *If X has an all-ones diagonal and $\epsilon = \pm 1$, then for distinct $i, j \in \{1, \ldots, n\}$, $X_{ij} = \epsilon$ if and only if $e_i - \epsilon e_j \in \ker X$.*

16.2 Comparing the Lovász–Schrijver and Lasserre Relaxations for Max-Cut

Let $K := \{x \in [-1, 1]^d \mid Ax \geq b\}$ be an explicitly given polytope and let $P := \mathrm{conv}(K \cap \{\pm 1\}^d)$ be the polytope whose linear description is to be found. (As we want to treat max-cut, it is more convenient for us to work with ± 1-polytopes rather than $0/1$-polytopes.)

The following notation will be used throughout. Write K as

$$K = \left\{ x \in [-1,1]^d \mid g_\ell^T \begin{pmatrix} 1 \\ x \end{pmatrix} \geq 0 \mid \ell = 1, \ldots, m \right\}, \tag{16.9}$$

where $g_1^T, \ldots, g_m^T \in \mathbb{R}^{d+1}$ are the rows of the matrix $(-b\ A)$. For a polytope $Q \subseteq \mathbb{R}^d$ the set

$$\tilde{Q} := \left\{ \lambda \begin{pmatrix} 1 \\ x \end{pmatrix} \mid x \in Q,\ \lambda \geq 0 \right\} \tag{16.10}$$

denotes the *homogenization* of Q; $\tilde{Q}$ is a cone in $\mathbb{R}^{d+1}$ (the additional coordinate is indexed by 0) and $Q = \{x \in \mathbb{R}^d \mid \binom{1}{x} \in \tilde{Q}\}$.

16.2.1 The Lovász–Schrijver construction

Let $M(K)$ denote the set of symmetric matrices $Y = (y_{ij})_{i,j=0}^d$ satisfying

$$y_{jj} = y_{00} \text{ for } j = 1, \ldots, d, \tag{16.11}$$

$$Y(e_0 + e_j),\ Y(e_0 - e_j) \in \tilde{K} \text{ for } j = 1, \ldots, d, \tag{16.12}$$

and let $M_+(K) := \{Y \in M(K) \mid Y \succeq 0\}$. Set

$$N(K) := \left\{ x \in \mathbb{R}^d \mid \begin{pmatrix} 1 \\ x \end{pmatrix} = Ye_0 \text{ for some } Y \in M(K) \right\}.$$

$$N_+(K) := \left\{ x \in \mathbb{R}^d \mid \begin{pmatrix} 1 \\ x \end{pmatrix} = Ye_0 \text{ for some } Y \in M_+(K) \right\}.$$

Then

$$P \subseteq N_+(K) \subseteq N(K) \subseteq K.$$

The inclusion $P \subseteq N_+(K)$ follows from the fact that the matrix $Y := \binom{1}{x}\binom{1}{x}^T$ belongs to $M_+(K)$ for all $x \in K \cap \{\pm 1\}^d$, the inclusion $N_+(K) \subseteq N(K)$ is obvious, and the inclusion $N(K) \subseteq K$ follows from property (16.12). The following consequence of (16.12) will be used in Section 16.3:

$$N(K) \subseteq \operatorname{conv}(K \cap \{x \mid x_j = \pm 1\}) \text{ for all } j = 1, \ldots, d. \tag{16.13}$$

The sets $N(K)$ and $N_+(K)$ are, respectively, linear and semidefinite relaxations of P. Define iteratively $N^1(K) := N(K)$, $N_+^1(K) := N_+(K)$, and, for $t \geq 2$, $N^t(K) := N(N^{t-1}(K))$, $N_+^t(K) := N_+(N_+^{t-1}(K))$. Then

$$P \subseteq N^d(K) \subseteq \cdots \subseteq N^{t+1}(K) \subseteq N^t(K) \subseteq \cdots \subseteq N(K) \subseteq K.$$

Lovász and Schrijver [28] show that

$$N^d(K) = P.$$

Hence the sequence $N_+^t(K)$ also converges to P in d steps. There are instances where it converges faster to P than the sequence $N^t(K)$. This is the case, for example, when $P = \text{ST}(G)$ is the stable set polytope of a graph G and $K = \text{FR}(G)$ is its fractional stable set polytope defined by

$$\text{FR}(G) := \{x \in \mathbb{R}_+^V \mid x_i + x_j \leq 1 \ (ij \in E)\}. \tag{16.14}$$

Lovász and Schrijver [28] show that $N_+(\text{FR}(K_n)) = \text{ST}(K_n)$, while the smallest t for which $N^t(\text{FR}(K_n)) = \text{ST}(K_n)$ is $t = n - 2$. On the other hand, there are also cases where the N_+ operator does not help. This is the case, for instance, for the polytope[1] $P = \{x \in \mathbb{R}^d \mid \sum_{i=1}^d x_i \geq 1\}$ if we start from its relaxation $K = \{x \in \mathbb{R}^d \mid \sum_{i=1}^d x_i \geq \frac{1}{2}\}$; then the same number d of iterations is needed to find P using the N or the N_+ operator [8]. Other examples are given in [8, 15]. Moreover, geometric conditions are studied in [15] under which the N_+ operator yields a tighter relaxation than the N operator.

If we apply the Lovász–Schrijver construction to the pair $P = \text{CUT}(G)$, $K = \text{MET}(G)$, we obtain the sequence of linear and semidefinite relaxations $N^t(\text{MET}(G))$ and $N_+^t(\text{MET}(G))$ for the cut polytope. As in (16.7), one can obtain at least as good relaxations by applying the Lovász–Schrijver construction to the metric polytope of K_n and then projecting back on the edgeset of the graph G, namely, set

$$N(G) := \pi_E(N(\text{MET}(K_n)))$$

and define $N_+(G)$ as in (16.6). These relaxations are studied in detail in [22], where the following results are shown: $N(G) \subseteq N(\text{MET}(G))$, $N_+(G) \subseteq N_+(\text{MET}(G))$ (with equality if $G = K_n$). If the graph G has t edges whose contraction produces a graph with no K_5 minor, then $N^t(\text{MET}(G)) = \text{CUT}(G)$. In particular, $N^{n-\alpha(G)-3}(\text{MET}(G)) = \text{CUT}(G)$ if G has a maximum stable set whose deletion leaves a graph with at most three connected components; $N^{n-\alpha(G)-3}(G) = \text{CUT}(G)$ for a graph G on n nodes with stability number $\alpha(G)$. No graph is known for which the sequence of relaxations converges faster to $\text{CUT}(G)$ when using the N_+ operator than when using the N operator. Note that optimizing over the relaxation $N_+(K_n)$ amounts to solving a semidefinite program having a matrix variable Y of order $1 + \binom{n}{2}$ and $2\binom{n}{2} \cdot 4\binom{n}{3}$ linear inequalities.

16.2.2 The Lasserre construction—general presentation

We first introduce some notation. In this subsection, we let $V = \{1, \ldots, d\}$ since we are looking for relaxations of the polytope P lying in $\mathbb{R}^d$. Let $\mathcal{P}(V)$ denote the collection of all subsets of V and, given an integer $t \geq 0$, let $\mathcal{P}_t(V)$ denote the collection of subsets of V with cardinality not greater than t. The components of a vector $y \in \mathbb{R}^{\mathcal{P}(V)}$ are denoted as y_I or $y(I)$; we also set $y_0 = y_\emptyset$, $y_{i_1 \ldots i_k} = y_{\{i_1, \ldots, i_k\}}$. Given $y \in \mathbb{R}^{\mathcal{P}(V)}$ and an integer $t \geq 0$, the matrices

$$M(y) := (y(I \Delta J))_{I,J \in \mathcal{P}(V)}, \quad M_t(y) := (y(I \Delta J))_{I,J \in \mathcal{P}_t(V)} \tag{16.15}$$

[1] In this example and the previous one of the stable set problem, P is a $0/1$-polytope and thus one should use the corresponding definition of the N and N_+ operators for the $0/1$-context, namely, replace (16.11) with $y_{jj} = y_{0j}$ and (16.12) with $Ye_j, Y(e_0 - e_j) \in \tilde{K}$ for $j = 1, \ldots, d$.

are known as the *moment matrices* of y (where $I \Delta J$ denotes the symmetric difference of the sets I, J). It is useful to realize that the principal submatrix of $M(y)$ indexed by a set $\mathcal{I} \subseteq \mathcal{P}(V)$ coincides with the principal submatrix of $M(y)$ indexed by the set $\mathcal{I}\Delta A := \{I\Delta A \mid I \in \mathcal{I}\}$ for any set $A \subseteq V$. Given $g, y \in \mathbb{R}^{\mathcal{P}(V)}$, define the vector $g * y \in \mathbb{R}^{\mathcal{P}(V)}$ with entries

$$g * y(I) := \sum_{J\in\mathcal{P}(V)} g_J y_{I\Delta J} \text{ (for } I \subseteq V),$$

that is, $g * y = M(y)g$. Given a subset $A \subseteq V$, let $\psi^A \in \{\pm 1\}^V$ denote its ± 1-incidence vector with entries $\psi^A(i) := -1$ if $i \in A$ and $\psi^A(i) := 1$ if $i \in V \setminus A$. The vectors e_A ($A \subseteq V$) denote the standard unit vectors in $\mathbb{R}^{\mathcal{P}(V)}$. We use the representation of K given in (16.9); it will be convenient to consider g_ℓ as a vector in $\mathbb{R}^{\mathcal{P}(V)}$, setting $g_\ell(I) := 0$ if $|I| \geq 2$.

Lemma 16.2. *Given $A \subseteq V$, if $\psi^A \in K$, then*

$$M(y) \succeq 0, \ M(g_\ell * y) \succeq 0 \ \textit{for } \ell = 1, \dots, m, \tag{16.16}$$

where $y := ((-1)^{|A\cap I|})_{I\subseteq V}$.

Proof. Indeed, $M(y) = yy^T$ and $M(g_\ell * y) = g_\ell^T y \cdot yy^T$, since $y(I\Delta J) = y(I)\cdot y(J)$ for all $I, J \subseteq V$. Moreover, $g_\ell^T y \geq 0$ for all ℓ, since $\psi^A \in K$ and the projection of y on the subspace indexed by the singletons is equal to ψ^A. □

Let Z denote the symmetric matrix indexed by $\mathcal{P}(V)$ with entries

$$Z_{I,J} = (-1)^{|I\cap J|} \text{ for } I, J \subseteq V. \tag{16.17}$$

This is the ± 1-analogue of the zeta matrix of the lattice $\mathcal{P}(V)$ considered in [28] in the 0/1-case. The next result is the ± 1-analogue of a result from [23] for the 0/1-case; we include the proof for completeness.

Lemma 16.3. *For $y \in \mathbb{R}^{\mathcal{P}(V)}$, equality $Z \operatorname{diag}(Zy)Z = 2^{|V|}M(y)$ holds. More generally, given $g \in \mathbb{R}^{\mathcal{P}(V)}$, $Z \operatorname{diag}(u)Z = 2^{|V|}M(g * y)$, where $u \in \mathbb{R}^{\mathcal{P}(V)}$ has entries $u_A := (Zy)_A \cdot g^T Ze_A$ (for $A \subseteq V$) (Ze_A denoting the Ath column of Z).*

Proof. Given $I, J \subseteq V$, the (I, J)th entry of the matrix $Z \operatorname{diag}(u)Z$ is equal to

$$\textstyle\sum_A Z_{IA}Z_{JA}u_A = \sum_A Z_{IA}Z_{JA}(Zy)_A \cdot g^T Ze_A$$

$$= \textstyle\sum_A Z_{IA}Z_{JA}\left(\sum_R Z_{AR}y_R\right)\cdot\left(\sum_S Z_{AS}g_S\right) = \sum_{R,S} y_R g_S \left(\sum_A Z_{IA}Z_{JA}Z_{AR}Z_{AS}\right).$$

Since $Z_{IA}Z_{JA}Z_{AR}Z_{AS} = (-1)^{|I\cap A|+|J\cap A|+|R\cap A|+|S\cap A|} = (-1)^{|A\cap(I\Delta J\Delta R\Delta S)|}$, the inner sum (over A) is equal to $2^{|V|}$ if $I\Delta J\Delta R\Delta S = \emptyset$ and to 0 otherwise. Therefore, the (I, J)th entry of $Z \operatorname{diag}(u)Z$ is equal to $2^{|V|}\sum_S g_S y_{I\Delta J\Delta S} = 2^{|V|}g*y(I\Delta J) = 2^{|V|}M(g*y)(I, J)$. This concludes the proof. □

For $t \geq 0$, let $P_t(K)$ denote the set of vectors $y \in \mathbb{R}^{\mathcal{P}_{2t+2}(V)}$ satisfying

$$M_{t+1}(y) \succeq 0, \ M_t(g_\ell * y) \succeq 0 \text{ for all } \ell = 1, \dots, m, \tag{16.18}$$

and let $Q_t(K)$ denote the projection of $P_t(K) \cap \{y \mid y_\emptyset = 1\}$ onto the subspace $\mathbb{R}^d$ indexed by the singletons. Then

$$P \subseteq Q_t(K) \subseteq K;$$

the first inclusion follows from Lemma 16.2 and the second follows from the fact that $g_\ell * y(\emptyset) \geq 0$ for all ℓ and $y \in Q_t(K)$. The hierarchy of relaxations $Q_t(K)$ was introduced by Lasserre [18, 20], who showed that P is found after d steps, that is, $P = Q_d(K)$. His construction is motivated by results about representations of positive polynomials as sums of squares, and his original presentation involves moment matrices indexed by integer sequences (rather than subsets of V). The presentation given here is taken from [23], where the following elementary proof for the convergence result is given, based on the above lemmas.

Let $\mathcal{C}_K$ denote the cone in $\mathbb{R}^{\mathcal{P}(V)}$ generated by the columns of Z corresponding to points in K, that is, $\mathcal{C}_K$ is generated by the vectors Ze_A for all the sets $A \subseteq V$ for which $\psi^A \in K$. Then, $\mathcal{C}_K$ is a simplicial cone and P is equal to the projection of the polytope $\mathcal{C}_K \cap \{y \mid y_\emptyset = 1\}$ on the subspace $\mathbb{R}^n$ indexed by the singletons.

Lemma 16.4. $P_d(K) = \mathcal{C}_K$.

Proof. By the definition, $y \in P_d(K)$ if and only if $M(y) \succeq 0$ and $M(g_\ell * y) \succeq 0$ for all $\ell = 1, \dots, m$. Using Lemma 16.3, this is equivalent to the conditions $Zy \geq 0$ and $(Zy)_A \cdot g_\ell^T Ze_A \geq 0$ for all $A \subseteq V$ and $\ell = 1, \dots, m$; this in turn holds if and only if $Zy \geq 0$ and, for all $A \subseteq V$, $(Zy)_A = 0$ whenever the vector ψ^A does not belong to K. Therefore, $y \in P_d(K)$ if and only if y belongs to the cone $\mathcal{C}_K$. $\square$

Corollary 16.5. $Q_d(K) = P$.

If we apply the above construction to the pair $P = \mathrm{CUT}(G)$, $K = \mathrm{MET}(G)$, then we obtain the sequence of semidefinite relaxations $Q_t(\mathrm{MET}(G))$ $(t = 0, \dots, |E|)$ for the cut polytope $\mathrm{CUT}(G)$. The definition of $Q_t(\mathrm{MET}(G))$ involves the semidefinite program (16.18), which contains as many semidefinite constraints as the number of circuits in G (which can therefore be exponentially large in terms of n); moreover, the program is in the variable $y \in \mathbb{R}^{\mathcal{P}_{2t+2}(E)}$ since $\mathrm{MET}(G)$ lies in the edge space $\mathbb{R}^E$. One way to get around the difficulty of the large number of constraints is to consider, instead of $Q_t(\mathrm{MET}(G))$, the set $\pi_E(Q_t(\mathrm{MET}(K_n)))$ whose definition now involves $1 + 4\binom{n}{3}$ constraints (corresponding to the triangles in K_n) and the variable $y \in \mathbb{R}^{\mathcal{P}_{2t+2}(E_n)}$. Although the number of constraints is now polynomial in n, we will see in the next subsection that, if we apply the Lasserre construction to the node model of max-cut, we obtain a much simpler relaxation, involving only one semidefinite constraint.

16.2.3 The Lasserre construction—the node model for max-cut

If we consider the formulation (16.2) for the max-cut problem, we arrive naturally at the following relaxations introduced by Lasserre [18] and obtained in the following way: Apply the Lasserre construction to the polytope $K := [-1, 1]^n$ and project on the edge subspace $\mathbb{R}^E$ (instead of projecting on the subspace $\mathbb{R}^n$ in which the starting polytope K lies). Thus in this subsection we let $V = \{1, \ldots, n\}$ since our starting polytope $K = [-1, 1]^n$ lies in the space $\mathbb{R}^n$. For $t \geq 0$ we have:

$$P_t(K) = \{y \in \mathbb{R}^{\mathcal{P}_{2t+2}(V)} \mid M_{t+1}(y) \succeq 0\}.$$

Let $Q_t(G)$ denote the projection of the set $P_t(K) \cap \{y \mid y_\emptyset = 1\}$ on the edge subspace $\mathbb{R}^E$. Then

$$\mathrm{CUT}(G) \subseteq Q_t(G).$$

We now mention a more concise formulation for the set $Q_t(G)$. For this let $\mathcal{E}(V)$ (resp. $\mathcal{E}_t(V)$) denote the collection of all even subsets (resp. all even subsets of size not greater than t) of V; $\mathcal{O}(V)$ and $\mathcal{O}_t(V)$ are the analogous families of odd subsets of V. It is also convenient to use the symbol $\mathcal{U}_t(V)$ to denote the collection of subsets of V whose cardinality is not greater than t and has the same parity as t. Given a vector $y \in \mathbb{R}^{\mathcal{E}_{2t}(V)}$, let us define its *reduced moment matrix* $\widetilde{M}_t(y)$ as

$$\widetilde{M}_t(y) := (y(I \Delta J))_{I,J \in \mathcal{U}_t(V)}. \tag{16.19}$$

Note that $\widetilde{M}_{n-1}(y) = \widetilde{M}_n(y)$ and that $\widetilde{M}_{n-1}(y)$ can be assumed to be indexed by either $\mathcal{E}(V)$ or $\mathcal{O}(V)$, since $\mathcal{O}(V) = \mathcal{E}(V)\Delta\{1\} := \{I \Delta \{1\} \mid I \in \mathcal{E}(V)\}$.

Lemma 16.6. *For $t \geq 0$, $Q_t(G)$ is equal to the projection on $\mathbb{R}^E$ of the set of vectors $y \in \mathbb{R}^{\mathcal{E}_{2t+2}(V)}$ satisfying*

$$\widetilde{M}_{t+1}(y) \succeq 0, \; y_\emptyset = 1. \tag{16.20}$$

Proof. Let $\widetilde{Q}_t(G)$ denote the projection on $\mathbb{R}^E$ of the solution set to (16.20). The inclusion $Q_t(G) \subseteq \widetilde{Q}_t(G)$ is obvious. Conversely, assume that $\widetilde{M}_{t+1}(y) \succeq 0$, where $y \in \mathbb{R}^{\mathcal{E}_{2t+2}(V)}$. Extend y to a vector $z \in \mathbb{R}^{\mathcal{P}_{2t+2}(V)}$ by setting $z_I := y_I$ if $|I|$ is even and $z_I := 0$ if $|I|$ is odd. Then the matrix $M_{t+1}(z)$ has the block configuration

$$\begin{array}{c} \\ \mathcal{E}_{t+1}(V) \\ \mathcal{O}_{t+1}(V) \end{array} \begin{array}{c} \begin{array}{cc} \mathcal{E}_{t+1}(V) & \mathcal{O}_{t+1}(V) \end{array} \\ \begin{pmatrix} A & 0 \\ 0 & B \end{pmatrix} \end{array}, \tag{16.21}$$

where $A = \widetilde{M}_{t+1}(y)$ and B is the principal submatrix of A indexed by $\{I\Delta\{1\} \mid I \in \mathcal{O}_{t+1}(V)\}$ if t is odd, and $B = \widetilde{M}_{t+1}(y)$ and A is the principal submatrix of B indexed by $\{I\Delta\{1\} \mid I \in \mathcal{E}_{t+1}(V)\}$ if t is even. From this follows that $M_{t+1}(z) \succeq 0$. Since y and z have the same even indexed entries, the reverse inclusion $\widetilde{Q}_t(G) \subseteq Q_t(G)$ follows. □

For $t = 0, 1, \ldots, n-2$ set

$$\mathcal{F}_t(n) := \{Y \succeq 0 \mid Y = \widetilde{M}_{t+1}(y) \;\text{ for some } y \in \mathbb{R}^{\mathcal{E}_{2t+2}(V)} \text{ with } y_\emptyset = 1\}. \tag{16.22}$$

Let us see what the matrix sets $\mathcal{F}_t(n)$ are for small values of t. For $t = 0$, $\mathcal{F}_0(n)$ is the basic semidefinite relaxation of the cut polytope $\mathrm{CUT}(K_n)$ consisting of the $n \times n$-symmetric positive semidefinite matrices with an all-ones diagonal. For $t = 1$, $\mathcal{F}_1(n)$ is equal to the set of symmetric positive semidefinite matrices Y indexed by $\{\emptyset\} \cup E_n$ having an all-ones diagonal and satisfying the two conditions

$$Y_{ij,ik} = Y_{\emptyset,jk} \text{ and } Y_{ij,rs} = Y_{ir,js} = Y_{is,jr}$$

for all distinct $i, j, k, r, s \in V$. If we remove in the definition of $\mathcal{F}_1(n)$ the second condition $Y_{ij,rs} = Y_{ir,js} = Y_{is,jr}$, then we obtain the larger matrix set $\mathcal{F}_n$ underlying the relaxation (SDP3) defined by Anjos and Wolkowicz [1] and their relaxation

$$F_n := \left\{ x \in \mathbb{R}^{E_n} \mid \begin{pmatrix} 1 \\ x \end{pmatrix} = Ye_\emptyset \text{ for some } Y \in \mathcal{F}_n \right\} \tag{16.23}$$

of the cut polytope $\mathrm{CUT}(K_n)$; thus

$$\mathcal{F}_1(n) \subseteq \mathcal{F}_n \text{ and } Q_1(K_n) \subseteq F_n.$$

A useful property of the matrix set $\mathcal{F}(n)$ (and thus of $\mathcal{F}_1(n)$) is that it implies the triangle inequalities.

Lemma 16.7 (Anjos and Wolkowicz [1]). $F_n \subseteq \mathrm{MET}(K_n)$.

Proof. Let $Y \in \mathcal{F}_n$ and set $y_{ij} := Y_{\emptyset,ij}$ for $ij \in E_n$. Given three nodes $1, 2, 3 \in V$, the principal submatrix X of Y indexed by the set $\{\emptyset, 12, 13, 23\}$ has the form

$$X = \begin{array}{c} \\ \emptyset \\ 12 \\ 13 \\ 23 \end{array} \begin{array}{c} \begin{array}{cccc} \emptyset & 12 & 13 & 23 \end{array} \\ \begin{pmatrix} 1 & y_{12} & y_{13} & y_{23} \\ y_{12} & 1 & y_{23} & y_{13} \\ y_{13} & y_{23} & 1 & y_{12} \\ y_{23} & y_{13} & y_{12} & 1 \end{pmatrix} \end{array}.$$

As $X \succeq 0$, we have $e^T Xe \geq 0$, which implies that $y_{12} + y_{13} + y_{23} \geq -1$. The other triangle inequalities are obtained by suitably flipping signs in X. □

The cuts of $\mathrm{CUT}(K_n)$ correspond to certain special matrices in $\mathcal{F}_t(n)$. Given $A \subseteq V$, the vector

$$y^A := ((-1)^{|I \cap A|})_{I \in \mathcal{E}(V)} \tag{16.24}$$

is called a *cut vector*; its projection on $\mathbb{R}^{E_n}$ is the cut $\delta(A)$. As $y^A = y^{V \setminus A}$, there are 2^{n-1} distinct cut vectors y^A obtained, for instance, for all $A \subseteq V \setminus \{n\}$. For convenience we often use below the same symbol y^A to denote the cut vector in $\mathbb{R}^{\mathcal{E}(V)}$ and its projection on a subspace of it. For $t = 0, \ldots, n-2$, the reduced-moment matrix $\widetilde{M}_{t+1}(y^A)$ is called a *cut matrix* of $\mathcal{F}_t(n)$ and is also denoted as $\widetilde{M}_{t+1}(A)$ for the sake of simplicity. Thus

$$\widetilde{M}_{n-1}(A) = y^A (y^A)^T$$

and every cut matrix $\widetilde{M}_{t+1}(A)$ has rank 1. The next lemma shows that the eigenvectors of any reduced-moment matrix $\widetilde{M}_{n-1}(y)$ are the cut vectors, which permits us to show that $\mathcal{F}_{n-2}(n)$ is a simplex with the cut matrices as vertices.

Lemma 16.8. *Let $Y = \widetilde{M}_{n-1}(y)$, where $y \in \mathbb{R}^{\mathcal{E}(V)}$. The eigenvectors of Y are the 2^{n-1} distinct vectors $y^A \in \mathbb{R}^{\mathcal{E}(V)}$ with respective eigenvalues $y^T y^A$.*

Proof. We verify that $Yy^A = (y^T y^A)y^A$, that is, $(Ye_S)^T y^A = y^T y^A \cdot (-1)^{|A\cap S|}$ for any $S \in \mathcal{E}(V)$. Indeed,

$$(Ye_S)^T y^A = \sum_{I\in\mathcal{E}(V)} y(S\Delta I)(-1)^{|A\cap I|} = \sum_{I\in\mathcal{E}(V)} y(S\Delta I)(-1)^{|A\cap S|}(-1)^{|A\cap(S\Delta I)|}$$
$$= (-1)^{|A\cap S|} \sum_{J\in\mathcal{E}(V)} y(J)(-1)^{|A\cap J|} = (-1)^{|A\cap S|} \cdot y^T y^A. \qquad \square$$

Corollary 16.9. *The set $\mathcal{F}_{n-2}(n)$ is the simplex in $\mathbb{R}^{\mathcal{E}(V)}$ whose vertices are the 2^{n-1} distinct cut matrices $\widetilde{M}_{n-1}(A)$ ($A \subseteq V$).*

Proof. By Lemma 16.8, any matrix $Y = \widetilde{M}_{n-1}(y) \in \mathcal{F}_{n-2}(n)$ can be written as $Y = \frac{1}{2^{n-1}} \sum_{A\subseteq V\setminus\{n\}} y^T y^A \cdot y^A (y^A)^T$, where $y^T y^A \geq 0$ for all A and $\frac{1}{2^{n-1}} \sum_A y^T y^A = 1$. $\square$

16.2.4 Comparing the Lasserre relaxation $Q_t(G)$ and the Lovász–Schrijver relaxation $N_+^{t-1}(G)$

We show here an inclusion relationship between the two semidefinite relaxations $N_+^{t-1}(G)$ and $Q_t(G)$ obtained earlier for the cut polytope using, respectively, the Lovász–Schrijver construction (applied to the edge model of max-cut) and the Lasserre relaxation (applied to the node model). As before, $G = (V, E)$ is a graph with $V = \{1, \ldots, n\}$. We begin with a preliminary result.

Lemma 16.10. *Given $y \in \mathbb{R}^{\mathcal{P}_{2t+2}(V)}$, $t \geq 0$, $M_{t+1}(y) \succeq 0$ implies $M_t((e_\emptyset + \epsilon e_{ij}) * y) \succeq 0$ for all $\epsilon = \pm 1$ and $ij \in E_n$.*

Proof. Let $ij = 12$ and set $u := (e_\emptyset + \epsilon e_{12}) * y$. Observe that $u(I) = \epsilon u(I\Delta\{1,2\})$ for all I. Let us consider the partition of $\mathcal{P}_t(V)$ into the following three sets: $\mathcal{I}_1 := \{I \in \mathcal{P}_t(V) \mid 1 \notin I, |\{I\Delta\{1,2\}\}| \leq t\}$, $\mathcal{I}_2 := \{I \in \mathcal{P}_t(V) \mid 1 \notin I, |\{I\Delta\{1,2\}\}| > t\}$, and $\mathcal{I}_3 := \{I\Delta\{1,2\} \mid I \in \mathcal{I}_1\}$. With respect to this partition, the matrix $M_t(u)$ has the block configuration

$$\begin{array}{c} \\ \mathcal{I}_1 \\ \mathcal{I}_2 \\ \mathcal{I}_3 \end{array} \begin{array}{c} \begin{array}{ccc} \mathcal{I}_1 & \mathcal{I}_2 & \mathcal{I}_3 \end{array} \\ \begin{pmatrix} A & B & \epsilon\cdot A \\ B^T & C & \epsilon\cdot B^T \\ \epsilon\cdot A & B & A \end{pmatrix} \end{array}.$$

Hence, $M_t(u) \succeq 0$ if and only if its principal submatrix X indexed by $\mathcal{I} := \mathcal{I}_1 \cup \mathcal{I}_2$ is positive semidefinite. Therefore, it suffices to show that $M_{t+1}(y) \succeq 0$ implies $X \succeq 0$. For this, consider the principal submatrix Y of $M_{t+1}(y)$ indexed by $\mathcal{I}_1' \cup \mathcal{I}_2'$, where $\mathcal{I}_i' := \{I\Delta\{i\} \mid I \in \mathcal{I}\}$ for $i = 1, 2$. Then Y has the block configuration

$$\begin{array}{c} \\ \mathcal{I}_1' \\ \mathcal{I}_2' \end{array} \begin{array}{c} \begin{array}{cc} \mathcal{I}_1' & \mathcal{I}_2' \end{array} \\ \begin{pmatrix} E & F \\ F^T & E \end{pmatrix} \end{array}.$$

Now $Y \succeq 0$ implies that $E + \epsilon F \succeq 0$ for $\epsilon = \pm 1$, since $(x^T\ \epsilon x^T) Y \binom{x}{\epsilon x} = 2 \cdot x^T (E + \epsilon F) x \geq 0$ for all x. Observe finally that $X = E + \epsilon F$. $\square$

Theorem 16.11. *For any graph G and $t \geq 1$, $Q_t(G) \subseteq N_+(Q_{t-1}(G))$.*

Proof. Let $x \in Q_t(G)$, that is, x is the projection on $\mathbb{R}^E$ of $y \in \mathbb{R}^{\mathcal{P}_{2t+2}(V)}$ satisfying $y_\emptyset = 1$ and $M_{t+1}(y) \succeq 0$. Let Y denote the principal submatrix of $M_{t+1}(y)$ indexed by $\{\emptyset\} \cup E$. Thus $\binom{1}{x} = Ye_0$ and $Y \succeq 0$. In order to show that $x \in N_+(Q_{t-1}(G))$, it suffices now to verify that the vector $z := Y(e_\emptyset + \epsilon e_{ij})$ belongs to $\widetilde{Q_{t-1}(G)}$ (the homogenization of $Q_{t-1}(G)$) for all $\epsilon = \pm 1$ and $ij \in E$. This follows from the fact that z is equal to the projection on $\mathbb{R}^{\{\emptyset\}\cup E}$ of the vector $u := (e_\emptyset + \epsilon e_{ij}) * y$ and that $M_t(u) \succeq 0$ by Lemma 16.10. □

Corollary 16.12. *For any graph G and $t \geq 1$, $Q_t(G) \subseteq N_+^{t-1}(G)$.*

Proof. The proof follows directly from Theorem 16.11 and Lemma 16.7 using induction on $t \geq 1$. □

16.3 Bounds on the Rank of the Lasserre Procedure

An interesting question is to determine the class $\mathcal{L}_t$ consisting of the graphs G for which $\mathrm{CUT}(G) = Q_t(G)$. Indeed, the max-cut problem can be solved in polynomial time (with an arbitrary precision) over the class $\mathcal{L}_t$ for any fixed t. The same holds for the class $\mathcal{G}_t$ of the graphs G for which $\mathrm{CUT}(G) = N_+^t(G)$. By Corollary 16.12, we have that $\mathcal{G}_{t-1} \subseteq \mathcal{L}_t$ for $t \geq 1$; inclusion is strict, for instance, for $t = 2$. The class $\mathcal{G}_t$ is closed under taking minors [22]. We show that the same holds for $\mathcal{L}_t$. A crucial tool for showing that $\mathcal{G}_t$ is closed under taking contraction minors is the fact that validity of an inequality for the set $N_+^t(G) \cap \{x \mid x_e = \pm 1\}$ can be expressed in terms of validity of a transformed inequality for the set $N_+^t(G/e)$. An analogous idea will be used to show that $\mathcal{L}_t$ is closed under taking contraction minors.

We begin with some definitions and preliminary observations. Let $G = (V, E)$ be a graph with $V = \{1, \ldots, n\}$ and let $e := uv$ be a given edge of G. Deleting e produces the graph $G\backslash e := (V, E \setminus \{e\})$, while contracting e produces the graph $G / e := (V \setminus \{u, v\} \cup \{w\}, F)$, where w is the new node created by the contraction of the edge e and F is the resulting edgeset (erasing multiple edges). A minor of G is any graph obtained from G by a sequence of deletions and contractions. For a node $i \in V$, $N_G(i)$ denotes the set of nodes that are adjacent to i in G. The following is a simple but powerful property of the metric polytope:

$$\text{If } y \in \mathrm{MET}(G) \text{ has } y_{uv} = \epsilon = \pm 1, \text{ then } y_{ui} = \epsilon y_{vi} \text{ for } i \in N_G(u) \cap N_G(v). \tag{16.25}$$

The same property holds if we replace $\mathrm{MET}(G)$ with its subset $\mathrm{CUT}(G)$ or with $Q_t(G)$ ($t \geq 0$) (in view of Lemma 16.1(ii)). Based on this property, let us define for $x \in \mathbb{R}^F$ its *ϵ-extension* $y \in \mathbb{R}^E$ by

$$\begin{aligned} &y_{uv} := \epsilon, \ y_{ij} := x_{ij} \text{ for } ij \in E \text{ with } i, j \neq u, v, \\ &y_{ui} := x_{wi} \text{ for } i \in N_G(u) \setminus \{v\}, \ y_{vi} := \epsilon x_{wi} \text{ for } i \in N_G(v) \setminus \{u\}. \end{aligned} \tag{16.26}$$

One can easily verify that

$$x \in \mathrm{CUT}(G/e) \Longleftrightarrow y \in \mathrm{CUT}(G). \tag{16.27}$$

In order to establish the analogous result for $Q_t(G)$, we need to extend the notion of ϵ-extension to reduced moment matrices. For convenience we set $u = w = 1$ and $v = n$.

Lemma 16.13. *Let $t \geq 0$, $X = \widetilde{M}_{t+1}(x)$, where $x \in \mathbb{R}^{\mathcal{E}_{2t+2}(V\setminus\{n\})}$, and $\epsilon = \pm 1$. Extend x to $y \in \mathbb{R}^{\mathcal{E}_{2t+2}(V)}$ by setting $y_I := \epsilon \cdot x_{I\Delta\{1,n\}}$ for all $I \in \mathcal{E}_{2t+2}(V)$ with $n \in I$ and set $Y := \widetilde{M}_{t+1}(y)$; Y is called an* ϵ-extension *of X. Then, $X \in \mathcal{F}_t(n-1)$ if and only if $Y \in \mathcal{F}_t(n)$.*

Proof. We have to show that $X \succeq 0$ if and only if $Y \succeq 0$. The "only if" part is obvious since X is a principal submatrix of Y. Assume now that $X \succeq 0$. Partition the index set $\mathcal{U}_{t+1}(V)$ of Y into $\mathcal{I}_1 \cup \mathcal{I}_2 \cup \mathcal{I}_3$, where $\mathcal{I}_1 := \{H \in \mathcal{U}_{t+1}(V \setminus \{n\}) \mid |H\Delta\{1,n\}| \leq t+1\}$, $\mathcal{I}_2 := \{H \in \mathcal{U}_{t+1}(V \setminus \{n\}) \mid |H\Delta\{1,n\}| > t+1\}$, and $\mathcal{I}_3 := \{H\Delta\{1,n\} \mid H \in \mathcal{I}_1\} = \{H \in \mathcal{U}_{t+1}(V) \mid n \in H\}$. Then X and Y have the following block configurations:

$$X = \begin{matrix} & \mathcal{I}_1 & \mathcal{I}_2 \\ \begin{matrix}\mathcal{I}_1\\ \mathcal{I}_2\end{matrix} & \multicolumn{2}{c}{\begin{pmatrix} B & C \\ C^T & D \end{pmatrix}} \end{matrix}, \quad Y = \begin{matrix} & \mathcal{I}_1 & \mathcal{I}_2 & \mathcal{I}_3 \\ \begin{matrix}\mathcal{I}_1\\ \mathcal{I}_2\\ \mathcal{I}_3\end{matrix} & \multicolumn{3}{c}{\begin{pmatrix} B & C & \epsilon\cdot B \\ C^T & D & \epsilon\cdot C^T \\ \epsilon\cdot B & \epsilon\cdot C & B \end{pmatrix}} \end{matrix}.$$

Indeed, for $H \in \mathcal{U}_{t+1}(V\setminus\{n\})$ and $K \in \mathcal{I}_3$, $Y(H,K) = y(H\Delta K)$ is equal to $\epsilon \cdot y(H\Delta K\Delta\{1,n\})$ (by the definition of y) and thus to $\epsilon \cdot Y(H, K\Delta\{1,n\})$, where $K\Delta\{1,n\} \in \mathcal{I}_1$. This implies that $Y \succeq 0$. □

Lemma 16.14. *Let $t \geq 0$ and $Y = \widetilde{M}_{t+1}(y) \in \mathcal{F}_t(n)$, and let X be the principal submatrix of Y indexed by $\mathcal{U}_{t+1}(V \setminus \{n\})$. If $y_{1n} = \epsilon = \pm 1$, then Y is the ϵ-extension of X.*

Proof. We have to show that $y_I = \epsilon \cdot y_{I\Delta\{1,n\}}$ for all $I \in \mathcal{E}_{2t+2}(V)$ containing n. For this, let $H \in \mathcal{U}_{t+1}(V)$ contain n; then $H\Delta\{1,n\} \in \mathcal{U}_{t+1}(V)$. As $Y(H, H\Delta\{1,n\}) = y_{1n} = \epsilon$, Lemma 16.1(ii) implies that $Ye_H = \epsilon \cdot Ye_{H\Delta\{1,n\}}$ and thus, for all $K \in \mathcal{U}_{t+1}(V)$, $Y(H,K) = \epsilon \cdot Y(H\Delta\{1,n\}, K)$, which yields $y(H\Delta K) = \epsilon \cdot y(H\Delta K\Delta\{1,n\})$, concluding the proof. □

Corollary 16.15. *Let $t \geq 0$, $\epsilon = \pm 1$, and $x \in \mathbb{R}^F$, and let $y \in \mathbb{R}^E$ be its ϵ-extension. Then, $x \in Q_t(G/e)$ if and only if $y \in Q_t(G)$.*

Proof. Say e is the edge $1n$ in order to match the notation in Lemmas 16.13 and 16.14. Suppose first that $x \in Q_t(G/e)$ and let $X \in \mathcal{F}_t(n-1)$, whose projection on $\mathbb{R}^F$ is x. Then the ϵ-extension Y of X belongs to $\mathcal{F}_t(n)$ (by Lemma 16.13) and its projection on $\mathbb{R}^E$ is equal to y, which shows that $y \in Q_t(G)$. Conversely, suppose that $y \in Q_t(G)$ and let $Y \in \mathcal{F}_t(n)$, whose projection on $\mathbb{R}^E$ is y. Then the principal submatrix X of Y indexed by $\mathcal{U}_{t+1}(V \setminus \{n\})$ belongs to $\mathcal{F}_t(n-1)$ and Y is the ϵ-extension of X by Lemma 16.14. As the projection of X on $\mathbb{R}^F$ is equal to x, we deduce that $x \in Q_t(G/e)$. □

As a first application, we can show that $\mathcal{L}_t$ is closed under taking minors.

Theorem 16.16. *Given $t \geq 0$, if* $\mathrm{CUT}(G) = Q_t(G)$, *then* $\mathrm{CUT}(G/e) = Q_t(G/e)$ *and* $\mathrm{CUT}(G\backslash e) = Q_t(G\backslash e)$.

Proof. Let $x \in Q_t(G/e)$ and let $y \in \mathbb{R}^E$ be its 1-extension. By Corollary 16.15, y belongs to $Q_t(G)$; hence $y \in \mathrm{CUT}(G)$, which, by (16.27), implies that x belongs to $\mathrm{CUT}(G/e)$. Equality $\mathrm{CUT}(G\backslash e) = Q_t(G\backslash e)$ follows from the fact that $\mathrm{CUT}(G\backslash e)$ (resp. $Q_t(G\backslash e)$) is the projection on $\mathbb{R}^{E\backslash\{e\}}$ of $\mathrm{CUT}(G)$ (resp. of $Q_t(G)$). □

As a second application, we can show the following result.

Theorem 16.17. *Given $t \geq 0$, if* $\mathrm{CUT}(G/e) = Q_t(G/e)$, *then* $\mathrm{CUT}(G) = Q_{t+1}(G)$.

Proof. By Theorem 16.11, $Q_{t+1}(G)$ is contained in $N_+(Q_t(G))$ which, in turn, is contained in $\mathrm{conv}(Q_t(G) \cap \{y \mid y_e = \pm 1\})$ by (16.13). Hence, it suffices to show that $Q_t(G) \cap \{y \mid y_e = \pm 1\} \subseteq \mathrm{CUT}(G)$. Let $y \in Q_t(G)$ with $y_e = \epsilon \in \{\pm 1\}$. By Corollary 16.15, y is the ϵ-extension of $x \in Q_t(G/e)$. By assumption, $x \in \mathrm{CUT}(G/e)$ which, by (16.27), implies that $y \in \mathrm{CUT}(G)$. □

We will see in Section 16.5 that

$$K_3 \in \mathcal{L}_1 \setminus \mathcal{L}_0, \; K_5 \in \mathcal{L}_2 \setminus \mathcal{L}_1, \; K_6 \in \mathcal{L}_2 \setminus \mathcal{G}_1, \; K_7 \notin \mathcal{L}_2.$$

The class $\mathcal{L}_0$ consists of the graphs with no K_3 minor (indeed, $K_3 \notin \mathcal{L}_0$ and, if G has no K_3 minor, then $\mathrm{CUT}(G) = [-1,1]^E$ is thus equal to $Q_0(G)$). The class $\mathcal{L}_1$ consists of the graphs having no K_5 minor (indeed, $K_5 \notin \mathcal{L}_1$ and, if G has no K_5 minor, then $\mathrm{CUT}(G) = \mathrm{MET}(G)$ is thus equal to $Q_1(G)$). The graph K_7 is a forbidden minor for the class $\mathcal{L}_2$ (we do not know whether K_7 is a *minimal* forbidden minor). There are other forbidden minors for the class $\mathcal{L}_2$ since the max-cut problem is known to be NP-hard for the class of graphs having no K_6 minor (and also for the graphs having a node whose deletion results in a planar graph) [5].

One can show that the class $\mathcal{L}_t$ is closed under taking clique k-sums ($k = 0, 1, 2, 3$); the same holds for the class $\mathcal{G}_t$ [22] (the proof for $\mathcal{L}_t$ is analogous to that for $\mathcal{G}_t$). The next result follows as an application of Theorem 16.17, Corollary 16.12, and the fact from [22] that $\mathrm{CUT}(G) = N^t(G)$ for $t := \max(0, n - \alpha(G) - 3)$.

Corollary 16.18. $\mathrm{CUT}(K_n) = Q_{n-4}(K_n)$ *for $n \geq 6$ and, for a graph G on n nodes with stability number $\alpha(G)$,* $\mathrm{CUT}(G) = Q_t(G)$ *for* $t := \max(1, n - \alpha(G) - 2)$.

16.4 Geometric Properties of the Matrix Sets $\mathcal{F}_t(n)$

In this section we study some geometric properties of the matrix sets $\mathcal{F}_t(n)$ underlying the Lasserre relaxations $Q_t(K_n)$.

Let us first recall some definitions. A convex subset F of a convex set $\mathcal{K}$ is called a *face* of $\mathcal{K}$ if, for all $x \in F$, $y, z \in \mathcal{K}$, $0 < \alpha < 1$, $x = \alpha y + (1-\alpha)z$ implies that $y, z \in F$. Given $x \in \mathcal{K}$, let $F(x)$ denote the smallest face of $\mathcal{K}$ containing x. A point

$x \in \mathcal{K}$ is an *extreme point* if $F(x) = \{x\}$ and a *vertex* if its normal cone is full dimensional. One says that the points $x_1, \ldots, x_k \in \mathcal{K}$ *form a face* of $\mathcal{K}$ if the set $\text{conv}(\{x_1, \ldots, x_k\})$ is a face of $\mathcal{K}$.

Consider a convex set $\mathcal{K}$ of the form

$$\mathcal{K} = \{X \in \text{PSD}_n \mid \langle A_i, X\rangle = b_i \;\text{ for } i = 1, \ldots, m\},$$

where the A_i's are symmetric matrices and $b_i \in \mathbb{R}$. It follows from a result in [11] that the smallest face $F(A)$ of $\mathcal{K}$ containing a given element $A \in \mathcal{K}$ is given by

$$F(A) = \{X \in \mathcal{K} \mid \ker X \supseteq \ker A\}.$$

This description of the faces applies in particular to any set $\mathcal{F}_t(n)$.

Analogously to $\text{CUT}(K_n)$, the set $\mathcal{F}_t(n)$ enjoys lots of symmetries. In particular, it is invariant under any permutation of the indices in $\{1, \ldots, n\}$ and under the following "switching" operation.

Lemma 16.19. *Given $A \subseteq V$, the* switching mapping

$$r_A : Y \mapsto r_A(Y) := \text{diag}(\zeta^A)\; Y\; \text{diag}(\zeta^A)$$

leaves the set $\mathcal{F}_t(n)$ invariant, where $\zeta^A := ((-1)^{|A\cap H|})_{H\in\mathcal{U}_{t+1}(V)}$.

Proof. Let $Y = \widetilde{M}_{t+1}(y) \in \mathcal{F}_t(n)$. Then $r_A(Y) = \widetilde{M}_{t+1}(z)$, where $z(I) := (-1)^{|A\cap I|}y(I)$ for all $I \in \mathcal{E}_{2t+2}(V)$. Therefore, $r_A(Y)$ is a reduced moment matrix and $r_A(Y) \succeq 0$ by the definition of r_A. $\square$

The matrix set $\mathcal{F}_t(n)$ permits us to formulate the following semidefinite relaxation for the max-cut problem:

$$\max\; \frac{1}{2}w(E) - \frac{1}{2}\sum_{ij\in E} w_{ij}y_{ij} \;\text{ subject to }\; \widetilde{M}_{t+1}(y) \succeq 0,\; y_\emptyset = 1. \tag{16.28}$$

Let $Y := \widetilde{M}_{t+1}(y) \in \mathcal{F}_t(n)$ be an optimum solution to the program (16.28) and let $X := \widetilde{M}_1(y)$; hence the off-diagonal entries of X are y_{ij} ($ij \in E_n$). If the rank of X is equal to 1, then X is a cut matrix and, therefore, the program (16.28) solves the max-cut problem exactly. An interesting question is to find conditions on the rank of X ensuring that X can be written as a convex combination of cut matrices and, therefore, that the relaxation (16.28) solves the max-cut problem.

For $t = 0$, there exist matrices $X \in \mathcal{F}_0(n)$ with rank 2 that cannot be written as a convex combination of cut matrices. For $t = 1$ Anjos and Wolkowicz [2] show that, if X has rank ≤ 2, then Y (and thus X) can be written as a convex combination of two cut matrices. This result can be generalized for any $t \geq 2$.

Theorem 16.20. *Let $t \geq 0$, $n \geq t + 2$, $Y = \widetilde{M}_{t+1}(y) \in \mathcal{F}_t(n)$, and $X := \widetilde{M}_1(y)$. If* rank $X \leq t + 1$, *then Y can be written as a convex combination of 2^t cut matrices and, therefore, the vector $(y_{ij})_{ij\in E_n}$ belongs to* $\text{CUT}(K_n)$.

The proof of Theorem 16.20 will be given in Section 16.4.1. We now examine some properties of the faces of the convex set $\mathcal{F}_t(n)$. All the 2^{n-1} cut matrices are vertices of $\mathcal{F}_t(n)$ (since they have rank 1). It is shown in [25] that the cut matrices are the *only* vertices of $\mathcal{F}_0(n)$. Moreover, it is shown in [24] that any two distinct cut matrices form a one-dimensional face of $\mathcal{F}_0(n)$. This adjacency property extends to each of the matrix sets $\mathcal{F}_t(n)$.

Proposition 16.21. *Let R and S be distinct subsets of $\{1, \ldots, n\}$ and $t \geq 0$. Then the set* $\mathrm{conv}(\{\widetilde{M}_{t+1}(R), \widetilde{M}_{t+1}(S)\})$ *is a face of $\mathcal{F}_t(n)$.*

Proof. In view of the switching symmetry from Lemma 16.19, we can assume, without loss of generality, that $R = \emptyset$. Set $Y_0 := \widetilde{M}_{t+1}(\emptyset)$, $Y_1 := \widetilde{M}_{t+1}(S)$, and $A := \frac{1}{2}(Y_0 + Y_1)$. Then A has the block decomposition $A = \left(\begin{smallmatrix} J & 0 \\ 0 & J \end{smallmatrix}\right)$ with respect to the partition of its index set $\mathcal{P}_{2t+2}(V)$ into the sets $\mathcal{I}_o$ and $\mathcal{I}_e$ consisting, respectively, of the sets I having odd and even intersections with S. We show that $F(A)$, the smallest face of $\mathcal{F}_t(n)$ containing A, is equal to the interval $[Y_0, Y_1]$. For this let $Y \in F(A)$, that is, $Y \in \mathcal{F}_t(n)$ and $\ker A \subseteq \ker Y$. Then all the columns of Y indexed by sets in $\mathcal{I}_o$ (resp. in $\mathcal{I}_e$) are identical. As Y is symmetric positive semidefinite with an all-ones diagonal, we find that Y has the block decomposition $Y = \left(\begin{smallmatrix} J & \alpha J \\ \alpha J & J \end{smallmatrix}\right)$ for some scalar $\alpha \in [-1, 1]$. Therefore, $Y = \frac{1+\alpha}{2} Y_0 + \frac{1-\alpha}{2} Y_1$. This concludes the proof. $\square$

Barahona and Mahjoub [6] had shown earlier that any two cuts form an edge of the cut polytope $\mathrm{CUT}(K_n)$. Padberg [29] showed moreover that any two cuts form an edge of the metric polytope $\mathrm{MET}(K_n)$, that is, $\mathrm{MET}(K_n)$ has the Trubin property with respect to $\mathrm{CUT}(K_n)$. This implies that any Lasserre relaxation $Q_t(K_n)$ also has the Trubin property. (For $t \geq 1$ this is true since $Q_t(K_n) \subseteq \mathrm{MET}(K_n)$ and, for $t = 0$, this is true by the above-mentioned result of [24] since $Q_0(n)$ is a linear bijective image of $\mathcal{F}_0(n)$.) In fact, $\mathrm{CUT}(K_n)$ and $\mathrm{MET}(K_n)$ have lots of higher dimensional common faces; for instance, any three cuts or any set $\delta(S_1), \ldots, \delta(S_k)$ of cuts in general position (meaning that each cell in the Venn diagram of the sets $S_1, \ldots, S_k$ is nonempty) form a face of $\mathrm{MET}(K_n)$ and thus of $\mathrm{CUT}(K_n)$ [10].

One may wonder whether some analogous result holds for the matrix set $\mathcal{F}_t(n)$. We saw above that any two cut matrices form a face of $\mathcal{F}_0(n)$; note that this does not extend to a set of three cut matrices (consider, e.g., $\mathcal{F}_0(3)$). On the other hand, Corollary 16.9 shows that $\mathcal{F}_{n-2}(n)$ is a simplex with the cut matrices as vertices. This suggests the following question.

Problem 16.22. Is it true that, for $t = 0, \ldots, n-2$, any set of 2^{t+1} cut matrices forms a face of $\mathcal{F}_t(n)$?

If this property holds, this suggests a "continuous" evolution from $\mathcal{F}_0(n)$ to the final simplex $\mathcal{F}_{n-2}(n)$. We saw above that the answer is yes for the two extreme values $t = 0$ and $t = n-2$, and the next result shows the answer is also positive for the next case $t = 1$. The proof of Theorem 16.23 is delayed until Section 16.4.2.

Theorem 16.23. *Any four cut matrices form a face of $\mathcal{F}_1(n)$.*

16.4.1 Proof of Theorem 16.20

A preliminary result. We begin with showing the auxiliary result from Proposition 16.25 that will play a central role in the proof of Theorem 16.20. The following lemma will be used as the base of induction.

Lemma 16.24. *Let* $Y = \widetilde{M}_2(y) \in \mathcal{F}_1(3)$ *and* $X := \widetilde{M}_1(y)$. *If* rank $X \leq 2$, *then* $y_{ij} = \pm 1$ *for some pair* ij *with* $1 \leq i < j \leq 3$.

Proof. As rank $X \leq 2$, there exists a nonzero vector $u \in \ker X$; thus $v := \binom{u}{0}$ belongs to $\ker Y$. Therefore, Y has at least two zero eigenvalues (since none of the eigenvectors y^A of Y has a zero coordinate). This implies that the vector (y_{12}, y_{13}, y_{23}) satisfies at least two of the equalities $y_{12} + y_{13} + y_{23} = -1$, $y_{12} - y_{13} - y_{23} = -1$, $y_{13} - y_{12} - y_{23} = -1$, and $y_{23} - y_{12} - y_{13} = -1$. Therefore, one of the components y_{ij} of y is equal to ± 1. □

Proposition 16.25. *Let* $Y = \widetilde{M}_{n-1}(y) \in \mathcal{F}_{n-2}(n)$ *and* $X := \widetilde{M}_1(y)$. *If* rank $X \leq n-1$, *then there exists a nonempty set* I *of even cardinality for which* $y_I = \pm 1$.

Proof. The proof is by induction on $n \geq 3$. The result holds for $n = 3$ by Lemma 16.24. Let $n \geq 4$ and suppose that the result holds for all $p \leq n-1$; we show that it also holds for n. We know from Lemma 16.8 that the eigenvectors of Y are the cut vectors y^A for $A \subseteq \{1, \ldots, n-1\}$ with corresponding eigenvalues $y^T y^A$. Set $V := \{1, \ldots, n\}$, $U := \{1, \ldots, n-1\}$, and

$$\mathcal{A} := \{A \subseteq U \mid y^T y^A \neq 0\}, \ \mathcal{B} := \{A \subseteq U \mid y^T y^A = 0\}. \tag{16.29}$$

Thus $\ker Y$ is spanned by the vectors y^A for $A \in \mathcal{B}$ and

$$y = \sum_{A \in \mathcal{A}} \lambda_A y^A, \tag{16.30}$$

where $\lambda_A = \frac{1}{2^{n-1}} y^T y^A > 0$ for all $A \in \mathcal{A}$.

Let $R_{\mathcal{A}}$ denote the matrix of order $|\mathcal{A}| \times n$ whose rows are the ± 1-incidence vectors ψ^A of the sets $A \in \mathcal{A}$ (viewing A as a subset of V). We claim that

$$\ker X = \ker R_{\mathcal{A}}. \tag{16.31}$$

Indeed, let $u \in \mathbb{R}^n$. Then, u belongs to $\ker X$ if and only if the extended vector $v := (u \ 0 \ \ldots \ 0) \in \mathbb{R}^{\mathcal{E}(V)}$ belongs to $\ker Y$ which, in turn, is equivalent to the fact that $v^T y^A = 0$ for all $A \in \mathcal{A}$. As the projection of y^A on the subspace $\mathbb{R}^n$ indexed by $\emptyset$ and the pairs $12, \ldots, 1n$ is equal to $\pm \psi^A$ (the sign $\pm$ depending on whether $1 \in A$), we have that $v^T y^A = \pm u^T \psi^A$. Therefore, $u \in \ker X$ if and only if $R_{\mathcal{A}} u = 0$, and thus (16.31) holds. We can assume that

$$\text{rank } X = n - 1.$$

Indeed, if rank $X \leq n-2$, then consider the principal submatrix Y' (resp. X') of Y (resp. of X) indexed by subsets of $\{1, \ldots, n-1\}$; as rank $X' \leq n-2$, the induction assumption implies that $y_I = \pm 1$ for some nonempty even set $I \subseteq \{1, \dots, n-1\}$, and thus Proposition 16.25 holds. Together with (16.31), this implies that

$$\text{rank } R_{\mathcal{A}} = n - 1. \tag{16.32}$$

In view of (16.30), we have $y_I = 1$ (resp. -1) for a set $I \in \mathcal{E}(V)$ if and only if $|I \cap A|$ is even (resp. odd) for all $A \in \mathcal{A}$. Thus we are left with the task of showing the existence of a nonempty set $I \in \mathcal{E}(V)$ and of $\epsilon \in \{0, 1\}$ satisfying $|I \cap A| \equiv \epsilon$ (modulo 2) for all $A \in \mathcal{A}$. Let $M_{\mathcal{A}}$ denote the matrix of order $|\mathcal{A}| \times (n-1)$ whose rows are the 0/1-incidence vectors $\chi^A \in \{0, 1\}^{n-1}$ of the sets $A \in \mathcal{A}$ (viewed as subsets of $U = \{1, \dots, n-1\}$). Equivalently, we have to show that at least one of the following two systems in the binary variable $x \in \mathbf{GF(2)}^{n-1}$:

$$\begin{cases} M_{\mathcal{A}} x = 0, \ x \neq 0, \\ M_{\mathcal{A}} x = e, \end{cases}$$

has a solution, where e denotes the all-ones vector. Indeed, if x is a solution of one of the above two systems and $I_0 := \{i \in \{1, \dots, n-1\} \mid x_i = 1\}$, then $I := I_0$ (resp. $I := I_0 \cup \{n\}$) is the required nonempty even set with $y_I = \pm 1$ if $|I_0|$ is even (resp. odd).

If the matrix $M_{\mathcal{A}}$ has rank less than or equal to $n-2$ over $\mathbf{GF(2)}$, then the system $M_{\mathcal{A}} x = 0$ has a nonzero solution over $\mathbf{GF(2)}$ and we are done. A collection of sets $A_1, \dots, A_p \in \mathcal{A}$ is said to form a $\mathbf{GF(2)}$-dependency if $\Delta_{i=1}^p A_i = \emptyset$. We claim the following:

$$\begin{array}{l} \text{If the rank of } M_{\mathcal{A}} \text{ over } \mathbf{GF(2)} \text{ is equal to } n-1 \text{ and every} \\ \mathbf{GF(2)}\text{-dependency in } \mathcal{A} \text{ involves an even number of sets,} \\ \text{then the system } M_{\mathcal{A}} x = e \text{ has a solution over } \mathbf{GF(2)}. \end{array} \tag{16.33}$$

To see this let $A_1, \dots, A_{n-1} \in \mathcal{A}$, whose incidence vectors are linearly independent over $\mathbf{GF(2)}$, and let x be a solution to the system

$$\chi^{A_i} x = 1 \text{ (modulo 2) } (i = 1, \dots, n-1).$$

Then $M_{\mathcal{A}} x = e$ holds. Indeed, if $A \in \mathcal{A} \setminus \{A_1, \dots, A_{n-1}\}$, then $A = \Delta_{i \in P} A_i$, where P is an odd subset of $\{1, \dots, n-1\}$. Therefore, $\chi^A x \equiv \sum_{i \in P} \chi^{A_i} x \equiv |P| \equiv 1$ (modulo 2). Thus (16.33) holds. In order to conclude the proof of Proposition 16.25, it suffices now to show the following result.

Lemma 16.26. *If there exists a $\mathbf{GF(2)}$-dependency in $\mathcal{A}$ involving an odd number of sets, then the rank of the matrix $M_{\mathcal{A}}$ over $\mathbf{GF(2)}$ is not greater than $n-2$.*

Proof. Let $A_0 = \Delta_{i=1}^p A_i$ be the smallest $\mathbf{GF(2)}$-dependency in $\mathcal{A}$ involving an odd number of sets, that is, p is even. Hence the vectors $\chi^{A_1}, \dots, \chi^{A_p}$ are linearly independent over $\mathbf{GF(2)}$. Suppose for a contradiction that $M_{\mathcal{A}}$ has rank $n-1$. Let $A_{p+1}, \dots, A_{n-1} \in \mathcal{A}$, whose incidence vectors together with those of $A_1, \dots, A_p$ are linearly independent over $\mathbf{GF(2)}$. We make the following claim:

$$\text{The vectors } \chi^{A_i \cup \{n\}} \ (i = 0, 1, \dots, n-1) \text{ are linearly independent over } \mathbf{GF(2)}. \tag{16.34}$$

Suppose not. Then $\Delta_{i\in H}(A_i \cup \{n\}) = \emptyset$ for some set $H \subseteq \{0, 1, \ldots, n-1\}$. Therefore, $|H|$ is even (in order to eliminate n) and $\Delta_{i\in H} A_i = \emptyset$, which implies that $0 \in H$ (since the vectors $\chi^{A_1}, \ldots, \chi^{A_{n-1}}$ are linearly independent over **GF(2)**). This combined with the fact that $A_0 = \Delta_{i=1}^{p} A_i$ implies that $H \setminus \{0\} = \{1, \ldots, p\}$. We reach a contradiction since p is even while $H \setminus \{0\}$ has an odd cardinality. Thus (16.34) holds. This implies the following:

$$\text{The vectors } \chi^{A_i \cup \{n\}} \ (i = 0, 1, \ldots, n-1) \text{ are linearly independent over } \mathbb{R}. \tag{16.35}$$

Indeed, suppose not. Then $\sum_{i=0}^{n-1} \lambda_i \chi^{A_i \cup \{n\}} = 0$ for some $\lambda_i \in \mathbb{R}$ not all equal to 0. Such λ_i exist that are rational valued and thus integer valued, not all of them even. Taking a reduction modulo 2 we find a linear dependency over **GF(2)** contradicting (16.34). Finally we show the following:

$$\text{The vectors } \psi^{A_i} \ (i = 0, 1, \ldots, n-1) \text{ are linearly independent over } \mathbb{R}. \tag{16.36}$$

For this note that $\psi^{A_i} = \begin{pmatrix} -2I & e \\ 0 & 1 \end{pmatrix} \chi^{A_i \cup \{n\}}$ (where I is the identity matrix of order $n-1$ and e is the all-ones vector of length $n-1$) and that the matrix $\begin{pmatrix} -2I & e \\ 0 & 1 \end{pmatrix}$ is nonsingular. We reach a contradiction since the matrix $R_{\mathcal{A}}$ has rank $n-1$ by (16.32). This concludes the proof of Lemma 16.26 and in turn the proof of Proposition 16.25. □

Corollary 16.27. *Let $Y = \widetilde{M}_{n-1}(y) \in \mathcal{F}_{n-2}(n)$ and $X := \widetilde{M}_1(y)$. If $\operatorname{rank} X \leq n-1$, then Y can be written as a convex combination of at most 2^{n-2} cut matrices.*

Proof. As in the proof of Proposition 16.25, let $\mathcal{A}$ be defined by (16.29) and, as in (16.30), let $y = \sum_{A\in\mathcal{A}} \lambda_A y^A$, $Y = \sum_{A\in\mathcal{A}} \lambda_A \widetilde{M}_{n-1}(y^A)$, where $\lambda_A > 0$ for all $A \in \mathcal{A}$ and $\sum_A \lambda_A = y_\emptyset = 1$. It suffices now to show that $|\mathcal{A}| \leq 2^{n-2}$. By Proposition 16.25, there exists a nonempty set $I \in \mathcal{E}(V)$ for which $y_I = \pm 1$. Therefore, the 0/1-incidence vectors of the sets $A \in \mathcal{A}$ are solutions of the equation $\chi^I x = \epsilon$ (modulo 2) where $\epsilon = 0$ if $y_I = 1$ and $\epsilon = 1$ if $y_I = -1$. This implies that $|\mathcal{A}| \leq 2^{n-2}$. □

Proof of Theorem 16.20. The proof of Theorem 16.20 is by induction on n. The case $n = t+2$ has been settled in Corollary 16.27. Hence we can assume that $n \geq t+3$ and that the result of Theorem 16.20 holds for $n-1$. We can also assume that $t \geq 1$ as the result holds obviously for $t = 0$. As before, $V = \{1, \ldots, n\}$ and $\mathcal{U}_{t+1}(V)$ is the set indexing the matrix Y, which consists of the sets $H \subseteq V$ with $|H| \leq t+1$ and $|H| \equiv t+1$ (modulo 2). Set

$$\mathcal{I} := \{I \in \mathcal{E}_{2t+2}(V) \mid I \neq \emptyset \text{ and } y_I = \pm 1\}.$$

Consider the relation on $\mathcal{U}_{t+1}(V)$: Given $H, K \in \mathcal{U}_{t+1}(V)$,

$$\begin{aligned} H \sim K \text{ if } y(H\Delta K) = \pm 1, \ \text{ i.e., } H\Delta K \in \mathcal{I} \cup \{\emptyset\} \\ \text{or, equivalently, if } Ye_H = \pm Ye_K, \end{aligned} \tag{16.37}$$

where Ye_H denotes the Hth column of Y (the last equivalence in (16.37) follows using Lemma 16.1(ii)). This is obviously an equivalence relation on $\mathcal{U}_{t+1}(V)$. We begin with some preliminary results about the collection $\mathcal{I}$.

Lemma 16.28. *If $I \neq J \in \mathcal{I}$ with $|I \setminus J|, |J \setminus I| \leq t+1$, then $I\Delta J \in \mathcal{I}$.*

Proof. As I, J have an even cardinality, the three sets $I \setminus J$, $J \setminus I$, and $I \cap J$ have the same parity. Say, $|I \setminus J| \geq |J \setminus I|$. Suppose first that $|I \cap J| \leq t+1$. If $|I \cap J| \equiv t+1$ (modulo 2), then $I \setminus J,\ J \setminus I,\ I \cap J \in \mathcal{U}_{t+1}(V)$ with $I \setminus J \sim I \cap J$ (since $I \in \mathcal{I}$) and $I \cap J \sim J \setminus I$ (since $J \in \mathcal{I}$). Therefore, $I \setminus J \sim J \setminus I$ implying that $(I \setminus J)\Delta(J \setminus I) = I \Delta J \in \mathcal{I}$. If $|I \cap J| \not\equiv t+1$ (modulo 2), let $a \in I \setminus J$. Then, $I \setminus (J \cup \{a\}),\ (I \cap J) \cup \{a\},\ (J \setminus I) \cup \{a\} \in \mathcal{U}_{t+1}(V)$ with $I \setminus (J \cup \{a\}) \sim (I \cap J) \cup \{a\}$ and $(I \cap J) \cup \{a\} \sim (J \setminus I) \cup \{a\}$, which implies that $I \Delta J \in \mathcal{I}$.

Suppose now that $|I \cap J| > t+1$. Let A be a subset of $I \cap J$ for which $|A| + |I \setminus J| = t+1$ and set $B := (I \cap J) \setminus A$. Then, $|A| + |J \setminus I| \leq t+1$, $|A| + |J \setminus I| \equiv t+1$ (modulo 2), $|B| = |I| - t - 1 \leq t+1$, and $|B| \equiv t+1$ (modulo 2). As $(I \setminus J) \cup A \sim B$ and $B \sim (J \setminus I) \cup A$, we deduce again that $I \Delta J \in \mathcal{I}$. □

As a consequence of Proposition 16.25 we have the following result.

Lemma 16.29. *For any subset $T \subseteq V$ with $|T| \geq t+2$, there exists a set $I \in \mathcal{I}$ that is contained in T; moreover, if $|T| \geq t+3$, such an I exists having cardinality not greater than $t+1$.*

Proof. The first part of the lemma is a direct application of Proposition 16.25. (Indeed, let $S \subseteq T$ with $|S| = t+2$ and let Y' (resp. X') be the principal submatrix of Y (resp. of X) indexed by subsets of S; thus $Y' \in \mathcal{F}_t(t+2)$ with rank $X' \leq t+1$ and Proposition 16.25 implies the existence of a set $I \in \mathcal{I}$ contained in S.) Suppose now that $|T| \geq t+3$ and let $T_1, T_2 \subseteq T$ with $|T_1| = |T_2| = t+2$ and $|T_1 \Delta T_2| = 2$. If T_1 or T_2 contains a member of $\mathcal{I}$ of size not greater than $t+1$, then we are done. Otherwise, both T_1 and T_2 belong to $\mathcal{I}$, which, using Lemma 16.28, implies that $T_1 \Delta T_2 \in \mathcal{I}$. Thus we have found a member of $\mathcal{I}$ contained in T of size $2 \leq t+1$. □

Choose a set $I_0 \in \mathcal{I}$ having the minimum cardinality among all sets in $\mathcal{I}$. Then, $|I_0| \leq t+1$ by Lemma 16.29 (since $|V| \geq t+3$). We can assume, without loss of generality, that the element $n \in V$ belongs to I_0. Set

$$\mathcal{I}(n) := \{I \in \mathcal{I} \mid n \in I\},\ \mathcal{I}(\overline{n}) := \mathcal{I} \setminus \mathcal{I}(n).$$

The rest of the proof of Theorem 16.20 can be sketched as follows. Let Y_0 denote the principal submatrix of Y indexed by $\mathcal{U}_{t+1}(V \setminus \{n\})$ and let y_0 denote the projection of y onto the subspace $\mathbb{R}^{\mathcal{E}_{2t+2}(V \setminus \{n\})}$; thus $Y_0 = \widetilde{M}_{t+1}(y_0)$. Using the induction assumption, we know that Y_0 can be decomposed as a convex combination of at most 2^t cut matrices or, equivalently, that y_0 can be written as a convex combination of at most 2^t cut vectors, that is,

$$Y_0 = \sum_{A \in \mathcal{A}} \lambda_A \widetilde{M}_{t+1}(A),\ y_0 = \sum_{A \in \mathcal{A}} \lambda_A y^A, \tag{16.38}$$

where $\lambda_A > 0$ for $A \in \mathcal{A}$, $\sum_A \lambda_A = 1$, and $\mathcal{A}$ is a collection of subsets of $V \setminus \{n\}$ with $|\mathcal{A}| \leq 2^t$. Our goal now is to show that the above decomposition of y_0 can be extended to a decomposition of y. Namely, we will show that for each set $A \in \mathcal{A}$ one can define $A' := A$ or $A \cup \{n\}$ in such a way that the identity $y = \sum_{A \in \mathcal{A}} \lambda_A y^{A'}$ holds, which implies then that Y can be written as a convex combination of at most 2^t cut matrices, concluding the proof.

Lemma 16.30. *For any $H \in \mathcal{U}_{t+1}(V)$, there exists a set $I \in \mathcal{I}(n)$ for which $|H \Delta I| \leq t+1$.*

Proof. Suppose not. Let $H \in \mathcal{U}_{t+1}(V)$ be a counterexample (that is, $|H \Delta I| \geq t+2$ for all $I \in \mathcal{I}(n)$) of minimum cardinality. Choose $I_1 \in \mathcal{I}(n)$ with $|I_1| \leq t+1$ and for which $|I_1 \Delta H|$ is minimum. Then, $|H \Delta I_1| \geq t+2$, implying $|H \Delta I_1| \geq t+3$ since $|H \Delta I_1| \equiv |H| \equiv t+1$ (modulo 2). By Lemma 16.29, there exists $I \in \mathcal{I}$ such that $I \subseteq (H \Delta I_1) \setminus \{n\}$. Then, the set $J := I \Delta I_1$ belongs to $\mathcal{I}$ by Lemma 16.28 (since $I \setminus I_1 \subseteq H$ and $I_1 \setminus I \subseteq I_1$ have size not greater than $t+1$) and thus $J \in \mathcal{I}(n)$. Suppose first that $|I \setminus I_1| \leq |I \cap I_1|$. Then, $|J| = |I \setminus I_1| + |I_1 \setminus I| \leq |I_1| \leq t+1$ and $|H \Delta J| = |H \Delta I_1| - |I| < |H \Delta I_1|$ which contradicts the minimality assumption made about the set I_1. Therefore, $|I \setminus I_1| > |I \cap I_1|$. This implies that $|H \Delta I| = |H| - |I \setminus I_1| + |I \cap I_1| < |H|$ and thus $H \Delta I \in \mathcal{U}_{t+1}(V)$. Therefore, by the minimality assumption made about H, the set $H \Delta I$ is not a counterexample to the lemma and thus there exists a set $J \in \mathcal{I}(n)$ for which $|H \Delta I \Delta J| \leq t+1$. Now, $H \Delta I \Delta J \in \mathcal{U}_{t+1}(V)$, $H \Delta I \Delta J \sim H \Delta I$ (since $J \in \mathcal{I}$), $H \Delta I \sim H$ (since $I \in \mathcal{I}$), which implies that $H \Delta I \Delta J \sim H$ and thus $I \Delta J \in \mathcal{I}$. As $I \Delta J \in \mathcal{I}(n)$ with $|H \Delta I \Delta J| \leq t+1$, we contradict our assumption that H is a counterexample to the lemma. □

Denote by $\mathcal{R}_1, \dots, \mathcal{R}_m$ the equivalence classes of the equivalence relation $\sim$ from (16.37) on $\mathcal{U}_{t+1}(V)$. By Lemma 16.30, each class $\mathcal{R}_j$ contains sets $H, K \in \mathcal{U}_{t+1}(V)$ with $n \in H \setminus K$. For $j = 1, \dots, m$ set

$$\mathcal{I}_j(n) := \{H \Delta K \mid H, K \in \mathcal{R}_j,\ n \in H \setminus K\}.$$

Thus, $\mathcal{I}(n) = \bigcup_{j=1}^m \mathcal{I}_j(n)$. Recall that I_0 is a member of $\mathcal{I}(n)$ having minimum cardinality. The following property of I_0 will play a crucial role in the rest of the proof.

Lemma 16.31. $I_0 \in \bigcap_{j=1}^m \mathcal{I}_j(n)$.

Proof. Fix $j = 1$; we show that $I_0 \in \mathcal{I}_1(n)$. Let H be a member of $\mathcal{R}_1$ containing n for which $|H \Delta I_0|$ is minimum. If $|H \Delta I_0| \leq t+1$, then $H \Delta I_0 \in \mathcal{U}_{t+1}(V)$, $H \Delta I_0 \sim H$, which implies that $I_0 \in \mathcal{I}_1(n)$, and we are done. Suppose now that $|H \Delta I_0| \geq t+2$ and thus $|H \Delta I_0| \geq t+3$. Let $I \in \mathcal{I}$ such that $I \subseteq H \Delta I_0$ and $|I| \leq t+1$ (apply Lemma 16.29). Then, $I \Delta I_0 \in \mathcal{I}(n)$ (by Lemma 16.28) and thus $|I \Delta I_0| \geq |I_0|$ by the choice of I_0. Therefore, $|I \setminus I_0| \geq |I \cap I_0|$. Then, $|H \Delta I| = |H| - |I \setminus I_0| + |I \cap I_0| \leq |H| \leq t+1$ and, hence, $H \Delta I \in \mathcal{U}_{t+1}(V)$. Now, $H \Delta I \sim H$ implying that $H \Delta I \in \mathcal{R}_1$. As $n \in H \Delta I$ and $|H \Delta I \Delta I_0| < |H \Delta I_0|$, we reach a contradiction with the way we have chosen H. □

As a consequence of (16.38) and of the definition (16.24) of y^A, we have that

$$y_I = (-1)^{|A \cap I|} \text{ for all } A \in \mathcal{A} \text{ and } I \in \mathcal{I}(\overline{n}). \tag{16.39}$$

We now observe that we can assume, without loss of generality, that

$$y_I = 1 \text{ for all } I \in \mathcal{I}(\overline{n}). \tag{16.40}$$

Indeed, fix $A \in \mathcal{A}$ and consider the matrix $Y' := r_A(Y)$ obtained using the switching symmetry described in Lemma 16.19. Then, $Y' = \widetilde{M}_{t+1}(y')$, where $y'_I = (-1)^{|A \cap I|} y_I$ for all $I \in \mathcal{E}_{2t+2}(V)$. Therefore, in view of (16.39), $y'_I = 1$ for all $I \in \mathcal{I}(\overline{n})$. If we can show that Y' is a convex combination of 2^t cut matrices, then the same holds for $Y = r_A(Y')$. Hence, replacing Y with Y', we can assume that (16.40) holds. Moreover, we can assume, without loss of generality, that

$$y_{I_0} = 1. \tag{16.41}$$

Indeed, if $y_{I_0} = -1$, then we can replace Y with its switching $r_{\{n\}}(Y)$.

Lemma 16.32. *For all* $I \in \mathcal{I}(n)$, $y_I = 1$ *and* $|A \cap I| \equiv |A \cap I_0|$ (modulo 2) *for all* $A \in \mathcal{A}$.

Proof. Let $I \in \mathcal{I}(n)$. Then $I, I_0 \in \mathcal{I}_j(n)$ for some $j = 1, \dots, m$ by Lemma 16.31. Hence, $I_0 = H_0 \Delta K_0$, $I = H \Delta K$, where $H, H_0, K, K_0 \in \mathcal{R}_j$ with $n \in H \cap H_0$ and $n \notin K \cup K_0$. As $H \Delta H_0$, $K \Delta K_0 \in \mathcal{I}(\overline{n})$, we deduce from (16.40) that $y(H \Delta H_0) = y(K \Delta K_0) = 1$. Together with $y(H_0 \Delta K_0) = 1$ from (16.41), this implies that $Ye_H = Ye_{H_0}$, $Ye_K = Ye_{K_0}$, $Ye_{H_0} = Ye_{K_0}$, and thus $Ye_H = Ye_K$, yielding $y_I = y_{H \Delta K} = 1$. This shows the first part of the lemma.

We have that $I \Delta I_0 = (H \Delta H_0) \Delta (K \Delta K_0)$, where $H \Delta H_0$, $K \Delta K_0 \in \mathcal{I}(\overline{n})$. As each $A \in \mathcal{A}$ has an even intersection with every $I \in \mathcal{I}(\overline{n})$ by (16.39) and (16.40), we deduce that A has an even intersection with $I \Delta I_0$ and thus $|A \cap I| \equiv |A \cap I_0|$ (modulo 2). ☐

Let $A \in \mathcal{A}$. Set $\epsilon_A := (-1)^{|A \cap I_0|}$. Then, $(-1)^{|A \cap I|} = \epsilon_A$ for all $I \in \mathcal{I}(n)$ by Lemma 16.32. If $\epsilon_A = 1$, set $A' := A$, and, if $\epsilon_A = -1$, set $A' := A \cup \{n\}$. Then,

$$(-1)^{|A' \cap I|} = 1 = y_I \quad \text{for all } I \in \mathcal{I}(n).$$

Define $z \in \mathbb{R}^{\mathcal{E}_{2t+2}(V)}$ by

$$z := \sum_{A \in \mathcal{A}} \lambda_A y^{A'}$$

and $Z := \widetilde{M}_{t+1}(z)$. To conclude the proof we show that $y = z$ or, equivalently, $Y = Z$. We already know that $z_I = y_I$ for all $I \in \mathcal{E}_{2t+2}(V \setminus \{n\})$ and $I \in \mathcal{I}(n)$. Hence the submatrices of Y and Z indexed by $\mathcal{U}_{t+1}(V \setminus \{n\})$ are identical and it suffices now to verify that $Y(H, K) = Z(H, K)$ for all $H, K \in \mathcal{U}_{t+1}(V)$ with $n \in H \setminus K$. Pick such H, K and let $K' \in \mathcal{U}_{t+1}(V \setminus \{n\})$ for which $H \sim K'$ (which exists by Lemma 16.30). Then, $Ye_H = Ye_{K'}$ and $Ze_H = Ze_{K'}$. Therefore, $Y(H, K) = Y(K, K') = Z(K, K') = Z(H, K)$. Thus we have shown that $Y = Z$. This concludes the proof of Theorem 16.20.

16.4.2 Proof of Theorem 16.23

In view of the switching symmetry it suffices to show Theorem 16.23 for four cut matrices $\widetilde{M}_2(S_i)$ ($i = 0, 1, 2, 3$), where $S_0 := \emptyset$. We proceed as follows. Set $A := \frac{1}{4} \sum_{i=0}^{3} \widetilde{M}_2(S_i)$. Then, $\ker(A) = \bigcap_{i=0}^{3} \ker\ \widetilde{M}_2(S_i)$. Our goal is to show that $F(A)$, the smallest face of $\mathcal{F}_1(n)$ containing A, is equal to the convex hull of the cut matrices $\widetilde{M}_2(S_i)$, that is, any $Y \in F(A)$ can be written as a convex combination of the matrices $\widetilde{M}_2(S_i)$ ($i = 0, 1, 2, 3$). The following result will be used in the proof.

Lemma 16.33. *If all the cuts $x := \delta(S_i)$ $(i = 0, 1, 2, 3)$ satisfy some linear equality $\sum_{jk \in E_n} u_{jk} x_{jk} = u_0$, then any matrix $Y \in F(A)$ satisfies the equation*

$$\sum_{jk \in E_n} u_{jk} Y e_{jk} = u_0 Y e_{\emptyset}, \tag{16.42}$$

that is, $\sum_{jk \in E_n} u_{jk} e_{jk} - u_0 e_{\emptyset} \in \ker Y$.

Proof. Note first that (16.42) holds for every $Y := \widetilde{M}_2(S_i)$; this follows from the fact that $Y_{jk,\emptyset} = \delta(S_i)_{jk}$ and $Y_{jk,rs} = \delta(S_i)_{jk} \cdot \delta(S_i)_{rs}$. Hence (16.42) holds for A and thus for $Y \in F(A)$, since the vector $\sum_{jk} u_{jk} e_{jk} - u_0 e_{\emptyset}$ belongs to $\ker A \subseteq \ker Y$. □

Theorem 16.23 holds for $n = 3$ since $\mathcal{F}_1(3)$ is a simplex. We now assume that $n \geq 4$. We first settle the case $n = 4$.

Lemma 16.34. *For $S_0 = \emptyset$, $S_i = \{1, i+1\}$ $(i = 1, 2, 3)$, the cut matrices $\widetilde{M}_2(S_i)$ $(i = 0, 1, 2, 3)$ form a face of $\mathcal{F}_1(4)$.*

Proof. Let $Y \in F(A)$, that is, $Y = \widetilde{M}_2(y) \succeq 0$ for some $y \in \mathbb{R}^{\mathcal{E}_4(V)}$ and $\ker Y \supseteq \ker A$. As each cut vector y^{S_i} satisfies the equation $y_{1234} = 1$, the same holds for y (since $e_{12} - e_{34} \in \ker A \subseteq \ker Y$ by Lemma 16.1(ii)) and thus $Y_{12,34} = Y_{13,24} = Y_{14,23} = 1$. The principal submatrix X of Y indexed by the set $\{\emptyset, 12, 13, 23\}$ belongs to the simplex $\mathcal{F}_1(3)$ and thus can be written as $X = \alpha_1 \widetilde{M}_2(\emptyset) + \alpha_2 \widetilde{M}_2(12) + \alpha_3 \widetilde{M}_2(13) + \alpha_4 \widetilde{M}_2(23)$ with $\alpha_i \geq 0$ and $\sum_i \alpha_i = 1$. Then, $Y = \alpha_1 \widetilde{M}_2(\emptyset) + \sum_{i=2}^{4} \alpha_i \widetilde{M}_2(1i)$, where the cut matrices are now matrices in $\mathcal{F}_1(4)$. (To see this note that $\widetilde{M}_2(23) = \widetilde{M}_2(14)$ and use the fact that $y_{1234} = 1$.) □

Proposition 16.35. *Any four cut matrices form a face of $\mathcal{F}_1(4)$.*

Proof. Consider four cut matrices $\widetilde{M}_2(S_i)$, where $S_0 = \emptyset$. The case when all S_i are even sets has been settled in Lemma 16.34; hence we can assume, without loss of generality, that $S_1 = \{1\}$. Let $Y = \widetilde{M}_2(y) \in F(A)$, that is, $Y \succeq 0$ and $\ker Y \supseteq \ker A$. The vector y can be decomposed as

$$y = \begin{pmatrix} y_{\emptyset} = 1 \\ \widetilde{y} \\ y_{1234} \end{pmatrix},$$

where $\widetilde{y}$ is indexed by the pairs of $V = \{1, \ldots, 4\}$. By Lemma 16.33, $\widetilde{y}$ satisfies any triangle equality that is satisfied at equality by all the cuts $\delta(S_i)$ $(i = 0, 1, 2, 3)$. Denote by $\mathcal{T}$ the set of common triangle equalities satisfied by all the cuts $\delta(S_i)$. As $Q_1(K_4) = \text{CUT}(K_4)$, $\widetilde{y}$ can be written as a convex combination of cuts. Let us assume that conditions (16.43) and (16.44) hold:

$$\mathcal{T} \neq \emptyset. \tag{16.43}$$

The only cuts satisfying all the equalities from $\mathcal{T}$ are $\delta(S_i)$ $(i = 0, 1, 2, 3)$. (16.44)

Then $\widetilde{y} = \sum_{i=0}^{3} \alpha_i \delta(S_i)$ for some $\alpha_i \geq 0$ with $\sum_i \alpha_i = 1$. Therefore, the matrices Y and $Y' := \sum_{i=0}^{3} \alpha_i \widetilde{M}_2(S_i)$ coincide everywhere except perhaps at their (12, 34), (13, 24), and

$(14, 23)$ entries. By (16.43), there exists a triangle equality that is satisfied by all the cuts $\delta(S_i)$. Suppose, to fix ideas, that this triangle equality is $-y_{12}+y_{13}-y_{23}=-1$. By Lemma 16.33, we deduce that $Y(-e_{12}+e_{13}-e_{23}+e_{\emptyset})=0$. Computing entry 34, we find that $Y_{12,34}=y_{14}-y_{24}+y_{34}$ can be expressed in terms of entries of $\widetilde{y}$ only. As the matrix Y' satisfies the same identity $Y'(-e_{12}+e_{13}-e_{23}+e_{\emptyset})=0$, it follows that $Y_{12,34}=Y'_{12,34}$. This shows, therefore, that $Y=Y'=\sum_{i=0}^{3}\alpha_i\widetilde{M}_2(S_i)$, which concludes the proof.

We now proceed to show the claims (16.43) and (16.44). Since $S_0=\emptyset$ and $S_1=\{1\}$, the only possible triangle equalities in $\mathcal{T}$ are of the form

$$y_{rs}-y_{rt}-y_{st}=-1 \text{ for } t\in\{2,3,4\} \text{ and } r\neq s\in\{1,2,3,4\}\setminus\{t\}. \tag{16.45}$$

We consider several cases depending on the number of singletons among the sets S_i. If $S_i=\{i\}$ for $i=1,2,3$, then $\mathcal{T}$ consists of the three triangle equalities in (16.45) for $t=4$. If $S_i=\{i\}$ for $i=1,2$ and $S_3=\{1,2\}$, then $\mathcal{T}$ consists of the triangle equalities from (16.45) for $(t=3, rs=14, 24)$ and for $(t=4, rs=13, 23)$. If $S_i=\{i\}$ for $i=1,2$ and, say, $S_3=\{1,3\}$, then $\mathcal{T}$ consists of the triangle equalities from (16.45) for $(t=3, rs=12, 14)$ and for $(t=4, rs=12, 23)$. Finally, if $S_2=\{1,2\}$ and $S_3=\{1,3\}$, then $\mathcal{T}$ consists of the triangle equalities from (16.45) for $(t=2, rs=14, 34)$, $(t=3, rs=14, 24)$, and $(t=4, rs=23)$. In each case one can verify that (16.44) holds. □

We now consider the general case $n\geq 4$. As before, $S_0=\emptyset$. If some cell in the Venn diagram of S_1, S_2, S_3 contains two distinct points, say 1 and n, then each cut matrix $\widetilde{M}_2(S_i)$ satisfies the equation $Y_{\emptyset,1n}=1$ and thus the same holds for any $Y\in F(A)$. Hence each matrix in $F(A)$ is the 1-extension of some matrix in $\mathcal{F}_1(n-1)$. Using Lemmas 16.13 and 16.14, one can verify that the cut matrices $\widetilde{M}_2(S_i)$ form a face of $\mathcal{F}_1(n)$ if and only if the cut matrices $\widetilde{M}_2(S_i\setminus\{n\})$ form a face of $\mathcal{F}_1(n-1)$. Repeating this argument, we arrive at the conclusion that one can assume that each cell in the Venn diagram contains at most one point and thus $n\leq 8$. We first settle the case when each cell contains exactly one point, that is, $n=8$ and, say,

$$S_1:=\{1,2,3,6\},\ S_2:=\{1,2,4,5\},\ S_3:=\{1,3,4,7\}, \tag{16.46}$$

with Venn diagram

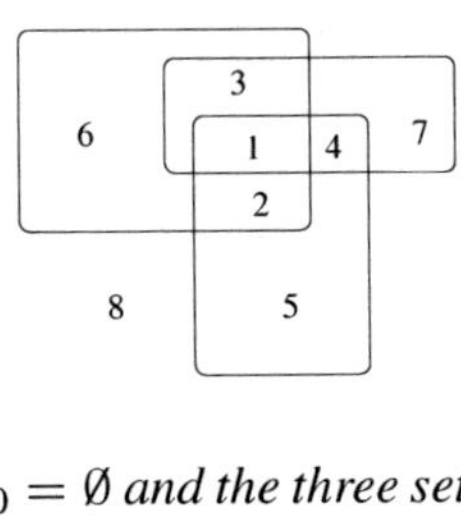

Proposition 16.36. *Assume that $S_0=\emptyset$ and the three sets S_1, S_2, S_3 are as in* (16.46). *Then the four cut matrices $\widetilde{M}_2(S_i)$ $(i=0,1,2,3)$ form a face of $\mathcal{F}_1(8)$.*

Proof. We use the following result from [25]:

$$\text{The cut matrices } \widetilde{M}_1(S_i)\ (i=0,1,2,3) \text{ form a face of } \mathcal{F}_0(8). \tag{16.47}$$

Set $A_1 := \frac{1}{4}\sum_{i=0}^{3}\widetilde{M}_1(S_i)$. Observe that $\widetilde{M}_1(S_i)$ (resp. A_1) is equal to the principal submatrix of $\widetilde{M}_2(S_i)$ (resp. A) indexed by the set $\mathcal{I} := \{\emptyset, 12, \ldots, 18\}$. Let $Y \in F(A)$ and let X be the principal submatrix of Y indexed by $\mathcal{I}$. Say, $Y = \widetilde{M}_2(y)$, where $y \in \mathbb{R}^{\mathcal{E}_4(V)}$. Then $X = \widetilde{M}_1(y)$ and $\ker X \supseteq \ker A_1$ (by Lemma 16.1(i)). Therefore, X belongs to $F(A_1)$, the smallest face of $\mathcal{F}_0(8)$ containing A_1, which, using (16.47), implies that $X = \sum_{i=0}^{3}\alpha_i\widetilde{M}_1(S_i)$ for some nonnegative α_i with $\sum_i \alpha_i = 1$. Set $Y' := \sum_{i=0}^{3}\alpha_i\widetilde{M}_2(S_i)$. Our task now is to show that $Y = Y'$. By construction, Y and Y' have the same (I, J)-entries for $I, J \in \mathcal{I}$. It suffices therefore to verify that $Y_{ab,cd} = Y'_{ab,cd}$ for any 4-tuple $abcd \subseteq V$.

Let $\mathcal{T}$ denote the set of 4-tuples $T := abcd$ for which $|T \cap S_i|$ is even for all i. Then $\widetilde{M}_2(S_i)_{ab,cd} = 1$ for all i implying $A_{ab,cd} = 1$ and thus $Y_{ab,cd} = 1$ (by Lemma 16.1). One can verify that $\mathcal{T}$ consists of the following 4-tuples: 1567, 2348, 1245, 3678, 1347, 2568, 1236, 4578, 1358, 2467, 1278, 3456, 1468, and 2357. The relation

$$ab \sim cd \text{ if } abcd \in \mathcal{T} \tag{16.48}$$

is an equivalence relation on the set E_8 of pairs of points of V and induces the partition of E_8 into the following seven equivalence classes:

$$\begin{gathered}\mathcal{L}_1 := \{12, 36, 45, 78\},\ \mathcal{L}_2 := \{13, 26, 47, 58\},\\ \mathcal{L}_3 := \{14, 25, 37, 68\},\ \mathcal{L}_4 := \{15, 24, 38, 67\},\\ \mathcal{L}_5 := \{16, 23, 48, 57\},\ \mathcal{L}_6 := \{17, 28, 34, 56\},\ \mathcal{L}_7 := \{18, 27, 35, 46\}.\end{gathered} \tag{16.49}$$

The following property of the set $\mathcal{T}$ holds:

$$\text{For every 4-tuple } T \notin \mathcal{T},\ \text{there exists } T_0 \in \mathcal{T} \text{ such that } |T \cap T_0| = 3. \tag{16.50}$$

Indeed, let $T := abcd$ and let $\mathcal{L}_i$ be the equivalence class containing the pair ab. Then, $\mathcal{L}_i$ contains the pair cc' for some $c' \in V \setminus \{a, b, c\}$, which implies that $T_0 := abcc'$ belongs to $\mathcal{T}$ and meets T in three elements.

Based on this, one can verify that $Y_{ab,cd} = Y'_{ab,cd}$ for any 4-tuple $T := abcd$. Indeed, if $T \in \mathcal{T}$, then $Y_{ab,cd} = Y'_{ab,cd} = 1$. Otherwise, let $T_0 \in \mathcal{T}$ such that $|T \cap T_0| = 3$ (which exists by (16.50)), say, $T_0 := abcd'$. Then, $Y_{ab,cd} = (Ye_{ab})_{cd} = (Ye_{cd'})_{cd} = Y_{\emptyset,dd'}$ and, similarly, $Y'_{ab,cd} = Y'_{\emptyset,dd'} = Y_{\emptyset,dd'}$. Thus $Y = Y'$, which concludes the proof. □

We finally consider the case when some cell in the Venn diagram of $S_1, \ldots, S_3$ is empty. In other words, given $W \subseteq V = \{1, \ldots, 8\}$, we have to show that the cut matrices $\widetilde{M}_2(S_i \cap W)$ form a face of $\mathcal{F}_1(n)$, where $n := |W| \le 8$.

Denote by E_W the set of pairs ij ($1 \le i < j \le 8$) that are contained in W. If $|W| \le 4$, then we are done, by Proposition 16.35. We now assume that $|W| \ge 5$. This implies that E_W contains at least one edge e_i from each class $\mathcal{L}_i$ in (16.49). Set $E_0 := \{e_i \mid i = 1, \ldots, 7\}$. Let A^W denote the principal submatrix of A indexed by $E_W \cup \{\emptyset\}$. By the definition of E_0, the principal submatrix of A^W indexed by $E_0 \cup \{\emptyset\}$ is equal to A_1 (defined in the proof of Proposition 16.36 as the principal submatrix of A indexed by $\{\emptyset, 12, \ldots, 18\}$).

Let $X \in \mathcal{F}_1(n)$ with $\ker X \supseteq \ker A^W$; we have to show that X is a convex combination of the cut matrices $\widetilde{M}_2(S_i \cap W)$. Say, $X = \widetilde{M}_2(x)$, where $x \in \mathbb{R}^{\mathcal{E}_4(W)}$. We extend x to a vector $y \in \mathbb{R}^{\mathcal{E}(V)}$ in the following way: For $ij \in E_8 \setminus E_W$, let $ab \in E_0$ such that $ij \sim ab$, and set $y_{ij} := x_{ab}$. Let T be a 4-tuple of elements of V. If $T \in \mathcal{T}$, then let $y_T := 1$.

Otherwise, let $T_0 \in \mathcal{T}$ such that $|T \cap T_0| = 3$ (recall (16.50)). Then $|T \Delta T_0| = 2$ and set $y_T := y_{T\Delta T_0}$. Finally, set $Y := \widetilde{M}_2(y)$.

Then, $Y \succeq 0$ (since Y is an extension of X). Moreover, $\ker Y \supseteq \ker A$. This follows from the fact that $\ker A$ is spanned by the vectors $(u\ 0\ \dots\ 0)^T$ for $u \in \ker A^W$ and the vectors $e_{ab} - e_{ij}$ for $ab \sim ij$ with $ab \in E_0, ij \in E_8$. We know from Proposition 16.36 that $Y = \sum_{i=0}^{3} \alpha_i \widetilde{M}_2(S_i)$ for some nonnegative scalars α_i with $\sum_i \alpha_i = 1$. Restricting to the entries in $E_W \cup \{\emptyset\}$, we deduce that $X = \sum_i \alpha_i \widetilde{M}_2(S_i \cap W)$. This concludes the proof of Theorem 16.23.

16.5 Numerical Comparison of the Various Relaxations for Small n

In this section we examine in detail how the Lasserre relaxations $Q_t(K_n)$ approximate the cut polytope $\mathrm{CUT}(K_n)$ for small n and t. In particular, we compare them with the Anjos–Wolkowicz relaxation F_n (defined in (16.23)) and with the Lovász–Schrijver relaxation $N_+^t(K_n)$. Some of our results have been obtained using the software package[2] SeDuMi for solving SDPs. Recall the inclusions

$$\mathrm{CUT}(K_n) \subseteq Q_t(K_n) \subseteq N_+(K_n) \subseteq F_n \subseteq \mathrm{MET}(K_n) \cap Q_0(K_n)$$

for $t \geq 1$. For $n = 3, 4$, one has

$$\mathrm{CUT}(K_n) = \mathrm{MET}(K_n) = Q_1(K_n) \subset Q_0(K_n).$$

Indeed, the matrix

$$X := \begin{pmatrix} 1 & -\frac{1}{2} & -\frac{1}{2} \\ -\frac{1}{2} & 1 & -\frac{1}{2} \\ -\frac{1}{2} & -\frac{1}{2} & 1 \end{pmatrix}$$

belongs to $\mathcal{F}_0(3)$, the matrix $\begin{pmatrix} X & 0 \\ 0 & 1 \end{pmatrix}$ belongs to $\mathcal{F}_0(4)$, while the vector $\left(-\frac{1}{2}, -\frac{1}{2}, -\frac{1}{2}\right)^T$ does not belong to $\mathrm{CUT}(K_3)$.

For $n = 5$, one has

$$\mathrm{CUT}(K_5) = N_+(K_5) = Q_2(K_5) \subset Q_1(K_5) \subset F_5.$$

The equality $\mathrm{CUT}(K_5) = N_+(K_5)$ is shown in [22]. The strict inclusion $\mathrm{CUT}(K_5) \subset Q_1(K_5)$ follows from the fact that the minimum of the linear objective function $\sum_{ij \in E_5} y_{ij}$ over $\mathrm{CUT}(K_5)$ is equal to -2, while its minimum over $Q_1(K_5)$ is equal to -2.5 attained at the matrix $Y = \widetilde{M}_2(y) \in \mathcal{F}_1(5)$, where $y_{ij} := -\frac{1}{4}$ $(ij \in E_5)$ and $y_{ijhk} := \frac{3}{8}$ $(1 \leq i < j < h < k \leq 5)$; note that the minimum over the relaxation F_5 is also equal to -2.5 [1]. We verified the strict inclusion $Q_1(K_5) \subset F_5$ using a computer. For instance, the minimum of the linear objective function $14y_{12} + 13y_{13} + 14y_{14} + 12y_{15} + 13y_{23} + 15y_{24} + 17y_{25} + 13y_{34} + 11y_{35} + 14y_{45}$ is equal to -34.833887 over F_5 and to -34.3402792 over $Q_1(K_5)$. Note, however, that for many random linear objective functions, one finds the same optimum over F_5 and $Q_1(K_5)$.

[2] This optimization software for semidefinite programming has been developed by J. Sturm and is accessible from his homepage http://fewcal.kub.nl/sturm.

For $n = 6$, one has

$$\mathrm{CUT}(K_6) = Q_2(K_6) \subset N_+(K_6).$$

Indeed, the minimum of the linear objective function $2\sum_{i=2}^{6} y_{1i} + \sum_{2\le i<j\le 6} y_{ij}$ is equal to -4 over both $\mathrm{CUT}(K_6)$ and $Q_2(K_6)$, while it is equal to $-\frac{49}{12} < -4$ over $N_+(K_6)$ (cf. [22]). The equality $\mathrm{CUT}(K_6) = Q_2(K_6)$ now follows from the fact that the cut polytope $\mathrm{CUT}(K_6)$ is determined by the triangle inequalities (16.5), the *pentagonal inequalities*

$$\sum_{1\le i<j\le 5} y_{ij} \ge -2, \tag{16.51}$$

the *hexagonal inequalities*

$$2\sum_{i=2}^{6} y_{1i} + \sum_{2\le i<j\le 6} y_{ij} \ge -4, \tag{16.52}$$

and the inequalities obtained from (16.51) and (16.52) by permutation of the nodes and switching by cuts.

We now treat the case $n = 7$. Grishukhin [17] has computed that all the facets of $\mathrm{CUT}(K_7)$ are, up to permutation and switching, induced by one of the following 11 inequalities:

(i) the triangle inequality (16.5);

(ii) the pentagonal inequality (16.51);

(iii) the hexagonal inequality (16.52);

(iv) the inequality $\sum_{ij\in E_7} y_{ij} \ge -3$;

(v) the inequality $4y_{12} + \sum_{j=3}^{7} 2y_{1j} + 2y_{2j} + \sum_{3\le i<j\le 7} y_{ij} \ge -6$;

(vi) the inequality $3\sum_{j=2}^{7} y_{1j} + \sum_{2\le i<j\le 7} y_{ij} \ge -7$;

(vii) the (bicycle odd-wheel) inequality $y_{12} + \sum_{j=3}^{7} y_{1j} + y_{2j} + \sum_{j=3}^{6} y_{j,j+1} + y_{37} \ge -4$;

(viii) the inequality $4y_{12} + \sum_{j=3}^{7} 2y_{1j} + 2y_{2j} + \sum_{3\le i<j\le 7} y_{ij} - y_{12} - y_{23} - y_{34} - y_{14} \ge -8$;

(ix) the inequality $5y_{12} + 5y_{13} + 3y_{23} + \sum_{j=4}^{7} 3y_{1j} + 2y_{2j} + 2y_{3j} \ge -9$;

(x) the (parachute) inequality $\sum_{j=1}^{5} y_{j,j+1} - \sum_{j=4,5,6} y_{1j} - \sum_{j=2,3} y_{j6} - \sum_{j=2}^{5} y_{j7} \ge -5$;

(xi) the (Grishukhin) inequality $\sum_{1\le i<j\le 4} y_{ij} + y_{56} + y_{57} - 2\sum_{j=1}^{4} y_{j5} - y_{16} - y_{36} - y_{27} - y_{47} - y_{67} \ge -5$.

(Inequalities (i) to (vi) belong to the class of hypermetric inequalities and (vii) to (ix) to the class of clique-web inequalities; cf. Section 30.5 in [9] for details).

It is shown in [22] that the inequalities (i), (ii), (vii), and (x) are valid for the Lovász–Schrijver relaxation $N_+(K_7)$ and thus for $Q_2(K_7)$ too (by Corollary 16.12). Using the computer program SeDuMi, we have computed the minimum of the linear objective function $c^T y$ over $Q_t(K_7)$ for $t \leq 2$, where $c^T y$ is the left-hand side of the remaining inequalities within (i) to (xi). Table 16.1 summarizes our results.

Table 16.1. *Comparing the facet-defining inequalities for* $\mathrm{CUT}(K_7)$.

Inequality	Min. over $\mathrm{CUT}(K_7)$	Min. over $Q_2(K_7)$	Min. over $Q_1(K_7)$	Min. over F_7	Min. over $Q_0(K_7)$	Min. over $N_+(K_7)$
triangle (i)	−1	−1	−1	−1	−1.5	−1
pentagonal (ii)	−2	−2	−2.5	−2.5	−2.5	−2
hexagonal (iii)	−4	−4	−4.5	−4.5	−4.5	−49/12 ∼ −4.0833
(iv)	−3	−3.5	−3.5	−3.5	−3.5	?
(v)	−6	−6.051882	−6.5	−6.5	−6.5	?
(vi)	−7	−7	−7.5	−7.5	−7.5	?
bicycle (vii)	−4	−4	−5	−5.0045	−5.8090	−4
(viii)	−6	−6	−6.5817	−6.6522	−7.9661	?
(ix)	−9	−9	−9.6433	−9.7036	−11.0166	?
parachute (x)	−4	−4	−4.7439	−4.8099	−5.9220	−4
Grishukhin (xi)	−5	−5	−5.6152	−5.7075	−6.9518	?

The set F_7 improves the relaxation $Q_0(K_7)$ for the inequalities (i) and (vii) to (xi) that altogether make up more than 96 percent of the total number of facets of $\mathrm{CUT}(K_7)$. On the other hand, the improvement of $Q_1(K_7)$ over F_7 does not seem to be very significant.

We know that $Q_3(K_7) = \mathrm{CUT}(K_7)$. Note that $Q_2(K_7)$ already approximates $\mathrm{CUT}(K_7)$ very well; indeed, the minimum over $Q_2(K_7)$ is strictly less than the minimum over $\mathrm{CUT}(K_7)$ only for the inequalities (iv) and (v), which represent less than 1.3 percent of the total number of facets of $\mathrm{CUT}(K_7)$.

Given $c \in \mathbb{Q}^{E_n}$ and $t \geq 0$, it is of interest to evaluate the *integrality ratio*

$$\rho_t := \frac{\sum_{ij} c_{ij} - \min(c^T y \mid y \in \mathrm{CUT}(K_n))}{\sum_{ij} c_{ij} - \min(c^T y \mid y \in Q_t(K_n))},$$

that is, the ratio of the maximum weight of a cut with respect to the weights c to the maximum obtained by optimizing over the relaxation $Q_t(K_n)$. Goemans and Williamson [16] showed that

$$\rho_0 \geq 0.878$$

Table 16.2. *The integrality ratios for the facets of K_7.*

Inequality	$\sum_{ij} c_{ij}$	ρ_2	ρ_1	ρ_F	ρ_0
triangle (i)	3	1	1	1	$\frac{8}{9} \sim$ **0.888**
pentagonal (ii)	10	1	0.96	0.96	0.96
hexagonal (iii)	20	1	0.979	0.979	0.979
(iv)	21	**0.979**	0.979	0.979	0.979
(v)	34	0.998	0.987	0.987	0.987
(vi)	33	1	0.987	0.987	0.987
bicycle (vii)	16	1	**0.952**	**0.952**	0.917
(viii)	30	1	0.984	0.982	0.948
(ix)	47	1	0.988	0.987	0.965

Table 16.3. *The antiweb graph AW_9^2.*

Inequality	Min. over CUT(K_9)	Min. over $Q_2(K_9)$	Min. over $Q_1(K_9)$	Min. over F_9	Min. over $Q_0(K_9)$
AW_9^2	-6	-6	-6.8282634	-6.9937	-9
integ. ratio		1	0.966	0.960	$\frac{8}{9} \sim 0.888$

when $c \geq 0$ and Feige and Schechtman [14] constructed graphs for which the integrality ratio ρ_0 attains the worst-case value 0.878. It is, however, known that, in practice, the integrality ratio ρ_0 is larger than the worst-case value. As an indication we have computed the integrality ratio for the facets of K_7 and the relaxations $Q_t(K_7)$ and F_7 (in which case the ratio is denoted by ρ_F). Table 16.2 gives the results. Observe that the worst-case value for ρ_0 is $\frac{8}{9} \sim 0.888$ (attained at the triangle inequality), while the worst-case values for ρ_F, ρ_1, and ρ_2 are, respectively, 0.952, 0.952, and 0.979.

Another example demonstrating the strict inclusion $Q_1(n) \subset F_n$ is as follows. Consider the circulant graph $G = AW_n^2$ on n nodes whose edgeset E consists of the pairs $(i, i+1)$ and $(i, i+2)$ $(i = 1, \ldots, n)$ (indices being taken modulo n). Table 16.3 gives the values of the minimum of $\sum_{ij \in E} y_{ij}$ over CUT(K_9), $Q_1(K_9)$, $Q_2(K_9)$, and F_9. The last row shows the corresponding integrality ratios. In fact, CUT(AW_n^2) $= Q_2(AW_n^2)$ for any odd n, since contracting edge 12 in AW_n^2 produces a planar graph (use Theorem 16.17).

16.6 Concluding Remarks

Application to the Boolean quadric polytope. In this chapter we have considered in detail the hierarchy of semidefinite relaxations $Q_t(G)$ $(t \geq 0)$ of the cut polytope CUT(G). All the results we have presented have counterparts for the analogous problem in 0/1-variables, namely, for the *unconstrained 0/1-quadratic programming problem*

$$\max\ x^T Qx \text{ subject to } x \in \{0, 1\}^n$$

and the associated *Boolean quadric polytope*

$$\mathrm{QP}_n := \mathrm{conv}(\{(x_i x_j)_{1\le i\le j\le n} \mid x \in \{0,1\}^n\})$$

studied in detail by Padberg [29]. Indeed, the mapping

$$x \in \{0,1\}^n \mapsto y := (1, 1-2x_1, \ldots, 1-2x_n)^T \in \{\pm 1\}^{n+1}$$

yields the correspondence $xx^T \mapsto yy^T$ between the vertices of QP_n and the vertices of $\mathrm{CUT}(K_{n+1})$. Therefore, as is well known, QP_n and $\mathrm{CUT}(K_{n+1})$ are in affine bijection.

The Lasserre construction can be applied for constructing semidefinite relaxations of QP_n. Namely, for $t \ge 0$, let $\mathcal{Q}_t(n)$ denote the set of positive semidefinite matrices of the form

$$M_{t+1}(y) := (y(I \cup J))_{\substack{I,J\subseteq V \\ |I|,|J|\le t+1}},$$

where $y \in \mathbb{R}^{\mathcal{P}_{2t+2}(V)}$ with $y_\emptyset = 1$ (comparing with (16.8), note that the symmetric difference is now replaced by the union). Then, the projection of $\mathcal{Q}_t(n)$ on the subspace indexed by the pairs ij with $1 \le i \le j \le n$ is a semidefinite relaxation of QP_n. The set $\mathcal{Q}_0(n)$ is the basic semidefinite relaxation for QP_n consisting of the symmetric positive semidefinite matrices Y of order $n+1$ having their main diagonal equal to their first row and $Y_{0,0} = 1$.

Given a graph $G = (V, E)$ with $V = \{1, \ldots, n\}$, Padberg [29] observed that the stable set polytope $\mathrm{ST}(G)$ of G arises as the projection of a face of the Boolean quadric polytope QP_n, namely, $d \in \mathbb{R}^V$ belongs to $\mathrm{ST}(G)$ if and only if $(d, y) \in \mathrm{QP}_n$ for some $y \in \mathbb{R}^{E_n}$ satisfying $y_{ij} = 0$ for all edges $ij \in E$. Therefore, each relaxation $\mathcal{Q}_t(n)$ for QP_n yields a semidefinite relaxation for $\mathrm{ST}(G)$. More precisely, the projection on $\mathbb{R}^V$ of the set

$$\{M_{t+1}(y) \in \mathcal{Q}_t(n) \mid y_{ij} = 0 \ (ij \in E)\}$$

is a semidefinite relaxation of $\mathrm{ST}(G)$, which, in the case $t = 0$, coincides with the basic semidefinite relaxation $\mathrm{TH}(G)$; moreover, this semidefinite relaxation coincides with the set $Q_t(\mathrm{FR}(G))$ obtained by applying the Lasserre construction to the fractional stable set polytope $\mathrm{FR}(G)$ (defined in (16.14)). See [23] for more details.

Lower bounds for the rank of the Lasserre procedure. It would be interesting to find lower bounds for the *Lasserre rank* of a graph G, which is defined as the smallest integer t for which $\mathrm{CUT}(G) = Q_t(G)$; the *LS rank* of G is defined analogously as the smallest t for which $\mathrm{CUT}(G) = N_+^t(G)$. As $Q_t(G) \subseteq N_+^{t-1}(G)$, the Lasserre rank is less than or equal to the LS rank plus one. In the case of the stable set problem, it has been shown in [28] that the smallest t for which equality $N_+^t(\mathrm{FR}(G)) = \mathrm{ST}(G)$ holds satisfies $t \le \alpha(G)$, with equality when G is the line graph of K_{2n+1} [31]. In the case of max-cut, the LS rank of K_n is conjectured to be equal to $n-4$; equality has been shown for $n = 4, 5, 6, 7$ [22]. We saw above that the Lasserre rank of K_n is equal to 1, 2, 2, 3 for $n = 4, 5, 6, 7$, respectively.

It is shown in [22] that, for n odd, the inequality

$$\sum_{1\le i<j\le n} y_{ij} \ge \frac{1-n}{2} \tag{16.53}$$

is valid for $N_+^{\frac{n-3}{2}}(K_n)$ and thus for $Q_{\frac{n-1}{2}}(K_n)$. We conjecture that the inequality (16.53) is *not* valid for $Q_{\frac{n-3}{2}}(K_n)$ for n odd. If true, this would imply that the Lasserre rank of K_n is at least $\frac{n-1}{2}$ and thus that the LS rank is at least $\frac{n-3}{2}$ for n odd.

In order to show the above conjecture, we have to find a positive semidefinite moment matrix $\widetilde{M}_{\frac{n-1}{2}}(y)$ with $\sum_{ij\in E_n} y_{ij} < \frac{1-n}{2}$. Set $a_0 := 1$ and, for $1 \leq r \leq \frac{n-1}{2}$,

$$a_{2r} := (-1)^r \prod_{i=0}^{r-1} \frac{2i+1}{n-2i-1},$$

and define $y \in \mathbb{R}^{\mathcal{E}(V)}$ by letting $y_I := a_{|I|}$ for all $I \in \mathcal{E}(V)$. Then, $\sum_{ij\in E_n} y_{ij} = \binom{n}{2} a_2 = -\binom{n}{2}\frac{1}{n-1} < \frac{1-n}{2}$. We conjecture that $\widetilde{M}_{\frac{n-1}{2}}(y) \succeq 0$ for all n odd. We verified that this fact is true for small $n = 3, 5, 7$.

Note Added in Proof. This conjecture has now been proved to hold for any odd $n \geq 3$ (Laurent [21]).

Bibliography

[1] M.F. Anjos and H. Wolkowicz. A strengthened SDP relaxation via a second lifting for the max-cut problem. *Discrete Applied Mathematics,* 119:79–106, 2002.

[2] M.F. Anjos and H. Wolkowicz. Geometry of semidefinite max-cut relaxations via ranks. *Journal of Combinatorial Optimization*, 6:237–270, 2002.

[3] E. Balas, S. Ceria, and G. Cornuéjols. A lift-and-project cutting plane algorithm for mixed 0-1 programs. *Mathematical Programming*, 58:295–324, 1993.

[4] F. Barahona. On cuts and matchings in planar graphs. *Mathematical Programming*, 60:53–68, 1993.

[5] F. Barahona. On the computational complexity of Ising spin glass models. *Journal of Physics* A, *Mathematical and General*, 15:3241–3253, 1982.

[6] F. Barahona and A.R. Mahjoub. On the cut polytope. *Mathematical Programming*, 36:157–173, 1986.

[7] V. Chvátal. Edmonds polytopes and a hierarchy of combinatorial problems. *Discrete Mathematics*, 4:305–337, 1973.

[8] W. Cook and S. Dash. On the matrix-cut rank of polyhedra. *Mathematics of Operations Research*, 26:19–30, 2001.

[9] M. Deza and M. Laurent. *Geometry of Cuts and Metrics*. Springer-Verlag, Berlin, 1997.

[10] M. Deza, M. Laurent, and S. Poljak. The cut cone III: On the role of triangle facets. *Graphs and Combinatorics*, 8:125–142, 1992.

[11] R.D. Hill and S.R. Waters. On the cone of positive semidefinite matrices. *Linear Algebra and Its Applications*, 90:81–88, 1987.

[12] F. Eisenbrand. On the membership problem for the elementary closure of a polyhedron. *Combinatorica*, 19:299–300, 1999.

[13] F. Eisenbrand and A.S. Schulz. Bounds on the Chvátal rank of polytopes in the 0/1 cube. In G. Cornuéjols, R.E. Burkard, and G.J. Woeginger, editors, Integer Programming and Combinatorial Optimization, *Lecture Notes in Computer Science* 1610, pages 137–150, Springer, Berlin, 1999.

[14] U. Feige and G. Schechtman. On the optimality of the random hyperplane rounding technique for MAX CUT. *Random Structures and Algorithms*, 20:403–440, 2002.

[15] M.X. Goemans and L. Tunçel. When does the positive semidefiniteness constraint help in lifting procedures? *Mathematics of Operations Research*, 26:796–815, 2001.

[16] M.X. Goemans and D.P. Williamson. Improved approximation algorithms for maximum cuts and satisfiability problems using semidefinite programming. *Journal of the Association for Computing Machinery*, 42:1115–1145, 1995.

[17] V.P. Grishukhin. All facets of the cut cone C_n for $n = 7$ are known. *European Journal of Combinatorics*, 11:115–117, 1990.

[18] J.B. Lasserre. *Optimality Conditions and LMI Relaxations for 0-1 Programs*. Technical Report N. 00099, LAAS, Toulouse, 2000.

[19] J.B. Lasserre. Global optimization with polynomials and the problem of moments. *SIAM Journal on Optimization*, 11:796–817, 2001.

[20] J.B. Lasserre. An explicit exact SDP relaxation for nonlinear 0-1 programs. In K. Aardal and A.M.H. Gerards, editors, *Lecture Notes in Computer Science* 2081, pages 293–303, Springer, Berlin, 2001.

[21] M. Laurent. Lower bound for the number of iterations in semidefinite hierarchies for the cut polytope. *Mathematics of Operations Research,* 28:871–883, 2003.

[22] M. Laurent. Tighter linear and semidefinite relaxations for max-cut based on the Lovász–Schrijver lift-and-project procedure. *SIAM Journal on Optimization*, 12:345–375, 2001.

[23] M. Laurent. A Comparison of the Sherali-Adams, Lovász-Schrijver and Lasserre Relaxations for 0-1 Programming. *Mathematics of Operations Research*, 28:470–496, 2003.

[24] M. Laurent and S. Poljak. On a positive semidefinite relaxation of the cut polytope. *Linear Algebra and Its Applications*, 223/224:439–461, 1995.

[25] M. Laurent and S. Poljak. On the facial structure of the set of correlation matrices. *SIAM Journal on Matrix Analysis and Applications*, 17:530–547, 1996.

[26] M. Laurent and F. Rendl. Semidefinite programming and integer programming. In K. Aardal, G. Nemhauser, and R. Weismantel, editors, *Handbook on Discrete Optimization*, to appear.

[27] L. Lovász. On the Shannon capacity of a graph. *IEEE Transactions on Information Theory*, IT-25:1–7, 1979.

[28] L. Lovász and A. Schrijver. Cones of matrices and set-functions and 0–1 optimization. *SIAM Journal on Optimization*, 1:166–190, 1991.

[29] M. Padberg. The Boolean quadric polytope: Some characteristics, facets and relatives. *Mathematical Programming*, 45:139–172, 1989.

[30] H.D. Sherali and W.P. Adams. A hierarchy of relaxations between the continuous and convex hull representations for zero-one programming problems. *SIAM Journal on Discrete Mathematics*, 3:411–430, 1990.

[31] T. Stephen and L. Tunçel. On a representation of the matching polytope via semidefinite liftings. *Mathematics of Operations Research*, 24:1–7, 1999.

Part VI

Computation

Chapter 17
The Steinberg Wiring Problem

*Nathan W. Brixius** *and Kurt M. Anstreicher*†

It is clear that much more effort is needed and should be expended to solve this interesting riddle posed to combinatorial optimizers well over 35 years ago.
—M.W. Padberg and M.P. Rijal (1996)

MSC 2000. 90C27, 90C09, 90C10
Key words. Quadratic assignment problem, branch-and-bound, Gilmore-Lawler bound

17.1 Introduction

In a 1961 paper [44], Leon Steinberg described a "backboard wiring" problem that has resisted solution for 40 years. The problem concerns the placement of computer components so as to minimize the total amount of wiring required to connect them. In the particular instance considered by Steinberg, 34 components with a total of 2625 interconnections are to be placed on a backboard with 36 open positions. The geometry of the backboard is illustrated in Figure 17.1.

To formulate the wiring problem mathematically it is convenient to add two dummy components, with no connections to any others, so that the numbers of components and locations are both $n = 36$. Let a_{ik} be the number of wires that connect components i and k and b_{jl} be the "distance" between locations j and l on the backboard. (There are several possible choices for the b_{jl}. In his paper Steinberg considered using 1-norm, 2-norm, and

*Microsoft Corporation, Redmond, WA.
†Department of Management Sciences, University of Iowa, Iowa City, IA.

P01 •	P02 •	P03 •	P04 •	P05 •	P06 •	P07 •	P08 •	P09 •
P10 •	P11 •	P12 •	P13 •	P14 •	P15 •	P16 •	P17 •	P18 •
P19 •	P20 •	P21 •	P22 •	P23 •	P24 •	P25 •	P26 •	P27 •
P28 •	P29 •	P30 •	P31 •	P32 •	P33 •	P34 •	P35 •	P36 •

Figure 17.1. *Backboard for Steinberg problem.*

squared 2-norm distances between the backboard locations.) Let $x_{ij} = 1$ if component i is placed at location j on the backboard and $x_{ij} = 0$ otherwise. Doubling the objective, the problem can then be written in the form

$$
\begin{aligned}
\text{(SWP)} \qquad \min \sum_{i,j,k,l} a_{ik} b_{jl} x_{ij} x_{kl} \text{ subject to} \\
\sum_{j} x_{ij} = 1, \quad i = 1, \dots, n, \\
\sum_{i} x_{ij} = 1, \quad j = 1, \dots, n, \\
x_{ij} \in \{0, 1\}, \quad i, j = 1, \dots, n.
\end{aligned}
$$

Note that the constraints of SWP are exactly that $X = \{x_{ij}\}$ is an $n \times n$-permutation matrix.

Steinberg devised a heuristic method to obtain (hopefully) a good solution for the wiring problem and applied it to the 2-norm and squared 2-norm versions of the problem. Most subsequent research has been directed to the 1-norm formulation.

In this chapter we describe the development of a branch-and-bound (B&B) algorithm to solve the 1-norm version of SWP to optimality. SWP is an example of a quadratic assignment problem (QAP), described in the next section. In Section 17.3 we describe lower-bounding schemes that have been proposed for QAP. In Section 17.4 we give some comparisons between bounds on SWP and similar problems, outline the construction of a complete B&B algorithm, and give computational results.

17.2 Quadratic Assignment Problems

The general QAP, introduced by Lawler [30], has the form

$$
\begin{aligned}
\text{(QAP)} \qquad \min \sum_{i,j,k,l} d_{ijkl} x_{ij} x_{kl} \text{ subject to} \\
X \in \Pi,
\end{aligned}
$$

where Π denotes the set of $n \times n$-permutation matrices. The problem SWP is an example of a "symmetric Koopmans–Beckmann" QAP (KBP). The term "Koopmans–Beckmann" denotes that the objective coefficient for $x_{ij} x_{kl}$ has the product form $a_{ik} b_{jl}$, and "symmetric" means that $a_{ij} = a_{ji}$ and $b_{ij} = b_{ji}$ for all i, j. The 1-norm, 2-norm, and squared 2-norm versions of SWP are now known as the ste36a, ste36c, and ste36b QAPs; these and all other problem names are taken from QAPLIB [11].

The QAP can be used to formulate a variety of interesting problems in location theory, manufacturing, data analysis, and other areas [9, 12, 40]. Unfortunately, the QAP is, typically, extraordinarily difficult to solve due to its size. Several well-known combinatorial optimization problems, such as the traveling salesman problem (TSP), can be formulated as QAPs, and therefore the QAP is NP-hard. However, while TSPs with thousands of cities are now tractable [6, 37], in general a QAP with $n = 30$ presents a formidable computational challenge. For example, the well-known nug30 problem, posed in 1968 [34], was only recently solved using the equivalent of approximately seven years of serial computation [2].

Because of the extreme difficulty of the QAP, many heuristic approaches have been proposed to generate what we hope are good quality solutions. These techniques include GRASP [39], genetic algorithms [16], simulated annealing [14], tabu search [43, 45], and ant systems [17]. The best known objective value for the 1-norm version of SWP, 9526, was first obtained in 1990 using a tabu search algorithm [43] and has been subsequently rediscovered many times. One permutation (assignment of components to locations) attaining this value is

$$(12, 19, 30, 11, 2, 3, 22, 20, 10, 21, 5, 4, 13, 15, 31, 32, 28, 29,$$
$$24, 14, 17, 18, 16, 9, 8, 7, 6, 23, 33, 34, 35, 25, 27, 26, 1, 36).$$

Note that in this assignment the two dummy components (numbers 35 and 36) are placed in corners of the grid that are diagonally opposite one another.

17.3 Solution Approaches for the Quadratic Assignment Problem

Most exact solution methods for the QAP have been of the B&B type. A key component in such algorithms is the choice of method used to obtain lower bounds. There are a variety of lower-bounding approaches for the QAP, some of which have been used successfully in complete B&B algorithms.

17.3.1 Gilmore–Lawler bound

The most widely used lower bound for the QAP is the Gilmore–Lawler bound (GLB) [18, 30]. Note that the objective in QAP can be written in the form

$$\sum_{i,j} \left(\sum_{k,l} d_{ijkl} x_{kl} \right) x_{ij}.$$

Let f_{ij} denote the solution value in the linear assignment problem (LAP)

$$\min \sum_{k,l} d_{ijkl} x_{kl} \text{ subject to}$$
$$X \in \Pi, \; x_{ij} = 1.$$

It is then clear that GLB := LAP(F) $\leq$ QAP, where LAP(F) denotes the LAP with cost matrix F, and for convenience we use the name of an optimization problem to also refer to its solution value. For the general QAP the computation of GLB requires the solution of $n^2 + 1$ LAPs. However, for a KBP the LAP associated with each f_{ij} is trivial to solve, and as a result F can be obtained in a total of only $O(n^3)$ operations.

Several successful B&B algorithms for the QAP have utilized the GLB [8, 10, 13, 33]. GLB-based algorithms have proved effective for problems up to about size $n = 24$, but for larger problems the growth in nodes may become overwhelming.

17.3.2 Eigenvalue and related bounds

A KBP, with an added linear term, can be written in the matrix form

$$\text{(KBP)} \quad \min_{X \in \Pi} \operatorname{tr}(AXB + C)X^T,$$

where tr(·) denotes the trace of a matrix. When A and B are symmetric, a bound for the quadratic term can be based on the fact that $X \in \Pi \Rightarrow X \in \mathcal{O}$, where $\mathcal{O}$ denotes the set of orthogonal matrices: $\mathcal{O} = \{X \mid XX^T = I\}$. For a symmetric matrix A let $\lambda(A) \in \mathbb{R}^n$ denote the vector of eigenvalues of A, and for vectors u and v let $\langle u, v \rangle_-$ denote the "minimal product"

$$\langle u, v \rangle_- := \min_{\pi} \sum_{i=1}^{n} u_i v_{\pi(i)},$$

where $\pi(\cdot)$ is a permutation of $1, 2, \ldots, n$. It is easy to show that $\langle u, v \rangle_-$ is obtained by putting the components of one of the vectors in nondecreasing order, and the components of the other in nonincreasing order, before taking the inner product. It can then be shown [15] that

$$\min_{X \in \mathcal{O}} \operatorname{tr}(AXBX^T) = \langle \lambda(A), \lambda(B) \rangle_-, \tag{17.1}$$

and therefore

$$\langle \lambda(A), \lambda(B) \rangle_- + \text{LAP}(C) \tag{17.2}$$

is a valid lower bound for a symmetric KBP. Unfortunately, the basic eigenvalue bound (17.2) is too weak to be computationally useful. Various schemes for improving the bound have been considered [15, 19, 41]. The most promising of these appears to be the projected eigenvalue bound (PB) of [19]. The construction of PB is based on enforcing the row and column sum constraints on X, in addition to orthogonality. Let V be an $n \times (n-1)$ matrix whose columns are an orthonormal basis for the nullspace of $e^T = (1, 1, \ldots, 1)$, and let $D = C + (2/n)Aee^T B$. Then

$$\text{PB} := \langle \lambda(V^T AV), \lambda(V^T BV) \rangle_- + \text{LAP}(D) - \frac{1}{n^2}(e^T Ae)(e^T Be).$$

As shown in [19], for many problems PB provides a good quality bound at modest computational cost.

A quadratic programming bound (QPB) for KBP that is related to PB was devised in [4]. By construction QPB $\geq$ PB, and evaluating QPB requires the approximate solution of a convex quadratic program (QP) in the n^2 variables X. In [4] QPB was evaluated by solving the QP using an interior-point algorithm. This approach provides a very accurate solution, but is too expensive to use in a B&B context. In [7] the Frank–Wolfe (FW) algorithm is used to approximately solve the QP associated with QPB. Although the asymptotic properties of the FW algorithm are known to be poor, this scheme is of interest in the context of QPB because the work on each iteration of the FW algorithm is dominated by the solution of an LAP. The resulting B&B algorithm exhibits state-of-the-art performance on many benchmark KPBs. In [2] the same QPB-based B&B algorithm, implemented using the "master-worker" distributed processing platform, obtains the first solution of several large problems including the nug30 QAP.

There has also been recent work devising bounds for KBP based on semidefinite programming. In [5] it is shown that there is a semidefinite programming interpretation for (17.1), and this interpretation is used in the derivation of QPB in [4]. Semidefinite programming bounds for KBP are described in [31] and [46]. In [3] it is shown that the basic semidefinite programming bound of [46] is closely related to PB. More complex semidefinite programming bounds described in [31] and [46] are also related to the linear programming bounds described below. These semidefinite programming bounds are often of excellent quality, but are obtained at a very high computational cost.

17.3.3 Linear programming and dual linear programming bounds

A large class of bounds for the QAP are related to linear programming relaxations of the problem. Defining new variables $y_{ijkl} = x_{ij}x_{kl}$ and dropping the integrality conditions results in a linear programming relaxation [1, 42]

$$
\begin{aligned}
\text{(LPQAP)} \quad & \min \sum_{i,j,k,l} y_{ijkl}d_{ijkl} \text{ subject to} \\
& \sum_j x_{ij} = 1, \quad i = 1,\dots,n, \\
& \sum_i x_{ij} = 1, \quad j = 1,\dots,n, \\
& \sum_l y_{ijkl} = x_{ij}, \quad i,j,k = 1,\dots,n, \\
& \sum_k y_{ijkl} = x_{ij}, \quad i,j,l = 1,\dots,n, \\
& y_{ijkl} = y_{klij}, \quad i,j,k,l = 1,\dots,n, \qquad (17.3) \\
& x_{ij} \geq 0, \quad y_{ijkl} \geq 0, \quad i,j,k,l = 1,\dots,n.
\end{aligned}
$$

The symmetry constraints (17.3) imply that LPQAP can be formulated using variables $y_{ijkl}, i \leq k$. Additional variables can be eliminated using the facts that $y_{ijij} = x_{ij}$ for all i and j, $y_{ijil} = 0$ for all i and $j \neq l$, and $y_{ijkj} = 0$ for all $i \neq k$ and j for X feasible in QAP. Taken together, these observations allow for a reformulation of LPQAP as a linear programming problem with $n^2 + n^2(n-1)^2/2$ variables. Further analysis [38, Section 7.1] can be used to reduce the number of equality constraints required in LPQAP to $2n(n-1)^2 - (n-1)(n-2), n \geq 3$. For a symmetric problem like SWP, LPQAP can be formulated using $n^2+n^2(n-1)^2/4$ variables and $n^2(n-2)+2n-1$ equality constraints, $n \geq 3$ [25], [38, Section 7.3]. For a symmetric problem with $n = 36$, for example, LPQAP can be written using 398,196 variables and 44,135 equality constraints. The solution of LPQAP using an interior-point method was investigated in [42]. This approach produces excellent bounds for many problems, but appears to be prohibitively costly for implementation in a B&B algorithm.

It is known [1] that, if the symmetry conditions (17.3) are dropped, then the solution value in LPQAP is exactly GLB. It can also be shown [28] that many bounding schemes for QAP can be viewed as Lagrangian procedures that attempt to approximately solve the dual of LPQAP. Computationally, the most successful of these is a method motivated by the Hungarian algorithm for LAP, due to P. Hahn and coworkers [20, 21, 22]. The B&B code of Hahn et al. recently obtained the first solution of the kra30a QAP, a hospital layout problem dating from 1972 [23].

In [28] a dual linear programming procedure similar to that proposed in [20] is used to obtain a lower bound of 7860 for the 1-norm version of SWP. To our knowledge this is the best known lower bound for the problem.

17.3.4 The polyhedral approach

The polyhedral approach to QAP is based on investigating the convex hull of 0/1-valued solutions to the linear programming relaxation LPQAP. This line of research was initiated by Padberg and Rijal [38] and has been further developed by Kaibel and Jünger [24, 25, 26, 27]. The convex hull of 0/1-valued solutions of LPQAP is a face of the Boolean quadric polytope, studied in [36]. An essential element of the polyhedral approach is the characterization of valid inequalities that can be added to LPQAP to tighten the relaxation. In [38, Section 1.5] the polyhedral approach is applied to a linear programming relaxation of SWP. (The relaxation is similar to LPQAP, but is specialized for a symmetric KBP and also exploits sparsity of the matrix A.) Solution of the resulting linear program took approximately one month on a 50MHz Sun workstation and obtained a lower bound of 7794. This was the best known lower bound for the problem prior to the dual linear programming bound obtained in [28].

The polyhedral approach to discrete optimization has resulted in very successful *branch-and-cut* algorithms for particular discrete optimization problems such as TSP [37, 6]. Branch-and-cut algorithms typically invest a large amount of time generating valid inequalities, and resolving subproblems, in an effort to reduce branching to a minimum. The development of branch-and-cut algorithms for QAP is still in its infancy, but recent results [24] indicate that the methodology promises to become a general purpose solution method.

17.4 Solving the Steinberg Problem

In this section we consider applying a B&B algorithm to solve the 1-norm version of SWP to optimality. In Table 17.1 we give the values for a number of different bounds applied to the problem. In the table "Sum" is the trivial bound obtained from the fact that there are 2625 interconnections[1] between components and all distances are at least one. TDB, the triangle decomposition bound of [29], is a parametric strengthening of PB that can be applied to problems with distance matrices arising from 1-norms on grids. QPB is computed using

[1]The number of interconnections is given as 2620 in [44]. This appears to be due to an error in computing the sum of the entries in row/column 29 of the matrix A; see [44, Figure 1].

Table 17.1. *Bounds for 1-norm wiring problem.*

Bound	Value	Gap
Dual-LP	7860	17%
Polyhedral	7794	18%
GLB	7124	25%
TDB	6997	27%
Sum	5250	45%
QPB	−10294	208%
PB	−11700	223%

500 FW iterations (see [7] for details), and all gaps are computed relative to the best-known value of 9526. It is clear that PB and the related QPB perform very poorly. The performance of GLB is reasonable, and although the dual linear programming and polyhedral bounds are better, the computational cost of these bounds is many orders of magnitude higher than that of GLB. The computation to obtain TDB is also much greater than that required for PB or GLB.

It is well known that eigenvalue bounds can be negative on instances of KBP for which zero is a trivial lower bound. In [29] it is suggested that this poor performance may be related to sparsity of the flow matrix A. In Figure 17.2 we give the sparsity (fraction of zero components) and coefficient of variation (CV, equal to the standard deviation of the components divided by their mean) for the flow matrices from a number of grid-based KBPs from QAPLIB [11]. It is clear that ste36a is very sparse, with a high CV. In the context of heuristics for QAP, CV is often termed "flow dominance" [17] and has been used as an algorithm control parameter.

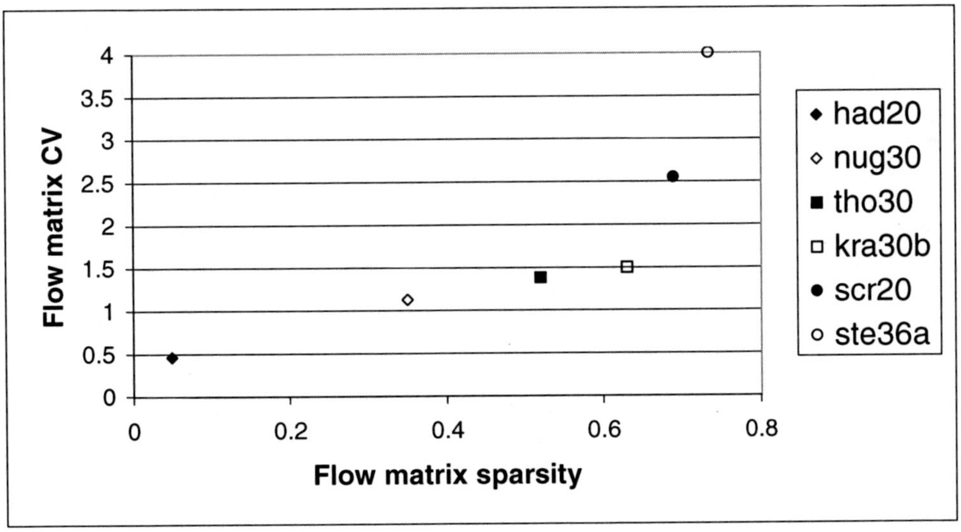

Figure 17.2. *Characteristics of flow matrices of grid-based QAPs.*

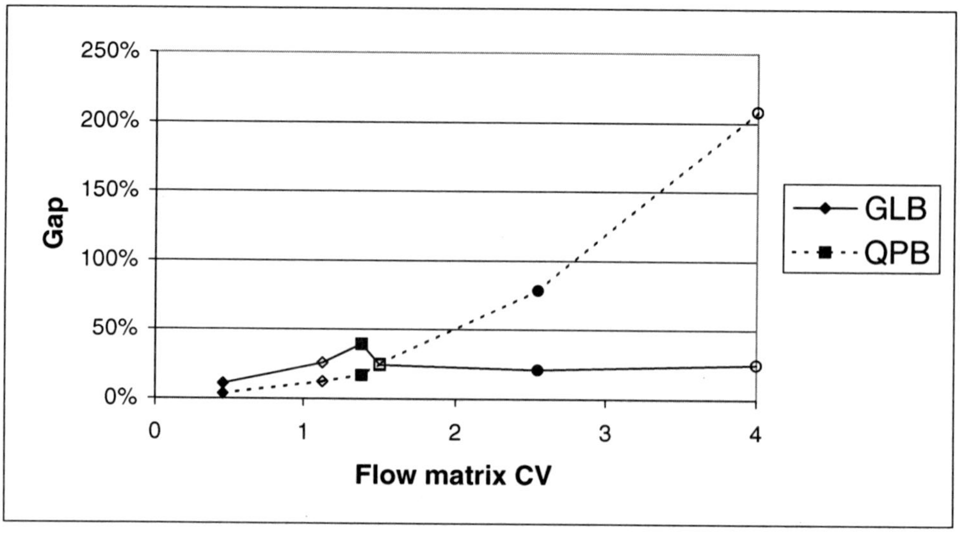

Figure 17.3. *Gaps for bounds on grid-based QAPs.*

In Figure 17.3 we give the gaps for GLB and QPB for the same problems considered in Figure 17.2. The markers used to denote the problems are the same as in Figure 17.2. The strong relationship between CV and the quality of QPB is evident. It is worth noting that the most successful applications of QPB reported in [2, 7] correspond to problems with relatively low CV values, such as had20, nug30, tho30, and kra30b. For had20, the problem with the lowest CV, solution using QPB is faster than the GLB-based algorithm of [8] by a factor of over 3000, after adjusting for hardware differences [7]. On the other hand, the equivalent time to solve scr20 using QPB is about a factor of 2.2 times that required in [8]. These observations suggest that QPB might not be a good candidate for the solution of ste36a, and consequently we consider the application of a GLB-based B&B algorithm.

17.4.1 Branching rules

As described in Section 17.3.1, the value of GLB for a QAP is obtained from LAP(F), where F is first derived from the original problem data. Associated with the solution of LAP(F) is a nonnegative reduced-cost matrix U such that

$$F \bullet X = z^* + U \bullet X$$

for any X with $Xe = X^T e = e$, where $F \bullet X = \mathrm{tr}(FX^T)$ and z^* is the solution value in GLB. (If X^* solves LAP(F), then $X^* \bullet U = 0$.) It follows that, if v is the value of a known solution to QAP, then

$$z^* + u_{ij} > v \Rightarrow x_{ij} = 0 \tag{17.4}$$

in any optimal solution X of QAP.

The use of (17.4) to eliminate children in the course of branching was introduced in [33], and it has been employed in many subsequent papers. Mautor and Roucairol [33] also introduced *polytomic branching*, where at any node candidate children are obtained by either (*row branching*) fixing one facility and assigning it to all available locations, or (*column branching*) fixing one location and assigning to it all available facilities. In our implementation we use polytomic row and column branching. We consider two branching rules, Rules 2 and 4, that are motivated by similar QPB-based branching rules from [7]. For simplicity we describe the rules here as they would be implemented at the root node, using row branching. The problem associated with an arbitrary node in the B&B tree is a lower dimensional QAP, on which the implementation of the rules is very similar. Let $N = \{1, 2, \ldots, n\}$.

Rule 2. Branch on the row i that produces the smallest number of children. In the event of a tie, choose the row with the largest value $\sum_{j \in N_i'} u_{ij}$, where $N_i' = \{j \in N \mid z + u_{ij} < v\}$.

Note that the set N_i' in Rule 2 consists exactly of the child problems $x_{ij} = 1$ that *cannot* be eliminated. Rule 2 is an extension of the branching rule used in [33] and is effective in reducing the size of the tree on small problems. Close to the root on larger instances, however, the information provided by the reduced-cost matrix U may be insufficient to make good branching decisions. Consequently, we consider obtaining more information about the effect of setting $x_{ij} = 1$ before actually deciding where to branch.

Rule 4. Let I denote the set of rows having the NBEST highest values of $\sum_{j \in N} u_{ij}$. For each $i \in I$, $j \in N$, compute the GLB z^{ij} for the QAP obtained by setting $x_{ij} = 1$. Let U^{ij} be the reduced-cost matrix associated with z^{ij}. Let v^{ij} be the maximal row sum of U^{ij}, and let $w^{ij} = (|N| - 1)z^{ij} + v^{ij}$. Branch on the row i having the highest value of $\sum_{j \in N} w^{ij}$.

In the context of B&B algorithms Rule 4 is an example of a *strong branching rule* [32]. Because of the use of the U^{ij} matrices, Rule 4 can also be viewed as a *look-ahead procedure* that tries to maximize the bounds two levels deeper in the tree.

In addition to the elimination of children based on bounds, described above, redundant children can be eliminated using symmetry of the grid on which the distance matrix B is based (see Figure 17.1). For example, the children of the root node can be based on assignments $x_{ij} = 1$, $j \in J_1 = \{1{:}5, 10{:}14\}$, regardless of the choice of i. (For integers $m < n$ we use $m{:}n$ to denote the collection of integers k with $m \le k \le n$.) In addition, if at any node the current assignments are all to locations contained in the set $J_2 = \{5, 14, 23, 32\}$, then the children can be restricted to $x_{ij} = 1$, $j \in J_3 = \{1{:}5, 10{:}14, 19{:}23, 28{:}32\}$, regardless of the choice of i. In cases where symmetry can be exploited we use row branching, with the index set N in Rules 2 and 4 replaced by a suitable $J \subset N$. In all *other* cases Rules 2 and 4 are implemented so as to consider column branching as well as row branching, with only minor modifications required. (For example, in Rule 2 we choose the row or column that produces the least number of children.)

17.4.2 Computational results

We implemented a GLB-based B&B algorithm, using the branching rules described above, to solve the 1-norm SWP. As in [2] the choice of branching rule to apply at a given node is

Table 17.2. *Branching strategy used to solve ste36a.*

Rule	Depth	Gap	NBEST
4a	5	0.35	36
4b	6	0.30	10
2	50	0.00	–

determined by depth in the tree and the relative gap. The relative gap for a node is defined to be

$$g = \frac{v - z'}{v - z_0},$$

where v is the incumbent value, z' is the lower bound at the current node, and z_0 is the root lower bound. The exact branching strategy used is given in Table 17.2. At a given node the rules are scanned from the top down until a rule is found whose maximum depth is greater than or equal to the node's depth, and whose minimum gap is less than the node's relative gap. The B&B tree was traversed using depth-first search.

The solution of the problem required a total of approximately 7.75×10^8 nodes in the B&B tree. The best known value of 9526 was verified as being optimal. In Figure 17.4 we give the number of nodes at each level of the tree. Note the logarithmic scale for the y-axis. Subproblems at level 33 of the tree correspond to QAPs of dimension three, which were solved by enumeration. The solution required approximately 186 hours of CPU time on a single 800 MHz Pentium III PC. (Based on a direct comparison this machine is approximately 40% faster on our application than the HP™ 9000 model C3000 used in [7].) In Figure 17.5 we give the cumulative CPU time (in hours) expended for the nodes

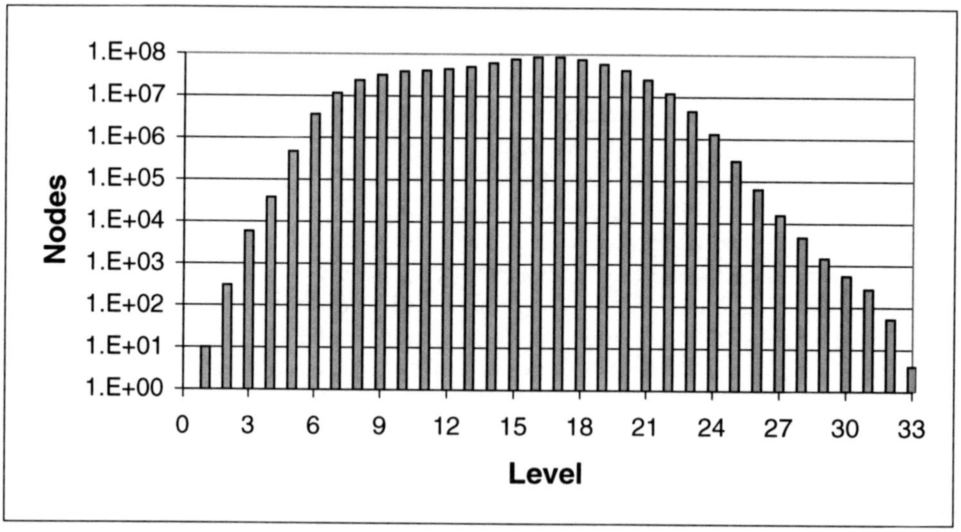

Figure 17.4. *Distribution of nodes in solution of ste36a.*

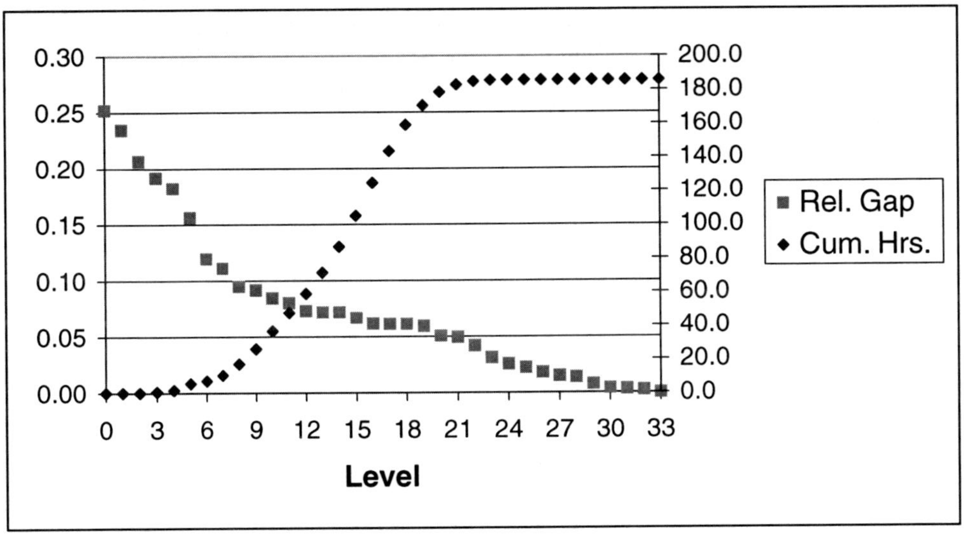

Figure 17.5. *Relative gap and cumulative time in solution of ste36a.*

up to each level of the tree. In the figure we also give the gap to optimality at each level, computed using the minimum bound obtained at that level. From the figure it is clear that it is relatively inexpensive to reduce the gap to about one-half its initial value. (The worst bound for a level 6 node was 8388, corresponding to a gap of 12%. The cumulative time to process all nodes at levels 0 to 6 is about 7 hours.)

To evaluate the effect of using Rule 4 at the top of the B&B tree we also ran the algorithm using only Rule 2, through level 7. In Table 17.3 we give comparitive statistics for the nodes through level 8 obtained from the solution run, and the run using only Rule 2. "L" denotes the level in the B&B tree. The "Fthm" and "Elim" columns report the fraction of nodes fathomed and the fraction of potential children of unfathomed nodes eliminated, respectively. "Gap" is the average gap to the optimal value for nodes on a given level,

Table 17.3. *Comparison of solution run with Rule 2 only.*

	Rule 2 only				Solution run			
L	Nodes	Gap	Fthm	Elim	Nodes	Gap	Fthm	Elim
0	1	2402.0	0.000	0.000	1	2402.0	0.000	0.000
1	10	2280.5	0.000	0.000	10	1953.7	0.000	0.054
2	318	1770.3	0.003	0.069	301	1218.4	0.000	0.421
3	9941	1239.8	0.080	0.549	5869	697.1	0.070	0.787
4	136112	727.7	0.209	0.580	38263	542.5	0.320	0.441
5	1445612	594.7	0.336	0.535	465182	354.3	0.404	0.569
6	13832243	438.4	0.445	0.629	3703103	260.8	0.579	0.752
7	85562934	322.4	0.546	0.613	11627541	183.0	0.641	0.806
8	436142577	266.5			23549921	132.2	0.730	0.821

computed using the lower bound inherited from the parent node. From the table it is clear that the use of Rule 4 at the top levels has an enormous effect on the subsequent evolution of the tree. Note that using only Rule 2 increases the number of nodes on level 8 by a factor of over 18. Moreover, the average gap for these level 8 nodes is approximately doubled, suggesting that the number of nodes at deeper levels will continue to worsen substantially compared to the solution run. We believe that the time to solve ste36a using only Rule 2 would be at least a factor of 100 higher than the time obtained here using Rules 4 and 2 together. Further evidence of the value of Rule 4 is provided by the results of a preliminary solution run that used Rule 4 only on levels 0, 1, and 2 of the tree. This earlier run required more than double the nodes (1.79×10^9) and time (435 hours) of the final solution run reported here.

It is interesting to compare some characteristics of the B&B tree for ste36a with the solution of nug30 obtained in [2]. For example, statistics like "Fthm" and "Elim" are substantially better near the top of the tree for ste36a than for nug30. On the other hand, the node distribution for ste36a, as shown in Figure 17.4, is much "flatter" than the corresponding distribution for nug30. Although the peak number of nodes is modest compared to the solution of nug30, there are 14 levels (8–21) where the number of nodes is within a factor of 5 of the peak number (8.7×10^7, on level 17). In the B&B tree for the nug30 problem only 6 levels had node counts within a factor of 5 of the peak (2.66×10^9, on level 10). We conclude that, while the use of the GLB with strong branching is effective in limiting the growth of the B&B tree for ste36a, there is still room for improvement in the overall time required to solve the problem.

After this chapter was written, we learned of a previously unreleased technical report by M. Nyström [35] that describes the solution of the ste36b/c problems. Nyström used a distributed B&B algorithm based on the GLB, implemented on twenty-two 200 MHz Pentium Pro CPUs. The serial time to solve the ste36b/c instances on one of these CPUs is estimated to be approximately 60 days and 200 days, respectively. (The time for ste36c is substantially higher because this problem was solved using an initial incumbent value of $+\infty$.)

Bibliography

[1] W.P. Adams and T. Johnson. Improved linear programming based lower bounds for the quadratic assignment problem. In P. Pardalos and H. Wolkowicz, editors, *Quadratic Assignment and Related Problems*, volume 16 of *DIMACS Series in Discrete Mathematics and Theoretical Computer Science*, pages 43–77. AMS, Providence, RI, 1994.

[2] K. Anstreicher, N. Brixius, J.-P. Goux, and J. Linderoth. Solving Large Quadratic Assignment Problems on Computational Grids. *Mathematical Programming Series* B, 91:563–588, 2002.

[3] K.M. Anstreicher. Eigenvalue bounds versus semidefinite relaxations for the quadratic assignment problem. *SIAM Journal on Optimization*, 11:254–265, 2000.

[4] K.M. Anstreicher and N.W. Brixius. A new bound for the quadratic assignment problem based on convex quadratic programming. *Mathematical Programming*, 89:341–357, 2001.

[5] K. Anstreicher and H. Wolkowicz. On Lagrangian relaxation of quadratic matrix constraints. *SIAM Journal on Matrix Analysis and Applications*, 22:41–55, 2000.

[6] D. Applegate, R. Bixby, V. Chvátal, and W. Cook. On the solution of traveling salesman problems. *Documenta Mathematica*, Extra Volume III:645–656, 1998.

[7] N.W. Brixius and K.M. Anstreicher. Solving quadratic assignment problems using convex quadratic programming relaxations. *Optimization Methods and Software*, 16:49–68, 2001.

[8] A. Brüngger, A. Marzetta, J. Clausen, and M. Perregaard. Solving large-scale QAP problems in parallel with the search library ZRAM. *Journal of Parallel and Distributed Computing*, 50:157–169, 1998.

[9] R.E. Burkard, E. Çela, P.M. Pardalos, and L.S. Pitsoulis. The quadratic assignment problem. In D.-Z. Du and P.M Pardalos, editors, volume 3 of *Handbook of Combinatorial Optimization*, pages 241–337. Kluwer Academic, Boston, 1998.

[10] R.E. Burkard and U. Derigs. *Assignment and Matching Problems: Solution Methods with Fortran Programs*, volume 184 of *Lecture Notes in Economics and Mathematical Systems*. Springer-Verlag, Berlin, 1980.

[11] R.E. Burkard, S.E. Karisch, and F. Rendl. QAPLIB—a quadratic assignment problem library. *Journal of Global Optimization*, 10:391–403, 1997. See also www.opt.math.tu-graz.ac.at/qaplib.

[12] E. Çela. *The Quadratic Assignment Problem: Theory and Algorithms*. Kluwer, Dordrecht, Boston, 1998.

[13] J. Clausen and M. Perregaard. Solving large quadratic assignment problems in parallel. *Computational Optimization and Applications*, 8:111–127, 1997.

[14] D.T. Connolly. An improved annealing scheme for the QAP. *European Journal of Operational Research*, 46:93–100, 1990.

[15] G. Finke, R.E. Burkard, and F. Rendl. Quadratic assignment problems. *Annals of Discrete Mathematics*, 31:61–82, 1987.

[16] C. Fleurent and J.A. Ferland. Genetic hybrids for the quadratic assignment problem. In P. Pardalos and H. Wolkowicz, editors, *Quadratic Assignment and Related Problems*, volume 16 of *DIMACS Series in Discrete Mathematics and Theoretical Computer Science*, pages 173–187. AMS, Providence, RI, 1994.

[17] L.M. Gambardella, É.D. Taillard, and M. Dorigo. Ant colonies for the quadratic assignment problem. *Journal of the Operational Research Society*, 50:167–176, 1999.

[18] P.C. Gilmore. Optimal and suboptimal algorithms for the quadratic assignment problem. *Journal of the Society for Industrial and Applied Mathematics*, 10:305–313, 1962.

[19] S.W. Hadley, F. Rendl, and H. Wolkowicz. A new lower bound via projection for the quadratic assignment problem. *Mathematics of Operations Research*, 17:727–739, 1992.

[20] P.M. Hahn and T. Grant. Lower bounds for the quadratic assignment problem based upon a dual formulation. *Operations Research*, 46:912–922, 1998.

[21] P.M. Hahn, T. Grant, and N. Hall. A branch-and-bound algorithm for the quadratic assignment problem based on the Hungarian method. *European Journal of Operational Research*, 108:629–640, 1998.

[22] P.M. Hahn, W.L. Hightower, T.A. Johnson, M. Guignard-Spielberg, and C. Roucairol. *Tree Elaboration Strategies in Branch and Bound Algorithms for Solving the Quadratic Assignment Problem.* Technical report, Systems Engineering, University of Pennsylvania, Philadelphia, 1999.

[23] P.M. Hahn and J. Krarup. A hospital facility layout problem finally solved. *The Journal of Intelligent Manufacturing,* 5/6:487–496, 2001.

[24] M. Jünger and V. Kaibel. Box-inequalities for quadratic assignment polytopes, *Mathematical Programming,* 91:175–197, 2001.

[25] M. Jünger and V. Kaibel. On the SQAP-polytope. *SIAM Journal on Optimization*, 11:444–463, 2000.

[26] V. Kaibel. Polyhedral combinatorics of quadratic assignment problems with less objects than locations. In R.E. Bixby, E.A. Boyd, and R.Z. Ríos-Mercado, editors, *Integer Programming and Combinatorial Optimization*, volume 1412 of *Lecture Notes in Computer Science*, pages 409–422. Springer-Verlag, Berlin, 1998.

[27] V. Kaibel. Polyhedral methods for the QAP. In P.M. Pardalos and L. Pitsoulis, editors, *Nonlinear Assignment Problems*. Kluwer Academic, Dordrecht, Boston, 2000.

[28] S.E. Karisch, E. Çela, J. Clausen, and T. Espersen. A dual framework for lower bounds of the quadratic assignment problem based on linearization. *Computing*, 63:351–403, 1999.

[29] S.E. Karisch and F. Rendl. Lower bounds for the quadratic assignment problem via triangle decompositions. *Mathematical Programming*, 71:137–152, 1995.

[30] E.L. Lawler. The quadratic assignment problem. *Management Science*, 9:586–599, 1963.

[31] C.-J. Lin and R. Saigal. *On Solving Large-Scale Semidefinite Programming Problems—A Case Study of Quadratic Assignment Problem.* Technical report, Department of Industrial and Operations Engineering, University of Michigan, Ann Arbor, MI, 1997.

[32] J.T. Linderoth and M.W.P. Savelsbergh. A computational study of branch and bound search strategies for mixed integer programming. *INFORMS Journal on Computing*, 11:173–187, 1999.

[33] T. Mautor and C. Roucairol. A new exact algorithm for the solution of quadratic assignment problems. *Discrete Applied Mathematics*, 55:281–293, 1994.

[34] C.E. Nugent, T.E. Vollman, and J. Ruml. An experimental comparison of techniques for the assignment of facilities to locations. *Operations Research*, 16:150–173, 1968.

[35] M. Nyström. *Solving Certain Large Instances of the Quadratic Assignment Problem: Steinberg's Examples*. Technical report, Department of Computer Science, California Institute of Technology, Pasadena, CA, 1999.

[36] M. Padberg. The Boolean quadratic polytope: Some characteristics, facets, and relatives. *Mathematical Programming*, 45:139–172, 1989.

[37] M. Padberg and G. Rinaldi. A branch-and-cut algorithm for the resolution of large-scale symmetric traveling salesman problems. *SIAM Review*, 33:60–100, 1991.

[38] M.W. Padberg and M.P. Rijal. *Location, Scheduling, Design and Integer Programming*. Kluwer Academic, Boston, 1996.

[39] P.M. Pardalos, L.S. Pitsoulis, and M.G.C. Resende. A parallel GRASP implementation for the quadratic assignment problem. In A. Ferreira and J. Rolim, editors, *Parallel Algorithms for Irregularly Structured Problems—Irregular '94*, pages 111–130. Kluwer Academic, Dordrecht, Netherlands, 1995.

[40] P.M. Pardalos, F. Rendl, and H. Wolkowicz. The quadratic assignment problem: A survey and recent developments. In P.M. Pardalos and H. Wolkowicz, editors, *Quadratic Assignment and Related Problems*, volume 16 of *DIMACS Series in Discrete Mathematics and Theoretical Computer Science*, pages 1–42. AMS, Providence, RI, 1994.

[41] F. Rendl and H. Wolkowicz. Applications of parametric programming and eigenvalue maximization to the quadratic assignment problem. *Mathematical Programming*, 53:63–78, 1992.

[42] M.G.C. Resende, K.G. Ramakrishnan, and Z. Drezner. Computing lower bounds for the quadratic assignment problem with an interior point algorithm for linear programming. *Operations Research*, 43:781–791, 1995.

[43] J. Skorin-Kapov. Tabu search applied to the quadratic assignment problem. *ORSA Journal on Computing*, 2:33–45, 1990.

[44] L. Steinberg. The backboard wiring problem: A placement algorithm. *SIAM Review*, 3:37–50, 1961.

[45] É.D. Taillard. Robust taboo search for the quadratic assignment problem. *Parallel Computing*, 17:443–455, 1995.

[46] Q. Zhao, S.E. Karisch, F. Rendl, and H. Wolkowicz. Semidefinite programming relaxations for the quadratic assignment problem. *Journal of Combinatorial Optimization*, 2:71–109, 1998.

Chapter 18

Mixed-Integer Programming: A Progress Report

Robert E. Bixby,[*] Mary Fenelon,[†] Zonghao Gu,[†] Ed Rothberg,[†] and Roland Wunderling[†]

Abstract. Over the last several decades, from the early 1970s to as recently as 1998, the underlying solution technology in commercial mixed-integer programming codes remained essentially unchanged. In spite of important advances in the theory, many of these advances have clear computational value. In the last several years, that situation has changed. The result has been a major step forward in our ability to solve real-world mixed-integer programming problems.

MSC 2000. 90C11, 90C05, 90C10
Key words. Linear programming, mixed-integer programming, simplex algorithms, barrier algorithms, cutting planes, branch-and-cut, preprocessing

18.1 Linear Programming

The focus of this chapter is on computational mixed-integer programming. However, advances in computational linear programming are a fundamental part of that story. We will devote the initial sections of this chapter to an overview of that subject.

A *linear program* (LP) is an optimization problem of the form

$$\begin{aligned} \text{minimize} \quad & c^T x \\ \text{subject to} \quad & Ax = b, \\ & l \le x \le u, \end{aligned}$$

*ILOG, Inc. and Rice University (bixby@rice.edu).
†ILOG, Inc.

where A is an $m \times n$ matrix, called the *constraint matrix*, x is a vector of *variables*, c is a vector of *objective function* coefficients, and l and u are vectors of *bounds*. The definition of an LP is very restrictive; in exchange there are powerful, scalable solution algorithms that, today, allow us to solve instances with hundreds of thousands of constraints and more.

Rather than attempting to give a complete list of linear programming computational improvements, we focus on the last roughly 15 years and give a list of what we consider to be the top four computational advances during that period:

- robust dual simplex algorithms;
- linear algebra improvements;
- interior-point algorithms;
- automatic problem simplification ("presolve").

It is perhaps surprising that we rank interior-point algorithms no higher than third. Karmarkar's paper [19] on this subject started a virtual revolution in the theory of linear programming; moreover, there is little doubt that the effect on computation has been substantial. Indeed, later results in this chapter will indicate that current barrier algorithms, the reintroduction of which was motivated by Karmarkar's work, are today the fastest CPLEX algorithms for linear programming. Nevertheless, it is mixed-integer programming that is the dominant application of linear programming in practice, and in this context the dominant computation is the solution of a sequence of LPs in which good starting-point information is known. Barrier algorithms have been unsuccessful in exploiting this starting-point information. Simplex algorithms, on the other hand, and most particularly the dual simplex algorithm, work very well in this context. As a result, these algorithms remain dominant in mixed-integer programming applications, and hence for linear programming in general.

Number one on our list is the development of robust implementations of the dual simplex algorithm. One apparent reason is their aforementioned central role in mixed-integer programming. In addition, for general linear programming, even when solving with no starting-point information, the current dual simplex implementations outperform the best primal simplex implementations. A key reason is the availability of fast, robust versions of so-called steepest-edge pricing for the dual, as introduced in Goldfarb [16] and Forrest and Goldfarb [14]. No similarly effective version of steepest edge is known for the primal.

Second on our list is linear algebra. Improvements in linear algebra algorithms are not central to the theory of linear programming, but are nevertheless central to computational progress in this subject. This effect has been felt by barrier and simplex algorithms alike.

For barrier algorithms, the computation of the Cholesky factorizations is the dominant computational step, often consuming more than 95% of the total computation time. Essential in computing these factorizations is the application of matrix-ordering algorithms to reduce the numerical "fill." The development of such algorithms specifically targeted at linear programming has led to significant overall improvements in barrier performance. See Rothberg and Hendrickson [25] for a description of the best current algorithms.

For simplex algorithms the improvements in linear algebra routines may be explained as follows. At each step of a simplex algorithm, several linear systems must be solved. The order of these systems is the number of rows in the underlying LP. As problem size grows, these computations dominate the total solution time. However, an examination of the solution characteristics reveals the following property: For each such computation there is a given input vector and a subsequent output vector produced by solving the system. The input vectors are almost always very sparse, with a number of nonzeros usually in the range from one to ten, independent of the number of rows. Most importantly, the output vector is often very sparse as well, with just a few additional nonzeros. Since it is unlikely that the sparsity of the output resulted from cancellation during the solve, this means that relatively little of the underlying matrix was touched during the solve. Thus it is essential to the speed of the computation that it be carried out without explicitly accessing any irrelevant parts of the matrix. As it happens, algorithms that solve exactly this problem have been known in the linear algebra community for some time (see Gilbert and Peierls [15]). Exploiting these algorithms has led to major improvements in speed for simplex algorithms.

The final item on our list is presolve. As most readers will be aware, presolve is a compendium of ideas for reducing model size before actually applying a given solution algorithm. The seminal paper on this subject is Brearley, Mitra, and Williams [7]. The effect of presolve on problem size can often be substantial, though the effect on actual solution times is mitigated to some extent, particularly for simplex algorithms, by the linear algebra improvements mentioned in the previous paragraph.

18.1.1 Linear programming speed improvements

We begin with an example. The patient distribution system (PDS) models were introduced in Carolan et al. [8]. The underlying model is a military logistics application with a multicommodity flow structure. Such structures have traditionally been difficult for simplex algorithms. In Table 18.1 we list PDS instances PDS02 through PDS100. Solution times are for CPLEX 1.0 (the first release of the CPLEX code) and CPLEX 8.0 running on a 667 MHz HP AlphaServer ES40.

The results for CPLEX 1.0 in Table 18.1 indicate that the PDS models were indeed difficult at the time they were introduced. According to the results reported in Bixby [6], today's fastest desktop computers are on the order of 800 times faster than the desktop computers available in 1990. Based on this factor and the solution time presented in Table 18.1,

Table 18.1. *PDS models—solution times.*

Instance	Rows	CPLEX 1.0	CPLEX 8.0	Speedup
pds02	2953	0.4	0.1	4.0
pds06	9881	26.4	0.9	29.3
pds10	16558	208.9	2.6	80.3
pds20	33874	5168.8	20.9	247.3
pds30	49944	15891.9	39.1	406.4
pds40	66844	58920.3	79.3	743.0
pds50	83060	122195.9	114.6	1066.3
pds60	99431	205798.3	160.5	1282.2
pds70	114944	335292.1	197.8	1695.1
pds80	129181	–	304.4	–
pds90	142823	–	320.3	–
pds100	156171	–	256.3	–

Table 18.2. *Speedups—1988 to 2002.*

Algorithm	
Simplex algorithms	960
Simplex and barrier algorithms	2360
Machines	
Simplex algorithms	800
Barrier algorithms	13000

PDS20 would have taken about 50 days to solve using CPLEX 1.0 in 1990, and PDS70 would have been unsolvable at that time, with an estimated solution time of approximately 8.5 years! The final column of this table is particularly interesting. For the largest of the models tested, the speedup in solution time due to algorithmic improvements alone is a factor of 1695, far exceeding machine improvements. (CPLEX 1.0 times for the largest PDS instances were not generated due to the excessive computation times that were expected.)

It seems appropriate here to give a general description of the way in which we carried out our computational comparisons for the linear programming results in this section, the mixed-integer programming results in later sections, and the more extensive linear programming results in Bixby [6]. Consider a testset containing N models for some N, and suppose that we want to compare two algorithms. Then each of the algorithms would be run on all N models, a total of $2N$ runs, with some fixed time limit. For any model and algorithm in which this time limit is exceeded, the time limit is used in further comparisons as the solution time. Note that this approach leads to conservative estimates of speedups where one code is consistently slower (as is the case here). Given these $2N$ numbers, we then compute N ratios of solve times, and from these N ratios the geometric mean. This geometric mean is taken as our measure of the "average" speedup. We remark that summarizing performance differences in this way, using a single average speedup, often does oversimplify the many subtleties in the differences. However, we have not found a better way to give a concise sense of what to expect when comparing two algorithms.

The results in Table 18.1, while interesting, are restricted to just one class of models. In Bixby [6] more extensive tests were reported in which, in the biggest of the tests, a suite of 680 models was used. The largest LP used in that test had over seven million constraints. Results of linear programming tests in Bixby [6] are summarized in Table 18.2. For algorithmic improvement alone we found that the best of the primal and dual simplex algorithms today is approximately 960 times faster than the primal simplex algorithm (the only algorithm available) in CPLEX 1.0. Taking the best of primal, dual, and barrier, we obtained an estimated speedup of a factor of 2360!

For our machine comparison, as documented in Bixby [6], we restricted our attention to desktop computing, including workstations, and obtained a speedup of a factor of 800 for simplex algorithms and a remarkable 13,000-fold speedup for barrier algorithms.[1] As is evident from these numbers, barrier algorithms have been much more effective in exploiting the properties of modern computing architectures. This fact is a principal reason these algorithms have emerged as such powerful tools in practical linear programming computations.

[1] In contrast to simplex algorithms, barrier algorithms can also be very effectively parallelized, leading to an even larger factor for machine speedup.

Combining the algorithmic and machine factors in Table 18.2, we obtain a speedup of 960×800 for simplex algorithms, and a 2360×800 speedup, a factor of nearly 1.9 million, for the best of the three algorithms compared to the primal simplex implementations available in the late 1980s.

Some readers may wonder how the three algorithms, primal and dual simplex and barrier, compare among themselves. Table 18.3 gives results for the various CPLEX 8.0 implementations, where ratios greater than one indicate the factor by which the algorithms in the denominators were better. Thus, for models with more than 100,000 rows, of which there were 73 in the test, the barrier algorithm was on average (using geometric mean) 1.6 times faster than the dual. This table indicates that the dual algorithm consistently outperformed primal by a factor of about two to one, while barrier was clearly the overall winner, especially for larger models. However, it is interesting to note that barrier algorithms and the best of the simplex algorithms produce roughly equivalent performance, even for large models.

Table 18.3. *Algorithm comparison.*

No. rows	No. models	Primal/ dual	Dual/ barrier	Barrier/ best simplex
$>$ 0	680	1.5	1.1	1.1
$>$ 10000	248	2.0	1.0	1.2
$>$100000	73	2.1	1.6	0.9

18.2 Mixed-Integer Programming

A *mixed-integer program (MIP)* is an optimization problem of the form

$$\begin{array}{ll} \text{minimize} & c^T x \\ \text{subject to} & Ax = b, \\ & l \le x \le u, \\ \multicolumn{2}{l}{\text{some or all } x_j \text{ integral.}} \end{array}$$

Thus, an MIP is an LP together with the additional condition that some or all of the variables must take integer values. This condition transforms a class of problems (LPs) for which we have powerful, robust, and scalable algorithms—linear programming—into a class that is NP-hard. In exchange we obtain a very powerful modeling paradigm. Indeed, the power of this paradigm has long been recognized. Unfortunately, what was also recognized for many years was that the power of the paradigm was not matched by the power of the available solution algorithms. It is the central theme of this chapter that this situation has changed dramatically in just the last two to three years.

The remainder of this chapter is taken up with describing our view of the current computational state of the art in mixed-integer programming. We begin with some examples. While the computational results at the end of the chapter illustrate major overall progress, we are not claiming that mixed-integer programming has become easy. For linear programming we can now reasonably assert that, in practice, small, unsolvable LPs simply do not occur, and that even larger LPs, some with several hundred thousand constraints, do not present major difficulties. We cannot make such a claim for mixed-integer programming. The mixed-integer programming landscape is and will remain unpredictable.

Example 18.1 (schedule generation module (SGM)). In spite of the progress that has been made in linear programming, there still remain MIPs that are unsolvable today primarily because the associated linear programming relaxations are just too difficult.

The SGM model was developed at Sabre Decision Technologies. In the standard fleet assignment problem, airplane types are assigned to already scheduled flight legs to meet given passenger demands over some fixed time horizon. The flight schedule and demands are traditionally taken as given inputs in this process. However, there are clear advantages to combining the optimization of both the equipment assignment and the schedule generation into a single module. SGM is an early prototype for this approach. The particular instance that we tested was not small, but also not particularly large: 157,323 constraints, 182,812 variables, and 6,348,437 nonzero constraint coefficients.

In spite of its size, the initial linear programming relaxation of SGM turns out to be quite difficult. Given the size of the Cholesky symbolic factorization, we estimated a solution time using the barrier algorithm of three to six days. The fastest algorithm we found was primal simplex using steepest-edge pricing. Total solution time was 64,000 seconds, almost one day. Worse yet, when we ran the MIP for a total of roughly two weeks, only 368 branch-and-cut nodes were processed. The MIP itself did not appear to be fundamentally difficult (for instance, the number of integer infeasibilities was steadily decreasing), but the difficulty of the LPs made the MIP practically unsolvable with today's technology. Note, however, that with a *mere* thousandfold increase in the LP solution speed, modest compared to what has happened over the last 15 years, it is not unlikely that this model would indeed fall into the realm of solvability.

Example 18.2 (MIP really is HARD). It is well known that mixed-integer programming is NP-hard, but the situation is actually worse than that. The specific class of algorithms we employ, linear programming–based branch-and-cut, can be proved to be *non-polynomial*. Thus it should be no surprise that we encounter small, difficult models in practice.

As an example of a difficult model we offer a small customer model with just 44 constraints, 51 variables, and 167 nonzeros in the constraint matrix. A key feature of the model is that all 51 variables are general integers, and all 51 have infinite bounds. When CPLEX 8.0 was applied, we immediately found an integer feasible solution of value -2136.0 and an initial upper bound of -1379.4 (the model is a maximization problem). The code was then allowed to run for 120,000 seconds (about 1.5 days), generating 32,000,000 branch-and-cut nodes and a 5.5 GB search tree. Unfortunately, after all that computation, the upper and lower bounds remained unchanged. To date, this model has not been solved.

We should note that, from a practical point of view, perhaps this model should be viewed as solvable, or at least as an instance of bad modeling. If bounds of 128 are placed on all variables, then the model easily solves in a few minutes. Guided by this solution, the judicious relaxation of 9 of the reduced bounds to infinity produces an instance of the original model that is solvable and in which no variables are binding at their bounds. There is of course no guarantee that the resulting solution is an optimal solution of the original model, but it seems unlikely that the user would have found this fact disturbing.

Example 18.3 (be careful what you conclude). The model *p2756* is taken from Crowder, Johnson, and Padberg [9]. It has 755 constraints and 2756 variables, all binary, and was apparently the hardest among the models studied in [9]. The solution time the authors reported for this model, using the results of their research, was 54.4 minutes. This same model is now solved by CPLEX 8.0 in 1.3 seconds (other modern codes solve *p2756* in similar times). A solution time of 1.3 seconds would seem to be a remarkable improvement, a factor of roughly 2500 times faster. However, that was 1983, and the difference in machine speeds between then and now is likely greater than the factor of 2500. The fact is, for this model, the key ideas that led to its solution in 1983 are probably still those that are key today. The difference is that these techniques are now available in standard codes, together with many other techniques, some related, some not.

Example 18.4 (supply-chain scheduling). Our final example illustrates what we consider a more typical situation from the point of view of solvability. It also demonstrates a growing trend toward using mixed-integer programming for day-of-operation, detailed-scheduling models, not just the more traditional, longer term planning models. This trend is a direct result of the increased power of the solvers.

The model in question is a weekly model using daily time buckets. The objective is to minimize end-of-day inventory. It includes production (at a single facility), inventory, shipping (via a dedicated trucking fleet), and demands from wholesale warehouses. The demands are given as inputs and may be treated as deterministic.

In an initial modeling phase, a simplified prototype was developed, and then complicating constraints were added to deal with certain consecutive-day production requirements and some rather complex minimum-use constraints for the truck fleet. The result of this process was not unexpected: The model was too difficult to solve in reasonable computation time. As a consequence, a "decomposition" approach was used. The model formulators talked to the schedulers who at the time were building the schedule by hand. The approach used by these schedulers was to first decide which products were assigned to which machines and make other decisions dependent upon these choices. This initial decision phase was easily simulated using constraint programming techniques (see Lustig and Puget [21] for a discussion of constraint programming), yielding fixings for a subset of the variables in the original formulation. The resulting "fixed" model was then solvable with available MIP technology, at that time CPLEX 5.0. The solution time on a 2 GHz Pentium IV using CPLEX 5.0 is 3466 seconds.[2] Running on the same machine, CPLEX 8.0 is able to solve the fixed problem in 1.4 seconds. Perhaps most interestingly, CPLEX 8.0 solves the original, unfixed model to optimality in 2 hours of computation time, and the resulting objective value is 20% lower than the optimal objective value for the fixed model!

18.3 A Short Computational History of Mixed-Integer Programming

Our object here is not to give the complete computational history of mixed-integer programming but only to mention a few highlights with the goal of providing historical context

[2]CPLEX versions 4.0 and earlier were unable even to solve the fixed model.

for a phenomenon that is at the heart of the remarkable advances that have occurred in the last few years.

The paper by Dantzig, Fulkerson, and Johnson [11] is now generally regarded as a seminal contribution in computational mixed-integer programming, perhaps *the* seminal contribution. By solving linear programming relaxations (largely by hand) and adding what were effectively "cutting planes," they managed to solve a 42-city instance of the traveling salesman problem (TSP) to optimality. The paper of Dantzig, Fulkerson, and Johnson was followed in the late 1950s and early 1960s by a body of work due to Gomory (see, for example, Gomory [17]) that laid the theoretical foundation for integer programming. While Gomory's work was long viewed as having limited computational value, we will see in later sections that the mixed-integer cuts he introduced are a very effective computational tool.

Linear programming–based branch-and-bound was introduced in 1960 by Land and Doig in [20]. Dakin [10] later introduced dichotomous branching. Using this basic approach, two excellent commercial mixed-integer programming codes were developed in the early 1970s: MPSX/370 (see Benichou et al. [5]) and the UMPIRE code (see Forrest, Hirst, and Tomlin [13]). These codes combined then state-of-the-art linear programming in a tight integration with linear programming–based branch-and-bound (B&B). The B&B parts of the codes, in turn, used carefully considered, common sense approaches to operations such as variable and node selection, approaches that are still in use today. These were the codes that first made mixed-integer programming a successful commercial optimization tool.

Following the introduction of MPSX/370 and UMPIRE there was a long period of stagnation in commercial mixed-integer programming technology stretching from the early 1970s into the late 1990s. The codes got better, but they got better largely because linear programming got better and computing machinery became more powerful. The structure of the codes themselves remained largely unchanged. However, during this same period, there was an unprecedented volume of successful research in integer programming and combinatorial optimization, both theoretical and computational.

In the early 1970s, Padberg wrote a sequence of papers (see, for example, Padberg [23], which introduced *sequential lifting*) evangelizing the use of cutting planes in the solution of MIPs. Balas [2, 3] promoted the idea of disjunctive arguments in deriving cutting planes, an idea that, in various guises (for example, Gomory mixed-integer cuts), has proved central in computation. The Crowder, Johnson, and Padberg [9] paper contained a beautiful and very influential computational study in which the MPSX commercial code was modified for pure 0/1-problems, adding cutting planes and clever preprocessing techniques. The resulting PIPEX code was used to solve a collection of previously unsolved, real-world MIPs. Van Roy and Wolsey [26] produced a similarly impressive modification of the Sciconic commercial code, a derivative of the aforementioned UMPIRE code, to deal with mixed 0/1-problems. However, strangely, neither of these modified codes ever had significant commercial impact. In addition to this work on general integer programming, through this entire period there was a steady stream of theoretical and computational results on the TSP by Grötschel (see, for example, Grötschel [18]), Padberg and Rinaldi [24], and others, which again demonstrated the efficacy of cutting planes in solving hard integer programs (IPs) arising in the context of combinatorial optimization.

What explains the phenomenon of so much fundamental progress in integer programming research, much of it computational, with relatively little effect on commercial practice? While we cannot claim to know the complete answer, at least part of the answer is probably the following. Academic research tends to focus on a single new idea. When studied computationally, such research often leads to implementations that focus on that single idea, to the exclusion of others. However, mixed-integer programming is NP-hard. As a result, general purpose algorithms for MIP seem to necessitate the application of a variety of, often disparate, ideas, each effective for certain kinds of structures and ineffective for others. This situation puts a premium on careful choices for the ways in which these ideas interact, and added emphasis on default settings that have the effect of bringing the right tools into play when needed and minimizing their use when not. Such a code involves a good measure of basic software engineering of the sort unlikely to be produced by an academic research project.

18.4 The New Generation of Codes

What are some of the important new features of the new generation of mixed-integer programming codes? Below is a by no means exhaustive list. This list is followed by a short discussion of some of these features.

- linear programming: stable, robust, dual simplex algorithms;
- variable/node selection: probing on dives, strong branching;
- primal heuristics: multiple heuristics applied within the search tree;
- node presolve: fast, incremental bound strengthening;
- presolve: probing in constraints;
- cutting planes: Gomory mixed-integer cuts, knapsack covers, flow covers, mixed-integer rounding (MIR) cuts, cliques, GUB covers, implied bounds cuts, path cuts, disjunctive cuts.

Advances in linear programming have clearly been important, particularly advances in the dual simplex algorithm. While not explicitly part of integer programming research, many of the developments in linear programming were motivated by this research (see Bixby [6]).

Simple aspects of branch-and-cut, such as variable and node selection, have greatly improved, sometimes in not-so-simple ways. As an example, probing on dives refers to the following idea (see Beale [4]). When a variable is chosen for branching, and a direction is chosen in which to branch, that decision is normally based upon some sort of estimate of the effect on the linear programming relaxation. However, it may happen that, when the actual branch is carried out, the effect on the LP is quite different from what was predicted. In that case, one can consider the option of branching in the other direction and comparing the two results. When diving (fixing a sequence of variables to move deeper into the branching tree), such a procedure can be very effective in avoiding dead ends, paths that eventually lead to integer infeasibility. Strong branching is another, very effective, idea for branch-variable selection, developed in the context of the TSP (see Applegate et al. [1]).

While academic research often focuses on finding optimal solutions, in practice the main goal is often to find "good" feasible solutions and find them as quickly as possible. Heuristics are thus extremely important. CPLEX 8.0 includes eight heuristics for finding integer feasible solutions. Six of these heuristics are of the more traditional type, using the solution of linear programming relaxations as the starting point in finding potentially better integer solutions. Two are different in character, starting with integer feasible solutions, not necessarily linear feasible.

Presolve is very important in mixed-integer programming, probably much more important than in linear programming. In linear programming, presolve reductions attempt to reduce the size of the model by effectively removing redundant or unnecessary parts. The result does not reduce the set of feasible solutions. For mixed-integer programming, presolve includes all "legal" linear reductions, together with some additional reductions, such as coefficient reduction and bound strengthening, that actually reduce the size of the feasible region for the linear programming relaxation, and thus tighten the formulation. Traditionally, these reductions have been applied only at the root node, but it was well known that such reductions could be very helpful if applied within the search tree. After some investigation, we concluded that the full application of presolve within the tree was simply too expensive computationally to be applied in general, but that reductions not changing the constraint matrix could be effectively applied. This investigation led to the introduction of incremental bound strengthening within the tree, similar to the "domain-reduction" methods that are standard in constraint programming (see Marriott and Stuckey [22]).

Probing within constraints is a simple idea. In general, probing refers to the idea of temporarily fixing 0/1-variables to 0 and 1, respectively, applying bound strengthening, and drawing logical conclusions about the model. For example, if a binary variable is fixed to 0 and bound strengthening concludes that the problem is infeasible, then we know that the probed variable can be permanently fixed to 1. The extensive application of probing can be quite expensive. However, applied within single constraints, the cost is limited. As an example of the benefits, consider the following typical sort of constraint:

$$\sum_{j\in J} x_j \leq \left(\sum_{j\in J} u_j\right) y,$$

where $0 \leq x_j \leq u_j$ $(j \in J)$ and $y \in \{0, 1\}$. By fixing y to 0, we readily deduce the validity of the "disaggregated" form of the above constraint, $x_j \leq u_j y (j \in J)$. It is well known that this disaggregated form yields a tighter linear programming relaxation, and hence a better approximation to the convex hull of integer feasible solutions. However, the additional constraints produced by disaggregation can be quite numerous and hence result in larger, more difficult LPs. When extracted by constraint probing, we treat these constraints as cuts to be added only when they help the formulation, that is, only when they are violated.

The final topic in our list is cutting planes. As our later computational results will show, cutting planes are by far the most important "single" feature of the new generation of codes. CPLEX now includes nine different types of cutting planes, eight of which are generated by default. Only disjunctive cuts are off by default.

The cuts that turn out to be most successful in our implementation are Gomory mixed-integer cuts. We now discuss these particular cuts in more detail, motivated in part by their effectiveness, but also by the fact that, while Gomory cuts are often mentioned, for example, in courses on integer programming, the Gomory fractional cuts are usually considered. These cuts are easy to describe and have beautiful theoretical properties. However, computational evidence suggests that they do not work well in practice. Gomory mixed-integer cuts are a bit more difficult to describe and do not possess the same elegant theory.

However, they are very effective in computation. In addition, they are among the easiest cuts to implement.

Gomory fractional cuts can be derived by using only integer rounding. The validity argument for mixed-integer cuts uses both rounding and disjunction. Suppose that we have solved the linear programming relaxation of an MIP using a simplex algorithm. To simplify the presentation, assume in addition that all variables are integral. The more general case, with continuous variables, is similar.

If all variables take integer values in the solution of the linear programming relaxation, then we are done: No cuts can or should be generated. In the alternative case, we may assume that some basic variable y takes a nonintegral value. Let the following be the corresponding row from an optimal simplex tableau:

$$y + \sum a_j x_j = d = \lfloor d \rfloor + f,\ f > 0,$$

where the x_j in this expression are the nonbasic variables in that row. Now let $f_j = a_j - \lfloor a_j \rfloor$ and write

$$t = y + \sum_{f_j \le f} \lfloor a_j \rfloor x_j + \sum_{f_j > f} \lceil a_j \rceil x_j,$$

noting that t is integral. It follows that

$$\sum_{f_j \le f} f_j x_j + \sum_{f_j > f} (f_j - 1) x_j = d - t.$$

It is now easy to see that

$$t \le \lfloor d \rfloor \Rightarrow \sum_{f_j \le f} f_j x_j \ge f \text{ and}$$

$$t \ge \lceil d \rceil \Rightarrow \sum_{f_j > f} (1 - f_j) x_j \ge 1 - f.$$

Since the left-hand side of each of these expressions is nonnegative, we have

$$\sum_{f_j \le f} (f_j / f) x_j + \sum_{f_j > f} (1 - f_j)/(1 - f) x_j \ge 1,$$

the corresponding Gomory mixed-integer cut.

Since the generation of Gomory mixed-integer cuts does introduce some numerical issues not arising for other, combinatorially motivated, cuts, we describe here in brief our procedure for generating these cuts. Starting with the solution of a linear programming relaxation, do the following:

(i) Make an ordered list of "sufficiently" fractional variables using Driebeek penalties [12].

(ii) Take the first 200 variables from the list, compute the corresponding tableau rows and cuts, and substitute out slacks, rejecting any of the resulting cuts for which the coefficient range is "too large."

(iii) Append the remaining cuts to the linear programming relaxation.

We apply the above procedure only at the root, twice by default.

18.5 Computational Results

The basis for our computational tests was an extensive library of customer and academic models assembled over the last 15 years. The library includes over 1500 models. A subset of 978 models was selected for testing. Where models were excluded, the typical criterion was the presence of multiple instances from the same source or having similar structure. In such cases, the instances included were typically those determined to be the most difficult or representative for a given size.

18.5.1 Feasibility

We ran CPLEX 5.0 and the current version, CPLEX 8.0, on the 978 models in the testset. CPLEX 6.0 would also have been a reasonable starting point for the comparison, that being the last "old generation" version for mixed-integer programming. The main difference between CPLEX 5.0 and 6.0 is in the linear algebra improvements for simplex algorithms, discussed earlier.

Defaults were used for both codes, CPLEX 5.0 and 8.0, and a 100,000 seconds time limit was imposed, a little over one day. All runs were made on a 667 MHz HP ES40 AlphaServer with 2 GB of physical memory. For CPLEX 5.0 this memory limit meant that a significant number of models, 228 in total, exhausted memory before reaching the time limit. For these models, a solution time of 100,000 seconds was recorded. CPLEX 8.0 includes a mechanism for using disk space to store, in compressed form, a specified portion of the branch-and-cut search tree. Since available disk space approximated 70 GB, no CPLEX 8.0 run terminated because of insufficient memory. For both of CPLEX 5.0 and 8.0, if early termination occurred because of numerical problems, the running time was again recorded as 100,000 seconds.

We begin with a summary of results that focuses on integer feasibility. This criterion is often the most important measure of performance in practice: finding "good" feasible solutions rather than proving optimality. In summary, the results were as follows:

- total models in test: 978;
- solved to optimality:
  - CPLEX 5.0: 569 (58%),
  - CPLEX 8.0: 755 (77%);
- among those not solved to optimality with CPLEX 8.0:
  - 116 had gap less than 10% (11.9%),
  - 32 had no integral solution (3.2%);
- using CPLEX 8.0 and "MIP emphasis feasibility" on the 32 models with no feasible solution:
  - 25 found no feasible solution (2.6%).

The *gap* is 100.0 times the absolute difference between the best integral solution found and the linear programming bound[3] divided by the absolute value of the best integral so-

[3]For a minimization problem the *linear programming bound* is the smallest value of an optimal solution of a linear programming relaxation at a currently active node in the branch-and-cut search tree.

lution plus 10^{-10} (to avoid dividing by zero). The MIP emphasis feasibility setting in CPLEX 8.0 emphasizes finding feasible solutions rather than proving optimality. While this setting is a nondefault setting, it is specifically designed for finding feasible solutions and so was a natural alternative in these tests. No other tuning of individual parameters was used.

It is interesting to analyze the 25 models for which no feasible solution was found, either using defaults or MIP emphasis feasibility. For 11 of these models, including the SGM instance discussed earlier, the linear programming relaxations were too difficult for the effective application of branch-and-cut: Within the 100,000 seconds time limit, it was possible to enumerate fewer than 1000 nodes,[4] and in some cases only the solution of the root-node relaxation was completed. Removing the 11 models for which the LPs were too difficult leaves 14. Of these 14, 5 included large numbers of general integer variables, many with large bound ranges (one instance had 109,000 such general integers). While such formulations may allow for the concise modeling of real-world phenomena, branch-and-cut is simply not a very effective method for handling such general integer variables.

We consider the above results to be quite good. The testset we used does contain some easy models, but it also contains many customer models that are in the testset precisely because customers found them difficult to solve. Nevertheless, running defaults we find integer feasible solutions in 97% of the cases. Moreover, of the 3% that are left (32 models), our analysis finds that only 9 models, about 1%, represent serious failures for the underlying technology.

18.5.2 Optimality

We next compare the relative performances of CPLEX 5.0 and 8.0 in proving optimality. As for linear programming, we used geometric means of ratios of solve times. It should, however, be pointed out that this measure can be misleading for mixed-integer programming. Unlike in linear programming, in mixed-integer programming a single new idea, such as a new kind of cutting plane, can convert a model that was unsolvable into one that solves in a fraction of a second. Clearly, such models can significantly affect mean ratios for a testset, and the effect will depend on the chosen time limit, though the potential impact of any one model on overall mean ratios is mitigated by our use of geometric means.

The specific comparisons we made were as follows. Our 978 model testset contains 220 models that were solved to optimality by neither CPLEX 5.0 nor CPLEX 8.0. To focus on optimality, we removed these models, leaving 758 models and yielding the results in Table 18.4.

The results in Table 18.4 may be read as follows. Consider the line beginning with 375. According to the table, there were 375 models from the original 758 for which CPLEX 5.0 took more than 100 seconds to prove optimality. For these 375 models, computing the ratio of the CPLEX 5.0 time divided by the CPLEX 8.0 solve time yielded 375 ratios, the geometric mean of which was 97, rounded to the nearest integer. This result represents a speedup of almost two orders of magnitude.

[4]It is worth noting that we did experiment with various alternative approaches on these models, such as applying the barrier algorithm at the nodes. While doing so typically increased the number of nodes processed within the time limit, it made no material difference in the outcomes.

Table 18.4. *Speedups for solvable models.*

No. models	CPLEX 5.0 time (seconds)	Geometric mean
758	> 0	12
551	> 1	33
463	> 10	59
375	> 100	97
294	> 1000	191
229	> 10000	357
189	>100000	528

18.5.3 Factors contributing to speedups

We have listed some of the new features in recent versions of CPLEX. A natural question is, which of these features is most important to the improved performance? To measure the effects of the different features, at least in terms of proving optimality, we performed the following test. From the 978 models in the original testset, there were 106 models that were solvable by CPLEX 8.0 in less than 1000 seconds and were not solvable by CPLEX 5.0 in 100,000 seconds or less.[5] For each of several features of the code, we turned off that feature, ran CPLEX 8.0 on the 106 models, and compared the running times with the default CPLEX 8.0 runs. The results are summarized in Table 18.5.

Table 18.5. *CPLEX 8.0–effects of individual features.*

Feature	Degradation
No cuts	53.7
No presolve	10.8
CPLEX 5.0 presolve	3.1
CPLEX 5.0 variable selection	2.9
No heuristics	1.4
No node presolve	1.3
No probing on dives	1.1

The "No cuts" entry in this table means that CPLEX 8.0 defaults were compared with CPLEX 8.0 run with all cuts disabled. The other entries headed by the word "No" have a similar meaning. The "CPLEX 5.0 presolve" entry means that CPLEX 8.0 defaults were compared with CPLEX 8.0 with presolve disabled but applied to the model produced by CPLEX 5.0 presolve. Finally, the "CPLEX 5.0 variable selection" entry refers to comparing CPLEX 8.0 defaults with running CPLEX 8.0 using the default variable selection rule from CPLEX 5.0.

[5]For this test we changed the CPLEX 5.0 default settings slightly. CPLEX 5.0 included methods for generating cliques and knapsack covers. To more accurately measure the effect of adding cutting planes, we disabled these cuts in the CPLEX 5.0 runs.

The clear winner in these tests was cutting planes. Disabling this feature resulted in a deterioration of a *factor* of almost 54 in overall performance, a remarkable difference. At the other extreme, we see that heuristics, node presolve, and probing on dives had a much smaller overall effect. However, it should be noted that heuristics and probing on dives are primarily aimed at more quickly and effectively generating good feasible solutions. Proving optimality often focuses on the effectiveness with which the linear programming bound can be moved.

18.5.4 A cut comparison

Our final test used the 106 models isolated for testing in the previous section to compare the 9 different cutting planes that are implemented in CPLEX. The test was performed as follows. For each of the 8 kinds of default cutting planes, we disabled that one kind of cut and compared the result with the default running time. For disjunctive cuts, off by default, these cuts were enabled and compared to defaults. The results are given in Table 18.6.

Table 18.6. *CPLEX 8.0–effects of individual cuts.*

Cut type	Factor
Gomory mixed-integer	2.52
MIR	1.83
Knapsack cover	1.40
Flow cover	1.22
Implied bound	1.19
Path	1.04
Clique	1.02
GUB cover	1.02
Disjunctive	0.53

Gomory cuts are the clear winner by this measure. The degradation in performance caused by disabling these cuts was a factor of roughly 2.5. At the other extreme, enabling disjunctive cuts resulted in a degradation in performance of a factor of almost two. Note that this latter result is not to be interpreted as meaning that disjunctive cuts are ineffective, but rather as a measure of their relative contribution and the fact that they are by far the most expensive to compute of the various cutting planes implemented in CPLEX. When applied in the absence of other cutting planes, disjunctive cuts can be demonstrated to be quite effective.

Bibliography

[1] D. Applegate, R. Bixby, V. Chvátal, and W. Cook. *Solving Traveling Salesman Problems*, 2002, forthcoming.

[2] E. Balas. *Disjunctive Programming: Properties of the Convex Hull of Feasible Points*. Technical report MSRR No. 330, Carnegie Mellon University, 1974.

[3] E. Balas. Disjunctive programming: Properties of the convex hull of feasible points. *Discrete Applied Mathematics*, 89:1–44, 1998.

[4] E.M.L. Beale. Branch and bound methods for mathematical programming systems. In P.L. Hammer, E.L. Johnson, and B.H. Korte, editors, *Annals of Discrete Mathematics 5: Discrete Optimization II*, pages 201–219, North Holland, 1979.

[5] M. Benichou, J.M. Gauthier, P. Girodet, G. Hentges, G. Ribiere, and O. Vincent. Experiments in mixed integer linear programming. *Mathematical Programming*, 1:76–94, 1971.

[6] R. Bixby. Solving real-world linear programs: A decade and more of progress. *Operations Research*, 50:3–15, 2002.

[7] A.L. Brearley, G. Mitra, and H.P. Williams. Analysis of mathematical programming problems prior to applying the simplex algorithm. *Mathematical Programming*, 8:54–83, 1975.

[8] W.J. Carolan, J.E. Hill, J.L. Kennington, S. Niemi, and S.J. Wichmann. An empirical evaluation of the KORBX algorithms for military airlift applications. *Operations Research*, 38:240–248, 1990.

[9] H.P. Crowder, E.L. Johnson, and M.W. Padberg. Solving large-scale zero-one linear programming problems. *Operations Research*, 31:803–834, 1983.

[10] R.J. Dakin. A tree search algorithm for mixed integer programming problems. *Computer Journal*, 8:250–255, 1965.

[11] G.B. Dantzig, D.R. Fulkerson, and S.M. Johnson. Solution of a large scale traveling salesman problem. *Operations Research*, 2:393–410, 1954.

[12] N.J. Driebeek. An algorithm for the solution of mixed integer programming problems. *Management Science*, 12:576–587, 1966.

[13] J.J. Forrest, J.P.H. Hirst, and J.A. Tomlin. Practical solution of large mixed integer programming problems with UMPIRE. *Management Science*, 20:736–773, 1974.

[14] J.J. Forrest and D. Goldfarb. Steepest-edge simplex algorithms for linear programming. *Mathematical Programming*, 57:341–374, 1992.

[15] J.R. Gilbert and T. Peierls. Sparse partial pivoting in time proportional to arithmetic operations. *SIAM Journal on Scientific and Statistical Computing*, 9:862–874, 1988.

[16] D. Goldfarb. Using the steepest-edge simplex algorithm to solve sparse linear programs. In J.R. Bunch and D.J. Rose, editors, *Sparse Matrix Computations*, pages 227–240. Academic Press, New York, 1976.

[17] R.E. Gomory. An algorithm for the integer solutions of linear programs. In R. Graves and P. Wolfe, editors, *Recent Advances in Mathematical Programming*, pages 269–302. McGraw-Hill, New York, 1963.

[18] M. Grötschel. On the symmetric traveling salesman problem: Solution of a 120-city problem. *Mathematical Programming Study*, 12:61–77, 1980.

[19] N. Karmarkar. A new polynomial-time algorithm for linear programming. *Combinatorica*, 4:373–395, 1984.

[20] A.H. Land and A.G. Doig. An automatic method for solving discrete programming problems. *Econometrica*, 28:497–520, 1960.

[21] I.J. Lustig and J.F. Puget. Program does not equal program: Constraint programming and its relationship to mathematical programming. *Interfaces*, 31:29–53, 2001.

[22] K. Marriott and P.J. Stuckey. *Programming with Constraints: An Introduction*, MIT Press, Cambridge, MA, 1999.

[23] M. W. Padberg. On the facial structure of set packing polyhedra. *Mathematical Programming*, 5:199–215, 1973.

[24] M. Padberg and G. Rinaldi. A branch-and-cut algorithm for the resolution of large-scale symmetric traveling salesman problems. *SIAM Review*, 33:60–100, 1991.

[25] E. Rothberg and B. Hendrickson. Sparse matrix ordering methods for interior point linear programming. *INFORMS Journal on Computing*, 10:107–113, 1998.

[26] T.J. Van Roy and L.A. Wolsey. Solving mixed 0-1 problems by automatic reformulation. *Operations Research*, 35:45–57, 1987.

Chapter 19

Graph Drawing: Exact Optimization Helps!

Petra Mutzel and Michael Jünger†*

MSC 2000. 90C57, 90C27, 90C10, 05C62, 05C85, 65K05

Key words. Graph drawing, planarization, crossing minimization, compaction, bend minimization

19.1 Introduction

Graph drawing deals with the design and implementation of algorithms for generating automatic layouts of graphs that can be read and understood easily. A good drawing should reveal the structure of the given graph. With applications in business process modeling, software (re-)engineering, and database design, the field of graph drawing is becoming increasingly important.

Figure 19.1(a) shows a diagram of the dependencies of the electric power industry as it appeared in a German newspaper [51], while Figure 19.1(b) shows an automatically generated layout of the same diagram. Figure 19.2(a) shows the original drawing of a unified modeling language (UML) diagram taken from [24], while Figure 19.2(b) shows an automatically generated layout of the same diagram.

It is difficult to model the *niceness* of a layout, since this often depends on the particular application. However, there exist some criteria that are commonly accepted as important. The vertices should be evenly distributed over the space, overlaps between nodes and other objects should be avoided, and the lengths of the edges and the drawing area should be small. Among the most important criteria is a small number of edge crossings. Users often

*Vienna University of Technology, Institute of Computer Graphics and Algorithms, Favoritenstraße 9-11, A-1040 Vienna.

†University of Cologne, Department of Computer Science, Pohligstraße 1, D-50969 Köln.

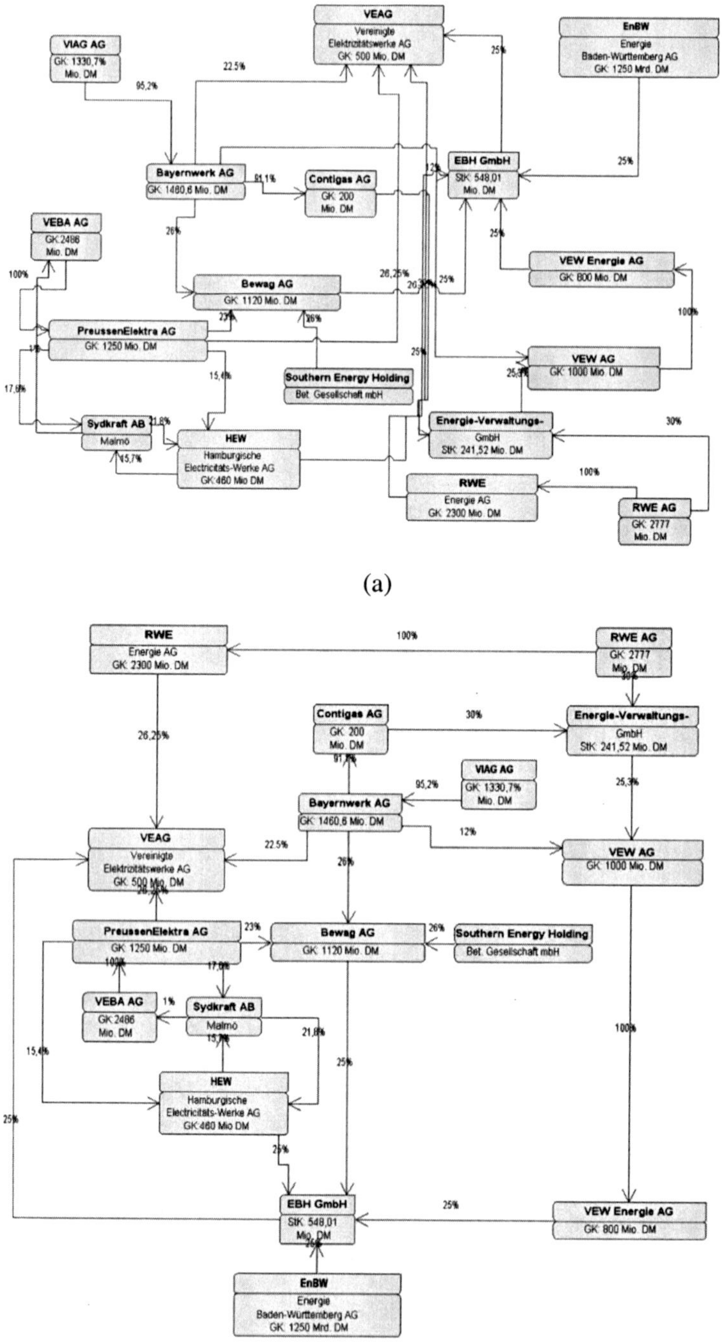

Figure 19.1. *A diagram showing interconnections in the electric power industry:* (a) *hand drawing,* (b) *automatic layout.*

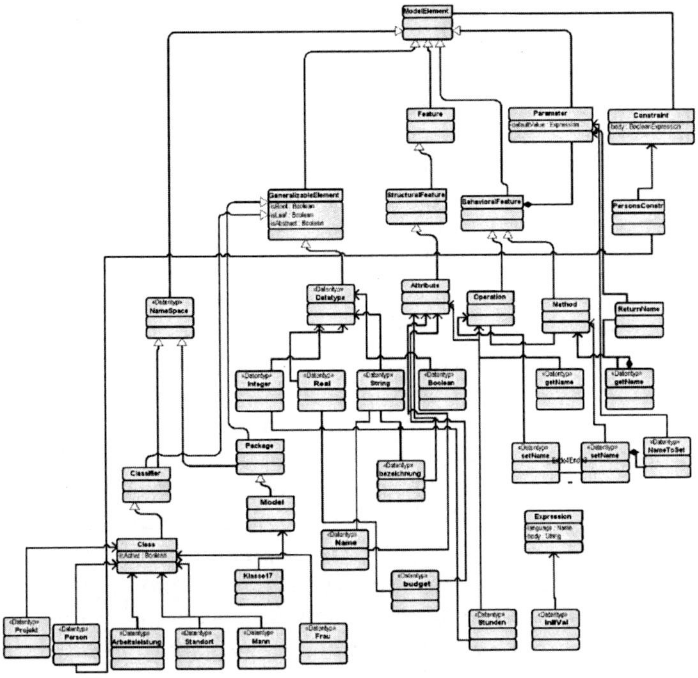

(a)

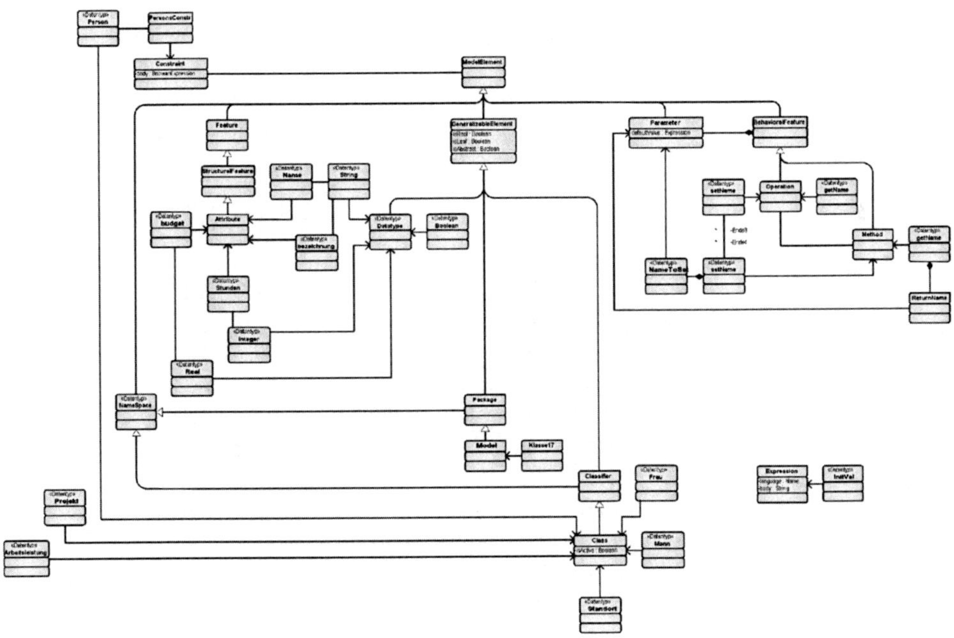

(b)

Figure 19.2. *A UML class diagram:* (a) *hand drawing,* (b) *automatic layout.*

prefer orthogonal drawings in which the edges are represented as paths of horizontal and vertical line segments. Orthogonal drawings with a small number of bends are preferred. If the data are symmetrical or hierarchical, this should be shown in the drawing.

Obviously, these criteria may be conflicting. A study by Purchase [46] has shown that, in general, crossing minimization is the most important criterion, followed by bend minimization. The topology-shape-metrics approach [3] suggests a stepwise approach that first fixes the topology of the drawing (the edge crossings), then the shape of the drawing (the bends), and in a third step the metrics (the edge lengths). Optimization problems arise in all three phases: crossing minimization, bend minimization, and compaction. For these three optimization problems, we will discuss the question of whether exact optimization does help in comparison with heuristic approaches. All three problems are NP-hard optimization problems. The latter two can be solved in many cases to optimality via a branch-and-cut approach, while no such approach is known for the first problem. We report on computational experiments that support the statement in the title of this chapter.

The crossing minimization problem is one of the crucial problems in graph drawing. However, despite the vast amount of published papers on this problem, no exact algorithm for solving even small instances (say 15 nodes) in reasonable computation time is known. In Section 19.3.1 we will briefly summarize the literature on crossing minimization. In Section 19.3.2 we will suggest mathematical programming approaches that might be able to attack the crossing minimization problem directly. In practice, the crossing minimization problem is solved heuristically using a two-step planarization approach, which we will discuss in Section 19.3.3.

Once the topology of a graph drawing is fixed, the substitution of the crossings with artificial vertices leads to a planar graph. Our next task is to determine the shape of an orthogonal drawing, i.e., the angles along the edges. In Section 19.4 we will show that this problem can be attacked with mathematical programming techniques.

In the last phase, the edge lengths are determined by solving the compaction problem that has already been studied in the context of VLSI-layout. In Section 19.5 we will report on new exact approaches.

19.2 Preliminaries

19.2.1 Crossings, planarity, and embeddings

In a drawing of a graph $G = (V, E)$ each vertex $v \in V$ is mapped to a distinct point p_v in the plane and each edge $(u, v) \in E$ is mapped to a closed simple curve that connects the points p_u and p_v and does not pass through the image of any other vertex. If two curves share an interior point p, we say that they cross at p. The crossing number $cr(G)$ is the minimal number of crossings in any drawing of G. The crossing number problem is the problem of finding the crossing number for a given graph G.

The graphs that can be drawn without any edge crossings are called *planar graphs*. A planar drawing of a graph divides the plane into regions called *faces*. Every drawing defines a *planar* and a *combinatorial embedding* of the graph G. Such an *embedding* essentially fixes the topology of the graph. A *combinatorial embedding* is defined as a clockwise ordered list of adjacent neighbors for each vertex $v \in V$. When, in addition, the outer face is fixed, the combinatorial embedding is also called a *planar embedding* of G. An alternative definition of a combinatorial embedding is an anticlockwise ordered list of the bordering edges for each face.

Given a planar graph, a combinatorial embedding can be computed in linear time [8, 35]. In general, a planar graph can have an exponential number of combinatorial embeddings. In the following section, we will use the name *embedding* for planar and combinatorial embeddings.

19.2.2 The SPQR-tree data structure

We give a brief overview of the SPQR-tree data structure for biconnected graphs. SPQR-trees have been suggested by Di Battista and Tamassia [11]. They represent a decomposition of a biconnected graph into triconnected components. A connected graph is triconnected if it does not contain a pair of vertices whose removal splits the graph into two or more components.

An SPQR-tree has four types of nodes, namely, S-, P-, Q-, and R-nodes. With each node μ of the SPQR-tree a biconnected graph, called the *skeleton* of μ, is associated. Each skeleton represents a "simplified" version of the original graph G'. Its vertices are associated with the vertices in the original graph, and each edge represents either an edge or a subgraph of the original graph. Moreover, each skeleton edge can be associated with an edge in the tree. The node types and their skeletons are as follows:

- Q-node: For each edge in G' we have exactly one Q-node. The skeleton consists of two vertices that are connected by two edges. One of the edges represents an edge e of the original graph and the other one the rest of the graph. There is exactly one Q-node for each edge.

- S-node: These nodes are associated with the components derived by a series decomposition of G'. The skeleton is a simple cycle with at least three vertices (associated with the cut-vertices of the decomposition).

- P-node: Each component derived from a parallel decomposition yields a P-node. The skeleton consists of two vertices connected by at least three edges.

- R-node: These nodes are associated with the real triconnected components of G'. The skeleton is a triconnected graph with at least four vertices.

All leaves of the SPQR-tree are Q-nodes, and all inner nodes S-, P-, or R-nodes. For any graph G, its SPQR-tree is unique when considered as unrooted. Figure 19.3 shows a graph and its SPQR-tree. The corresponding skeletons are shown in Figure 19.4. The size of the SPQR-tree is linear in the size of the original graph. An SPQR-tree of a graph G can be constructed in linear time [22, 18].

The most interesting property for this application is that SPQR-trees can be used to represent the set of all combinatorial embeddings of a biconnected planar graph. Every combinatorial embedding of the original graph defines a unique combinatorial embedding for each skeleton of a node in the SPQR-tree, and vice versa. Hence, the set of all combinatorial embeddings of a planar graph can be enumerated straightforwardly using the SPQR-tree data structure. Moreover, the number of embeddings can be directly computed from the tree (in linear time). The skeletons of S- and Q-nodes are simple cycles, so they

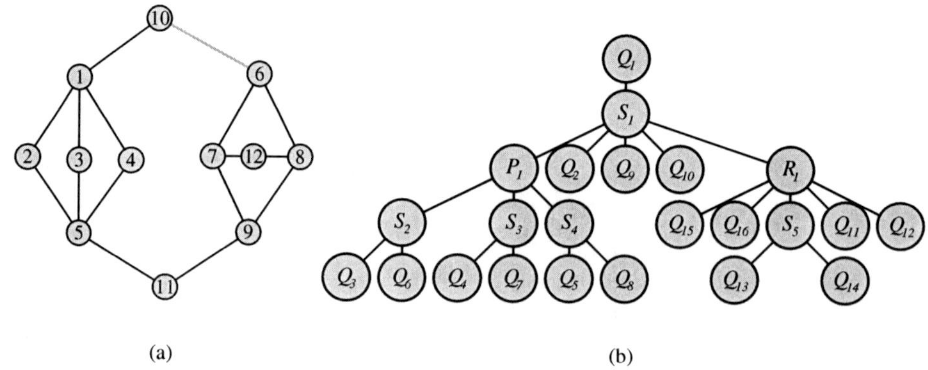

Figure 19.3. *A graph and its SPQR-tree.*

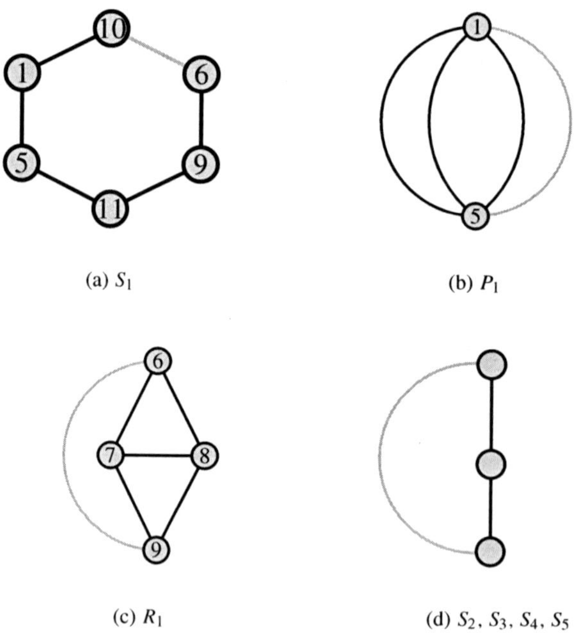

Figure 19.4. *The skeletons of the inner nodes of the SPQR-tree in Figure 19.3. Since the skeletons of the nodes S_2, S_3, S_4, and S_5 are isomorphic, their structure is shown in Figure 19.4(d) without vertex numbers.*

have only one embedding. The skeletons of R-nodes are triconnected graphs. The number of different embeddings of a P-node skeleton is $(k-1)!$, where k is the number of edges in the skeleton.

19.3 Topology: Crossing Minimization

19.3.1 Bounds from the literature

The crossing number represents a fundamental measure of nonplanarity of graphs and has been studied for more than 40 years by graph theorists. The algorithmical problem of computing the crossing number has also been studied in the context of VLSI-layout. Nevertheless, so far there are only a few infinite classes of graphs for which the crossing number is known. We do not even know the asymptotic value for the complete graph K_n with n vertices and for the complete bipartite graph $K_{n,n}$ with $2n$ vertices as n tends to infinity [47]. So far the correctness of the conjectures

$$cr(K_n) = \frac{1}{4}\left\lfloor\frac{n}{2}\right\rfloor\left\lfloor\frac{n-1}{2}\right\rfloor\left\lfloor\frac{n-2}{2}\right\rfloor\left\lfloor\frac{n-3}{2}\right\rfloor, \tag{19.1}$$

$$cr(K_{m,n}) = \left\lfloor\frac{m}{2}\right\rfloor\left\lfloor\frac{m-1}{2}\right\rfloor\left\lfloor\frac{n}{2}\right\rfloor\left\lfloor\frac{n-1}{2}\right\rfloor \tag{19.2}$$

has been verified only for K_n with $n \le 10$ and $K_{m,n}$ with $m \le 6$, as well as a few more special cases. The smallest complete bipartite graphs with unknown crossing numbers are $K_{7,11}$ and $K_{9,9}$.

We do know, however, that the crossing number problem and several of its variants are NP-hard [25, 6]. While many other prominent NP-hard problems have been successfully attacked with integer programming and branch-and-cut techniques, most notably the traveling salesman problem (TSP) for which Manfred Padberg played the key role after the classical paper of Dantzig, Fulkerson, and Johnson [9], no similar approach to the crossing number problem is known to date. To our knowledge, no exact algorithm exists that is able to solve even small instances of the crossing number problem to provable optimality within reasonable computation time.

The main problem with exact approaches is the lack of tight lower bounds for the crossing number. Most known lower bounds depend only on the number of vertices and edges of the graph, but not on its particular structural properties; e.g., an improved version by Pach and Tóth [43] of a bound introduced by Ajtai et al. [1] yields

$$cr(G) \ge \frac{1}{33.75}\frac{|E|^3}{|V|^2} - 0.9\,|V|.$$

A bound based on the structure of the graph can, e.g., be obtained from its skewness. The *skewness* of a graph G is the minimum number of edges that must be deleted from G in order to obtain a planar subgraph. The skewness problem is also NP-hard, but for this problem, approaches based on branch-and-cut are successful for many practical instances of moderate sizes up to 80 edges [26]. However, this bound is not tight in general.

An alternative lower bound is based on the bisection width of the graph and has been given by Leighton [33], Pach, Shahrokhi, and Szegedy [42], and Sýkora and Vrťo [49].

The *bisection width* $bw(G)$ of a graph G is the minimum number of edges whose removal partitions G into two parts having at most $2|V|/3$ vertices each. Recently, Djidjev and Vrťo [12] found a similar relation using the *cutwidth* $cw(G)$ of the graph defined as follows. Consider an injection of the vertices of G into points on a horizontal line. Draw the edges above the line using semicircles. Find a vertical line between a pair of consecutive points that intersects the maximum number of edges. Minimize the maximum over all injections. This minmax value is called the cutwidth of G. Notice that $cw(G) \geq bw(G)$, and there exist connected graphs with $bw(G) = 1$ but with arbitrarily large cutwidth. Let d_v denote the degree of vertex v. From [42, 49, 12] we know that

$$cr(G) + \frac{1}{16} \sum_{v \in V} d_v^2 \geq \frac{1}{40} bw^2(G), \tag{19.3}$$

$$cr(G) + \frac{1}{16} \sum_{v \in V} d_v^2 \geq \frac{1}{1176} cw^2(G). \tag{19.4}$$

Unfortunately, both the cutwidth problem and the bisection width problem are NP-hard.

For graphs with bounded degrees, Even, Guha, and Schieber [14] have recently suggested an approximation algorithm in which the sum of the numbers of vertices and crossings is $O(\log^3 |V|)$ times the minimum sum, thus improving the results of $O(\log^4 |V|)$ by Bhatt and Leighton [5] and Leighton and Rao [32]. It is based on a decomposition tree obtained by recursively bisecting the graph. Grohe [17] has given an exact algorithm that works in quadratic time if the crossing number is fixed. Both algorithms are of a rather theoretical nature and have so far not been useful for solving practical instances.

In the following we will discuss mathematical programming approaches to the crossing minimization problem.

19.3.2 Discussion of exact approaches

An intuitive approach to exact crossing minimization via mathematical programming could be the following. We introduce 0/1-variables x_{ef} for all pairs of edges $e, f \in E$ coding the crossings in a drawing: $x_{ef} = 1$ if the two edges e and f cross in the associated drawing and $x_{ef} = 0$ otherwise. We could add constraints forcing one crossing variable to 1 whenever a subdivision of K_5 or $K_{3,3}$ is detected. There are some problems with this formulation.

Unfortunately, no exact polynomial-time separation algorithm for these constraints is known. Moreover, these constraints are not strong enough. Crossing-critical graphs may be helpful in the search for stronger constraints. A graph is *crossing-critical* if deleting an arbitrary edge decreases its crossing number. Let M_k be the family of crossing-critical graphs with crossing number at least k. By Kuratowski's theorem we have that $M_1 = \{K_5, K_{3,3}\}$ up to subdivisions. But already the family M_2 is infinite. Recently, Ding et al. [13] have characterized 2-crossing-critical graphs that satisfy certain simple assumptions. Hlinený [20] has shown that the class of graphs in M_k has bounded pathwidth (this is not true for noncrossing-critical graphs with crossing number k). But still more research in the area of crossing-critical graphs is needed in order to come up with results that can be used in practical branch-and-cut algorithms.

However, a severe problem with this formulation is the fact that the *realizability problem*, "Given a vector $x \in \{0, 1\}^{\binom{E}{2}}$, does there exist a drawing consistent with x?," is

NP-complete [31]. It follows that the pure crossing information (which edge crosses which other edge) is not sufficient to compute a crossing-minimum drawing from x.

We need some additional information. A possibility could be to introduce variables that encode for all $e \in E$ the order in which other edges cross e. This would allow us to solve the realizability problem via planarity testing. Planarity testing can be done in linear time (e.g., [21]). However, it is not clear how to link the new variables with the crossing variables.

An alternative approach is to use skeletons. This approach is based on the observation that any graph drawn in the plane with some crossing structure X can be redrawn with an equivalent crossing structure such that the resulting drawing has the following properties [41]:

- All vertices are placed on a horizontal line (in arbitrary order).
- All edges are drawn as a series of semiellipses in which successive semiellipses lie on different sides of the horizontal line.

It can be shown [40] that the crossing configuration can be reconstructed in polynomial time from the crossing vector x and the information where and in which direction the semiellipses cross the horizontal line. We can get this information by inserting artificial edges between two consecutive vertices on the line and between the first and the last vertex. We call the resulting Hamilton cycle the *skeleton* of the new graph G'.

The skeleton approach could be attacked via the *constrained crossing minimization problem*. This problem asks for the minimum number of crossings obtained by a set of edges F when inserted into a planar graph P, while the embedding of P is not changed. The crossing minimization problem for G is equivalent to the constrained crossing minimization problem for the edge set of G inserted into the Hamilton cycle. Ziegler [53] has investigated this problem. He presented an integer linear programming formulation and developed a branch-and-cut algorithm. So far, the experiments have only worked for an edge set F of cardinality up to 10. This is useful within a planarization approach, but not yet for general crossing minimization.

However, there is still one more problem with our formulation: The vector x as defined above will give us the *pairwise crossing number* $cr_p(G)$, i.e., the minimum number of pairs of edges that cross over all drawings of G. It is not known whether $cr_p(G) = cr(G)$. Obviously, we have $cr_p(G) \leq cr(G)$, and it is known that there always exists a crossing-minimal drawing in which each pair of edges crosses at most once. Pach and Tóth [44] have written a paper entitled "Which crossing number is it anyway?," in which they discuss various types of crossing numbers. They have also shown that

$$cr_o(G) \leq cr_p(G) \leq cr(G) \leq 2(cr_o(G))^2,$$

where $cr_o(G)$ denotes the odd-crossing number of G, which is defined as the minimum number of pairs of edges that cross an odd number of times.

It might be easier to find the *rectilinear crossing number* $cr_l(G)$, which is the minimum number of crossings in any drawing of G, in which every edge is represented by a straight line segment. However, it is known that $cr_l(G) \neq cr(G)$ in general.

Concerning our question of whether exact optimization helps in the field of crossing minimization, we conjecture yes. It would be important not only for finding better drawings, but also for the field of crossing theory. Exact algorithms may help us to understand the theory and may provide good hints for new results. In Section 19.3.3 we describe how the crossing minimization problem has been attacked in practice so far.

19.3.3 The planarization approach

In practice, the crossing minimization problem is solved heuristically using a two-step planarization approach. In the first step, a minimum cardinality step of edges is deleted from G in order to obtain a planar graph G_P. In the second step, the edges are reinserted into the planar graph G_P while trying to keep the number of crossings small. In the final drawing, there are no edge crossings between edges in G_P.

The first problem is called the *maximum planar subgraph problem* and has been shown to be NP-hard [34]. If the number of edges to be deleted is small, the exact branch-and-cut algorithm suggested in [26] is able to provide a provably optimal solution quite fast. If this number exceeds 10, the algorithm usually needs far too much time to be acceptable for practical computation.

Ziegler [53] did computational studies on a benchmark set of graphs widely used in graph drawing. This set contains 11,529 graphs and has been generated from a core set of 112 graphs used in "real-life" software engineering and database applications with number of vertices ranging from 10 to 100. For the 4654 nonplanar graphs in the benchmark set with no more than 65 vertices, Ziegler has computed a maximum planar subgraph using the branch-and-cut algorithm. He found provably optimal solutions for 4458 of the graphs within 1 hour running time on a Sun Enterprise™ 10000 (with 12 GB of main memory). For 1426 (which is approximately 32%) of those graphs, the algorithm deleted only one edge. The maximum number of deleted edges was 9.

Ziegler also investigated heuristics for the maximum planar subgraph problem. While the standard heuristics produce solutions that are far from the optimum solution most of the time, they can be changed (e.g., by introducing random events and calling them 100 times) so that their quality improves greatly. However, due to the absence of an efficient exact algorithm for handling graphs with larger skewness, we cannot say anything about the quality of the heuristics.

The edge reinsertion step is also an NP-hard optimization problem. The standard algorithm used in practice reinserts the edges $e_1, e_2, \ldots, e_l$ iteratively. The approach is based on the observation that an edge e_i crosses an edge in G_P if and only if it uses an edge in the geometric dual graph of G_P. Hence, the problem of reinserting only one edge into G_P can be solved via a simple shortest path computation in the extended dual graph of G_P. (We need to extend the dual graph in order to connect the end-vertices of e_i with the dual graph.) After each insertion step i, the crossings generated by edge e_i are substituted with artificial vertices so that the resulting graph $G_P \cup \{e_1, \ldots, e_i\}$ becomes planar again ($i = 1, \ldots, l$).

However, the quality of the resulting drawing highly depends on the chosen embedding for G_P. Figure 19.5(a) shows an optimal solution of the edge reinsertion problem for the shown embedding of G_P, while Figure 19.5(b) shows the optimal solution over the set of all possible combinatorial embeddings of G_P. Recently, Gutwenger, Mutzel, and Weiskircher [19] have given a linear-time algorithm based on SPQR-trees for inserting one edge into a planar graph G_P so that the number of crossings in $G_P \cup \{e\}$ over the set of all possible planar embeddings of G_P is minimized.

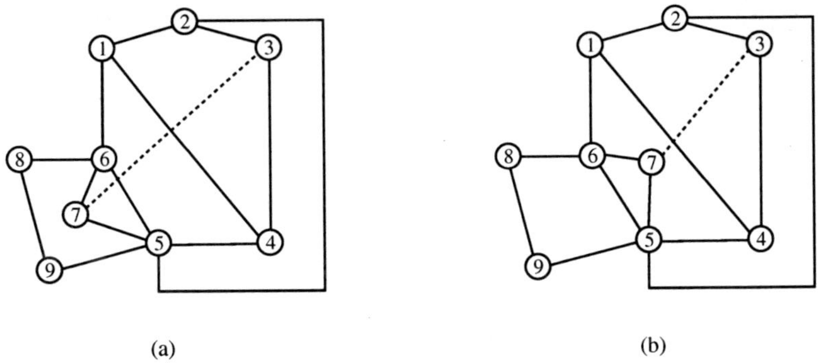

Figure 19.5. *Quality of edge reinsertion depends on the chosen embedding of G_P: (a) fixed embedding, (b) optimal embedding.*

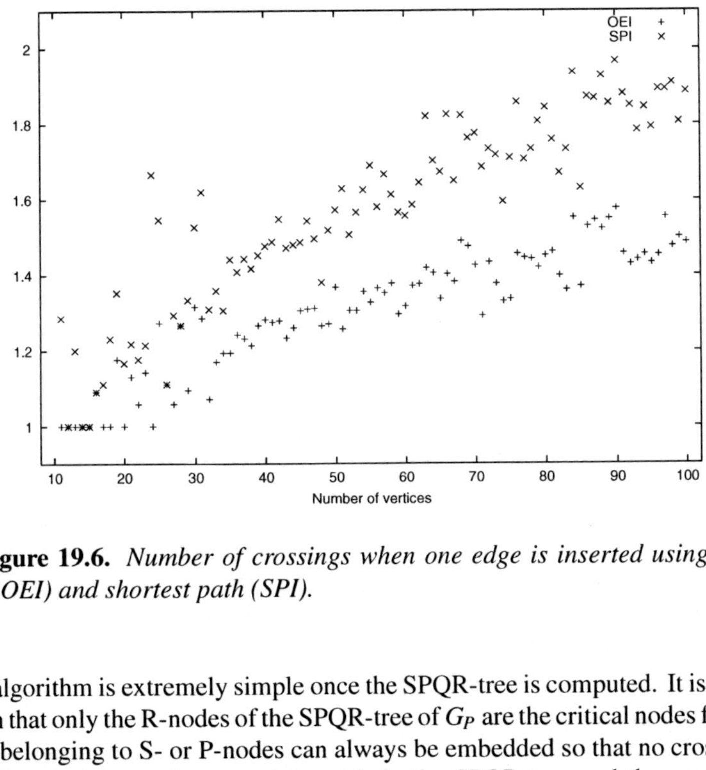

Figure 19.6. *Number of crossings when one edge is inserted using the optimal algorithm (OEI) and shortest path (SPI).*

The algorithm is extremely simple once the SPQR-tree is computed. It is based on the observation that only the R-nodes of the SPQR-tree of G_P are the critical nodes for crossings. Subgraphs belonging to S- or P-nodes can always be embedded so that no crossing occurs. The algorithm first determines a unique path in the SPQR-tree and then only keeps the R-nodes on this path. For each of these R-nodes μ, a shortest path in the skeleton of μ is computed. The number of crossings is the sum of the lengths of the shortest paths.

Computational experiments on the 8249 nonplanar graphs in the benchmark set (mentioned above) have shown that iterative use of the new algorithm for the set of all deleted edges $e_1, e_2, \ldots, e_l$ leads to a smaller number of crossings. Figure 19.6 shows the number of

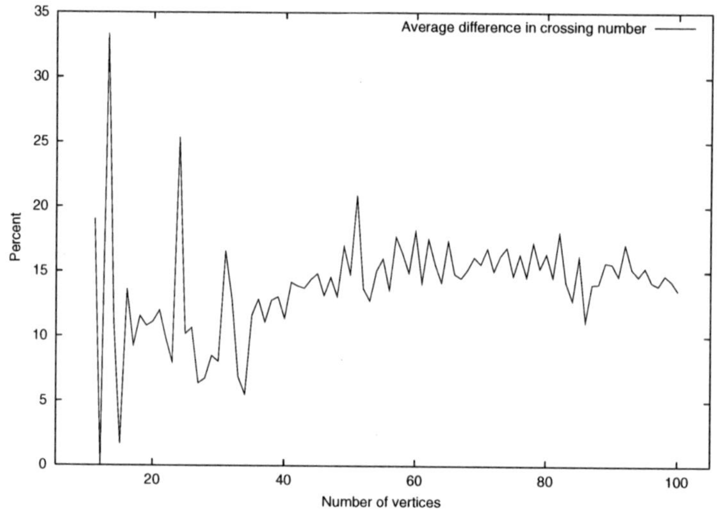

Figure 19.7. *Improvement of the exact algorithm over the shortest path inserter for the benchmark set using the iterative approach.*

crossings generated by using the standard approach (SPI) and the optimal approach (OEI) when only one edge needs to be inserted. The difference in quality is rather high. The average improvement on the standard benchmark set of graphs of the iterative method via the new optimal 1-edge insertion algorithm compared to the standard shortest path approach is shown in Figure 19.7. The average relative improvement is 14.4%, while the maximum improvement is 85.71% [52]. The average number of crossings produced by the standard algorithm ranges from 1.3 to 57.9 compared to the range from 1.3 to 49.8 for the 1-optimal algorithm.

Obviously, reinsertion of all edges at the same time will improve the solution. However, no practically efficient algorithm is known. When keeping the embedding of the planar graph G_P fixed, we have the constrained crossing minimization problem (see Section 19.3.2). It can be formulated as a k-shortest paths problem in the extended dual graph of G_P, where the objective is the sum of the paths plus the sum of crossings between the paths. Moreover, the nodes and edges of the paths need not be disjoint. The problem has been investigated in [40, 39, 53]. Experiments show that this approach is able to solve instances for which fewer than 10 edges need to be reinserted to provable optimality.

19.4 Shape: Bend Minimization

Once the topology of the graph is fixed, we substitute each crossing c with an artificial vertex and obtain a planar graph $G' = (V \cup C, E')$. We define $n := |V| + |C|$. The task considered in this section is to find a planar orthogonal drawing of G' with the minimum number of edge bends. This problem is also NP-hard [15].

For the restricted problem where a fixed planar embedding of G' is part of the input and the vertex degree is bounded by four, Tamassia [50] has given an $O(n^2 \log n)$ time

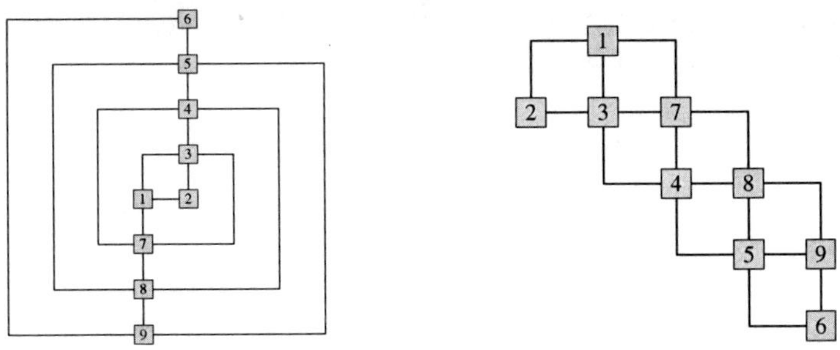

(a) (b)

Figure 19.8. *Bend minimum drawings* (a) *for a given fixed embedding and* (b) *over the set of all embeddings.*

algorithm based on a transformation of the problem into a minimum cost flow problem. Garg and Tamassia [16] have shown that this particular network flow problem can be solved in time $O(n^{\frac{7}{4}}\sqrt{\log n})$. The algorithm computes a so-called orthogonal representation of the graph G', which is essentially a list of the angles along the edges for each face.

In general, the choice of the planar embedding can strongly affect the number of bends of the drawing. Figure 19.8(a) shows a bend minimum drawing for the given embedding, while Figure 19.8(b) shows a bend minimum drawing over the set of all planar embedding for the same graph.

Again, we investigate the following question: Can exact optimization help to solve the bend minimization problem? So far, two different exact approaches for bend minimization exist: The approach in [4] consists of a branch-and-bound (B&B) algorithm that essentially enumerates over the set of all planar embeddings and solves the corresponding network flow problem. Moreover, it contains new methods for computing lower bounds. Recently, Weiskircher [52] presented a branch-and-cut algorithm based on an integer linear programming formulation for optimization over the set of all planar embeddings as suggested in [37]. Both approaches are based on the SPQR-tree data structure and are not restricted to maximal vertex degree four.

In the following, we describe the integer linear programming approach to the bend minimization problem. As already mentioned, for a fixed embedding, a bend minimum orthogonal representation can be constructed by solving a minimum cost flow problem in a network constructed from the geometric dual graph. Each unit of flow represents a 90 degree angle. A flow between a primal vertex v and a dual vertex c_f represents the angle at v in the face f. Each unit of flow from one face vertex c_f into the other via an edge e in the dual graph represents a 90 degree bend within the corresponding edge in the original graph. The network obviously depends on the chosen embedding.

In order to generalize it to the set of all embeddings, we set up a network that contains edges between all cycles sharing an edge that could be face cycles in at least one planar embedding. For this we first need a description of the variables and constraints for the set of all planar embeddings. Such an integer linear programming formulation has been suggested in [36, 37].

19.4.1 An integer linear programming formulation for the set of embeddings

We start by describing the recursive construction of the integer linear program (ILP) that represents all combinatorial embeddings of a graph.

The variables in our program correspond to directed cycles in the graph that are face cycles in at least one planar embedding. We can guarantee this by our recursive construction based on SPQR-trees. In a feasible solution of the ILP, a variable x_c has value 1 if the associated cycle is a face cycle in the represented embedding and has value 0 otherwise.

We use a recursive approach to construct the variables and constraints of the ILP. In order to generate the ILP for the whole graph, we start with the SPQR-tree and use a *splitting operation* recursively, thus producing subproblems that are smaller than the original problem. The splitting process stops when the corresponding trees contain only one S-, P-, or R-node. A graph whose SPQR-tree has only one inner node is isomorphic to the skeleton of this inner node. The ILPs for SPQR-trees with only one inner node are defined as follows:

- S-node: When the only inner node of the SPQR-tree is an S-node, the whole graph is a simple cycle. Thus it has two directed cycles and both are face cycles in the only combinatorial embedding of the graph. So the ILP consists of two variables, both of which must be equal to one.

- R-node: In this case, the whole graph is triconnected. According to our definition of a planar embedding, every triconnected graph has exactly two embeddings, which are mirror images of each other. When the graph has m edges and n nodes, we have $k = 2(m - n + 2)$ variables and two feasible solutions. The constraints are given by the convex hull of the points in k-dimensional space that correspond to the two solutions. If we order the variables $x_1, x_2, \ldots, x_k$ for the face cycles such that the variables with odd index represent the face cycles of the first embedding and the variables with even index represent the face cycles of the second embedding, the ILP has a very simple structure:

$$x_{2i-1} + x_k = 1 \qquad \text{for} \quad 1 \le i \le k/2, \tag{19.5}$$

$$x_{2i} - x_k = 0 \qquad \text{for} \quad 1 \le i \le k/2 - 1, \tag{19.6}$$

$$x_k \ge 0, \tag{19.7}$$

$$x_k \le 1. \tag{19.8}$$

- P-node: The whole graph consists of only two nodes connected by k edges with $k \ge 3$. Every directed cycle in the graph is a face cycle in at least one embedding of the graph, so the number of variables is equal to the number of directed cycles in the graph. The number of cycles is $l = 2\binom{k}{2}$. In order to construct an ILP for a P-node, we consider the problem of finding the embedding of P that minimizes the sum of the weights of the cycles that are face cycles. It is not hard to see that this problem is equivalent to the asymmetric TSP (ATSP). For the set of face cycles of an embedding of the P-node we can identify the set of edges in the associated ATSP tour, and vice versa. This observation enables us to use the same integer linear programming formulation as for the ATSP, which contains the degree constraints and the subtour elimination constraints. Since the number of subtour elimination constraints is exponential, we store with each P-node skeleton the corresponding ATSP graph. For each edge in the ATSP graph, we store the corresponding cycle in the P-node skeleton. In our integer linear programming approach, these inequalities are separated via minimum cut computations.

Now we describe how to construct the ILP of an SPQR-tree T from the ILPs of the split trees of T. Let G' be the graph that corresponds to T and $T_1, T_2, \ldots, T_k$ the split trees representing the graphs $G_1, G_2, \ldots, G_k$. Every directed cycle c in a graph G_i represented by a split tree represents a set of cycles $R(c)$ in the original graph. These cycles are either *local cycles* (only appearing in one of the graphs $G_1, G_2, \ldots, G_k$), or *global cycles* (not completely contained in any of the G_i). Global cycles are the cycles that have been split into several smaller cycles by the splitting operation.

The variables of the split trees that represent local cycles will also be variables of the ILP of the original graph G'. This is not always true for global cycles. The constraints of the ILP consist of lifted constraints, choice constraints, and a center graph constraint. The lifted constraints come from constraints in the subproblems. The choice constraints code the fact that only one of the cycles in $R(c)$ can be a face cycle in any combinatorial embedding of G'. The center graph constraint essentially fixes the number of cycles that can be face cycles. For further details, see [36, 37, 52].

19.4.2 An exact algorithm for bend minimization

The ILP so far is able to optimize an objective function over the set of all combinatorial embeddings of the given graph. In order to solve the bend minimization problem, we need to combine this with additional variables and constraints that model the flow in the network. We use essentially the same network as for a fixed planar embedding; the only difference is that, instead of only having flow between adjacent face cycles, here we have a flow between every adjacent pair of cycles that could be present in at least one combinatorial embedding. Moreover, we need to introduce additional variables representing the outer face. (This is needed to get the planar embedding from the combinatorial embedding.)

In the following, we report on computational experiments on both exact algorithms, the B&B and the integer linear programming–based branch-and-cut algorithm, and the network flow approach for our benchmark set of graphs (see [38, 52]). We use a standard planarization approach for fixing the topology of the nonplanar graphs. This leads to a set of 11,529 planar (or planarized) graphs G' having vertices in the range from 10 to 300. Figures 19.9 and 19.10 show the distribution of the graphs with respect to their number of vertices and number of embeddings.

Due to the high number of test instances, we restricted the running time to 1 h on a SUN Enterprise 450 (4 GB main memory, four 400 MHz processors). The B&B approach exceeded the time limit for 197 instances, while this happened only for 25 instances in the case of the branch-and-cut algorithm. Figure 19.11 shows the running times of the two algorithms. The B&B algorithm is faster for the graphs with only a few embeddings, while the strength of the branch-and-cut algorithm takes over for the graphs with an exponential number of embeddings. It seems that the time needed by the branch-and-cut algorithm grows more slowly with the size of the graphs compared to that of the B&B algorithm. Figure 19.12

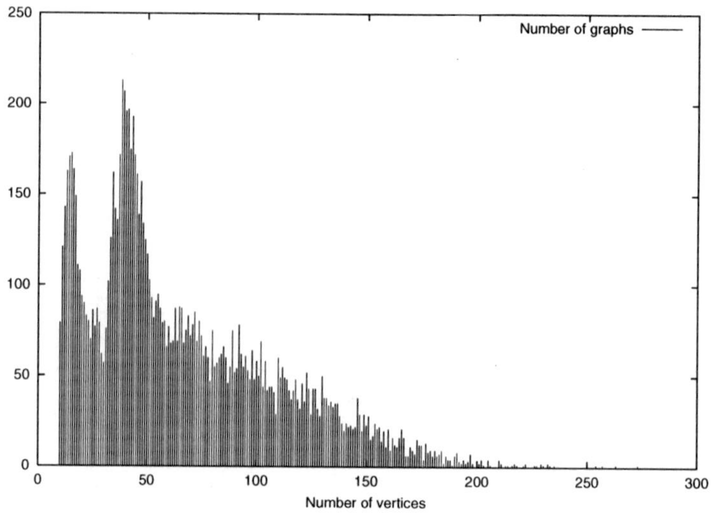

Figure 19.9. *Distribution of graphs with respect to number of vertices.*

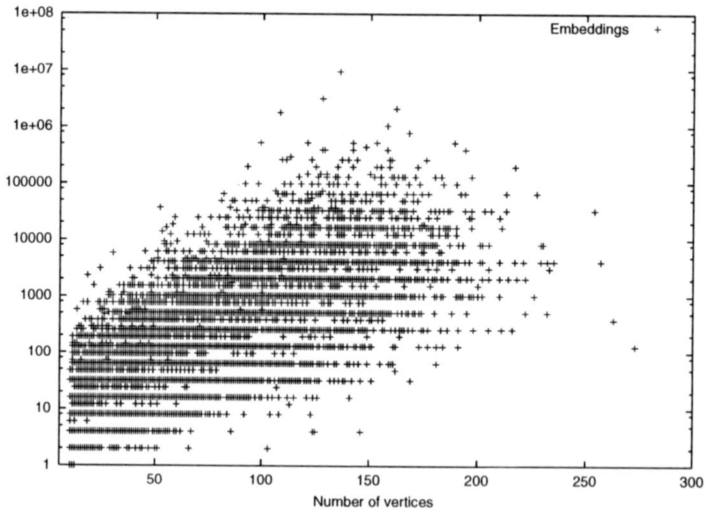

Figure 19.10. *Distribution of graphs with respect to number of embeddings.*

shows the average running time compared to the average number of embeddings. Notice that the figure is in log-scale.

Figure 19.13 shows the average number of variables and constraints in the integer programming formulation. We find it interesting that these numbers grow only linearly with the size of the graphs, although the numbers of embeddings grow exponentially.

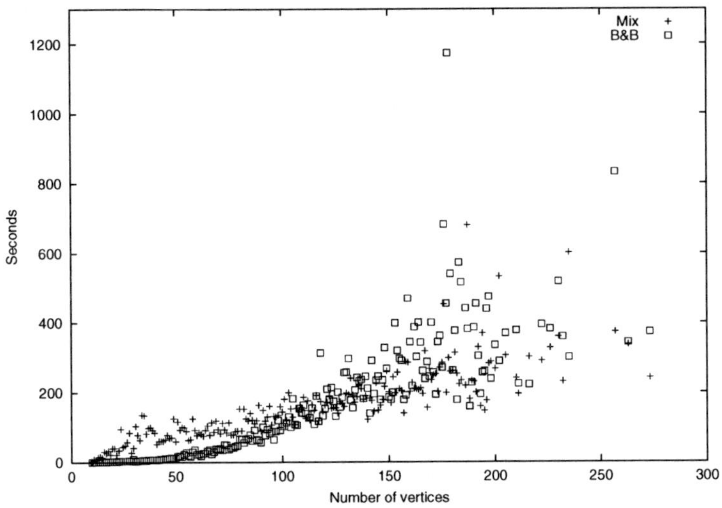

Figure 19.11. *Running times of the branch-and-cut (Mix) and the B&B algorithms.*

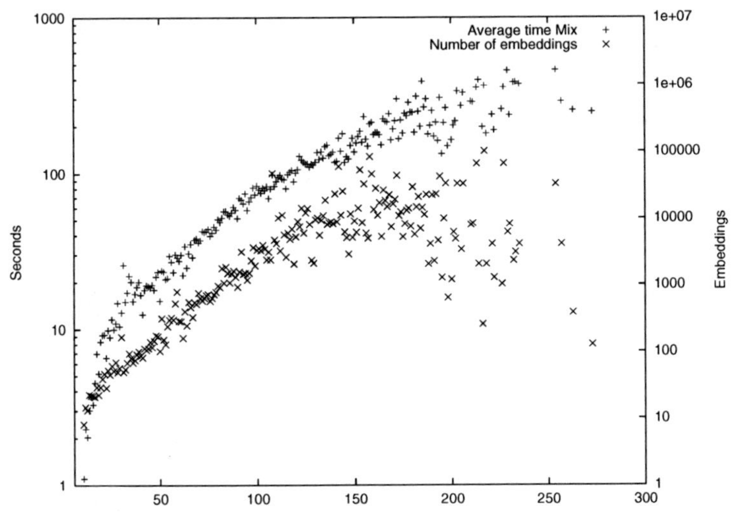

Figure 19.12. *Average running time of the branch-and-cut algorithm compared to the average number of embeddings.*

Finally, we compared our optimum solutions to those obtained by standard heuristics for bend minimization. Figure 19.14 shows the average percentage of improvement of the optimum solution compared to the min-cost-flow solution for a fixed planar embedding. Exactly 5224 graphs (which is about half of the graphs) had average improvement of more than 10%. The falling curve does not mean that the improvement gets smaller with increasing

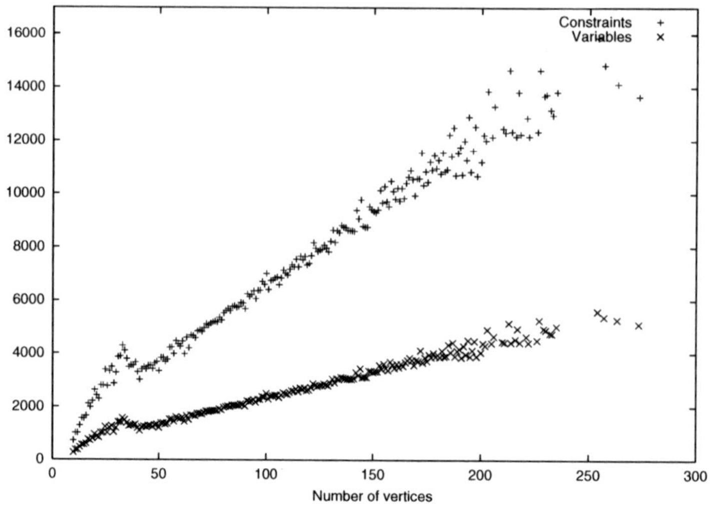

Figure 19.13. *Average number of variables and constraints in the integer linear programming formulation.*

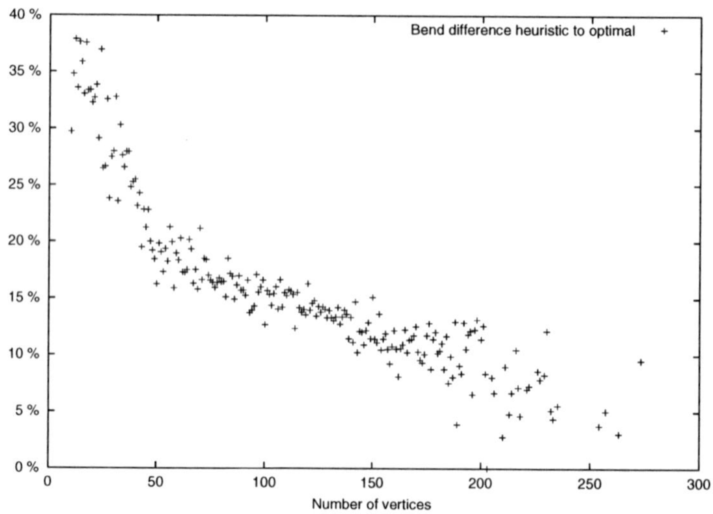

Figure 19.14. *Average percentage of improvement compared to min-cost-flow.*

graph sizes: The original number of vertices of the benchmark graphs was bounded by 100. This means that all the planarized graphs with $100 + c$ vertices have at least c artificial vertices coming from planarization. An increase in the number of artificial vertices (i.e., crossings) induces an increase in the number of faces of size four and an increase in the sizes of the triconnected components. As a consequence, the number of bends does not further increase with the sizes of the graphs (indicated by our experiments), and also the heuristics behave better.

For bend minimization, we can clearly answer our question positively. We conclude that not only is exact optimization helpful, but also integer programming techniques are superior to simple B&B algorithms.

19.5 Metrics: Compaction

The next step is to provide edge lengths to the graph with fixed topology and shape. The compaction phase deals with the transformation of an orthogonal representation into an orthogonal drawing with small total edge length or little area. Again, these are NP-hard problems [45]. Previous algorithmic research for this problem can be divided into construction and improvement heuristics. The standard method used in graph drawing has been suggested by Tamassia [50] and is based on a rectangular dissection of the original orthogonal representation. Recently, Bridgeman et al. [7] improved this technique by introducing the concept of *turn-regularity*.

The improvement methods *compression ridge* (e.g., Akers, Geyer, and Roberts [2]) as well as longest path and flow-based compaction techniques (e.g., Hsueh [23]) are known from VLSI-design. They consider the one-dimensional subproblems of reducing the horizontal or vertical edge length while keeping the other fixed. The flow-based method is optimal in one dimension. In many cases, iterative usage of these heuristics with alternating directions in a one-dimensional compaction scheme yields considerable improvements.

From VLSI-design, we also know two exact algorithms based on B&B. They have been suggested by Schlag, Liao, and Wong [48] and by Kedem and Watanabe [27]. However, the first integer linear programming formulation (for edge length minimization) was given in [30]. The formulation is based on transforming the compaction problem to a new combinatorial optimization problem based on constraint graphs. The task is to extend the two so-called shape graphs (essentially coding the shape and topology) by additional edges in which the compaction problem reduces to two separate one-dimensional problems for which optimal solutions lead to an optimal solution of the problem in two dimensions.

Figure 19.15 shows an instance for which different algorithms have been applied. Figure 19.15(a) has been generated using the standard rectangular dissection, Figure 19.15(b) using turn-regular dissection, and Figure 19.15(c) using rectangular dissection first and then using the flow-based improvement algorithm. Figure 19.15(d) shows the optimal solution.

We ask the question again: Does exact optimization help? We report on the computational results in [29] and [28], which have been conducted on a Sun Enterprise 450 with 1 GB of main memory and two 400 MHz CPUs. The benchmark set of graphs is the same as the one described in the previous sections. For all graphs, the standard planarization step and the orthogonalization step have been executed. In addition, all the bends have been substituted with artificial vertices in order to obtain the instances for the compaction problem.

The running time of the exact algorithm described in [30] was below 1 second for 99% of all benchmark instances and below 10 seconds for all of them, while the running times of all the heuristics stayed below 1 second. In comparison to the standard approach mostly used within the topology-shape-metrics approach, the improvement of the optimal solution

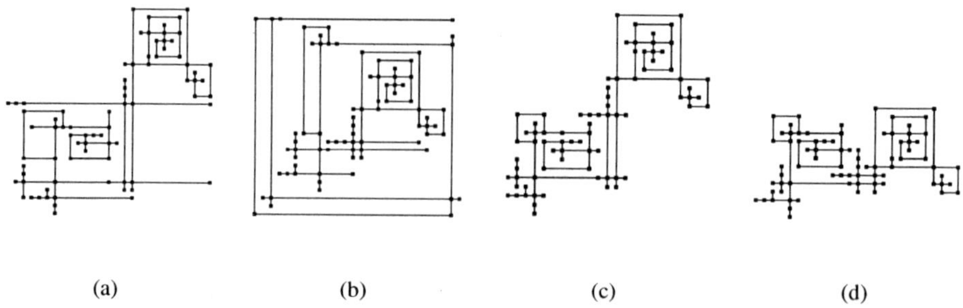

Figure 19.15. (a) *Rectangular dissection;* (b) *turn-regular dissection;* (c) *rectangular dissection with flow postprocessing;* (d) *optimal.*

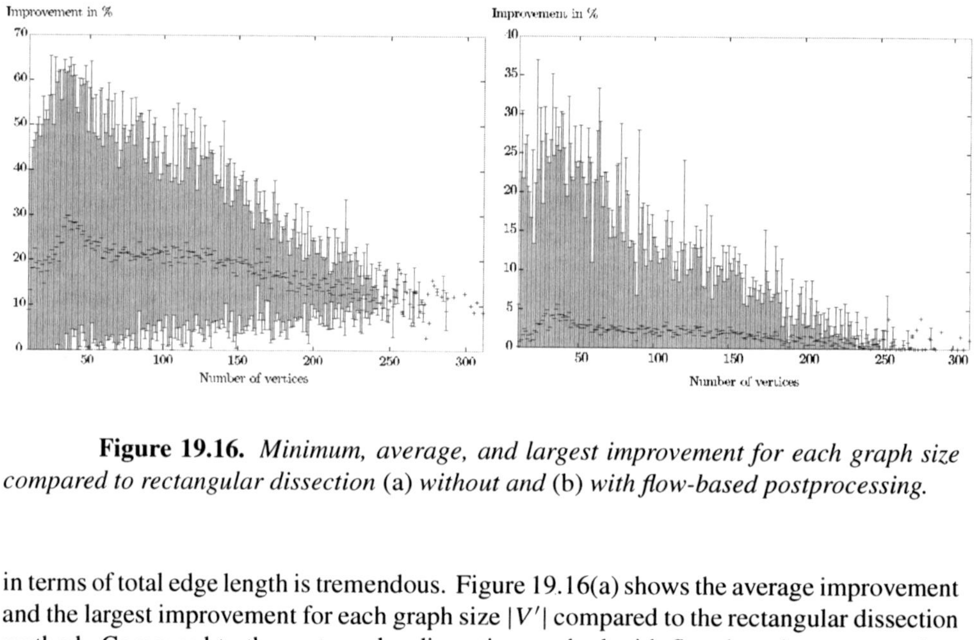

Figure 19.16. *Minimum, average, and largest improvement for each graph size compared to rectangular dissection* (a) *without and* (b) *with flow-based postprocessing.*

in terms of total edge length is tremendous. Figure 19.16(a) shows the average improvement and the largest improvement for each graph size $|V'|$ compared to the rectangular dissection method. Compared to the rectangular dissection method with flow-based postprocessing, the average improvement is only about 3%; but still, there exist many graphs for which the improvement is more than 20% (see Figure 19.16(b)).

Figure 19.17 groups the graphs according to size using a step size of 50 vertices and shows the quality of the solution for some construction heuristics with and without the flow-based improvement step. The construction heuristics considered are the rectangular dissection method based on longest paths (lp) and on flow (fl), and two heuristics based on turn-regularity dissection with flow (t1, t2). It can be observed that the choice of the constructive heuristics in the first step does not have a big impact on the final quality, while it is important to use the flow-based method in a postprocessing step. Figure 19.17 also

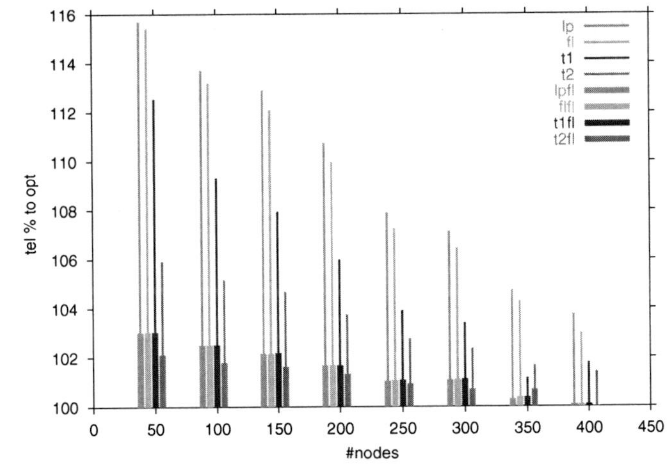

Figure 19.17. *Quality of heuristics with and without flow-based improvement step.*

indicates that combined constructive and improvement heuristics lead to almost optimal solutions in terms of the total edge length.

Figures 19.16 and 19.17 are slightly misleading, giving the impression that the quality of the heuristics improves with the sizes of the graphs. As already remarked in the previous section, the sizes (number of vertices) of the original graphs are bounded by 100. The numbers belonging to graphs beyond the 100 vertex limit contain, in general, many crossing vertices. A planarized graph with many crossing vertices contains many rectangular faces that are easy to compact.

To answer our question: We are not sure if exact optimization is really necessary for this drawing step. On the other hand, the exact approach does not take much longer than the heuristics, and there exist some cases where it indeed improves greatly. Moreover, the exact approach suggested in [30] is open to handling additional constraints and alternative variants of the problem. Sometimes it may be advantageous to trade a few more bends or crossings for a big decrease in the total edge length or the area (see [28]). Using the above approach, this can be handled quite easily, while efficient heuristic approaches in this direction are unclear.

19.6 Conclusion

We have considered optimization problems arising in the topology-shape-metrics approach. As we have seen, exact optimization helps in automatic graph drawing in different senses. In crossing minimization, we are still far from attacking the problem directly, and exact optimization helps in evaluating planarization heuristics and in exploring the strengths of edge deletion and insertion approaches. In bend minimization, our current conclusion is that exact optimization is far superior to known heuristic approaches. Of course, this does not necessarily mean that graph drawing systems should contain exponential-time algorithms whose running time could explode sometimes, but more likely that more research time should be invested in the search for efficient good quality heuristics. In compaction, we have almost the opposite situation. On the average, the known heuristics are fast and effective, and exact optimization helped us find out.

The question arises whether there is a better approach for drawing general graphs. The experimental study [10] includes four different algorithms for orthogonal graph drawing on the benchmark set described above. Two of these algorithms were based on the topology-shape-metrics approach (in one version the network flow algorithm was substituted with a simple "stretch" heuristic), and the others are incremental algorithms that focus on a small area and a small number of bends. The results show that the former two algorithms are superior in terms of the criteria number of crossings, number of bends, area, and edge length. The best algorithm had up to eight times fewer crossings, up to three times fewer bends, up to three times smaller total edge length, and up to four times smaller area in comparison to the worst algorithm. The performance of the other algorithms is worse except that the crossing number of the stretch heuristic is close to the best.

We believe that it is a good idea to separate the crossing minimization phase from the remaining phases of the algorithm. However, in our opinion, it would be better not to separate the bend minimization phase from the compaction phase. Our experience shows that the layouts produced by the current approach do need more space than necessary. Often, the area of the layout (hence also its resolution on the screen) can be reduced considerably by introducing a few more bends. However, so far no better method is known.

In practice, there are requirements on minimum distances between the objects in the drawing (including the edges). Also, usually only a small number of anchor points where edges may be attached to nodes are available. These and other constraints (e.g., sizes of vertices) force software designers to make certain compromises; see Figures 19.1 and 19.2 for examples, where clearly neither the number of bends nor the area is minimized. Nevertheless, the theory that assumes idealized models helps to produce good drawings in practice. Recent research tries to take more such side conditions into account, and, of course, the integer programming approach is especially suited for more refined models.

Acknowledgements. We thank our Ph.D. students Gunnar Klau, René Weiskircher, and Thomas Ziegler for providing figures and numbers from the experimental results in their theses.

Bibliography

[1] M. Ajtai, V. Chvátal, M.M. Newborn, and E. Szemerédi. Crossing-free subgraphs. *Annals of Discrete Mathematics*, 12:9–12, 1982.

[2] S.B. Akers, M.E. Geyer, and D.L. Roberts. IC mask layout with a single conductor layer. In *Proceedings of the 7th Design Automation Workshop*, pages 7–16. ACM/IEEE, 1970.

[3] C. Batini, E. Nardelli, and R. Tamassia. A layout algorithm for data-flow diagrams. *IEEE Transactions on Software Engineering*, SE-12:538–546, 1986.

[4] P. Bertolazzi, G. Di Battista, and W. Didimo. Computing orthogonal drawings with the minimum number of bends. In *WADS '97*, volume 1272 of *Lecture Notes in Computer Science*, pages 331–344. Springer-Verlag, Berlin, 1998.

[5] S.N. Bhatt and F.T. Leighton. A framework for solving VLSI layout problems. *Journal of Computer and System Sciences*, 28:300–343, 1984.

[6] D. Bienstock. Some provably hard crossing number problems. *Discrete and Computational Geometry*, 6:443–459, 1991.

[7] S.S. Bridgeman, G. Di Battista, W. Didimo, G. Liotta, R. Tamassia, and L. Vismara. Turn-regularity and planar orthogonal drawings. In *Graph Drawing (Proceedings of GD '99)*, volume 1731 of *Lecture Notes in Computer Science*, pages 8–26. Springer-Verlag, Berlin, 1999.

[8] N. Chiba, T. Nishizeki, S. Abe, and T. Ozawa. A linear algorithm for embedding planar graphs using PQ-trees. *Journal of Computer and System Sciences*, 30:54–76, 1985.

[9] G. Dantzig, D. Fulkerson, and S. Johnson. Solution of a large-scale traveling-salesman problem. *Operations Research*, 2:393–410, 1954.

[10] G. Di Battista, A. Garg, G. Liotta, R. Tamassia, E. Tassinari, and F. Vargiu. An experimental comparison of four graph drawing algorithms. *Computational Geometry Theory and Applications*, 7:303–326, 1997.

[11] G. Di Battista and R. Tamassia. On-line graph algorithms with SPQR-trees. In *Automata, Languages and Programming (Proceedings of 17th ICALP)*, volume 442 of *Lecture Notes in Computer Science*, pages 598–611. Springer-Verlag, Berlin, 1990.

[12] H. Djidjev and I. Vrťo. An improved lower bound for crossing numbers. In M. Jünger, S. Leipert, and P. Mutzel, editors, *Graph Drawing (Proceedings of GD '01)*, volume 2265 of *Lecture Notes in Computer Science*, pages 96–101. Springer-Verlag, Berlin, 2002.

[13] G. Ding, B. Oporowski, R. Thomas, and D. Vertigan. Large Four-Connected Nonplanar Graphs. In preparation.

[14] G. Even, S. Guha, and B. Schieber. Improved approximations of crossings in graph drawing and VLSI layout area. In *Proceedings of the 32nd ACM Symposium on Theory of Computing (STOC'00)*, pages 296–305. ACM Press, New York, 2000.

[15] A. Garg and R. Tamassia. On the computational complexity of upward and rectilinear planarity testing. In R. Tamassia and I.G. Tollis, editors, *Graph Drawing (Proceedings of GD '94)*, volume 894 of *Lecture Notes in Computer Science*, pages 286–297. Springer-Verlag, Berlin, 1995.

[16] A. Garg and R. Tamassia. A new minimum cost flow algorithm with applications to graph drawing. In S. North, editor, *Graph Drawing (Proceedings of GD '96)*, volume 1190 of *Lecture Notes in Computer Science*, pages 201–216. Springer-Verlag, Berlin, 1997.

[17] M. Grohe. Computing crossing numbers in quadratic time. In *Proceedings of the 32nd ACM Symposium on Theory of Computing (STOC'00)*, pages 231–236. ACM Press, 2000.

[18] C. Gutwenger and P. Mutzel. A linear time implementation of SPQR trees. In J. Marks, editor, *Graph Drawing (Proceedings of GD 2000)*, volume 1984 of *Lecture Notes in Computer Science*, pages 77–90. Springer-Verlag, Berlin, 2001.

[19] C. Gutwenger, P. Mutzel, and R. Weiskircher. Inserting an edge into a planar graph. In *Proceedings of the 12th Annual ACM-SIAM Symposium on Discrete Algorithms (SODA 2001)*, pages 246–255. ACM Press, New York, SIAM, Philadelphia, 2001.

[20] O. Hlinený. Crossing-critical graphs and path-width. In M. Jünger, S. Leipert, and P. Mutzel, editors, *Graph Drawing (Proceedings of GD '01)*, pages 102–114. Volume 2265 of *Lecture Notes in Computer Science*. Springer-Verlag, Berlin, 2002.

[21] J. Hopcroft and R.E. Tarjan. Efficient planarity testing. *Journal of the Association for Computing Machinery*, 21:549–568, 1974.

[22] J.E. Hopcroft and R.E. Tarjan. Dividing a graph into triconnected components. *SIAM Journal on Computing*, 2:135–158, 1973.

[23] M.Y. Hsueh. *Symbolic Layout and Compaction of Integrated Circuits*. Ph.D. thesis, University of California at Berkeley, 1979.

[24] M. Jeckle, Unified Modelling Language (UML). http://www.jeckle.de.

[25] M.R. Garey and D.S. Johnson. Crossing number is NP-complete. *SIAM Journal on Algebraic Discrete Methods*, 4:312–316, 1983.

[26] M. Jünger and P. Mutzel. Maximum planar subgraphs and nice embeddings: Practical layout tools. *Algorithmica, Special Issue on Graph Drawing*, 16:33–59, 1996.

[27] G. Kedem and H. Watanabe. Graph optimization techniques for IC—layout and compaction. *IEEE Transactions on Computer-Aided Design of Integrated Circuits and Systems*, CAD-3:12–20, 1984.

[28] G. Klau. *A Combinatorial Approach to Orthogonal Placement Problems*. Ph.D. thesis, Max-Planck-Institut für Informatik, Saarbrücken, 2001.

[29] G.W. Klau, K. Klein, and P. Mutzel. An experimental comparison of orthogonal compaction algorithms. In *Graph Drawing (Proceedings of GD 2000), Lecture Notes in Computer Sciences*. Springer-Verlag, Berlin, 2001.

[30] G.W. Klau and P. Mutzel. Optimal compaction of orthogonal grid drawings. In G.P. Cornuéjols, editor, *Integer Programming and Combinatorial Optimization (Proceedings of IPCO '99)*, volume 1610 of *Lecture Notes in Computer Science*, pages 304–319. Springer-Verlag, Berlin, 1999.

[31] J. Kratochvíl. String graphs. II. Recognizing string graphs is NP–hard. *Journal of Combinatorial Theory* (B), 52:67–78, 1991.

[32] F.T. Leighton and S. Rao. Multicommodity max-flow min-cut theorems and their use in designing approximation algorithms. *Journal of the Association for Computing Machinery*, 46:787–832, 1999.

[33] F.T. Leighton. New lower bound techniques for VLSI. *Mathematical Systems Theory*, 17:47–70, 1984.

[34] P.C. Liu and R.C. Geldmacher. On the deletion of nonplanar edges of a graph. In *Proceedings of the 10th Southeastern Conference on Combinatorics, Graph Theory, and Computation*, Congress. Numer. xxiii–xxiv, Utilitas Math., Winnipeg, Manitoba, 1979.

[35] K. Mehlhorn and P. Mutzel. On the embedding phase of the Hopcroft and Tarjan planarity testing algorithm. *Algorithmica*, 16:233–242, 1996.

[36] P. Mutzel and R. Weiskircher. Optimizing over all combinatorial embeddings of a planar graph. In G.P. Cornuéjols, editor, *Integer Programming and Combinatorial Optimization (Proceedings of IPCO '99)*, volume 1610 of *Lecture Notes in Computer Science*, pages 361–376. Springer-Verlag, Berlin, 1999.

[37] P. Mutzel and R. Weiskircher. Computing optimal embeddings for planar graphs. In D.-Z. Du, P. Eades, V. Estivill-Castro, X. Lin, and A. Sharma, editors, *Computing and Combinatorics, Proceedings of the Sixth Annual International Conference (COCOON 2000)*, volume 1858 of *Lecture Notes in Computer Science*, pages 95–104. Springer-Verlag, Berlin, 2000.

[38] P. Mutzel and R. Weiskircher. Bend minimization in orthogonal drawing using integer programming. In O.H. Ibarra and L. Zhang, editors, *Computing and Combinatorics, Proceedings of the Eighth Annual International Conference (COCOON 2002)*, volume 2387 of Lecture Notes in Computer Science, Springer-Verlag, Berlin, 2002.

[39] P. Mutzel and T. Ziegler. The constrained crossing minimization problem. In J. Kratochvíl, editor, *Graph Drawing (Proceedings of GD '99)*, volume 1731 of *Lecture Notes in Computer Science*, pages 175–185. Springer-Verlag, Berlin, 1999.

[40] P. Mutzel and T. Ziegler. The constrained crossing minimization problem—a first approach. In P. Kall and H.-J. Lüthi, editors, *Operations Research Proceedings 1998*, pages 125–134. Springer-Verlag, Berlin, 1999.

[41] T.A.J. Nicholson. Permutation procedure for minimising the number of crossings in a network. *IEE Proceedings*, 115:21–26, 1968.

[42] J. Pach, F. Shahrokhi, and M. Szegedy. Applications of the crossing number. *Algorithmica*, 16:111–117, 1996.

[43] J. Pach and G. Tóth. Graphs drawn with few crossings per edge. *Combinatorica*, 17:427–439, 1997.

[44] J. Pach and G. Tóth. Which crossing number is it anyway? *Journal of Combinatorial Theory, Series* B, 80:225–246, 2000.

[45] M. Patrignani. On the complexity of orthogonal compaction. In F. Dehne, A. Gupta, J.-R. Sack, and R. Tamassia, editors, *Proceedings of the 6th International Workshop on Algorithms and Data Structures (WADS '99)*, volume 1663 of *Lecture Notes in Computer Science*, pages 56–61. Springer-Verlag, Berlin, 1999.

[46] H. Purchase. Which aesthetic has the greatest effect on human understanding? In G. Di Battista, editor, *Graph Drawing (Proceedings of GD '97)*, volume 1353 of *Lecture Notes in Computer Science*, pages 248–261. Springer-Verlag, Berlin, 1997.

[47] R.B. Richter and C. Thomassen. Relations between crossing numbers of complete and complete bipartite graphs. *American Mathematical Monthly*, pages 131–137, February 1997.

[48] M. Schlag, Y.-Z. Liao, and C. K. Wong. An algorithm for optimal two-dimensional compaction of VLSI layouts. *Integration, the VLSI Journal*, 1:179–209, 1983.

[49] O. Sýkora and I. Vrťo. On VLSI layouts of the star graph and related networks. *The VLSI Journal*, 17:83–93, 1994.

[50] R. Tamassia. On embedding a graph in the grid with the minimum number of bends. *SIAM Journal on Computing*, 16:421–444, 1987.

[51] G. Henschel. Die Verflechtung der Stromwirtschaft (nach Michael Stelte) Mysterien und Wurstbrote oder: Die wirrsten Graphiken der Welt (8), TAZ, November 2, 1999, page 24.

[52] R. Weiskircher. *New Applications of SPQR-Trees in Graph Drawing*. Ph.D. thesis, Max-Planck-Institut für Informatik, Saarbrücken, 2002.

[53] T. Ziegler. *Crossing Minimization in Automatic Graph Drawing*. Ph.D. thesis, Max-Planck-Institut für Informatik, Saarbrücken, 2000.

Part VII

Appendix

Chapter 20
Reflections

20.1 Banquet Speech at the Celebration of Manfred Padberg's 60th Birthday by Egon Balas

Dear friends and colleagues,

It is a great pleasure to participate in this wonderful event so perfectly organized by Martin and his team. How fitting it is to have the banquet celebrating Manfred Padberg's 60th birthday on a cruise ship. How happy he must feel that so many of his friends and colleagues came to participate. I must admit that upon my arrival, marked by relentlessly pouring rain, for a fleeting second the thought crossed my mind that maybe Martin had lately lost his clout and his connections with the Higher Ups, but next day the sun came out and everything turned beautiful—the organizer's prestige was reestablished. But it was only tonight, at this banquet, that we found out what a thorough job they did in preparing this event, looking back through the years and digging up history, old events, half-forgotten connections and bringing them all here.

I was asked to be a banquet speaker because three decades ago I was Manfred's professor and thesis advisor. This happened at GSIA, the Graduate School of Industrial Administration of Carnegie Mellon University, where Manfred studied for his Ph.D. between 1968 and 1971. Those were heady days for me. I had joined the place just one year earlier, as a fresh immigrant from behind the Iron Curtain and a Ford Distinguished Research Professor, and was thoroughly enjoying my new freedom and the wonderful research environment. One day a young German showed up at my office, with a *Diplom* in Mathematics from Münster, and with a Ford Foundation Fellowship to study in the U.S. for a doctorate, for which he chose the area of Operations Research. The two of us hit it off well from the beginning, although I must say that if I had known at the time what I found out from Martin tonight, namely, that some of Manfred's ancestors were in the armed robbery business and one of them had even fought against Attila the Hun, that famous though distant relative of the Hungarians, I for sure would have refused to have anything to do with him. Fortunately I did not know these things at the time and so, luckily for me, Manfred became my first doctoral student.

It seems that we entered each other's lives at the appropriate moment. We worked closely together for three years and developed an interaction from which I certainly benefited a lot, given Manfred's talent and enthusiasm for our subject. As to Manfred, it came back to me a few years later that when asked about his experience at GSIA, he said among other things, "I certainly learnt how to work my head off." Whether he has learned that at GSIA is debatable, but he definitely had fire in his belly. If the proof of a theorem or a step in the proof was open or in doubt at midnight, Manfred would not go to sleep until the question was settled.

Since I presented no paper at this conference, I had a chance to relax, listen, and philosophize. As a result, I came across some intriguing thoughts. Throughout all my academic career, my basic credo centered around my research. Of course, like every academic, I would dutifully say when quizzed about it, that research and teaching are the two equally important aspects of academic life, but in my innermost value system—why deny it—research always took precedence. I always felt that discovering something new must take priority over passing on existing knowledge. But at this conference on the 60th anniversary of my student, listening to his students and the students of his students, and some of the students of the students of his students, I had a strange feeling of mixed-up, if not reversed, priorities. In the preface to his very nice book on Linear Programming and Extensions, Manfred talks about "the 'invisible hand' of Egon Balas of Carnegie Mellon University, whose enthusiasm and superb teaching of the subject literally got me hooked on linear and combinatorial optimization." If anybody else had called my teaching superb, I would have taken it as a routine compliment, but Manfred can certainly not be accused of throwing around praise and kind words too easily. But if I take his statement seriously, if I got Manfred Padberg hooked on combinatorial optimization, and Manfred got Martin Grötschel hooked, and Martin got Michael Jünger and Gerd Reinelt hooked, and Michael and Gerd got several others hooked who are presenting interesting results at this conference, then who knows what was—what is—ultimately more important: my research and the facts or properties that it established, or my teaching, which "got hooked" Manfred, whose teaching in turn got hooked Martin, and so on, so that today we have all these people whom I, through several others, "got hooked" on mathematical programming and combinatorial optimization? I frankly no longer know.

On another note, I would like to tell you how my work with Manfred got me invited in the 1980s to the leading research institute of our field in the Soviet Union.

In 1970 Manfred and I showed that an optimal integer solution of the set partitioning problem can be reached from any other integer solution through a sequence of at most p pivots in the associated simplex tableau, each solution in the sequence being integer and of no greater cost than the previous one. Here p is the number of components nonbasic in the starting solution and equal to 1 in the final solution. One implication of our result was that the edges of the set partitioning polytope are edges of the corresponding linear programming polytope, a fact discovered half a year earlier—unbeknownst to us—by the Ukrainian mathematician Trubin. Our result seemed to suggest that set partitioning problems can be solved by linear programming, restricting pivots to integer ones. This, however, does not follow, because the sequence of pivots that one needs may include degenerate integer pivots on entries equal to -1, which kills any of the known schemes meant to eliminate cycling in degeneracy. Several years later, in 1979, I was participating in an international conference where a Russian mathematician, the head of a research group at the Central

Economic-Mathematical Institute in Moscow, gave a survey of Soviet results in *Discrete Mathematics*. He prominently mentioned Trubin's result and claimed that it had reduced the apparently difficult problem of set partitioning to the easy problem of linear programming. I had a discussion with the lecturer after his talk and told him that while Trubin's discovery is very important indeed, it did not reduce set partitioning to the easy problem of linear programming, and pointed out the reason. The Russian replied that what he said was a well-established fact, and in the Soviet Union "every student knows it." Since Manfred and I had spent a considerable effort in trying to come up with a pivoting procedure that would do the trick, the framework was sufficiently fresh in my mind to allow me to put together that evening a small example of a set partitioning problem with a nonoptimal integer solution from which the optimal solution could not be reached by any known pivoting scheme. I gave it to the Russian and next day he came to see me: he finally understood the point the example was illustrating. His conclusion was an emphatic "you must come visit us in Moscow." An official invitation arrived soon after my return to the States.

For the last three years I had the honor to serve on the selection committee of the John von Neumann Theory Prize, and in that capacity in May of last year, at the Salt Lake City meeting of INFORMS, I had the great pleasure to award this most important prize of our profession to Ellis Johnson and Manfred Padberg, as a shared award. Five years earlier, when I had been awarded that prize myself, somebody asked me after a lecture how did it feel to get the von Neumann Prize. I was somewhat embarrassed, did not quite know what to say, and mumbled something to the effect that there is no greater honor in my profession than this. Upon which somebody from the other end of the room yelled "Yes, there is." Surprised, I asked "What is that?" And the answer came "It's when your student gets it!"

So let me finish this talk by reading the citation which accompanied the awarding to Manfred of the John von Neumann Theory Prize of INFORMS in May 2000:

Since receiving his Ph.D. from Carnegie Mellon University in 1971, Manfred Padberg has made fundamental contributions to both the theoretical and computational side of integer programming and combinatorial optimization. His early work on facets of the vertex packing polytope and their lifting, and on vertex adjacency on the set partitioning polytope, paved the way towards the wider use of polyhedral methods in solving integer programs. His characterization of perfect 0/1-matrices reinforced the already existing ties between graph theory and 0/1-programming.

Padberg is the originator and the main architect of the approach known as branch-and-cut. Concentrating on the traveling salesman problem as the main testbed, Padberg and Rinaldi successfully demonstrated that if cutting planes generated at various nodes of a search tree can be lifted so as to be valid everywhere, then interspersing them with branch and bound yields a procedure that vastly amplifies the power of either branch and bound or cutting planes by themselves. This work had and continues to have a lasting influence.

One of the basic discoveries of the 1980s in the realm of combinatorial optimization, arrived at by three different groups of researchers in the wake of the advent of the ellipsoid method for convex programming, was the equivalence of optimization and separation: Padberg and M.R. Rao formed one of these groups.

Padberg's work combines theory with algorithm development and computational testing in the best tradition of Operations Research and the Management Sciences. In his joint work with Crowder and Johnson as well as in subsequent work with others, Padberg set an example of how to formulate and handle efficiently very large scale practical 0/1-programs with important applications in industry and transportation."

Thanks for your attention.

20.2 Speech of Claude Berge, Read at the Workshop in Honor of Manfred Padberg, Berlin, October 13, 2001

Since Manfred is an old friend, I am extremely sorry for not being fit enough (physically, that is: the brain still ticks over occasionally) to present this speech myself as my tribute to him on his birthday. I suspect that for some of you, the fact that another person will be reading this out may be somewhat preferable. My own English has been distorted by various exposures to pidgin English in Papua New Guinea or in Irian Jaya ..., and, in addition, laced with an unshakable, though devastatingly seductive, French accent.

Manfred himself is a master of Italian, French, English, and, naturally, German. He has even been known to wax eloquent in Latin on certain occasions, when late in the evening he has found himself in the presence of colleagues talking about subjects that bore him: a useful method for changing the subject that I wish I could emulate. One of his subjects, for which he is unpeacheable, is the age of most of our friends. For many years, it was also the life of Charlemagne (Karl the great): the tomb of his father, Pepin, is in Saint Denis, near Paris, but if a rash interlocutor thinks that Charlemagne was more French than German, such an imprudent conviction may generate hours of harsh discussions....

One may bump into Manfred here, there, and everywhere, Berlin, Bonn, Lausanne, New York, Tampa, Hawaii, Grenoble, Paris, but do not interpret his work on the Traveling Salesman Problem in the context of his own peregrinations. If you meet him on the beach of Saint-Tropez, he will be very likely working on a portable, without a look to the sea or to a group of attractive ladies! My personal opinion is that Manfred Padberg is a perfect specimen of a new type of man, one who prefers spending his time in front of a computer. Maybe after *Homo Erectus, Neanderthals, Cro-Magnons, Homo Sapiens*, we are confronting a new breed of *Homo Mathematicus*?

This is the question we have to answer today!

Happy birthday, Manfred!
Claude

20.3 Banquet Speech in Honor of Manfred Padberg's 60th Birthday by Harold Kuhn

The question that I propose to answer this evening is, What has Manfred Padberg been doing for the last 60 years and 3 days? After Lawrence Wolsey's sparkling summary of Manfred's work in the opening session of this celebration, I can largely ignore Manfred's scientific work and concentrate on his poor choices of places of employment.

As I look out at the people present on this occasion, I see a large number who were in Atlanta last year at the International Symposium on Mathematical Programming, when

I gave a banquet talk with the title "Being in the Right Place at the Right Time." I was very tempted to call this talk "Being in the Wrong Place at the Wrong Time." That may seem too sad for this happy occasion. However, rest assured: There is a happy ending!

From 1941 to 1964, Manfred didn't seem to do much. He appears to have spent his time studying pure mathematics, which has stood him in good stead ever since. In 1965, he made his first bad move: he joined an Institute for Industrial Administration and Operations Research in Münster. There he produced his first scientific work. Just listen to this title (you will not find it among his published papers!):

"Reduktion durch Suffizienz
im Fall sequentieller statistischer Entscheidungsfunktionen."

As usual, it is longer in English (13 words instead of 8):

"Reduction by Means of Sufficiency
in the Case of Statistical Sequential Decision Functions."

This title is longer than any other title in the 105 published works listed on Manfred's website.

In 1967–68, while spending a very happy year as an Assistant Professor of Industrial Administration in Mannheim, he succumbed to WANDERLUST, and allowed his name to be entered for a Ford Foundation Fellowship in Management Education. The late Richard M. Cyert, then Dean of the Business School at Carnegie Mellon University, read his application, recognized his brilliance, and had the wisdom to choose him for the fellowship.

Manfred was supposed to be studying Industrial Administration but drifted off to study with Egon Balas, who wouldn't know how to administer an industry even if threatened by a penalty of death if he failed.

In 1971, with his Ph.D. Thesis (*Essays in Integer Programming*, clearly irrelevant to Industrial Administration) under his belt, Manfred went back to Germany for three years—again to an Institute of Management. Another bad choice.

In 1974, WANDERLUST took over again and he returned to the United States to accept an appointment in the distinguished business school of an illustrious university. It was a poor match for someone who was at heart a pure mathematician. For the next 26 years he had various titles and taught a variety of courses—but has always stayed in the closet, pursuing his true love, combinatorial optimization.

A measure of how happy Manfred was in this position is given by the statistic that he arranged to be on leave for 8 of those 26 years. In spite of teaching courses far from his true interests, he managed to make fundamental contributions to both the theoretical and the computational side of integer programming and combinatorial optimization, serving as a main architect of the approach known as branch-and-cut.

However, the story has a happy ending! Finally, Manfred made a correct professional decision. He decided to leave the United States and return to Europe. He now has the freedom to come out of the closet and pursue his true professional love, combinatorics as a branch of pure mathematics. Living in Marseille, where he spent summer vacations as a boy, and with the solid emotional support of his wife, Suzy, we all expect to see a new burst of creativity from Manfred.

Happy Birthday!

Index